Imperial College London
2406941027

AF598225

Plasmonic
Nanoguides and Circuits

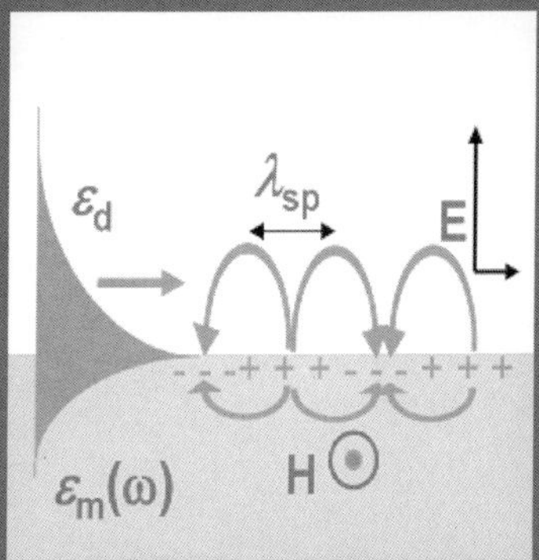

Plasmonic
Nanoguides and Circuits

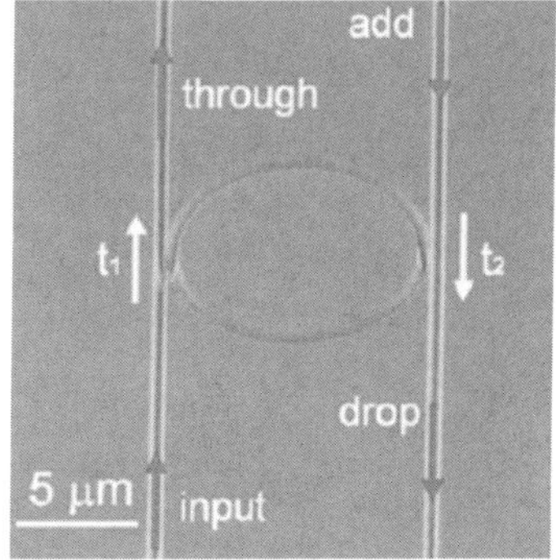

Editor

SERGEY I BOZHEVOLNYI
University of Southern Denmark, Denmark

PAN STANFORD PUBLISHING

Published by

Pan Stanford Publishing Pte. Ltd.
5 Toh Tuck Link
Singapore 596224

Distributed by

World Scientific Publishing Co. Pte. Ltd.
5 Toh Tuck Link, Singapore 596224
USA office: 27 Warren Street, Suite 401-402, Hackensack, NJ 07601
UK office: 57 Shelton Street, Covent Garden, London WC2H 9HE

British Library Cataloguing-in-Publication Data
A catalogue record for this book is available from the British Library.

PLASMONIC NANOGUIDES AND CIRCUITS

Copyright © 2009 by Pan Stanford Publishing Pte. Ltd.

All rights reserved. This book, or parts thereof, may not be reproduced in any form or by any means, electronic or mechanical, including photocopying, recording or any information storage and retrieval system now known or to be invented, without written permission from the Publisher.

For photocopying of material in this volume, please pay a copying fee through the Copyright Clearance Center, Inc., 222 Rosewood Drive, Danvers, MA 01923, USA. In this case permission to photocopy is not required from the publisher.

ISBN-13 978-981-4241-32-8
ISBN-10 981-4241-32-6

Printed in Singapore by B & JO Enterprise
E-mail: sankaran@rpsonline.com.sg

PREFACE

Modern communication systems dealing with vast amounts of data at ever increasing speed started to progressively stronger experience the fundamental limitations imposed on the existing electronic and optical technologies. Specifically, nm-sized electronic (metallic) circuits are inherently slow due to RC-delay times, whereas optical (dielectric) circuits having enormous information capacities are essentially μm-sized due to diffraction effects. It is fascinating to note that electromagnetic modes supported by metal-dielectric interfaces were found to be beating the diffraction limit while upholding the optical bandwidth. These modes, in which optical fields are coupled to electron oscillations, are usually called surface plasmons (SPs) and are characterized by the wavelength that is always shorter than the wavelength in the dielectric, exhibiting thereby subwavelength confinement. It should be emphasized that not all SP modes allow for *unlimited* squeezing in the cross section perpendicular to their propagation (i.e., going "nano"), a property that is imperative for realization of *plasmonic* circuits combining the compactness of electronics with huge bandwidth of optical networks.

In this book, we concentrate on SP guiding geometries ensuring truly nanoscale field confinement and review the current status of this very young but rapidly growing area of nanoplasmonics. Both fundamental physics and application aspects are overviewed in detail by world-leading experts highlighting the most significant achievements and reporting on the latest advances within nanoscale plasmonic guides and circuits.

SP modes that persist to exist in the limit of infinitely small waveguide cross sections (being correspondingly squeezed to progressively smaller transverse dimensions) are introduced in Chapters 1 and 2. The two chapters provide the main dispersion relations as well as explain the underlying physics of guiding beyond the diffraction limit. The chapters that follow are self-contained and they consider particular SP guiding configurations summing up previous developments and outlining future research directions. SP excitations supported by chains of nanoparticles are analyzed and experimentally investigated in Chapter 3. Chapters 4 through 8 consider the different aspects of SP guiding between two metal surfaces, with recent experiments on all-optical plasmonic modulators being presented in Chapter 7. Propagation of SP modes along metal grooves (channels) and wedges is analyzed in Chapters 9 and 10, followed by the experimental demonstration of nanophotonic components utilizing channel SP modes in Chapter 11. Adiabatic concentration of the optical energy and coherent control of SP propagation in tapered nanoplasmonic waveguides are treated in Chapter 12. Finally Chapter 13 reviews some of the key developments in the rapidly emerging field of nanoplasmonics as well as discusses challenges that need to be addressed before the dream of integrating plasmonic, electronic and optical components can be realized.

It has been a great pleasure to work with some of the leading scientists in nanoplasmonics during the preparation of this book. I am deeply grateful to all contributing authors for their efforts and enthusiasm in sharing the very latest results in this exciting field. I would also like to acknowledge Elena Bozhevolnaya for taking care of technical issues associated with actual collection and administration of the book chapters.

Sergey I. Bozhevolnyi
Editor
Odense, Denmark
June 2008

CONTENTS

Chapter 1

INTRODUCTION TO SURFACE PLASMON-POLARITON WAVEGUIDES

Sergey I. Bozhevolnyi

Institute of Sensors, Signals and Electrotechnics (SENSE)
University of Southern Denmark
Niels Bohrs Allé 1, DK-5230 Odense M, Denmark
E-mail: seib@sense.sdu.dk

Fundamental properties of surface plasmon polariton (SPP) modes propagating along metal-dielectric interfaces, such as short- and long-range SPPs and gap SPPs, are reviewed, paying attention to the underlying physical phenomena involved and providing useful explicit formulas approximating the exact (implicit) dispersion relations for these modes. Usage of the effective-index method for analyzing various waveguide configurations that ensure the lateral confinement of SPP modes, such as stripe, gap and channel waveguides, is also described.

1. Propagating and evanescent waves

Surface plasmon polaritons (SPPs) are electromagnetic (EM) excitations that, being coupled to surface collective oscillations of free electrons in a metal, are bound to and propagate along metal-dielectric interfaces[1-3]. Topology of the metal surface and adjacent dielectric determines characteristics of various propagating SPP modes existing at flat and curved, single and multiple surfaces, including SPP modes of complex particle arrays and metal nanostructures. SPP modes are renowned for their ability to concentrate electromagnetic fields beyond the diffraction limit (i.e., on the nanoscale), while enhancing local field strengths by several orders of magnitude. In this introductory chapter, basic properties of the main SPP modes are briefly overviewed leaving out the details that can be found elsewhere[1-3], including a very recent book on plasmonics[4].

Light-matter interactions in nanostructured materials give rise to nanostructured optical fields, implying that EM fields of SPP modes supported by metal nanostructures exhibit subwavelength spatial frequencies. In the integral plane-wave representation of the total field the latter correspond to evanescent field components. The propagating SPP modes being confined in the plane perpendicular to the propagation direction comprise necessarily evanescent fields as well. It is therefore instructive to briefly review the main properties of the propagating and evanescent waves formed at interfaces between different media[5,6]. Let us begin with the equation for EM waves propagating in a nonmagnetic (μ = 1), isotropic and homogeneous medium characterized by the dielectric permittivity $\varepsilon(\omega)$. The corresponding wave equation for the electric field $\mathbf{E}(\mathbf{r}; t)$ is given by

$$\Delta\mathbf{E} - \frac{\varepsilon(\omega)}{c^2}\frac{\partial^2\mathbf{E}}{\partial t^2} = 0 \quad , \tag{1}$$

with c being the speed of light in vacuum. Substituting in Eq. (1) the plane-wave trial solution in the form $\mathbf{E}(\mathbf{r};t) = \mathbf{E}_0 \exp[i(\mathbf{k}\cdot\mathbf{r} - \omega t)]$ results in the dispersion relation for the wave vector $\mathbf{k}(\omega)$

$$|\mathbf{k}(\omega)|^2 = k^2 = k_x^2 + k_y^2 + k_z^2 = \varepsilon(\omega)\frac{\omega^2}{c^2} = \varepsilon k_0^2 \quad . \tag{2}$$

Hereafter we consider the two-dimensional (2D) configuration, which exhibits invariance along the y-axis, i.e., the wave propagation directions are confined to the x-z plane (k_y = 0). The 2D geometry allows us to express the electric field of a plane wave propagating into the half-space $z \geq 0$ as follows [Eq. (2)]

$$\mathbf{E}(\mathbf{r},t) = \mathbf{E}_0 e^{-i\omega t} \exp[i(k_x x + k_z z)] \quad , \tag{3}$$

with $k_x \in (-\infty, +\infty)$, $k_z = \sqrt{\varepsilon k_0^2 - k_x^2}$.

It transpires from the above relations that, in lossless dielectric media with $\varepsilon = \varepsilon_d > 0$, all plane waves can be divided into propagating (homogeneous) waves with $k_x \leq \varepsilon k_0$ and evanescent (inhomogeneous) waves with $k_x > \varepsilon k_0$. The propagating waves are characterized by the propagation direction determined by the real wave-vector projections k_x and k_z [Eq. (3)] and the spatial period of oscillations, i.e., the wavelength, given by [Eq. (2)]: $\lambda = 2\pi/k = \lambda_0/n$ (where $\lambda_0 = 2\pi/k_0$ and $n^2 = \varepsilon$). The evanescent waves propagate along the x-axis exhibiting the spatial period $\lambda_x = 2\pi/|k_x|$ and exponentially decay along the z-axis with the decay length given by: $|k_z|^{-1}=(k_x^2-\varepsilon k_0^2)^{-0.5}$. The spatial periodicity of evanescent waves (in the propagation direction) can be infinitely small and the corresponding spatial frequencies – extremely high, suggesting thereby their usage for ultra-compact nanophotonic circuits, in general, and for optical interconnects on board, in particular.

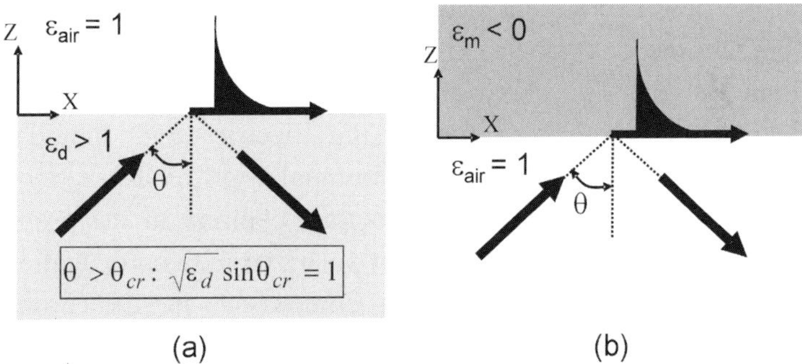

Fig.1. Schematic picture of the occurrence of evanescent waves excited (a) in dielectrics by total internal reflection of light and (b) in metals at any angle of light incidence.

An evanescent wave can be formed in the half-space $z \geq 0$, for example, in the process of total internal reflection (TIR), when a wave propagating in the half-space $z < 0$ occupied by a dielectric with $\varepsilon_d > 1$ impinges the interface $z = 0$ between the dielectric and air with $\varepsilon_{air} = 1$ under the angle θ of incidence larger than the critical one θ_{cr} determined by: $\sqrt{\varepsilon_d}\sin\theta_{cr} = 1$. The boundary conditions require the phase matching of the total electric fields on both sides of the interface, i.e., the conservation of k_x, implying thereby the formation of the evanescent wave (in the air half-space) with $k_x > \varepsilon_{air}k_0 = k_0$ [Fig. 1(a)]:

$$k_x = \sqrt{\varepsilon_d}\,k_0 \sin\theta > k_0 \Rightarrow \mathbf{E} = \mathbf{E}_0\, e^{i(k_x x - \omega t)} \exp\left(-zk_0\sqrt{\varepsilon_d \sin^2\theta - 1}\right). \quad (4)$$

Contrary to the above case, in lossless metals with $\varepsilon = \varepsilon_m < 0$, there are no propagating waves because the wave-vector magnitude (wave number) k is pure imaginary [Eq. (2)]. For example, considering the half-space $z > 0$ occupied by the metal having the interface $z = 0$ with air, one finds that any wave propagating in air (towards the metal) forms an evanescent wave in the metal [Fig. 1(b)]:

$$k_x^2 + k_z^2 = \varepsilon_m k_0^2 < 0 \Rightarrow \mathbf{E} = \mathbf{E}_0 \, e^{i(k_x x - \omega t)} \exp\left(-z k_0 \sqrt{\sin^2\theta - \varepsilon_m}\right). \quad (5)$$

Note that, for typical metals, the magnitude of dielectric permittivity is very large, $|\varepsilon_m| >> 1$, implying that the field penetration in metals, $|k_z|^{-1} \approx \lambda_0 / (2\pi |\varepsilon_m|^{0.5})$, is rather small, typically on the nanometer scale.

Concluding this Section we would like to emphasize that, even though the field expressions [Eqs. (3)-(5)] are valid for all electric field components, the components themselves are not independent because they should satisfy Maxwell's equations. More specifically, in homogeneous (charge-free) media, the electric field should satisfy Coulomb's law, which is in the differential form reads: $\nabla \cdot \mathbf{E} = 0$. For example, it can be immediately seen that, contrary to the propagating waves, the evanescent waves [Eq. (4)] polarized in the propagation plane (p-polarized) feature the nonzero component in the propagation direction, i.e., $E_x \neq 0$ (Fig. 1). The appropriate implications to the properties of SPP modes will be discussed in the following Sections.

2. Interface electromagnetic excitations

Metal-dielectric interfaces can support a number of different SPP modes bound to and propagating along the interfaces[1-4]. Even within the framework of 2D geometry, one can realize many different SPP modes by setting up various multiple-interface systems, i.e., alternating metal and dielectric films[7]. SPPs represent interface EM excitations, in which evanescent EM waves are coupled to surface collective oscillations of free electrons, propagating along metal-dielectric interfaces. In 2D multilayer geometry, plane waves [Eq. (3)] can be divided into the TM (transverse magnetic) and TE (transverse electric) waves that are polarized in (p-polarized) and perpendicular to (s-polarized) the propagation plane, respectively. The continuity of the tangential (parallel to an interface) components of EM fields across the interface required by Maxwell's equations implies that, in the case of nonmagnetic ($\mu = 1$) media, the electric field of the TE wave should be continuous along with its lateral derivative ($\partial E_y/\partial z$) across the interface (Fig. 1). These boundary conditions together with the requirement of field confinement to the interface (i.e., field vanishing infinitely far from the interface) signify that SPP modes cannot be *s*-polarized[1-3]. Conversely, the magnetic field of the TM wave should be continuous along with the product $(1/\varepsilon)\cdot(\partial H_y/\partial z)$, implying the change in its derivative sign across the interface between the dielectric ($\varepsilon_d > 0$) and metal ($\varepsilon_m < 0$) media and allowing thereby the existence of a confined solution. In passing we note that the interface between a dielectric and a negative refractive-index material, which is characterized with the dielectric permittivity and magnetic permeability both being negative[8]: $\varepsilon_{nim} < 0$ and $\mu_{nim} < 0$, would obviously support both *p*- and *s*-polarized surface polariton modes whose excitation can be advantageously exploited for subwavelength imaging through a slab of such a material[9].

One can also take up the physical argument by considering SPPs as hybrid modes comprising the features of both the propagating EM waves in dielectrics, which are transverse waves, and free electron (plasma) oscillations in metals, which are longitudinal. Since the TE waves are purely transverse waves (in the sense that $\nabla \cdot \mathbf{E} = 0$ everywhere) these waves cannot be coupled to the longitudinal electron oscillations in metals[7]. In the following Section, we consider only the TM (*p*-polarized) waves and the corresponding SPP modes in single- and double-interface configurations.

3. Surface plasmon polaritons (SPPs)

Let us consider monochromatic (with angular frequency ω) TM waves in a system consisting of a dielectric ($\varepsilon_d > 0$) material in the half-space $z > 0$ and a metal characterized with a complex, frequency-dependent permittivity $\varepsilon_m(\omega)$ in the half-space $z < 0$ (Fig. 2). Bound to the interface $z = 0$ TM modes, i.e., SPPs, can be viewed as consisting of two evanescent waves similar to those expressed by Eqs. (4) and (5) and shown in Fig. 1. Since the tangential field components are continuous across the interface (implying thereby that the wave vector component k_x parallel to the interface is conserved), the electric field of such a mode can be written as follows

$$\mathbf{E}(z>0) = \left(E_x^0, 0, E_z^d\right) \cdot e^{i\left(k_{sp}x - \omega t\right)} \exp\left(-z\sqrt{k_{sp}^2 - \varepsilon_d k_0^2}\right), \tag{6a}$$

$$\mathbf{E}(z<0) = \left(E_x^0, 0, E_z^m\right) \cdot e^{i\left(k_{sp}x - \omega t\right)} \exp\left(z\sqrt{k_{sp}^2 - \varepsilon_m k_0^2}\right), \tag{6b}$$

where E_x^0 and $E_z^{d(m)}$ are the amplitudes of the corresponding electric field components in the dielectric (metal) medium, and $k_{sp} = k_x$ is the mode propagation (complex) constant yet to be determined. Since the electric field components should satisfy Coulomb's law $\nabla \cdot \mathbf{E} = 0$ (both in the upper and lower half-spaces), the normal (perpendicular to the interface) field components can be related to the tangential one:

$$E_z^d = \frac{ik_{sp}}{\sqrt{k_{sp}^2 - \varepsilon_d k_0^2}} E_x^0 \quad \text{and} \quad E_z^m = -\frac{ik_{sp}}{\sqrt{k_{sp}^2 - \varepsilon_m k_0^2}} E_x^0 . \tag{7}$$

Finally, using the appropriate boundary condition for the normal electric field components: $\varepsilon_d E_z^d = \varepsilon_m E_z^m$, results in the SPP dispersion relation:

$$k_{sp}(\omega) = \frac{\omega}{c}\sqrt{\frac{\varepsilon_m(\omega)\,\varepsilon_d}{\varepsilon_m(\omega) + \varepsilon_d}} . \tag{8}$$

The necessary conditions for an SPP (as a bound and propagating SP mode) to exist can be obtained from the requirement of the square root expressions in Eq. (6) and (8) to have a positive real part, leading to

$$\mathrm{Re}\{\varepsilon_d \cdot \varepsilon_m(\omega)\} < 0 \text{ and } \mathrm{Re}\{\varepsilon_d + \varepsilon_m(\omega)\} < 0 , \Rightarrow \mathrm{Re}\{\varepsilon_m(\omega)\} < -\varepsilon_d . \tag{9}$$

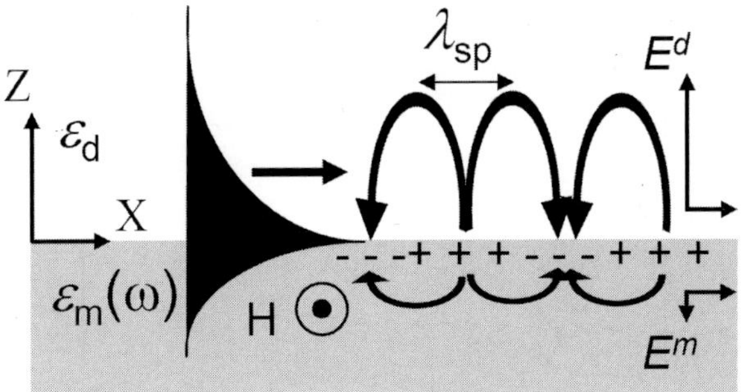

Fig. 2. Schematic representation of an intensity distribution and electromagnetic field components of a surface plasmon polariton supported by a metal-dielectric interface.

For most metals and dielectrics, this condition is satisfied in the long-wavelength part of the visible and in the infrared. The real part of the propagation constant determines the SPP wavelength $\lambda_{sp} = 2\pi / \mathrm{Re}(k_{sp})$, which is always smaller than the light wavelength in the dielectric: $\lambda_{sp} < \lambda = \lambda_0 / \sqrt{\varepsilon_d}$, and the SPP (intensity) propagation length $L_{sp} = 1/\mathrm{Im}(2k_{sp})$. Using the free-electron (Drude) approximation for the dielectric permittivity: $\varepsilon_m(\omega) = 1 - (\omega_p / \omega)^2$ with ω_p being the plasma frequency, one can work out the frequency dependencies of the SPP characteristics[1-3], observing, for example, that the SPP wavelength approaches zero at the high-frequency (short-wavelength) limit of the SPP dispersion, i.e., when $\omega \to \omega_p / \sqrt{1+\varepsilon_d}$. In practice, however, the SPP damping caused by ohmic losses of the electron oscillations associated with the SPP is a very important factor, limiting both the extent of the decrease of SPP wavelength (for short wavelengths) and the SPP propagation length (Fig. 3). The latter circumstance, i.e., a rather limited SPP propagation length, poses a serious problem when developing SPP-based waveguide components for practical applications. For example, the normalized SPP propagation length for the gold-air interface at the telecom wavelength of 1.55 μm can be estimated using the permittivity[10] ε_m = -132 + 12.7i as $L_{sp}/\lambda_{sp} \approx 220$, representing essentially the upper limit for the quality factor of corresponding SPP-based resonators or interferometers.

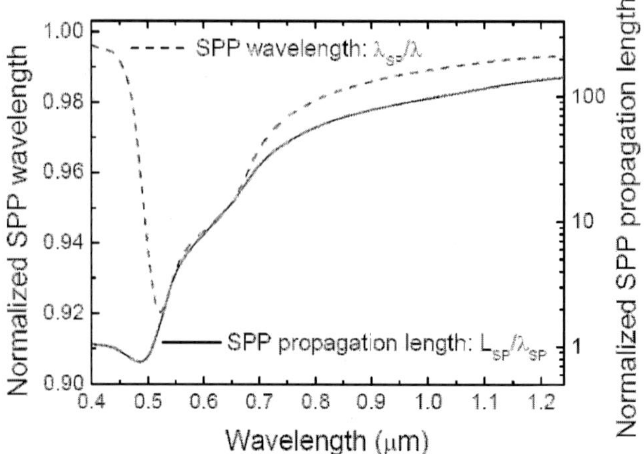

Fig. 3. Normalized surface plasmon-polariton (SPP) wavelength and propagation length calculated for a gold-air interface using the gold permittivity[10] as functions of the light wavelength.

The SPP dispersion relation [Eq. (8)] allows one to explicitly relate the amplitudes of normal and tangential electric field components in the dielectric and metal [Eq. (7)]:

$$E_z^d = i\sqrt{-\varepsilon_m / \varepsilon_d}\, E_x^0 \quad \text{and} \quad E_z^m = -i\sqrt{-\varepsilon_d / \varepsilon_m}\, E_x^0 \,. \tag{10}$$

Typically, metals exhibit large magnitudes of the dielectric permittivity, i.e. $|\varepsilon_m| >> \varepsilon_d$, resulting in the normal component being dominant in the dielectric and the tangential component - dominant in the metal (Fig. 2). The SPP electric field is thereby primarily transverse in the dielectric and longitudinal in the metal, reflecting the hybrid nature of SPPs that combine the features of both propagating EM waves in dielectrics and free electron oscillations in metals. Since the SPP damping occurs due to ohmic losses in the metal, it is the *longitudinal* electric field component of the SPP in the metal that determines the SPP damping, a circumstance which is important for understanding the properties of SPP modes supported by thin metal films considered in the next Section.

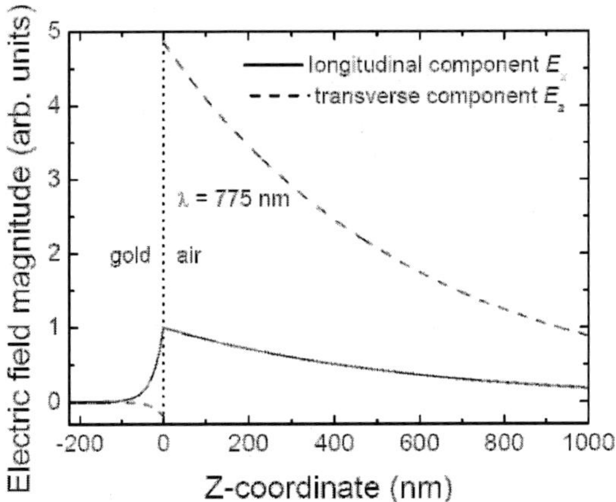

Fig. 4. Spatial distributions of transverse and longitudinal electric field components of a surface plasmon polariton supported by a gold-air interface at the wavelength of 775 nm.

The penetration depths (or the decay lengths) of the SPP electric field in the dielectric d_d and metal d_m are also different as seen from the corresponding SPP field distributions [Eq. (6)], and can be explicitly expressed by making use of the dispersion relation [Eq. (8)] as follows (for the 1/e amplitude decrease):

$$d_{d(m)} = \frac{\lambda}{2\pi} \cdot \sqrt{-\frac{\varepsilon_m + \varepsilon_d}{\varepsilon_{d(m)}^2}} , \qquad (11)$$

so that $\varepsilon_d d_d = |\varepsilon_m| d_m$. The latter relation implies that the field penetration in the metal is typically much shorter than that in the dielectric. It is interesting that the penetration depth in metals depends rather weakly on the wavelength staying at the level of a few tens of nanometers, while that in dielectrics increases fast and nonlinearly with the wavelength. These trends can be accounted for by using the Drude expression for the dielectric permittivity $\varepsilon_m(\omega)$ together with Eq. (11), resulting in the following asymptotic relations for the penetration depths in the long-wavelength limit:

$$d_m \approx c/\omega_p, \quad \text{and} \quad d_d \approx (\lambda/2\pi)^2 \cdot (\omega_p / c\varepsilon_d) . \qquad (12)$$

Concluding this Section let us consider the properties of the SPP supported by the air-gold interface at the excitation wavelength of 775 nm. Using the gold permittivity[10] ε_m = -23.6 + 1.69i and the above relations [Eqs. (8) and (11)] results in the following SPP characteristics: $\lambda_{sp} \approx 759$ nm, $L_{sp} \approx 38.2$ μm ($L_{sp}/\lambda_{sp} \approx 50$), $d_{air} \approx 587$ nm, $d_{gold} \approx 24.8$ nm. The corresponding electric field distributions (Fig. 4) demonstrate, as discussed above, the dominance of the transverse field component in air and that of the longitudinal one in gold. Finally, using the above relations [Eqs. (10) and (11)], one can approximate the fraction of the SPP electric field energy in the dielectric as simply being equal to $(|\varepsilon_m|/\varepsilon_d)^2$, which is somewhat counterintuitive as, normally, light is drawn towards regions with higher refractive indexes. The obtained expression signifies that the SPP electric field is progressively being forced out of the metal for longer wavelengths since, in the Drude approximation, $|\varepsilon_m| \propto \omega^{-2}$.

4. Short- and long-range SPP modes

Many different SPP modes can be found in multiple-interface systems[7], when the SPPs associated with individual metal-dielectric interfaces start interacting with each other. Considering the SPP modes associated with two metal-dielectric interfaces one finds that the SPP modes can be supported by either a thin metal film abutted with dielectrics or a thin dielectric layer surrounded by metals, often called the insulator-metal-insulator (IMI) or metal-insulator-metal (MIM) structures, correspondingly[11]. Let us consider the SPP modes in the *symmetric* IMI configuration of a thin metal film with the thickness t being embedded in the dielectric (extending indefinitely on both sides of the film). When two identical SPP modes start overlapping with each other for small thicknesses of the metal film, the propagation constants of the symmetric and anti-symmetric (individual) mode combinations become different[7,12]. Note, that the symmetric (with respect to the film mid-plane) configuration of the transverse component E_z corresponds to the anti-symmetric one for the longitudinal components E_x and *vice versa* [see Eq. (7)]. Hereafter the E_z-symmetric configuration is called the symmetric SPP mode[13]. Since the SPP damping is determined by the longitudinal SPP component (see Section 3 of this chapter), the symmetric SPP mode exhibiting the odd symmetry of E_x, which thereby crosses zero (changing its sign) at the mid-plane of the metal film, experiences considerably smaller attenuation than the anti-symmetric SPP mode, which is the E_x-symmetric configuration. The symmetric SPP mode is therefore conventionally called the long-range SPP (LR-SPP) and the anti-symmetric one – the short-range SPP (SR-SPP) [12,13].

Using the appropriate boundary conditions for the normal and tangential electric field components as in the case of SPPs and the aforementioned symmetry of the corresponding field component distributions, allows one to obtain the LR-SPP dispersion relation:

$$\tanh\left(k_z^{(m)} t/2\right) = -\frac{\varepsilon_m k_z^{(d)}}{\varepsilon_d k_z^{(m)}} \quad \text{with} \quad k_z^{(m,d)} = \sqrt{k_{lrsp}^2 - \varepsilon_{m,d} k_0^2} \; . \tag{13}$$

Note that, unlikely the case of SPPs, the LR-SPP propagation constant k_{lrsp} cannot be presented in an explicit form. However, for sufficiently thin metal films ($t \rightarrow 0$), one can use the approximation $\tanh x \approx x$ resulting in the following simple expression:

$$k_{lrsp} \approx k_0 \sqrt{\varepsilon_d + \left(t\, k_0 \varepsilon_d / 2\right)^2 \cdot \left[1 - \left(\varepsilon_d / \varepsilon_m\right)\right]^2} \; . \tag{14}$$

It is seen that, in the limit of very thin films, the LR-SPP propagation constant approaches the light line in the dielectric, implying that the LR-SPP field being forced out of the metal (the main component E_x crosses zero) spreads progressively into the dielectric [Eq. (6a)]. Note that both the real and imaginary part of the difference between the LR-SPP propagation constant and that of light in the dielectric decreases quadratically with the decrease in the film thickness [Eq. (14)]. The latter means also that, in the same manner, the LR-SPP propagation loss decreases to zero and the LR-SPP propagation length increases infinitely, a remarkable feature that allows one to employ the LR-SPPs for developing practical plasmonic components for integrated optics[14].

Following the same procedure for the SR-SPPs one obtains the corresponding dispersion relation:

$$\tanh\left(k_z^{(m)} t/2\right) = -\frac{\varepsilon_d k_z^{(m)}}{\varepsilon_m k_z^{(d)}} \quad \text{with} \quad k_z^{(m,d)} = \sqrt{k_{srsp}^2 - \varepsilon_{m,d} k_0^2} \; , \tag{15}$$

that can be simplified for thin metal films as follows:

$$k_{srsp} \approx k_0 \sqrt{\varepsilon_d + \left[2\varepsilon_d / (t k_0 \varepsilon_m)\right]^2} . \quad (16)$$

Contrary to the above case of LR-SPPs, the SR-SPP propagation constant k_{srsp} increases indefinitely with its real and imaginary parts: $k_{srsp} \approx -(2\varepsilon_d)/(t\varepsilon_m) \rightarrow \infty$, when the metal film thickness t decreases to zero. This means that both the SR-SPP wavelength and propagation length decrease, approaching zero for infinitely thin films. It can also be shown that, in this limit, their ratio tends to a constant determined by the ratio between the real and imaginary parts of the metal susceptibility: $L_{srsp} / \lambda_{srsp} \rightarrow \left|\mathrm{Re}(\varepsilon_m)\right| / \left[4\pi \, \mathrm{Im}(\varepsilon_m)\right]$, a ratio that can be rather large in the infrared. Furthermore, the SR-SPP field filling efficiently the metal (the main component E_x is nearly constant) is also tightly bound to the metal film [Eq. (6a)]. Since, at the same time, the main SR-SPP component in the dielectric (E_z) is anti-symmetric, the SR-SPP should exhibit a high reflectivity at the film termination due to impedance mismatch with the propagating EM waves (due to asymmetry and a very short field penetration in the dielectric). This non-trivial feature can be successfully exploited when designing various plasmonic resonator structures[15].

Concluding this Section let us consider the properties of the LR- and SR-SPP supported by a thin gold film surrounded by air at the excitation wavelength of 775 nm. Using the gold permittivity[10] $\varepsilon_m = -23.6 + 1.69i$ and the appropriate dispersion relations [Eqs. (13) and (15)], the effective mode index ($N_{eff} = k_{sp}/k_0$) and propagation length were calculated for both SPP modes (Fig. 5). Note that, in both cases, one has to solve a transcendent equation searching for a complex solution, and that the approximate values obtained directly from Eqs. (14) and (16) can be used only for very thin films (thinner than 40 nm). It turned out, however, that a very good approximation can be found in the closed form by making use of the following simplification: $(k_{l(s)rsp}^2 - \varepsilon_m k_0^2)^{0.5} \approx k_0 \sqrt{\varepsilon_d - \varepsilon_m}$ in both Eqs. (13) and (15). The corresponding dispersion relations can be then written down as follows[16]

$$k_{l(s)rsp} \approx k_0 \sqrt{\varepsilon_d + (\varepsilon_d - \varepsilon_m)(\varepsilon_d / \varepsilon_m)^2 \cdot \tanh^{(-)2}\left(0.5 k_0 t \sqrt{\varepsilon_d - \varepsilon_m}\right)} . \quad (17)$$

It is seen that the explicit solutions above result in accurate values of the mode effective index and propagation loss for both LR- and SR-SPPs in the whole range of film thicknesses, down to that of 10 nm (Fig. 5).

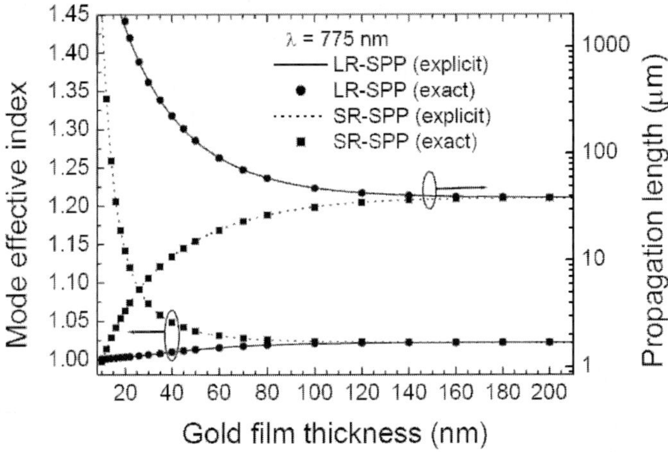

Fig. 5. The effective indexes of short- and long-range surface plasmon-polariton modes (SR-SPP and LR-SPP, respectively) and their propagation lengths at the excitation light wavelength of 775 nm for a thin gold film surrounded by air (adapted from Ref. 16).

5. Gap SPP modes

Let us now consider the SPP modes in the *symmetric* MIM configuration of a thin dielectric layer with the thickness t being sandwiched between two metal surfaces (with metal extending indefinitely on both sides of the dielectric). When two identical SPP modes start overlapping with each other for small layer thicknesses, the propagation constants of the symmetric and anti-symmetric (individual) mode combinations become different[7]. The only SPP mode surviving for all values t of the gap between metal surfaces, the so called gap SPP (G-SPP), is the mode exhibiting odd symmetry of the longitudinal electric field component E_x and, consequently, even symmetry of the transverse field component E_z.

Using the appropriate boundary conditions for the normal and tangential electric field components as in the case of SPPs and the aforementioned symmetry of the corresponding field component distributions, allows one to obtain the G-SPP dispersion relation[7,11]:

$$\tanh\left(k_z^{(d)} t/2\right) = -\frac{\varepsilon_d k_z^{(m)}}{\varepsilon_m k_z^{(d)}} \quad \text{with} \quad k_z^{(m,d)} = \sqrt{k_{gsp}^2 - \varepsilon_{m,d} k_0^2} \,, \qquad (18)$$

where k_{gsp} denotes the propagation constant of the fundamental G-SPP mode with the transverse field component E_z having the same sign across the gap. The G-SPP properties determine the SPP guiding in different channel structures, including the SPP guiding in V-grooves and trenches[17], and are worth considering in detail[18].

For sufficiently small gap widths ($t \rightarrow 0$), one can use the approximation $\tanh x \approx x$ resulting in the following expression:

$$k_{gsp} \approx k_0 \sqrt{\varepsilon_d + 0.5\left(k_{gsp}^0 / k_0\right)^2 + \sqrt{\left(k_{gsp}^0 / k_0\right)^2 \left(\varepsilon_d - \varepsilon_m + 0.25\left(k_{gsp}^0 / k_0\right)^2\right)}}$$

$$\text{with } k_{gsp}^0 = -\frac{2\varepsilon_d}{t\varepsilon_m} \,. \qquad (19)$$

Here, k_{gsp}^0 represents the G-SPP propagation constant in the limit of very narrow gaps ($t \rightarrow 0$), when the real part of the correspondent effective index, $\mathrm{Re}\{k_{gsp}\lambda/(2\pi)\}$, becomes much larger than the dielectric refractive index. At the same time, the imaginary part of the G-SPP propagation constant increases rapidly as well resulting in the decrease of the G-SPP propagation length. In this limit, the propagation constants of the SR-SPP (see §4) and G-SPP modes become *equal* for *complementary* structures (i.e. obtained one from another by exchanging metal and dielectric).

For relatively large gaps, when the two SPP modes are only starting to interact, one can use the approximation $\tanh(x) \approx 1 - 2\exp(-2x)$ resulting in the following first-order corrected expression for the G-SPP propagation constant:

$$k_{gsp}^{(1)} \approx k_{sp} \sqrt{1 - \frac{4\varepsilon_d \varepsilon_m}{\varepsilon_m^2 - \varepsilon_d^2} \exp\left(-k_{z,d}^{(1)} t\right)}$$

$$\text{with} \quad k_{z,d}^{(1)} = k_{z,d}^{(0)} \sqrt{1 + \frac{4\varepsilon_m^2}{\varepsilon_m^2 - \varepsilon_d^2} \exp\left(-k_{z,d}^{(0)} t\right)} \,, \qquad (20)$$

$$\text{where} \quad k_{z,d}^{(0)} = \sqrt{k_{sp}^2 - \varepsilon_d k_0^2} = \frac{k_0 \varepsilon_d}{\sqrt{-\varepsilon_m - \varepsilon_d}} \,.$$

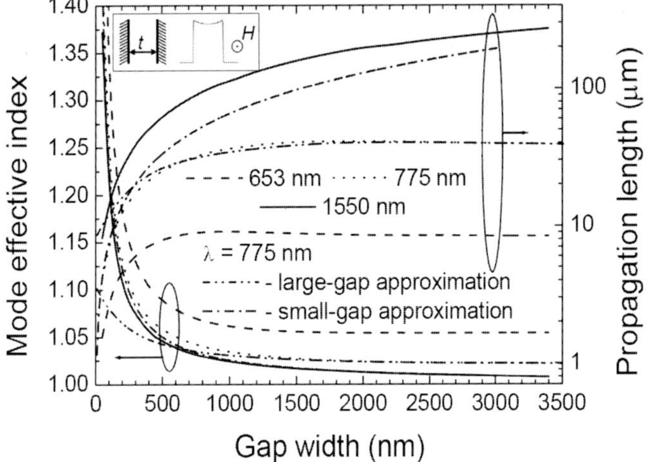

Fig. 6. The G-SPP effective index and propagation length as a function of the width t of air gap in gold calculated [Eq. (18)] for several light wavelengths along with the small [Eq. (19)] and large [Eq. (20)] gap approximations for $\lambda = 775$ nm. The insert shows the gap configuration and the G-SPP magnetic field orientation along with an example of the field distribution across a 1-μm-wide gap for $\lambda = 1550$ nm (adapted from Ref. 18).

It should be mentioned that one can proceed further with high-order corrections by using the above expression for $k_{gsp}^{(1)}$ for calculating the next correction of the G-SPP transverse wave-vector component $k_{z,d}^{(2)} = \sqrt{(k_{gsp}^{(1)})^2 - \varepsilon_d k_0^2}$. The latter value should in turn be used to determine the next correction $k_{gsp}^{(2)}$ in accordance with the corresponding expression [Eq. (20)]. The iteration process can be repeated, producing progressively accurate results.

The G-SPP characteristics were calculated for air gaps in gold and several wavelengths in the interval between visible and telecom wavelengths (Fig. 6). The following dielectric constants of gold were used in the simulations[10]: $n = 0.166 + 3.15i$ ($\lambda = 653$ nm), $0.174 + 4.86i$ (775 nm), $0.272 + 7.07i$ (1033 nm) and $0.55 + 11.5i$ (1550 nm). It is interesting that that the G-SPP propagation length first increases when the gap width t decreases from large values corresponding to uncoupled individual SPPs (Fig. 6). This surprising fact indicates that *G-SPP modes with better confinement can exhibit longer propagation lengths*[17]. The longest G-SPP propagation length is ≈ 8% larger than the SPP propagation length (for $t \to \infty$) independently on the wavelength, though it is achieved for different (gap) width-to-wavelength ratios: $t_{opt}/\lambda \approx 1.4$ ($\lambda = 653$ nm), 2.6 ($\lambda = 775$ nm), 3.3 ($\lambda = 1033$ nm) and 5.7 ($\lambda = 1550$ nm). It was found[17] that, with the accuracy of a few percents (for the above wavelengths), the optimum width t_{opt} can be evaluated simply as $t_{opt} \approx 0.5\lambda\,|\,\mathrm{Re}(\varepsilon_m) + 1\,|^{0.5}$. The explanation of this remarkable effect is related to the fact that, with the decrease of the gap width, the electric field of the G-SPP (composed of two exponents) approaches quickly to the electrostatic (capacitor) mode, which is constant across the gap (see the insert in Fig. 6). Consequently, the fractional electric field energy concentrated in the gap first increases (and the damping in metal decreases) when the gap width decreases, reaches its maximum and then starts to decrease with the field being squeezed from the gap into the metal. The exact analysis of the process of energy redistribution is quite complicated because the corresponding dispersion relation is implicit [Eq. (18)]. If one considers only the main electric field components and evaluates the G-SPP propagation constant using its large-gap approximation [Eq. (20)], one finds[18] that the normalized (to the corresponding SPP value) fraction of the electric field energy in the gap can be expressed as a function of the normalized gap width $u = t/d_d$, where $d_d = 1/\left|k_{z,d}^{(0)}\right|$ is the SPP penetration depth in dielectric [Eq. (11)]:

$$R(u = \frac{t}{d_d}) \propto \frac{\exp(-\alpha t)\left[2\alpha t + \exp(\alpha t) - \exp(-\alpha t)\right]}{(\alpha d_d)\left[1 + \exp(-\alpha t)\right]^2}\,, \qquad (21)$$

$$\text{with} \quad \alpha = (1/d_d)\sqrt{1 + 4\exp(-t/d_d)}\,.$$

Here α is in fact an approximation of $k_{z,d}^{(1)}$ [Eq. (20)] when $|\varepsilon_m| >> \varepsilon_d$, as is typically the case. This function reaches its maximum value, which is by 6% larger than that at $t \to \infty$, for $u = t/d_d \approx 3.48$, resulting in the following expression for the optimum width: $t_{opt} \approx 0.55\lambda\,|\,\mathrm{Re}(\varepsilon_m) + 1\,|^{0.5}$, a value which is quite close to that found above from the numerical simulations.

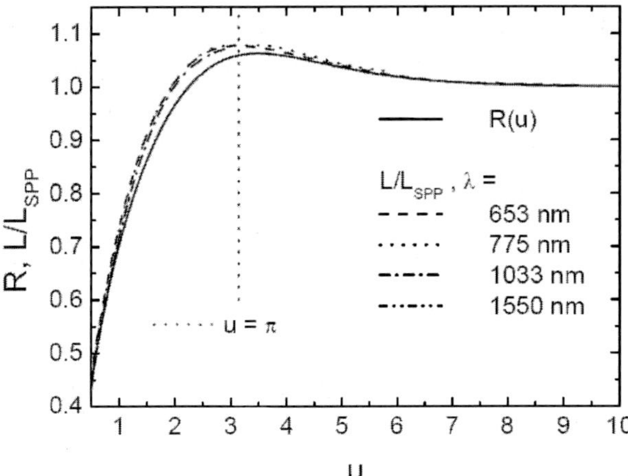

Fig. 7. The normalized fraction of the electric field energy in the gap [Eq. (21)] and G-SPP propagation length as a function of the normalized gap width $u = t/d_{\mathrm{d}}$ for the air gap in gold for several light wavelengths (adapted from Ref. 18).

It is interesting that within a wide range of the gap widths the normalized G-SPP propagation length seems dependent on only one parameter, the normalized gap width $u = t/d_{\mathrm{d}}$ (Fig. 7). This feature allows one to determine the G-SPP propagation length for any gap widths and wavelengths using the universal dependence $L_{\mathrm{gsp}} = L_{\mathrm{sp}}R^*(u)$, where the function $R^*(u)$ can be approximated by the above expression [Eq. (20)] with a few percents of error (Fig. 7). Finally, it should be noted that the condition for gap width ensuring the longest G-SPP propagation length, $u_{\mathrm{opt}} = t_{\mathrm{opt}}/d_{\mathrm{d}} \approx \pi$, resembles the resonator condition since $d_d^{-1} = k_{z,d}^{(0)}$ represents the SPP transverse wave-vector component [Eq. (20)].

6. Effective-index modelling of SPP waveguides

Let us now make use of the explicit relations established for different SPP modes in the previous sections and analyze various waveguide configurations ensuring the lateral, i.e. two-dimensional (2D), confinement of SPP modes such as stripe, groove and trench waveguides with the help of the effective-index method (EIM). The main attractive feature of the EIM is that it allows one to combine the results of modeling conducted for one-dimensional (1D) waveguide configurations so that the characteristics of 2D (channel) waveguides can be described[19]. For a rectangular-core waveguide, one should first analyze a planar (slab) waveguide obtained by letting one dimension of the original 2D waveguide approach infinity. Thus obtained mode propagation constant(s) is then used to define the corresponding effective dielectric index(s) assigned to the core index(s) of another 1D waveguide considered in the perpendicular direction. The propagation constant(s) of this second waveguide are taken to represent those of the original rectangular waveguide[19]. The EIM is one of the most extensively used approaches for modeling of dielectric channel waveguides in integrated optics[20], where its validity has been scrutinized and various corrections have been introduced[21]. The EIM has also been successfully applied to modeling of weakly guided SPP modes supported by thin metal stripes of finite width[14,22] and strongly confined SPP modes supported by V-grooves[17,23] and trenches[17] in metals. In this Section, several 2D metal waveguide configurations that are able of supporting laterally confined SPP modes (Fig. 8) are considered by making use of the EIM.

6.1. Symmetric thin-film optical waveguides

We start with introducing normalized waveguide parameters for 1D symmetric thin-film optical waveguides and recalling useful expressions[24] established for the mode effective index N and effective waveguide thickness W. An optical waveguide consisting of a film (core) with the thickness w and refractive index n_1 embedded in the dielectric medium (cladding) with the refractive index n_2 can be conveniently characterized with the normalized frequency (or film thickness) V and mode index b as follows[24]:

$$V = w k_0 \sqrt{n_1^2 - n_2^2} \quad \text{and} \quad b = \left(N^2 - n_2^2\right)/\left(n_1^2 - n_2^2\right). \tag{22}$$

It turned out that, for TE modes (the electric field is parallel to the film surfaces and perpendicular to the mode propagation direction), solutions of the corresponding (implicit) waveguide dispersion relation can be approximated for modes being very far from cutoff ($b \sim 1$) by a simple explicit relation[24]

$$b \cong 1 - \left\{ \frac{(m+1)\pi}{V+2} \right\}^2 , \quad m = 0, 1, 2, \ldots \tag{23}$$

where m is the mode index ($m = 0$ for the fundamental mode). In passing we note that, for modes being very close to cutoff ($b << 1$), the normalized mode index can be expressed by $b \cong 0.25(V - m\pi)^2$.

The single-mode waveguide design is often aimed at achieving the best lateral mode confinement because this confinement strongly influences the maximum density of waveguide components in integrated optical circuits[17]. It is convenient to introduce the effective waveguide thickness W that includes the core thickness and the mode field penetration depth into the cladding normalized with respect to the light wavelength and index contrast[24]:

$$W = k_0 \sqrt{n_1^2 - n_2^2} \left(w + \frac{2}{k_0 \sqrt{N^2 - n_2^2}} \right) = V + \frac{2}{\sqrt{b}} . \tag{24}$$

It is seen that the effective waveguide thickness diverges both for modes being very far from cutoff ($b \to 1$ implying $V \to \infty$) and for those being very close to cutoff ($b \to 0$), reflecting the fact that the waveguide mode cannot be indefinitely squeezed by decreasing the waveguide core thickness. The dependence $W(V)$ features quite a broad minimum[24], reaching the value $W_{min} = 4.93$ at $V_0 = 1.73$. Such a waveguide supports the fundamental mode with $b_0 \cong 0.39$ that carries ~ 60% of the mode power within the waveguide core, with the rest of power propagating in the cladding[25]. The parameter space for the single-mode symmetric waveguide, within which the fundamental TE mode is well confined, can thereby be determined by the following unequality: $1.73 \leq V \leq \pi$. The fundamental (even) mode at the onset of the second (odd) mode (i.e., when $V_1 = \pi$) is characterized with $b_1 \cong 0.65$ and carries ~ 84% of the mode power within the waveguide core[25].

6.2. Stripe SPP waveguides

The SPP waveguide configuration representing a finite-width metal stripe placed on a dielectric substrate [Fig. 8(a)] is probably the most natural (it is realized immediately by limiting the lateral extent of a metal film supporting SPPs) and certainly the best studied one to date[26]. The main problem with this configuration is that the stripe SPP modes are very close to the light line, even closer than SPPs supported by metal surfaces (Section 3), and prone therefore to experience radiation losses, leaking out into a dielectric substrate with a relatively large refractive index [Fig. 8(a)]. This radiation loss increases drastically for modes approaching cutoff (since their fields spill progressively out of the stripe) and becomes dominant, exceeding the "natural" SPP loss by absorption in the metal[27].

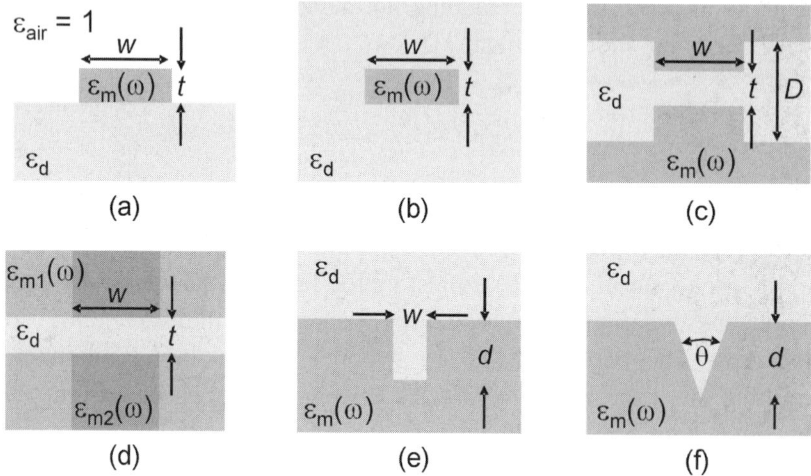

Fig. 8. Schematic of the considered SPP waveguide configurations ensuring the 2D lateral mode confinement.

Careful analysis of stripe SPP waveguides requires elaborate numerical modelling[27,28], but quite accurate design guidelines can be easily worked out using the EIM[22] especially when making use of the waveguide description with the normalized parameters introduced above. Within the EIM framework guided modes of the stripe SPP waveguide [Fig. 8(a)] can be approximated by TE modes of a symmetric dielectric waveguide consisting of the core with the thickness w and refractive index $n_1 = k_{sp}/k_o$ [Eq. (8)] embedded in the dielectric with the refractive index $n_2 = 1$. Here it is also assumed that the metal stripe thickness t is sufficiently large to decouple the SPPs supported by the upper and lower stripe surfaces, otherwise one should use a suitably corrected expression for the SPP propagation constant[1-3] instead of Eq. (8). The corresponding normalized frequency can then be found simply by combining Eqs. (8) and (22):

$$V_{sp} = wk_0 / \sqrt{|1+\varepsilon_m|} \quad . \tag{25}$$

Taking into account evolution of the effective waveguide thickness described by Eq. (24) along with the above relation, the best lateral mode confinement should be expected to occur for the stripe widths close to $w_0 \cong 0.275\lambda\sqrt{|1+\varepsilon_m|}$. Note that, since at this value ~ 40% of mode power is carried outside the stripe (contributing thereby to the radiation leakage into the substrate), this width should also be considered as the (cutoff) width, below which the mode propagation loss would drastically increase. Considering gold stripes and the light wavelength of 780 nm one obtains (using the gold susceptibility $\mathrm{Re}(\varepsilon_m) \cong -24.13$) the minimal stripe width $w_0 \cong 1.1\ \mu$m, a value which is in good agreement with the numerical result obtained using the full-vectorial finite-difference method[28]. Concluding this subsection, I would like to emphasize that, while the stripe SPP waveguides are conceptually appealing and easy to fabricate, their operation is diffraction limited, a circumstance that severely decreases their application potential.

6.3. Symmetric stripe SPP waveguides

The symmetric SPP waveguide configuration representing a finite-width metal stripe embedded in a dielectric [Fig. 8(b)] is probably the second-best studied configuration, at least with respect to supporting the LR-SPP modes[14,29,30]. Similarly to the previous configuration, careful analysis of these waveguides requires elaborate numerical modelling[29], but the EIM has been found adequate for evaluating the mode characteristics including the propagation length for modes being far from cutoff[14,22]. On the contrary, the symmetric stripe SPP modes are not leaky and the only loss mechanism, at least in the idealized configuration, is related to the radiation absorption, i.e. to ohmic losses, as is the case with conventional SPPs (Section 3). This means that the propagation length of symmetric stripe SPP modes is always larger than that of the corresponding plane (LR- or SR-) SPP and increases with the decrease of the stripe width, as the mode fields progressively spread further away into the dielectric.

Following the same procedure as in the previous subsection and using the approximation valid for thin metal films [Eq. (14)], one obtains for the normalized frequency of LR-SPP modes supported by finite-width stripes the following simple expression:

$$V_{lrsp} \cong 2\pi^2 \frac{wt}{\lambda^2} \varepsilon_d \left| 1 - \frac{\varepsilon_d}{\varepsilon_m} \right| . \tag{26}$$

Considering that very often $|\varepsilon_m| >> \varepsilon_d$, the condition for the best lateral mode confinement can be readily found: $w_0 \cong 0.088 \lambda^2 /(t\varepsilon_d)$. The onset of the second mode (i.e., when $V_1 = \pi$) determines the upper limit $w_1 \cong 0.16 \lambda^2 /(t\varepsilon_d)$ for stripe widths ensuring the single-mode operation. Considering 15-nm-thick gold stripes embedded in the dielectric with n = 1.535 and the light wavelength of 1550 nm one obtains the stripe width $w_0 \cong 6.0$ μm to achieve the best mode confinement and the second-mode cutoff width $w_1 \cong 11$ μm, both values being in good agreement with the reported results[14]. It transpires that the operation of symmetric stripe LR-SPP waveguides is also diffraction limited [Eq. (26)]. However, due to exceedingly small propagation (< 1 dB/mm) and coupling (< 1 dB) losses (when using single-mode fibers) achieved for nm-thin metal stripes at telecom wavelengths[14], a number of promising practical photonic components have already been successfully realized[14,30].

The symmetric SPP waveguide configuration [Fig. 8(b)] supports also the propagation of SR-SPP modes that are tightly bound to the metal surface (Section 4). Using the appropriate approximation for thin metal films [Eq. (16)], one obtains for the normalized frequency of SR-SPP modes supported by finite-width stripes the following simple expression:

$$V_{srsp} \cong 2 \frac{w\varepsilon_d}{t|\varepsilon_m|} . \tag{27}$$

Contrary to the previous cases [Eqs. (25) and (26)], the above expression does not contain *explicitly* the light wavelength, signifying the fact that the operation of symmetric stripe SR-SPP waveguides is not limited by diffraction. This remarkable feature is directly related to the SR-SPP property of being progressively confined to a metal film whose thickness t decreases to zero, with the SR-SPP wavelength and propagation distance being correspondingly decreased (Section 4). In this respect, the stripe SR-SPP configuration is similar to other SPP waveguide configurations ensuring truly subwavelength mode confinement (i.e., not limited by diffraction) discussed in detail in other chapters of this book. Considering the same configuration parameters as above, i.e. t = 15 nm, λ = 1550 nm, $\varepsilon_d \cong 2.36$, Re(ε_m) ≅ - 132, results in the width $w_0 \cong 726$ nm to achieve the best mode confinement and the second-mode cutoff width $w_1 \cong 1.32$ μm. A well-guided stripe SR-SPP mode exhibits the propagation length and effective index that are quite similar to those of the plane SR-SPP [29], and these can be readily evaluated using Eq. (17) as $L_{srsp} \cong 6.0$ μm and $N_{eff} \cong 1.65$. The latter value can in turn be used to determine the mode confinement in depth (perpendicular to the stripe surface) via the field evanescent decay length in the dielectric, $(\lambda / 2\pi)\cdot(N_{eff}^2 - \varepsilon_d)^{-0.5}$, as being equal to 816 nm, which makes the stripe SR-SPP mode featuring nearly symmetrical (and subwavelength) lateral confinement. Note that the stripe geometrical parameters (and thereby the mode confinement) can be further decreased by decreasing the stripe thickness [Eq. (27)], though at the expense of decreasing the propagation length (Section 4). Finally, since the main SR-SPP component in the dielectric (E_z) is anti-symmetric, the stripe SR-SPP mode should exhibit a high reflectivity at the stripe termination due to impedance mismatch with the propagating EM waves, a circumstance that can be successfully exploited when designing resonant nano-antennas[15,16]. In this respect, it can be advantageous to make use of *square* metal rods embedded in a dielectric that support various SPP modes, including the corresponding SR-SPP mode exhibiting radial orientation of the transverse electric field component[31].

6.4. G-SPP waveguides

The G-SPP characteristics discussed above (Section 5), especially a non-trivial dependence of the G-SPP propagation length on the gap width

(Fig. 7), indicate that the G-SPPs exploit in the *most efficient* way the available dielectric space (gap) between the metal walls minimizing thereby the absorption loss[17]. This remarkable feature stimulates investigations of various G-SPP-based waveguide geometries. Since the G-SPP effective index is strongly dependent on the gap width (increasing with its decrease) it is natural to exploit this dependence for achieving the 2D lateral mode confinement, e.g. by laterally varying the gap width[32] as shown in Fig. 8(c). The G-SPP mode in such a configuration is laterally confined to a dielectric stripe between the closest metal surfaces, where the G-SPP effective index reaches its largest value. This configuration is considered in detail in Chapter 4, and I present here only simple design relations in the spirit of the previous subsections.

For this purpose, it is desirable to further simplify the small-gap approximation given by Eq. (19), which is somewhat cumbersome to handle. Considering various terms in this approximation one notices that for not too small gaps, i.e. when $\left|k^0_{gsp}\right| < k_0 \Leftrightarrow t > (\lambda\varepsilon_d)/(\pi|\varepsilon_m|)$, it can be approximated as follows:

$$k_{gsp} \approx k_0 \sqrt{\varepsilon_d + \frac{2\varepsilon_d\sqrt{\varepsilon_d - \varepsilon_m}}{k_0 t(-\varepsilon_m)}} \quad . \tag{28}$$

The G-SPP characteristics were calculated for air gaps in gold and several wavelengths in the interval between visible and telecom wavelengths (for the parameters listed in Section 5) using the exact (implicit) dispersion relation [Eq. (18)] and the above explicit (analytic) formula (Fig. 9). It is seen that the above approximation gives fairly accurate values for both the G-SPP propagation length and effective index, in fact progressively more accurate for longer wavelengths because of larger values of $|\varepsilon_m|$. Considering the system parameters used for simulations of G-SPP waveguides[32], i.e. $\lambda = 532$ nm, $\varepsilon_d = 1$, Re(ε_m) ≅ - 12.6 (silver), t = 34 and 85 nm, results when using Eq. (28) in the effective G-SPP index $N_{eff} \cong 1.57$ and 1.26, respectively, values that agree well with the values $N_{eff} \cong 1.61$ and 1.28 obtained directly from the dispersion relation[32].

The normalized frequency corresponding to the variable-gap (G-SPP) waveguide configuration [Fig. 8(c)] can be readily obtained by using Eqs. (22) and (28):

$$V_{gsp} \cong 2w\sqrt{\frac{\pi\varepsilon_d\sqrt{|\varepsilon_d-\varepsilon_m|}}{\lambda|\varepsilon_m|}\left(\frac{1}{t}-\frac{1}{D}\right)}\,. \tag{29}$$

It seen that the above normalized frequency depends explicitly on the light wavelength, reflecting the fact that the gap width values considered are relatively large. In the limit of very narrow gaps, i.e. when $|k^0_{gsp}| >> k_0 \Leftrightarrow t << (\lambda\varepsilon_d)/(\pi|\varepsilon_m|)$, the G-SPP dispersion becomes the same as that of the SR-SPP (Section 5), and the corresponding normalized frequency becomes equal to that given by Eq. (27) if $D >> t$ [Fig. 8(c)]. The wavelength dependence disappears in this limit as expected for the SPP waveguides whose operation is not limited by diffraction. Note that the above condition is rather stringent requiring extremely narrow gaps, e.g. the gap width t << 13 nm is required for the system parameters used in the previous example.

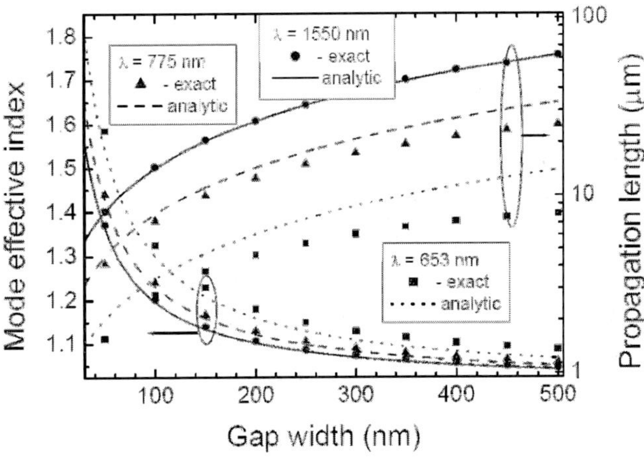

Fig. 9. The G-SPP mode effective index and its propagation length as a function of the width t of air gap in gold for several light wavelengths calculated exactly [Eq. (18)] and using the analytic (moderate gap) approximation [Eq. (28)].

The G-SPP modes can be confined in the lateral direction not only by varying the gap width[32] but also by using two metals with sufficiently different susceptibilities[33] as shown in Fig. 8(d) and considered in detail in Chapter 5. The corresponding normalized frequency can be easily worked out essentially in the same manner as that used above for G-SPP waveguides with the variable gap width. Considering the system parameters used for simulations of G-SPP hetero-waveguides[33], i.e. λ = 539.1 nm, ε_d = 2.25, Re(ε_{m1}) ≅ - 42.13 (aluminum), Re(ε_{m1}) ≅ - 10.55 (silver), t = 15 nm, results when using Eq. (19) in the effective G-SPP indexes $N_{eff} \cong 2.55$ and 3.80, respectively, and thereby in the normalized frequency $V \cong 0.033w$ [Eq. (22)]. The latter relation implies that the best lateral mode confinement should be expected for the silver width $w_0 \cong 53$ nm (corresponding to $V_0 \cong 1.73$), a value that agrees well with the reported numerical simulations regarding the mode width dependence upon the silver width[32]. Similarly to the previous case, the wavelength dependence of the normalized frequency of G-SPP hetero-waveguides disappears in the limit of very narrow gaps:

$$V_{hgsp} \cong 2\frac{w\varepsilon_d}{t}\left|\varepsilon_{m1}^{-1} - \varepsilon_{m2}^{-1}\right| , \qquad (30)$$

signifying the fact that their operation is not limited by diffraction as well. It should be emphasized that careful analysis of G-SPP waveguides requires accurate simulations using advanced numerical techniques (Chapters 4 and 5) and the relations introduced above can serve only as guidelines for narrowing down the parameter space to be explored in theoretical and experimental investigations.

6.5. Channel SPP waveguides

The possibility of guiding of electromagnetic waves by a channel cut into an otherwise planar surface of a solid (metal or polar dielectric) characterized by a negative dielectric function was first considered by Maradudin and co-workers (in the electrostatic limit[34] and with retardation being taken into account[35]) and then demonstrated experimentally for telecom wavelengths[23]. Considering a straight groove cut into metal and having a rectangular [Fig. 8(e)] or gradually varying [Fig. 8(f)] cross section, one realizes that, since light tends to be confined in regions with higher refractive indexes (because of total internal reflection), sufficiently deep grooves should support bound modes formed by G-SPPs, whose effective index increases with the decrease in the gap width (Section 5).

The SPP modes supported by rectangular trenches [Fig. 8(e)] can be described within the EIM framework by considering TE-modes in a three-layer structure, in which a dielectric core having the effective index of the corresponding G-SPP is sandwiched between the air cladding and the gold substrate[17]. Using the approximation of moderately narrow trenches, i.e. under the condition $\left|k_{gsp}^{0}\right|<k_0 \Leftrightarrow w>(\lambda\varepsilon_d)/(\pi|\varepsilon_m|)$, one can express the normalized frequency of trench SPP (TPP) waveguides as follows [Eqs. (22) and (28)]:

$$V_{tpp} \cong 2d\sqrt{\frac{\pi\,\varepsilon_d\sqrt{|\varepsilon_d-\varepsilon_m|}}{\lambda\,w|\varepsilon_m|}} \quad . \tag{31}$$

The TPP mode fields decrease to (nearly) zero at the trench bottom because of a large magnitude of the dielectric constant of the metal substrate and the boundary condition for the electric field[17]. For this reason, the TPP modes are nearly identical to the corresponding (odd) modes of the symmetrical (G-SPP) waveguide having the double width, a circumstance that allows one to deduce a simple relation for the single-mode TPP guiding [17]: $0.5\pi < V_{tpp}(w, d) < 1.5\pi$. Using as a compromise $V_{tpp} = \pi$, one obtains the following design formula for the single-mode TPP waveguides [Eq. (31)]:

$$\frac{d^2}{\lambda w}\left(V_{tpp}=\pi\right) \cong \frac{\pi}{4}\frac{|\varepsilon_m|}{\varepsilon_d\sqrt{|\varepsilon_d-\varepsilon_m|}} \quad . \tag{32}$$

Considering the system parameters used for simulations of TPP waveguides in gold at telecom wavelengths, i.e. w = 500 nm, λ = 1.55 μm, ε_d = 1, Re(ε_m) $\cong$ - 132, results when using Eq. (32) in the trench depth $d \cong 2.64$ μm, a value that ensures indeed good confinement of the TPP mode inside the trench[17].

The SPP modes supported by metal grooves, in particular V-grooves [Fig. 8(f)], are conventionally called channel SPPs (CPPs)[35] and considered in detail in Chapters 9 and 10 (experimental results on CPP-based nanophotonic components can be found in Chapter 11). Careful analysis of CPP waveguides requires elaborate numerical modelling[35-37], but the design guidelines can be worked out using the EIM[17,23]. Within the framework of the EIM, one can find the CPP modes supported by a V-groove through analyzing a one-dimensional layered (in depth) guiding structure, in which the top layer of air and the bottom layer of metal abut a stack of layers having refractive indexes determined by the layer depth: an index is equal to the G-SPP effective index for a gap width corresponding to the groove width at this depth[23]. The normalized frequency of CPP waveguides can be obtained by integrating the index contrast over the groove depth, a straightforward procedure when using the moderate gap approximation [Eq. (28)] that results in the following expression:

$$V_{cpp} \cong 2\sqrt{\frac{k_0 d\, \varepsilon_d \sqrt{|\varepsilon_d - \varepsilon_m|}}{|\varepsilon_m| \tan(\theta/2)}}\,. \tag{33}$$

It is seen that, since $\tan(\theta/2) = 0.5w/d$ (w being the groove width at the top of the groove), the CPP waveguide parameter is exactly twice that of TPP [Eq. (31)]: $V_{\rm cpp} = 2V_{\rm tpp}$. The latter relation is somewhat surprising as the cross section of a CPP waveguide is twice smaller than that of the TPP waveguide [cf. Figs. 8(e) and 8(f)]. On the other hand, the G-SPP effective index increases with the decrease in the gap width and, thereby, *narrower* channels (grooves) represent *stronger* waveguides (i.e., with larger normalized frequencies), as is also transparent from Eqs. (31) and (33), which accounts for the relation above. Note that the above relation for the CPP waveguide parameter [Eq. (33)] involves additional approximations as compared to Eq. (31). Thus, when using the EIM for CPP waveguides, it is implicitly assumed that the CPP electric field is polarized parallel to the sample surface, an approximation that can be justified only for narrow ($\theta \ll 1$) grooves[37]. Furthermore, when obtaining Eq. (33), the moderate gap approximation was applied for the gap widths varying all the way from 0 to the (maximum) groove width at the top of the groove. Conducting accurate simulations with a finite-element method it has recently been directly verified, that the CPP field magnitude distributions calculated for two different V-groove waveguides (operating at different wavelengths) characterized with the same parameter $V_{\rm cpp}(\lambda, \theta, d) = 1.34\pi$ [Eq. (33)] bear indeed a close resemblance to each other, exhibiting a rather similar CPP confinement to the groove region[38].

Concluding this Section I would like to note that, for the TPP and CPP waveguides being far from cutoff, the lateral (in width) mode confinement is simply determined by the groove width w at the groove top. It should be borne in mind that, as the TPP or CPP modes approach the cutoff and the mode fields become progressively larger at the groove edges, one should expect the occurrence of mode coupling to wedge SPP modes leading to their hybridization[37]. In general, for all considered SPP waveguide configurations ensuring the 2D lateral mode confinement (Fig. 8) such a hybridization of the SPP waveguide and wedge modes becomes important when approaching the cutoff. The correspondent behavior cannot be described with the EIM and requires accurate modeling using advanced numerical techniques described in detail in the following Chapters. On the other hand, the expressions for the normalized frequency of different SPP waveguides obtained in this Section can still be used as *practical guidelines* to design waveguides for operation within a given wavelength range, so as to ensure the single-mode operation with a well-confined fundamental mode being far away from the cutoff. The latter is very important since, in practice, any structural irregularities would result in the coupling of SPP waveguide modes (especially those close to the cutoff) to plane SPPs propagating away from the waveguide and thereby in additional propagation loss. Finally, I would like to remark that it is also possible to estimate the propagation length and effectives index of a well-guided fundamental mode simply by considering the corresponding LR-, SR- or G-SPP modes. Typically, such estimation would slightly overestimate the mode propagation loss and effective index as compared to the values obtained by using the EIM for determining the mode properties[17].

Acknowledgments

I am grateful to E. Bozhevolnaya for help in preparing this chapter, and acknowledge the support by the European Network of Excellence, PLASMO-NANO-DEVICES (FP6-2002-IST-1-507879) and by the

Danish Agency for Science, Technology and Innovation (contract No. 274-07-0258).

References

1. V.M. Agranovich and D.L. Mills, Eds., *Surface Polaritons* (North-Holland, Amsterdam, 1982).
2. H. Raether, *Surface Plasmons* (Springer-Verlag, Berlin, 1988).
3. A.D. Boardman, Ed., *Electromagnetic Surface Modes* (John Wiley & Sons, NewYork, 1982).
4. S. A. Maier, *Plasmonics: Fundamentals and Applications* (Springer, Berlin, 2007).
5. M. Born and E. Wolf, *Principles of Optics* (New York, Cambridge University Press, 7th ed., 1999).
6. M. Nieto-Vesperinas, *Scattering and Diffraction in Physical Optics* (John Wiley & Sons, New York, 1991).
7. E.N. Economou, *Phys. Rev.* **182**, 539 (1969).
8. V.G. Veselago, *Sov. Phys. Uspekhi* **10**, 509 (1968).
9. J.B. Pendry, *Phys. Rev. Lett.* **85**, 3966 (2000).
10. E.D. Palik, *Handbook of Optical Constants of Solids* (Academic, New York, 1985).
11. R. Zia, M.D. Selker, P.B. Catrysse, M.L. and Brongersma, *J. Opt. Soc. Am. A*, **21**, 2442 (2004).
12. D. Sarid, *Phys. Rev. Lett.* **47**, 1927 (1981).
13. J.J. Burke, G.I. Stegeman and T. Tamir, *Phys. Rev. B* **33**, 5186 (1986).
14. A. Boltasseva, T. Nikolajsen, K. Leosson, K. Kjaer, M.S. Larsen and S.I. Bozhevolnyi, *J. Lightwave Techn.* **23**, 413 (2005).
15. T. Søndergaard and S.I. Bozhevolnyi, *Phys. Rev. B* **75**, 073402 (2007).
16. S.I. Bozhevolnyi and T. Søndergaard, *Opt. Express* **15**, 10869 (2007).
17. S.I. Bozhevolnyi, *Opt. Express* **14**, 9467 (2006).
18. S.I. Bozhevolnyi, V.S. Volkov, E. Devaux, J.-Y. Laluet and T.W. Ebbesen, *Appl. Phys. A* **89**, 225 (2007).
19. G.B. Hocker and W.K. Burns, *Appl. Opt.* **16**, 113 (1977).
20. S.I. Bozhevolnyi, K.S. Buritskii, E.M. Zolotov, A.M. Prokhorov and V.A. Chernykh, *Sov. J. Quantum Electron.* **11**, 1517 (1981).
21. K.S. Chiang, C.H. Kwan and K.M. Lo, *J. Lightwave Technol.* **17**, 716 (1999).
22. R. Zia, A. Chandran and M. L. Brongersma, *Opt. Lett.* **30**, 1473 (2005).
23. S.I. Bozhevolnyi, V.S. Volkov, E. Devaux and T.W. Ebbesen, *Phys. Rev. Lett.* **95**, 046802 (2005).
24. H. Kogelnik and V. Ramaswamy, *Appl. Opt.* **13**, 1857 (1974).
25. H. Kogelnik and H.P. Weber, *J. Opt. Soc. Am.* **64**, 174 (1974).
26. J.R. Krenn and J.C. Weeber, *Philos. Trans. R. Soc. London* **362**, 739 (2004) ; and references therein.

27. R. Zia, M.D. Selker and M.L. Brongersma, *Phys. Rev. B* **71**, 165431 (2005).
28. R. Zia, J.A. Schuller and M.L. Brongersma, *Phys. Rev. B* **74**, 165415 (2006).
29. P. Berini, *Phys. Rev. B* **61**, 10484 (2000).
30. S.I. Bozhevolnyi, in *Nanophotonics with Surface Plasmons*, edited by V. M. Shalaev and S. Kawata (Elsevier, Amsterdam, 2007); P. Berini, in *Surface Plasmon Nanophotonics*, edited by P. G. Kik and M.L. Brongersma (Springer, Berlin, 2007); and references therein.
31. J. Jung, T. Søndergaard and S.I. Bozhevolnyi, *Phys. Rev. B* **76**, 035434 (2007).
32. K. Tanaka and M. Tanaka, *Appl. Phys. Lett.* **82**, 1158 (2003).
33. B. Wang and G.P. Wang, *Appl. Phys. Lett.* **85**, 3599 (2004).
34. J.Q. Lu and A.A. Maradudin, *Phys. Rev. B* **42**, 11159 (1990).
35. I.V. Novikov and A.A. Maradudin, *Phys. Rev. B* **66**, 035403 (2002).
36. D.K. Gramotnev and D.F.P. Pile, *Appl. Phys. Lett.* **85**, 6323 (2004).
37. E. Moreno, F.J. Garcia-Vidal, S.G. Rodrigo, L. Martin-Moreno and S.I. Bozhevolnyi, *Opt. Lett.* **31**, 3447 (2006).
38. S.I. Bozhevolnyi and J. Jung, *Optics Expr.* **16**, 2676 (2008).

Chapter 2

NEGATIVE DIELECTRIC OPTICAL WAVEGUIDES FOR NANO-OPTICAL GUIDING

Junichi Takahara

Graduate School of Engineering Science, Osaka University, Toyonaka, Osaka 560-8531, Japan

E-mail: takahara@ee.es.osaka-u.ac.jp

Negative dielectric optical waveguides are described from the viewpoint of low-dimensional optical waves. Principles are presented for making narrow optical beams beyond diffraction limit. Analytic solutions and fundamental properties of surface plasmon polariton modes propagating along plane and cylindrical metal-dielectric interfaces are reviewed.

1. Introduction

Dielectric optical waveguides are the basic elements for transmission of light. In dielectric optical waveguides, optical waves are confined and guided by total internal reflection at a dielectric interface. It is primarily the diffraction limit of light that determines how small a waveguide can be. To achieve optical waveguides on the nanometer scale –1 to 10 nm in size– a new approach is needed.

In this chapter, we will describe metal waveguides, which are widely known and have been extensively studied. Metal waveguides and metallic nano-structures are seen today as key devices in development of subwavelength optical technology, which is known as plasmonics.[1,2] The metal waveguides described in this chapter are those known in current terminology as "negative dielectric waveguides". We introduce the concept of "low-dimensional optical waves" and discuss a new method that enables us to confine and guide optical waves in nanostructures.[3,4] This approach can be used in the realization of nano-optical circuits.

In metals, the real part of the permittivity ϵ is negative ($Re[\epsilon] < 0$) at optical frequencies.[5] Materials with $Re[\epsilon] < 0$ are called negative dielectrics (NDs). In recent years, new type of optical metamaterials with simultaneously negative permittivity and permeability has been realized: i.e. negative index (NI) materials.[6–9] Now we are extending optical materials with only $\epsilon > 0$ and $\mu > 0$ to wide range of ϵ-μ space including negative sign of ϵ and μ. Thus, we can categorized ND waveguides within ϵ-μ diagram including NI materials as shown in Fig.1. In dielectric and NI materials, optical waves can propagate through them. Optical waves, however, cannot penetrate into NDs; hence we need a new approach for wave guiding. We are able to confine and guide optical waves as self-sustaining mode along the interface between ND and dielectric. For example, optical waves coupled to free-electron system in metals, i.e. surface plasmon polariton (SPP), can propagate at a metal surface. These ND waveguides are called plasmonic waveguides, which is the most promising candidate of nano-optical waveguides.

In this chapter, we introduce the concept of low-dimensional optical waves and wavenumber surface. We discuss the origin of diffraction limit in geometrical approach in wavenumber space. Then, we describe principles to confine and guide optical waves in nanostructures beyond the diffraction limit using low-dimensional optical waves. We next present structures and calculated transmission properties in various types of negative dielectric waveguides. Finally, we expand these schemes to novel interface including negative index materials.

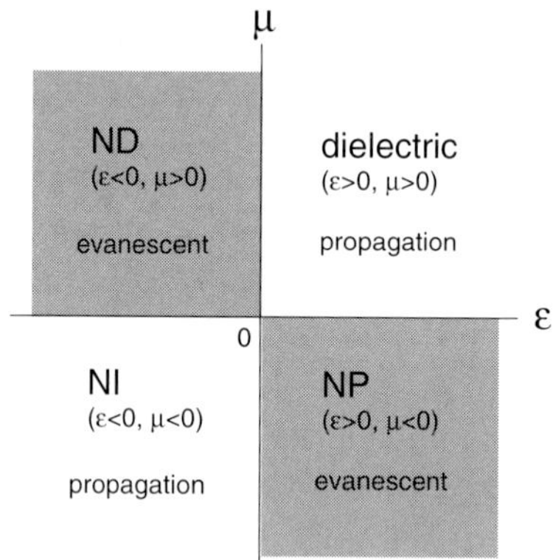

Fig. 1. NDs in ϵ-μ diagram for both signs of ϵ and μ. Dielectric, ND, materials with negative permeability (NP), negative-index materials are shown. Propagation of optical waves is prohibited in the bulk of ND and NP (gray areas).

2. Low-dimensional optical waves

In nano-optics, an essential role is played by optical waves, the spatial field of which decays exponentially. This is different from what occurs in conventional optics, where extended modes of optical waves play a dominant role. We can model this type of wave based on the concept of low-dimensional optical waves.[4,10]

We define the dimension of an optical wave as the number of real components in the wavevector $\boldsymbol{k}$. An optical wave having three real components in $\boldsymbol{k}$ is defined as a 3-dimensional (3D) optical wave; light propagating in free-space is an example of a 3D optical wave. An optical wave having one or two real components in $\boldsymbol{k}$ (the other components are imaginary) is defined, respectively, as a 1D- or a 2D optical wave. An optical wave that has no real components in $\boldsymbol{k}$ (because all the components are imaginary) is defined as a zero-dimensional (0D) optical wave. Two-dimensional, 1D, and 0D optical wave are all low-dimensional optical waves. In this chapter, we demonstrate that 1D and 2D optical waves can physically exist and that a low-dimensional optical waveguide is the key to the realization of nano-optical circuits.

3. Diffraction limit of 3D optical waves

As a basis for our discussion, let us first describe the origin of the diffraction limit of optical beams. Since an optical beam can be synthesized by the superposition of 3D optical plane waves, we consider a plane wave of angular frequency ω in a uniform dielectric (D) medium with refractive index n. The dispersion relation of light is

$$|\boldsymbol{k}|^2 = n^2 \frac{\omega^2}{c^2}, \tag{1}$$

where $|\boldsymbol{k}| = k$ is the wavenumber in the medium and c is the light speed in vacuum. In Cartesian coordinates, the components of $\boldsymbol{k} = (k_x, k_y, k_z)$ must satisfy

$$k_x^2 + k_y^2 + k_z^2 = (nk_0)^2 = \epsilon\mu(\frac{\omega}{c})^2, \tag{2}$$

where λ_0 and $k_0 = 2\pi/\lambda_0$ are the wavelength and wavenumber in vacuum, respectively, and ϵ and μ are the relative permittivity and the permeability of the medium, respectively. Eq.(2) is the surface of a sphere in 3D k-space

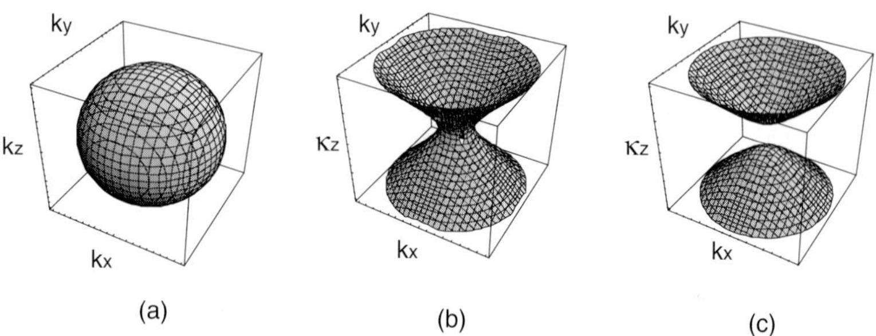

Fig. 2. Wavenumber surface for (a) 3D optical wave in D, (b) 2D optical wave in D, (c) 2D optical wave in ND.

$\boldsymbol{k} = (k_x, k_y, k_z)$ as shown in Fig. 2 (a). Here we call this surface in k-space an "wavenumber surface", which is an analogy to Fermi surface.[10–12]

Since all the components of the 3D optical wave are real, each component k_j (j=x,y,z) takes values satisfying $-k \leq k_j \leq k$. This means that the range of spatial frequency Δk is $2k$ ($\Delta k = 2nk_0$). According to the uncertainty relation of Fourier transforms Δk and the range of real space Δr satisfy $\Delta r \Delta k \geq \pi$. Therefore, Δr has a minimum given by the following relation

$$\Delta r \geq \frac{\pi}{\Delta k} = \frac{\lambda_0}{4n}. \tag{3}$$

Eq.(3) means that the minimum size of a beam synthesized by the superposition of 3D optical waves is limited to the order of λ_0 given as

$$\Delta r_{min3D} = \frac{\lambda_0}{4n}. \tag{4}$$

We underline the fact that this is the origin of the diffraction limit; it is inevitable as long as the optical wave is 3D. It is worth noting that optical waves in conventional dielectric waveguides, the optical waves are 3D in despite the 1D shape of the waveguides. Even in photonic crystal waveguides or high refractive index waveguides, the optical waves are 3D. Thus, the width of optical beams in such waveguides is limited by diffraction to the order of λ_0.

4. 2D optical waves

To break through the diffraction limit, we need a low-dimensional optical wave having larger Δk than a 3D optical wave. If Δk is larger than $2k$ (that is, $\Delta k > 2nk_0$), Δr becomes smaller than the diffraction-limited size $\lambda_0/(4n)$.

A typical example of a 2D optical wave is the evanescent wave generated at a dielectric interface by total internal reflection (TIR)(see Fig. 3(a)). Because $\boldsymbol{k}$ of the evanescent wave obeys $k_x^2 + k_y^2 > (nk_0)^2$, k_z is imaginary from Eq.(2); hence the optical wave is 2D. The $\boldsymbol{k}$ components of a 2D optical wave $(k_x, k_y, i\kappa_z)$ satisfies the relation

$$k_x^2 + k_y^2 - \kappa_z^2 = \epsilon_2 \mu (\frac{\omega}{c})^2, \tag{5}$$

where $\epsilon_2 (> 0)$ is the relative permittivity of the medium. As shown in Fig. 2 (b), Eq.(5) is the surface of one-sheeted circular hyperboloid (1SCH) in 2D k-space (k_x, k_y, κ_z).

An evanescent wave can also be excited by reflection inside a metal surface (Fig. 3(b)), i.e. a D/ND interface; "D/ND" represents an interface between D and ND, such kind of representation is used again in this chapter. Such an evanescent wave in an ND is also a 2D optical wave. The $\boldsymbol{k}$ components of a 2D optical wave $(k_x, k_y, i\kappa_z^{'})$ satisfies the relation

$$k_x^2 + k_y^2 - \kappa_z^{'2} = \epsilon_{\mathrm{ND}} \mu (\frac{\omega}{c})^2, \tag{6}$$

where $\epsilon_{\mathrm{ND}} (< 0)$ is the relative permittivity of the medium. As shown in Fig. 2 (c), Eq.(6) is the surface of two-sheeted circular hyperboloid (2SCH) in 2D k-space (k_x, k_y, κ_z). Here we define penetration depths of the 2D optical wave as $\delta_{\mathrm{D}} = 1/\kappa_z$ and $\delta_{\mathrm{ND}} = 1/\kappa_z^{'}$.

We can excite these evanescent waves in both dielectrics and NDs by the irradiation of 3D optical waves; in other words, evanescent waves always accompany 3D optical waves. Thus, k of a 2D optical wave is restricted by the wavenumber of the accompanying 3D optical wave $(\Delta k < 2nk_0)$.

Another kind of 2D optical wave propagates along a D/ND interface as a surface wave (Fig. 3(c)). This type of wave is self-sustaining mode that exists without accompanying 3D optical waves. When an ND is metal, this is known as a SPP. An SPP is the coupled mode of an electromagnetic (EM) field and the surface mode of the collective excitation of a free-electron system, or surface plasmon.[13] The EM field of an SPP is a Transverse Magnetic (TM) field mode and is localized at the D/ND interface as shown in Fig. 3(c). The field of an SPP decreases exponentially on both sides of D and ND; hence, the SPP is 2D optical waves on both sides.

5. Topology of wavenumber surfaces

Let us consider unique properties of an SPP from viewpoint of a wavenumber surface.[10–12] Wavenumber at an interface must coincide due to the

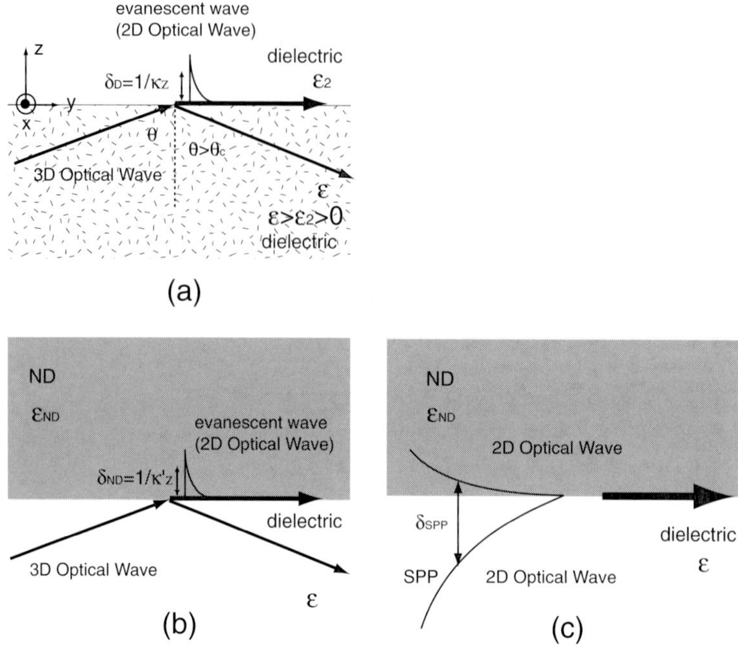

Fig. 3. Examples of 2D optical waves: (a) TIR and evanescent wave at dielectric interface, (b) reflection and evanescent wave at metal surface, (c) surface wave at metal surface.

boundary condition; hence we have to consider two wavenumber surfaces on both side of the interface. The boundary condition can be represented in a cylindrical surface as $k_x^2 + k_y^2 = \text{constant}$.

Figure 4 shows two wavenumber surfaces and the boundary condition at the interface. We underline the fact that surface topology in k-space is open for both 2SCH and 1SCH in contrast to a sphere as closed surface. Hence, as shown in Fig. 4(a) and (b), Δk of 2D optical waves is restricted by the radius of a sphere. For the case of an SPP, however, wavenumber surfaces are 2SCH and 1SCH; hence, Δk of 2D optical waves can be increased to infinity without any limitations. Thus, Δr goes to zero when $\Delta k \to \infty$. This implies that SPP can overcome the diffraction limit, and 2D optical waves play a crucial role in nanophotonics, as described below.

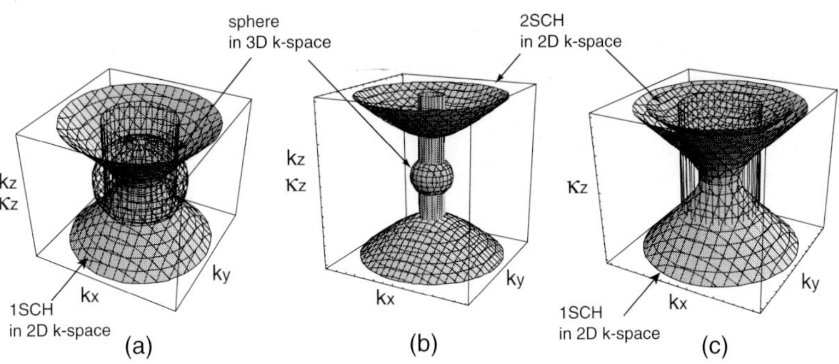

Fig. 4. Wavenumber surfaces and the boundary condition in k-space in (a)TIR at D/D, (b) reflection at D/ND, (c) SPP at D/ND.

6. Diffraction limit of 2D optical waves

As discussed in Sec.5, SPP has a potential ability to make very large k. Let us consider SPP on a plane D/ND interface as shown in Fig.5(a): the interface between an isotropic ND half-space with relative permittivity $\epsilon_1 = \epsilon_{\mathrm{ND}} < 0$ and, an isotropic dielectric half-space with $\epsilon_2 = \epsilon > 0$. We assume that the materials are nonmagnetic, so that relative permeability $\mu_1 = \mu_2 = 1$. EM field propagating in y direction along the interface is assumed in the form of $e^{i(\omega t - \beta y)}$, where β is a phase constant (wavenumber) along the interface. The characteristic equation is given by

$$\frac{\epsilon_1}{\gamma_1} + \frac{\epsilon_2}{\gamma_2} = 0, \tag{7}$$

where γ_j $(j = 1, 2)$ is defined by

$$\gamma_j = \sqrt{\beta^2 - \epsilon_j (\frac{\omega}{c})^2}. \tag{8}$$

The solution β of Eq.(7) is the wavenumber of SPP given by

$$k_{\mathrm{SPP}} \equiv \beta = \frac{\omega}{c}\sqrt{\frac{\epsilon\epsilon_{\mathrm{ND}}}{\epsilon + \epsilon_{\mathrm{ND}}}} = nk_0\sqrt{\frac{\epsilon_{\mathrm{ND}}}{\epsilon + \epsilon_{\mathrm{ND}}}}. \tag{9}$$

Eq. (9) implies that SPP feels effective permittivity of both materials as $1/\epsilon + 1/\epsilon_{\mathrm{ND}}$. Eq. (9) can be obtained under the condition (the condition of SPP propagation) as

$$\epsilon_{\mathrm{ND}} < -\epsilon < 0 \quad (|\epsilon_{\mathrm{ND}}| > \epsilon). \tag{10}$$

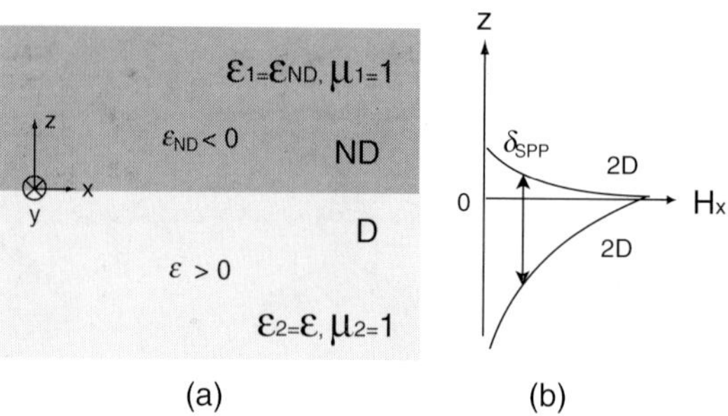

Fig. 5. Coordinates and magnetic field distribution $H_x(z)$ of SPP: (a) single D/ND interface, (b) magnetic field (H_x) profile and the width of 2D optical wave δ_{SPP}.

From Eq. (10), one can find that k_{SPP} is larger than nk_0 as

$$k_{\mathrm{SPP}} = nk_0 \frac{1}{\sqrt{1 - \frac{|\epsilon|}{|\epsilon_{\mathrm{ND}}|}}} > nk_0. \tag{11}$$

The EM field of SPP is confined at the interface as a 2D optical wave as shown in Fig. 5(b). Hence we can define the "width" of SPP by adding $\delta_{\mathrm{D}} = 1/\kappa_z$ and $\delta_{\mathrm{ND}} = 1/\kappa_z^{'}$ as

$$\delta_{\mathrm{SPP}} = \delta_{\mathrm{D}} + \delta_{\mathrm{ND}} = \frac{\lambda_0}{2\pi}\left(\frac{\sqrt{|\epsilon_{\mathrm{ND}}| - \epsilon}}{\epsilon} + \frac{\sqrt{|\epsilon_{\mathrm{ND}}| - \epsilon}}{|\epsilon_{\mathrm{ND}}|}\right). \tag{12}$$

In this chapter, we use permittivity of silver (Ag) as an typical example of NDs. A numerical example is $k_{\mathrm{SPP}} = 1.03k_0$ for $\epsilon = 1$ and $\epsilon_{\mathrm{ND}} = -19$ in ideal lossless silver at λ_0=633 nm (lossless values are selected for the sake of simplicity and used again below). In Sec.6 – 8, we consider lossless ND for the sake of simplicity. We will discuss lossy ND in Sec.9.

Fig. 6 (a) shows k_{SPP} with respect to ϵ_{ND} in complex plane for $\epsilon = 1$. For $|\epsilon_{\mathrm{ND}}| \to \epsilon$, k_{SPP} diverges to infinity; k_{SPP} has a pole at $\epsilon_{\mathrm{ND}} = -\epsilon$. Hence, k_{SPP} can be increased by changing material constant of ϵ or ϵ_{ND}. Fig. 6 (b) shows δ_{SPP} with respect to ϵ_{ND}; it can de decreased rapidly near the pole.

We can overcome the diffraction limit using SPP because k_{SPP} is larger than nk_0. The minimum beam size of 2D optical waves $\Delta r_{\mathrm{min2D}}$ is given as

$$\Delta r_{\mathrm{min2D}} = \frac{\lambda_{\mathrm{SPP}}}{4} = \frac{\lambda_0}{4n}\sqrt{\frac{\epsilon + \epsilon_{\mathrm{ND}}}{\epsilon_{\mathrm{ND}}}}. \tag{13}$$

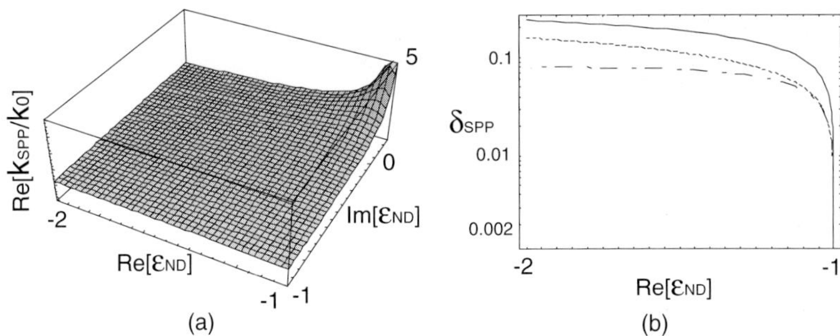

Fig. 6. Wavenumber and width of SPP: (a) $k_{\rm SPP}$ and (b) $\delta_{\rm SPP}$ (solid line) vs. the relative permittivity of ND for $\epsilon = 1$. $\delta_{\rm D}$ (dotted line) and $\delta_{\rm ND}$ (dash-dotted line) are also shown.

From Eq. (10), one can find that $\Delta r_{\rm min2D}$ is smaller than $\Delta r_{\rm min3D}$ as

$$\Delta r_{\rm min2D} = \frac{\lambda_0}{4n}\sqrt{1 - \frac{|\epsilon|}{|\epsilon_{\rm ND}|}} < \frac{\lambda_0}{4n}. \tag{14}$$

Thus, we can overcome diffraction limit of 3D optical wave by 2D optical wave; nevertheless the minimum beam size synthesized by 2D plane waves of SPP is still limited to the subwavelength order of $\Delta r_{\rm min2D}$. This is what we call the "diffraction limit of 2D optical waves".

Although Δr_{min2D} is smaller than $\Delta r_{\rm min3D}$, $\Delta r_{\rm min2D}$ still depends on permittivity of materials. This implies that controlling the minimum width of 2D optical beams is difficult on metal surface. SPP waveguides have been studied as possible nano-optical circuits, but the minimum beam width of SPP is still limited to the order of the subwavelength. If we can somehow increase k of 2D optical waves instead of by changing the material constants, we can make Δr much smaller than the diffraction limit of 3D optical waves; $\Delta r_{\rm min2D} \ll \Delta r_{\rm min3D}$. The coupled system of an SPP is suitable for this scheme.

7. 2D optical waveguides

7.1. *Structures*

Let us next consider the coupled system of SPP. We place two D/ND interfaces close together, as shown in Fig. 7, and consider the complementary coupled systems of SPPs: a thin dielectric layer sandwiched between semi-infinite ND layers (ND gap) and a thin ND layer sandwiched between

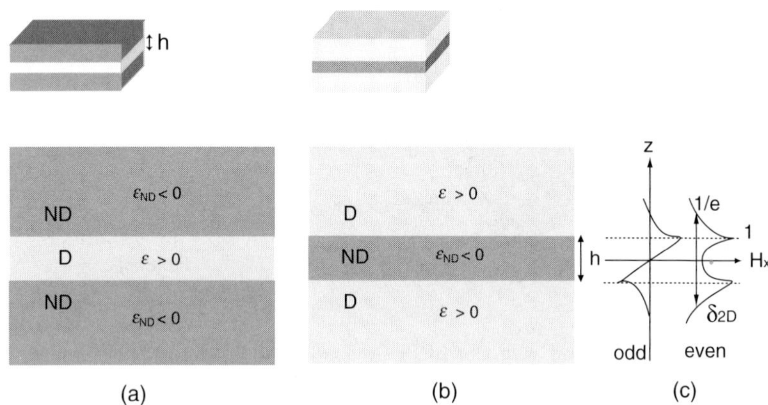

Fig. 7. Schematic view of 2D optical waveguides (double D/ND interface structures) and coordinate: (a) ND gap, (b) ND film, (c) schematic magnetic field (H_x) profile and the width of 2D optical wave.

dielectric layers (ND film). In these structures, the SPPs at each D/ND interface couple to form a new TM propagation mode when we reduce the thickness of the middle layer. This mode, known as the Fano mode, is a 2D optical wave.[14,15]

7.2. *Characteristic equations of ND gap and ND film*

We consider characteristic equations of the ND gap and film. Fig. 7(c) shows Cartesian coordinates of symmetric ND gap and film: in ND gap two ND half-spaces are separated by a D region at the distance of h, in an ND film two D half-spaces are separated by an ND region at the distance of h. EM field propagating in y direction along the waveguide is assumed in the form of $e^{i(\omega t-\beta y)}$, where β is a phase constant. The EM fields of the coupled SPP mode are shown in Fig. 7(c).

Characteristic equations of TM mode of 2D optical waves in symmetric ND gap and ND film are given by

$$\tanh(\gamma_1 \frac{h}{2}) + \frac{\epsilon_1 \gamma_2}{\epsilon_2 \gamma_1} = 0 \quad \text{(even mode)}, \tag{15}$$

$$\tanh(\gamma_1 \frac{h}{2}) + \frac{\epsilon_2 \gamma_1}{\epsilon_1 \gamma_2} = 0 \quad \text{(odd mode)}, \tag{16}$$

where ϵ_1 and ϵ_2 are relative permittivity in core and clad layer, respectively.

γ_j $(j = 1, 2)$ is given by

$$\gamma_j = \sqrt{\beta^2 - \epsilon_j(\frac{\omega}{c})^2}. \tag{17}$$

For the ND film, we take $\epsilon_1 = \epsilon_{\mathrm{ND}}$ and $\epsilon_2 = \epsilon$. For the ND gap, we take $\epsilon_1 = \epsilon$ and $\epsilon_2 = \epsilon_{\mathrm{ND}}$. One can obtain characteristic equation of single D/ND interface (Eq.(7)) from Eqs.(15) or (16) by taking the limit of $h \to \infty$ using $\tanh x \to 1$ for $x \to \infty$.

We define width δ_{2D} of 2D optical wave, as shown in Fig. 7(c). The beam width of a 2D optical wave δ_{2D} can be represented as

$$\delta_{\mathrm{2D}} = h + \frac{2}{\gamma_2}. \tag{18}$$

Here we underline the fact that δ_{2D} depends on structural parameter h; hence we can control width of 2D optical waves by structures instead of material constants.

Note that parity (even or odd) is defined by the distribution of magnetic field H_x, as shown in Fig. 7(c), since SPP is a TM mode. In the literature, even and odd mode are called antisymmetric and symmetric mode, respectively, due to electric charge symmetry induced by the tangential components of electric fields at the interfaces. These are opposite parities from our definition.

We can calculate β of 2D optical waves by solving Eqs.(15) – (17) numerically. In the case of ND gap, we have another propagation modes as 3D optical waves, which are conventional propagation modes that are known to exist in metal-clad waveguides. We can also calculate β for 3D optical waves by solving the conventional characteristic equations of the TM and transverse electric (TE) modes.

7.3. *Coupled mode of SPP*

Figure 8 shows the propagation constant β of a 2D optical wave with respect to the distance between the two D/ND interfaces for $\epsilon_{\mathrm{ND}} = -19$ and $\epsilon = 1$. As shown in Fig. 8(a), propagation modes in the ND gap split into two branches as the gap distance decreases. The propagation constant β of the upper branch increases to infinity, while the lower branch has a cutoff level. Propagation modes in the ND film also split into two branches as the film thickness decreases, as shown in Fig. 8(b). The propagation constant β of the upper branch increases to infinity while β of the lower branch

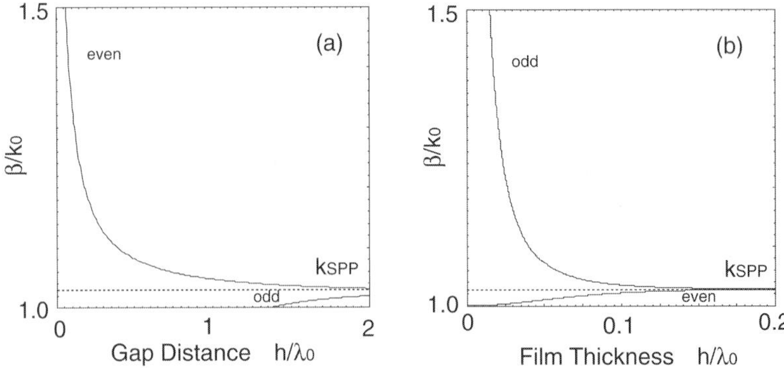

Fig. 8. Propagation modes of 2D optical wave in ND gap and ND film: (a) propagation constant vs. gap distance, (b) propagation constant vs. film thickness. The relative permittivities of the dielectric layer and ND layer are 1 and −19, respectively. Dotted lines indicate k_{SPP} in a single D/ND interface. β and h are normalized to k_0 and λ_0=633nm, respectively.

asymptotically approaches k_0. The lower branch in the ND film, which has long propagation length even in lossy ND, is called long-range SPP.[16]

For the case of the ND gap, we have additional propagation modes in 3D optical waves, which are conventional propagation modes in metal clad waveguides. We show all propagation modes (both 2D and 3D optical wave) in Fig.9(a). The modes of 3D optical wave are seen in $\beta/k_0 < \sqrt{\epsilon}$; both TM and TE modes in 3D optical wave have a cutoff level. We discuss here only propagation modes of 2D optical waves, and 3D optical wave is not a scope of this review, because these modes have a cutoff.

Let us consider EM field of the even-mode of 2D optical wave in the ND gap (the upper branch). Fig.9(b) shows δ_{2D} versus gap distance. δ_{2D} decreases to zero when h approaches zero because, based on Eq.(18), γ_2 also goes to infinity. This suggests that the EM field of 2D optical wave in the upper branch can be confined into thin ($\ll 0$) structures beyond diffraction limit.

These unique properties of 2D optical waves are very useful when it comes to producing extremely narrow optical beams. We have proposed a planar index-guided nano-optical waveguide based on ND gaps, which has been reported in the literature.[17–19]

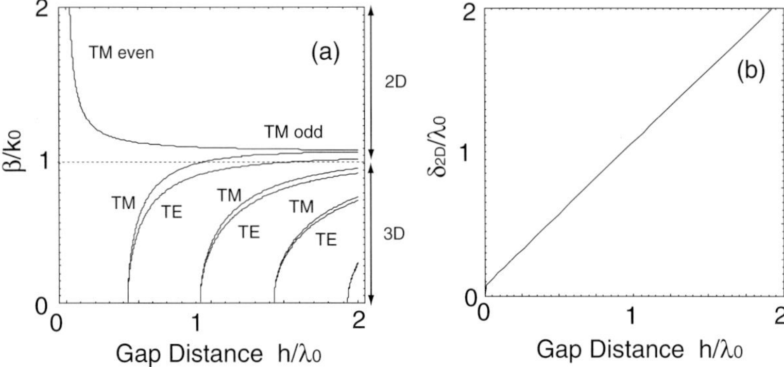

Fig. 9. All propagation modes in ND gap: (a) propagation constant vs. gap distance, (b) δ_{2D} vs. gap distance. The relative permittivities of the dielectric layer and ND layer are 1 and −19, respectively. Dotted line indicates separation between 2D and 3D optical wave. β and h are normalized to k_0 and λ_0=633nm, respectively.

8. 1D optical waveguides

8.1. *structures*

As indicated above, we can decrease the width of the 2D optical wave to zero by decreasing the distance between the two interfaces. But to confine the optical wave in a nanometer-size region, we have to confine it not only in the vertical direction but also in the horizontal direction. For this reason we turn our attention to waveguides with a cylindrical D/ND surface, as shown in Fig. 10. These waveguides can be considered rolled structures of ND gaps, or films. We have shown theoretically that some of the propagation modes in these waveguides are 1D optical waves and the cylindrical surfaces can be considered 1D optical waveguides. A 1D optical wave has only one real component of $\boldsymbol{k}$ along the axis, propagates along the axial direction and decays in the radial direction. While such kind of mode has been reported since 1970s, it has been considered to be useless for the application to waveguides due to large propagation loss.[20,21] We have first pointed out that 1D optical wave can be confined and guided beyond diffraction limit and applied to nano-optical waveguides.[3]

8.2. *Characteristic equations of ND rod and ND hole*

We consider a propagation mode in an ND rod and an ND hole (see Fig. 10(a) and (b)), where there is one D/ND interface with a radius of

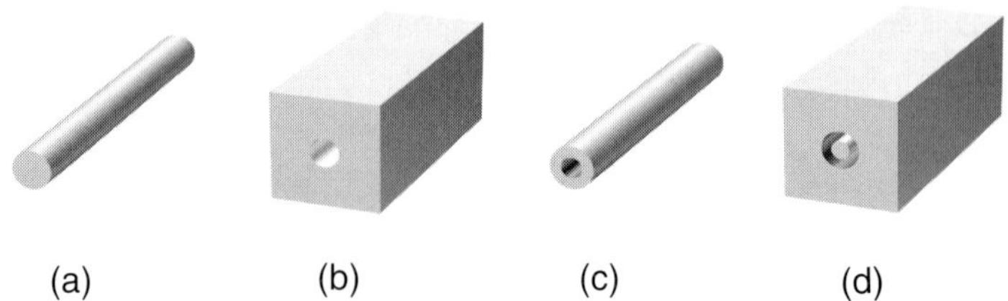

Fig. 10. Schematic view of 1D optical waveguides: (a) ND rod, (b) ND hole, (c) ND tube, and (d) ND coaxial hole.

a. In the analysis, cylindrical coordinates (r, ϕ, z) are selected as shown in Fig. 11(a). The relative permittivity of the core and the clad is ϵ_1 and ϵ_2, respectively. EM field propagating in z direction is assumed in the form of $e^{i(\omega t-\beta z)}$, where β is a phase constant.

Maxwell's equations in cylindrical coordinate are given by[22]

$$E_r = \frac{i}{\gamma^2}[\beta\frac{\partial E_z}{\partial r} + \frac{\omega\mu}{r}\frac{\partial H_z}{\partial \phi}], \tag{19}$$

$$E_\phi = \frac{i}{\gamma^2}[\frac{\beta}{r}\frac{\partial E_z}{\partial \phi} - \omega\mu\frac{\partial H_z}{\partial r}], \tag{20}$$

$$H_r = \frac{i}{\gamma^2}[-\frac{\omega\epsilon}{r}\frac{\partial E_z}{\partial \phi} + \beta\frac{\partial H_z}{\partial r}], \tag{21}$$

$$H_\phi = \frac{i}{\gamma^2}[\omega\mu\frac{\partial E_z}{\partial r} + \frac{\beta}{r}\frac{\partial H_z}{\partial \phi}], \tag{22}$$

$$\frac{1}{r}\frac{\partial}{\partial r}(r\frac{\partial E_z}{\partial r}) + \frac{1}{r^2}\frac{\partial^2 E_z}{\partial \phi^2} = \gamma^2 E_z, \tag{23}$$

$$\frac{1}{r}\frac{\partial}{\partial r}(r\frac{\partial H_z}{\partial r}) + \frac{1}{r^2}\frac{\partial^2 H_z}{\partial \phi^2} = \gamma^2 H_z, \tag{24}$$

where ϵ and μ is permittivity and permeability of the medium, respectively, γ is defined by $\gamma^2 = \beta^2 - \omega^2\epsilon\mu$. Since Eqs. (23) and (24) are separable for r and ϕ, one can obtain general solution of E_z and H_z given by

$$\{C_1 I_\nu(\gamma r) + C_2 K_\nu(\gamma r)\}\{C_3\cos(\nu\phi+\phi_1) + C_4\sin(\nu\phi+\phi_2)\}, \tag{25}$$

where $\gamma^2 > 0$, ν is integer, C_i $(i = 1,2,3,4)$ and ϕ_i $(i = 1,2)$ are any constants. I_ν and K_ν are the ν-th order modified Bessel functions. Then, one can derive other components of the EM fields form Eqs. (19) – (22).

Let us first consider the most simple case of $\nu = 0$ and seek physically meaning solutions. Here, we present only TM mode ($H_z = 0$) because TE mode ($E_z = 0$) is not a propagation mode in a 1D optical wave. The EM field in TM mode in a core ($r < a$) is given by

$$E_{z1} = AI_0(\gamma_1 r), \quad E_{r1} = \frac{i\beta}{\gamma_1^2} AI_1(\gamma_1 r), \tag{26}$$

$$H_{\phi 1} = \frac{i\omega\epsilon_1\epsilon_0}{\gamma_1^2} AI_1(\gamma_1 r), \tag{27}$$

and the EM field in a clad ($r > a$) is

$$E_{z2} = BK_0(\gamma_2 r), \quad E_{r2} = -\frac{i\beta}{\gamma_2^2} BK_1(\gamma_2 r), \tag{28}$$

$$H_{\phi 2} = -\frac{i\omega\epsilon_2\epsilon_0}{\gamma_2^2} BK_1(\gamma_2 r), \tag{29}$$

where ϵ_0 is permittivity in vacuum, A and B are any constants. γ_j ($j = 1, 2$) is defined by

$$\gamma_j = \sqrt{\beta^2 - \epsilon_j(\frac{\omega}{c})^2}. \tag{30}$$

¿From boundary conditions at $r = a$ of $E_{z1} = E_{z2}$ and $H_{\phi 1} = H_{\phi 2}$, we finally obtain the ratio of A/B and the characteristic equation of TM mode as[3]

$$\frac{A}{B} = \frac{K_0(\xi_2)}{I_0(\xi_1)}, \tag{31}$$

$$\frac{\epsilon_1 I_1(\xi_1)}{\xi_1 I_0(\xi_1)} + \frac{\epsilon_2 K_1(\xi_2)}{\xi_2 K_0(\xi_2)} = 0, \tag{32}$$

where $\xi_1 = \gamma_1 a$, $\xi_2 = \gamma_2 a$. Taking the limit of $a \to \infty$ in Eq.(32), one can obtain SPP dispersion in plane D/ND as Eq.(7).

For the case of hybrid mode ($H_z \neq 0$ and $E_z \neq 0$) of the order of ν, the EM field can be taken as

$$E_{z1} = CI_\nu(\gamma_1 r)\cos(\nu\phi), \; H_{z1} = C'I_\nu(\gamma_1 r)\sin(\nu\phi) \;\; \text{for } r < a, \tag{33}$$

$$E_{z2} = DK_\nu(\gamma_2 r)\cos(\nu\phi), \; H_{z2} = D'K_\nu(\gamma_2 r)\sin(\nu\phi) \;\; \text{for } r > a, \tag{34}$$

where C, C', D, D' are constants. Then, one can derive solutions of the other components of EM field similarly to the TM mode. We finally obtain the characteristic equation of ν-th order hybrid mode from the boundary condition as[23]

$$\{\frac{1}{\xi_1}\frac{I_\nu'(\xi_1)}{I_\nu(\xi_1)} - \frac{1}{\xi_2}\frac{K_\nu'(\xi_2)}{K_\nu(\xi_2)}\}\{\frac{\epsilon_1}{\xi_1}\frac{I_\nu'(\xi_1)}{I_\nu(\xi_1)} - \frac{\epsilon_{r2}}{\xi_2}\frac{K_\nu'(\xi_2)}{K_\nu(\xi_2)}\} = \nu^2(\frac{\beta}{k_0})^2(\frac{1}{\xi_1^2} - \frac{1}{\xi_2^2})^2. \tag{35}$$

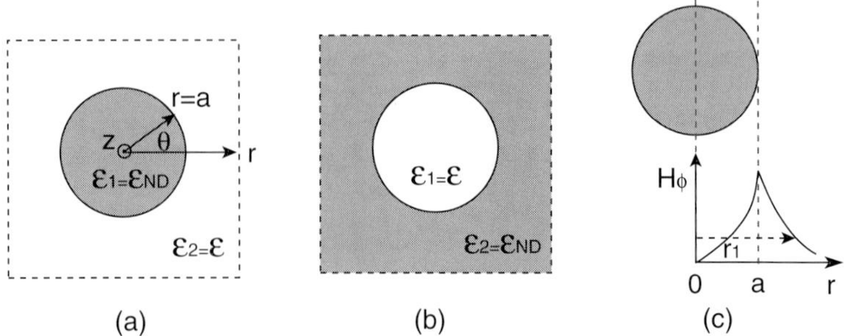

Fig. 11. Coordinates and parameters in one cylindrical D/ND interface: (a) ND rod, (b) ND hole, (c) the magnetic field distribution and the beam radius.

One can derive Eq.(32) by taking $\nu = 0$ in Eq.(35) using $I_0^{'}(x) = I_1(x)$ and $K_0^{'}(x) = -K_1(x)$. TM mode is considered to be the lowest (0th) order hybrid mode. Note that Eqs. (32) and (35) have similar forms to the characteristic equation of step index dielectric optical fiber.

EM fields of 1D optical wave in the ND rod and hole are localized at the interface and decays in both dielectric and ND as shown in Fig. 11(c). Thus, we define the beam radius r_1 of TM mode by the equation of $H_{\phi 2}(r_1) = \frac{1}{e}H_{\phi 2}(a)$. This definition gives

$$K_1(\gamma_2 r_1) = \frac{1}{e}K_1(\gamma_2 a). \tag{36}$$

We have calculated a phase constant of a 1D optical wave by solving Maxwell's equations; β and r_1 as a function of a can be calculated by solving Eqs.(32), (35) and (36) numerically. In the numerical calculations, the relative permittivity of ND and dielectric are taken to be $\epsilon_{\mathrm{ND}} = -19$ and $\epsilon = 1$, respectively. We note that propagation modes in an ND rod are only 1D optical waves while both 1D and 3D optical wave exist in an ND hole.

8.3. *Characteristic equations of ND tube and ND coaxial hole*

Next let us consider an ND tube and an ND coaxial hole (see Fig. 10(c) and (d)), where there are two D/ND interfaces.[11,23] The cross sectional views and structural parameters of these waveguides are shown in Fig.12. The relative permittivity of core, middle clad and outer clad are ϵ_1, ϵ_2 and ϵ_3, respectively. We define an inner core radius a and an outer radius

b. Because these two waveguides have two radii, we have an additional parameter $p \equiv b/a$, the ratio of the inner radius to the outer radius.

We can derive the characteristic equations of a TM and hybrid mode by the same procedure of the ND rod. ¿From the boundary conditions at $r = a$ and $r = b$, we finally obtain characteristic equation of the TM mode as determinant

$$\begin{vmatrix} -I_0(\xi_1) & I_0(\xi_2) & K_0(\xi_2) & 0 \\ 0 & I_0(\xi_3) & K_0(\xi_3) & -K_0(\xi_4) \\ \frac{\epsilon_1}{\gamma_1}I_1(\xi_1) & -\frac{\epsilon_2}{\gamma_2}I_1(\xi_2) & \frac{\epsilon_2}{\gamma_2}K_1(\xi_2) & 0 \\ 0 & \frac{\epsilon_2}{\gamma_2}I_1(\xi_3) & -\frac{\epsilon_2}{\gamma_2}K_1(\xi_3) & \frac{\epsilon_3}{\gamma_3}K_1(\xi_4) \end{vmatrix} = 0, \tag{37}$$

where $\xi_1 = \gamma_1 a$, $\xi_2 = \gamma_2 a$, $\xi_3 = \gamma_2 b$ and $\xi_4 = \gamma_3 b$. The expansion of Eq. (37) is given by

$$\begin{aligned} &\frac{\epsilon_1\epsilon_2}{\gamma_1\gamma_2}I_1(\xi_1)I_1(\xi_3)K_0(\xi_2)K_0(\xi_4) + \frac{\epsilon_2^2}{\gamma_2^2}I_0(\xi_1)I_1(\xi_3)K_0(\xi_4)K_1(\xi_2) + \\ &\frac{\epsilon_1\epsilon_2}{\gamma_1\gamma_2}I_0(\xi_2)I_1(\xi_1)K_0(\xi_4)K_1(\xi_3) - \frac{\epsilon_2^2}{\gamma_2^2}I_0(\xi_1)I_1(\xi_2)K_0(\xi_4)K_1(\xi_3) + \\ &\frac{\epsilon_1\epsilon_3}{\gamma_1\gamma_3}I_0(\xi_3)I_1(\xi_1)K_0(\xi_2)K_1(\xi_4) - \frac{\epsilon_1\epsilon_3}{\gamma_1\gamma_3}I_0(\xi_2)I_1(\xi_1)K_0(\xi_3)K_1(\xi_4) + \\ &\frac{\epsilon_2\epsilon_3}{\gamma_2\gamma_3}I_0(\xi_1)I_1(\xi_2)K_0(\xi_3)K_1(\xi_4) + \frac{\epsilon_2\epsilon_3}{\gamma_2\gamma_3}I_0(\xi_1)I_0(\xi_3)K_1(\xi_2)K_1(\xi_4) = 0. \end{aligned} \tag{38}$$

The characteristic equation of hybrid mode is 8×8 determinant of which expansion is so complicated that we do not show it here.

The beam radius r_2 of the TM mode in the ND tube and ND coaxial hole is defined in similar way to the ND rod. We define r_2 by the equation of $H_{\phi 3}(r_2) = \frac{1}{e}H_{\phi 3}(b)$ as shown in Fig.12(c). This definition gives

$$K_1(\gamma_3 r_2) = \frac{1}{e}K_1(\gamma_3 b). \tag{39}$$

β and r_2 can be calculated as a function of a or b by solving Eqs.(38) and (39) numerically. In the numerical calculations, the relative permittivity of ND and dielectric are taken to be $\epsilon_{\mathrm{ND}} = -19$ and $\epsilon = 1$, respectively.

8.4. *Propagation modes in ND rod*

Figure 13(a) shows the calculated propagation modes, with respect to core radius, of a 1D optical wave in an ND rod.[3,4] The propagation modes in the 1D optical wave include TM and hybrid modes, but not the TE mode. Here, three propagation modes are shown: the TM mode, and hybrid modes of modified Bessel functions of the orders $\nu = 1$ and $\nu = 2$.

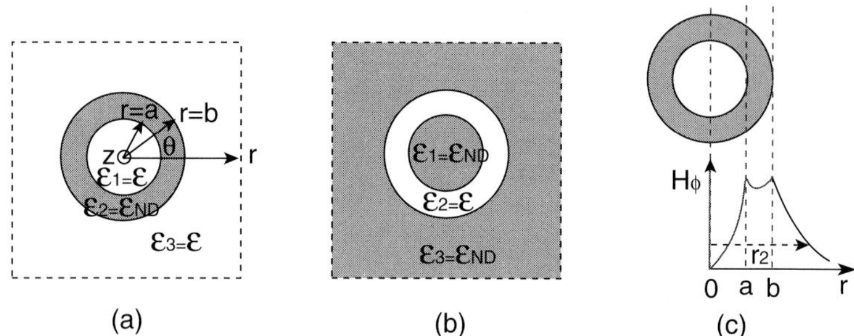

Fig. 12. Coordinates and parameters in two cylindrical D/ND interfaces: (a) ND tube, (b) ND coaxial hole, (c) the magnetic field distribution and the beam radius.

As shown in Fig. 13(a), the β of the TM mode diverges as the core radius a decreases to zero. This property is similar to those of the ND film denoted above. We underline the fact that such a dispersion relation of the TM mode is important for the confinement of an optical wave at nanometer-scale. The propagation constant of the $\nu = 1$ hybrid modes asymptotically approaches k_0 and the cutoff effect is never observed, whereas the $\nu = 2$ hybrid modes have a cutoff at $a/\lambda_0 = 0.7$.

The EM field in the ND rod is localized in the rod and decays rapidly in the dielectric. Figure 14 shows the field distribution of the ND rod at a=5 nm. As can be seen in Fig. 14, the field is localized at the D/ND interface and decays exponentially in the dielectric. Thus, we define the beam radius r_1 by the magnetic field of the TM mode, as shown in Fig. 11(c). For this example, r_1=22nm at λ_0=633nm: 1D optical beams can be confined within the diameter of 44nm by using visible light. We can obtain λ_{SPP}=120nm, which corresponds to the wavelength of UV light even in optical frequency.

Figure 13(b) shows the beam radius of the TM mode with respect to the core radius. As shown in Fig. 13 (b), r_1 decreases to zero as the core radius decreases beyond the diffraction limit. In contrast to TM mode, beam radius of hybrid mode of $\nu = 1$ diverge as the core radius decreased to zero although cutoff effect is never observed in this mode. It is therefore possible to create and propagate an optical beam with nanometer-order diameter by using the TM mode of the ND rod.

8.5. *Propagation modes in ND hole*

We have just seen that the upper branch of TM mode plays crucial role in the ND rod from the viewpoint of nano-optical waveguides. But in the case

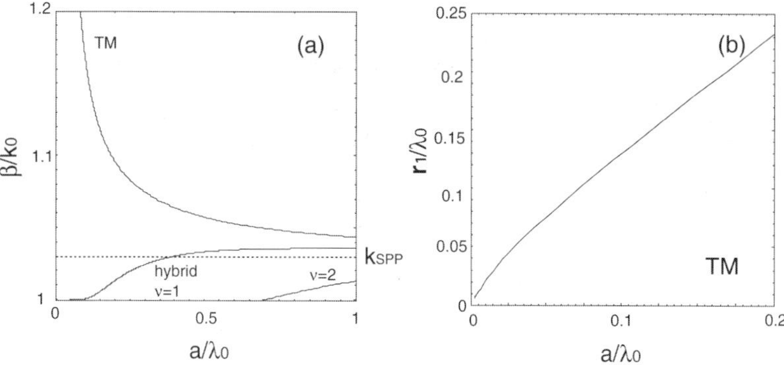

Fig. 13. Propagation modes and beam radius of 1D optical wave in ND rod: (a) propagation constant of TM and hybrid modes vs. core radius, (b) beam radius of TM mode vs. core radius. The relative permittivities of the dielectric and ND are 1 and −19, respectively. β is normalized to k_0, a and r_1 are normalized to λ_0=633nm.

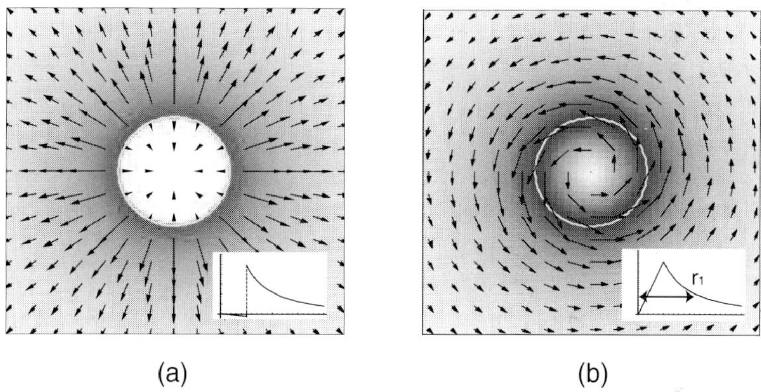

Fig. 14. Field distribution of TM mode in ND rod at a radius of 5 nm ($a/\lambda_0 = 0.0079$): (a) electric field, (b) magnetic field. Arrows show vector field in cross sectional plane. The color density changes from light to dark in proportion to the total amplitude of the field. Inset: the total amplitude of electric and magnetic field vs. core radius. The definition of the beam radius r_1 is shown.

of an ND hole, such kind of mode is not as usual. Figure 15(a) shows the propagation modes, with respect to core radius, in an ND hole. We have both 1D (TM mode) and 3D (TM and TE mode) optical waves in the ND hole, while only 1D optical wave can propagate in the ND rod. Note that the dimension of the optical waves is 1D above the horizontal dotted line ($\beta/k_0 > \sqrt{\epsilon}$), and 3D below the line ($\beta/k_0 < \sqrt{\epsilon}$). The TM mode in 1D

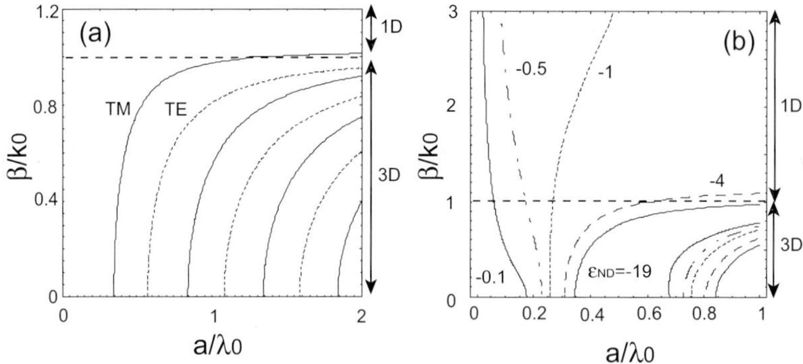

Fig. 15. Propagation modes of ND hole: (a) propagation constant of TM and TE modes vs. core radius. The relative permittivities of the dielectric and ND are 1 and -19, respectively. (b) propagation constant of TM modes for various permittivities of the ND of $-0.1, -0.5, -1, -4, -19$. β and a are normalized to k_0 and λ_0=633nm, respectively.

optical wave changes to 3D optical wave seamlessly as in Fig. 15(a). The propagation mode in the ND hole shows the same type of cutoff behavior as conventional metal waveguides.

Figure 15(b) shows the TM modes for various $\epsilon_{\rm ND}$ in an ND hole of $\epsilon = 1$. Under the condition of SPP propagation (i.e. $|\epsilon_{\rm ND}| > \epsilon$), the propagation mode shows the cutoff effect. But the dispersion curves are different for the case of $|\epsilon_{\rm ND}| < \epsilon$: β diverges as the core radius approaches zero and does not have a cutoff.[23] Furthermore, we note that in this case, the group velocity is negative.

We can fulfill the condition of $|\epsilon_{\rm ND}| < \epsilon$ at specific optical frequency range by using high refractive index materials such as a silicon (Si). Figure 16 shows typical two examples in the ND hole embedded by lossless Si as a dielectric material: β of TM mode under the condition of $|\epsilon_{\rm ND}| > \epsilon$ (λ_0=633nm) and $|\epsilon_{\rm ND}| < \epsilon$ (λ_0=500nm).[11,24] As shown in Fig. 16(a), TM mode shows a cutoff effect under the condition of $|\epsilon_{\rm ND}| < \epsilon$. We can satisfy the condition of $|\epsilon_{\rm ND}| < \epsilon$ by changing the wavelength. At λ_0=500nm, β does not have a cutoff even at $a/\lambda_0 < 0.1$, but diverges as the core radius approaches zero. Thus, an optical beam can pass through an ND hole of nanometer-order diameter only under the condition of $|\epsilon_{\rm ND}| < \epsilon$.

8.6. *Propagation modes in ND tube and ND coaxial hole*

Now, let us examine alternative structures of 1D optical waveguides: an ND tube and an ND coaxial hole (Fig. 10(c) and (d)). Figure 17(a) shows the

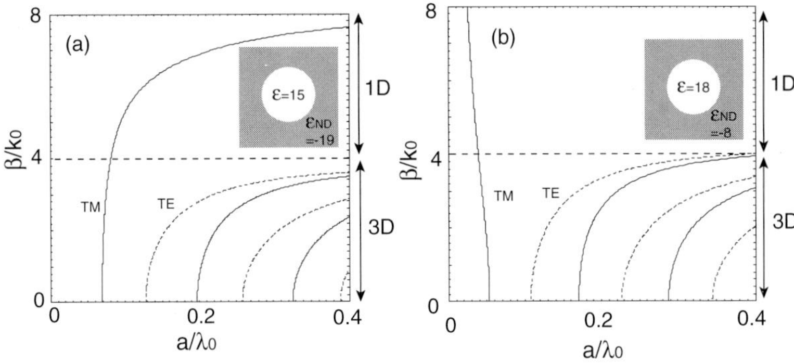

Fig. 16. Propagation modes in ND hole with Si core: (a) $\epsilon_{\rm ND} = -19$, $\epsilon = 15$ ($|\epsilon_{\rm ND}| > \epsilon$) λ_0=633nm (b) $\epsilon_{\rm ND} = -8$, $\epsilon = 18$($|\epsilon_{\rm ND}| < \epsilon$) λ_0=500nm. β and a are normalized to k_0 and λ_0, respectively. The dimension of optical wave is 1D above horizontal dashed-lines, and 3D below the lines.

calculated propagation modes, with respect to inner core radius a, of a 1D optical wave in an ND tube in the fixed ratio of $p = 2$. Propagation modes of the TM mode split into two branches as the core radius decreases, as shown in Fig. 17(a). The propagation constant β of the upper branch increases to infinity while β of the lower branch has a cutoff level. This is similar to the ND gap. Figure 17(b) shows β of TM mode, with respect to inner core radius a, in the fixed outer radius of $b/\lambda_0 = 2$. Two solutions are obtained only under the condition that the ND layer is thin ($(b - a)/\lambda_0 < 0.6$), because the mode splitting is attributed to the coupling of 1D optical waves between two interfaces.

Figure 18(b) shows beam radius of the upper branch in the TM mode r_2 with respect to inner core radius a. r_2 decreases linearly as a decreases and approaches to zero. Thus, the optical beam diameter in the ND tube can be shrunk in nanometer-order by decreasing a and b simultaneously.

Figure 19 shows the calculated propagation modes, with respect to inner core radius a, of a 1D and 3D optical wave in an ND coaxial hole in the fixed ratio of $p = 2$. As shown in Fig. 19, β of the TM mode only increases to infinity as a decreases, while hybrid modes have a cutoff level. The TM mode of an ND coaxial hole shows a similar property to that of the ND rod: β diverges as the inner core radius decreases while keeping p constant.

The TM mode of the ND tube and coaxial hole shows a property similar to that of the ND rod: β diverges as the inner core radius decreases, while the ratio of the inner and outer radii remains constant. This means that

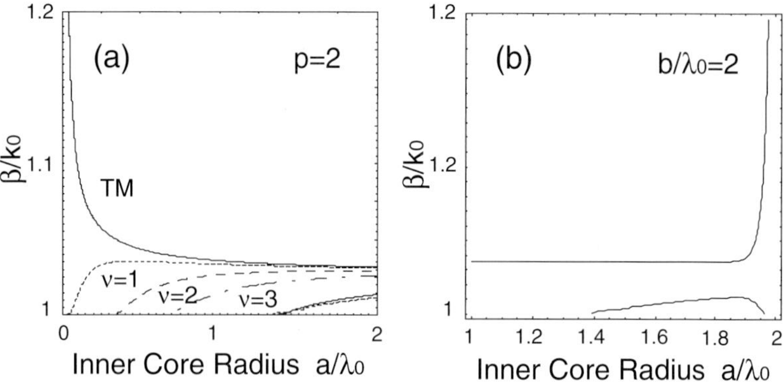

Fig. 17. Propagation modes of 1D optical wave in ND tube: (a) propagation constant of TM and hybrid modes vs. inner core radius for $p = 2$, (b) propagation constant of TM modes vs. inner core radius for fixed outer radius $b/\lambda_0 = 2$. The relative permittivities of the dielectric and ND are 1 and -19, respectively. β and a are normalized to k_0 and λ_0=633nm, respectively.

we can decrease the beam radius of a 1D optical wave in the ND tube and coaxial hole by decreasing the two radii simultaneously. This is a useful property as far as the creation of nano-optical waveguides is concerned. As denoted in Sec.8.5, mode properties of the ND hole are complicated; 1D optical waves exist under limited condition of material constants. Instead of an ND hole, we can use an ND coaxial holes as a high-transmission aperture for 1D optical waves, but 3D optical waves will be unable to pass through it.

9. Lossy 1D optical waveguides

9.1. *Propagation modes in lossy waveguides*

In Sec.7 and 8, we have treated metals as ideal lossless ND material. In the case of actual metals at optical frequencies, however, permittivity is a complex number with a small imaginary component, e.g. $\epsilon_{\rm ND} = -19 - 0.53i$ at λ_0=633 nm for silver.[15] The imaginary component of the permittivity induces Ohmic loss, which causes some transmission loss. We have calculated propagation constant ($\beta = \beta_{\rm R} - i\beta_{\rm I}$) for lossy ND rod and tube in order to estimate propagation length of 1D optical wave.

Figure 20 shows imaginary part of propagation constant $\beta_{\rm I}$ of TM mode with respect to core radius for lossy ND rod, ND tube and ND coaxial hole.

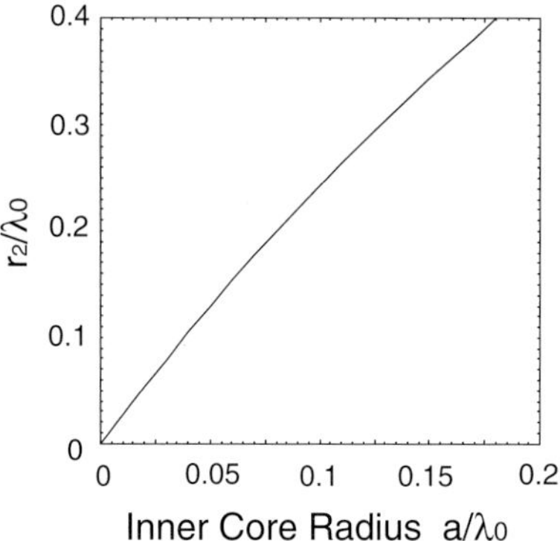

Fig. 18. Beam radius of 1D optical wave in ND tube for $p = 2$: beam radius of TM mode vs. inner core radius. The relative permittivities of the dielectric and ND are 1 and -19, respectively. r_2 and a are normalized to λ_0=633nm.

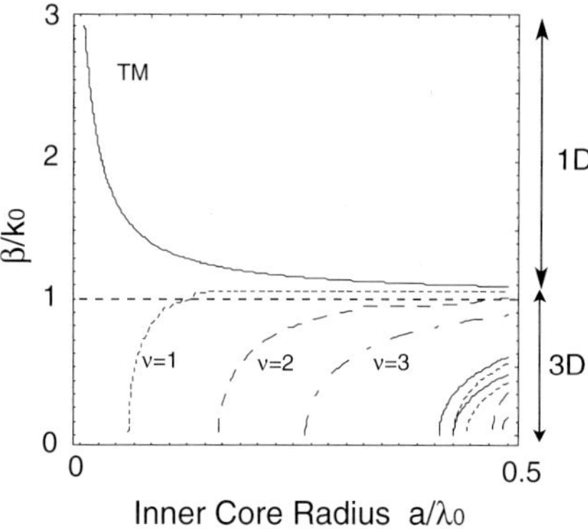

Fig. 19. Propagation modes of 1D and 3D optical wave in ND coaxial hole for $p = 2$: propagation constant of TM and hybrid modes vs. inner core radius. The relative permittivities of the dielectric and ND are 1 and -19, respectively. β and a are normalized to k_0 and λ_0=633nm, respectively.

Here we show only the results of β_I, since β_R does not change qualitatively between lossy and lossless ND as long as β_I is sufficiently small (i.e. $|\beta_I| << |\beta_R|$). In all 1D waveguides, β_I of the TM mode increases as the core radius decreases as shown in Fig.20. This directly implies that a 1D optical beam with smaller diameter has higher transmission loss.

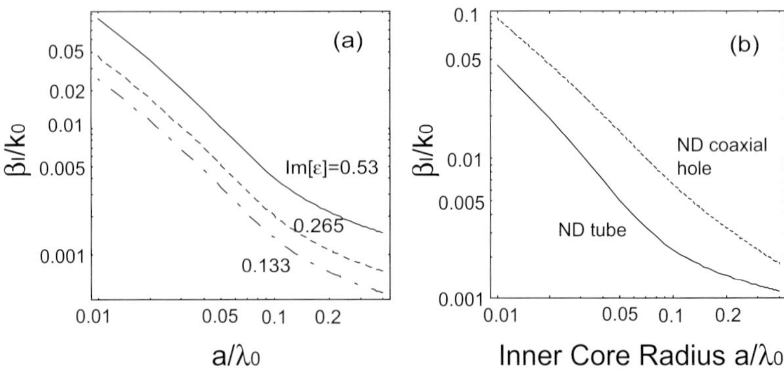

Fig. 20. Propagation loss of 1D optical waveguides: (a) Imaginary part of β vs. core radius in ND rod at $\epsilon' = 0.53$ (solid), 0.265 (dotted) and 0.133 (dashed-dotted) for $\epsilon_{\mathrm{ND}} = -19 - i\epsilon'$. (b) Imaginary part of β vs. inner core radius in ND tube (solid) and ND coaxial hole (dotted) for $p = 2$ and $\epsilon_{\mathrm{ND}} = -19 - 0.53i$. β_{I} and a are normalized to k_0 and λ_0=633nm, respectively.

9.2. *Propagation length*

Let us next evaluate propagation length of 1D optical waves for applications. Propagation length L is defined as $1/e$ decay length of the power by

$$L = \frac{1}{2\beta_{\mathrm{I}}}. \tag{40}$$

Figure 21 shows the propagation length with respect to wavelength for lossy ND rod in various diameters. The propagation length strongly depends on the rod diameter because β_{I} depends on the rod diameter as shown in Fig.20. The propagation length decreases as wavelength decreases due to Ohmic loss; it rapidly decrease near plasma frequency.[12,23]

For a silver-core ND rod of 10nm diameter at λ_0=633 nm in vacuum (ϵ =1), the beam diameter is 22 nm and the propagation length is 0.5 μ m. In addition, we point out that loss of the ν=1 hybrid mode approaches zero as the core radius decreases while its beam radius diverges. The propagation length of this mode reaches 1m at $a \sim$ 40nm; we can call this mode a "long-range 1D optical wave".[4] We underline that the transmission loss is not a very serious problem for short-range transmission ($< \lambda_0$) when applications to optical devices of nanometer size being considered.

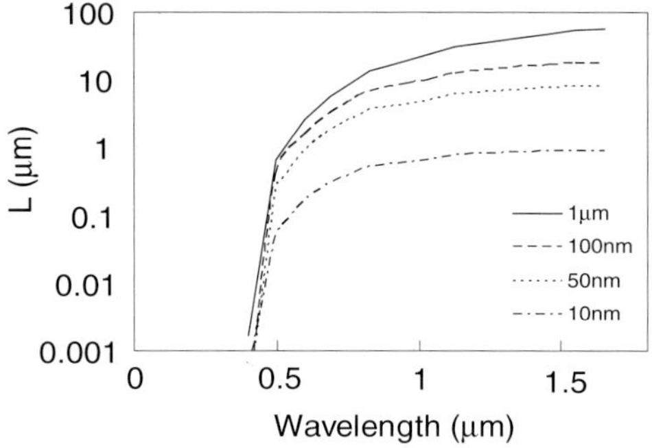

Fig. 21. Propagation length vs. wavelength for ND rod in vacuum for various rod diameters.

9.3. *Tapered waveguides*

We have seen that low-dimensional optical waves have very large wavenumber in the previous sections. In order to couple 3D optical waves in macroscopic system to low-dimensional optical waves in nanometer-sized system, we need efficient nano-optical couplers. At the early stage of the research in plasmonic waveguides, this was a serious problem for applications to nano-optical waveguides because the end-fire method or grating couplers have been used for excitaion of SPP. Such coupling problem has been resolved in principle by steep tapered ND waveguides.[25] Figure 22 shows 2D and 1D tapered ND waveguides. There are no reflections of 2D and 1D optical waves in tapered waveguides if taper angle is sufficiently small (adiabatic limit). These tapered structures can be used as efficient coupler from 3D optical waves in macroscopic dielectric waveguides to low-dimensional optical waves in nano-optical waveguides seamlessly.

Analytic solutions of EM fields at the tip of wedged and conical ND waveguides have been obtained theoretically.[26,27] The upper branch of TM mode in 1D optical wave plays central roles in tapered waveguides; they call the unique properties of the mode as "superfocusing effect". Remarkable properties of superfocusing is large field enhancement: i.e. the EM field of the mode increases as it propagates towards the tip because incident power conserves. Because superfocusing can be observed even in lossy ND, the field enhancement can degrade the field decreasing by Ohmic loss as discussed above, and even cancel it.[25,28,29]

The exact solutions of superfocusing can be applied only at the tip of

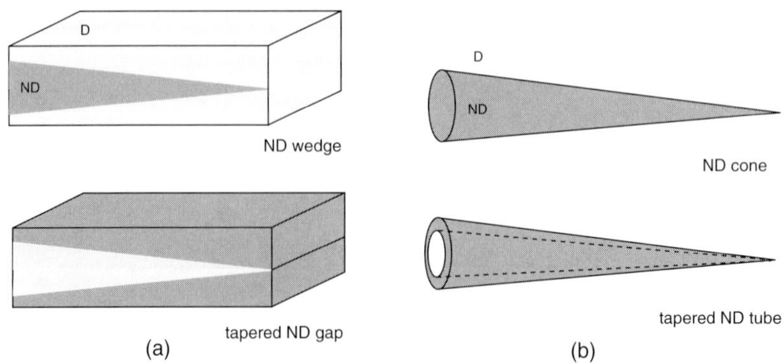

Fig. 22. Schematic view of tapered ND waveguides: (a) 2D ND waveguide, (b) 1D ND waveguide.

the taper. We have obtained the 0th-order perturbation solutions, which satisfy boundary conditions at all points of tapered geometries, using a new method called the quasi-separation of variables.[30]

9.4. *Applications*

An important potential application of 1D optical waveguides in the near future could be a high-efficiency optical probe with high spatial resolution. We could use such a device for SNOM, optical recording and nano-manipulation. Other potential applications include such as nano-optical circuits and devices. We have proposed 3D-2D optical couplers to nano-optical circuits based on the ND gap using superfocusing effects.[19,28] Low-dimensional optical waveguides could play a crucial role in constructing nano-optical integrated circuits: they could serve as a bridge linking the subwavelength world to a truly nanometer world.

10. Optical hetero-interfaces

Finally, it is interesting to note that low-dimensional optical waves can propagate not only along D/ND interfaces but also along the interfaces between dielectric and materials with negative permeability (NP). At D/NP interfaces, 2D optical waves are TE modes, which is interpreted as surface magnon polariton (SMP). Antiferromagnets such as MnF_2 or FeF_2 are NP at microwave frequency.[7] What's more, we can expand these arguments systematically to interfaces, which include negative-index materials.[4] Figure 23 shows six possible interfaces in parameter space for both sign of ϵ

and μ (see also Fig. 1).[4,31] Here we call these interfaces with opposite sign of ϵ and μ an "optical hetero-interface".

Let us expand SPP dispersion relation (Eq.(9)) to another 2D optical waves. As shown in Fig.5(a), we consider 2D optical waves propagating at the interface between an isotropic half-space 1 with ϵ_1, μ_1, and an isotropic half-space 2 with ϵ_2, μ_2, where ϵ_i and μ_i is relative permittivity and permeability in half-space i, respectively. It is natural to predict that a 2D optical wave becomes propagation mode under the conditions:

$$\epsilon_1\epsilon_2 < 0 \quad \text{and/or} \quad \mu_1\mu_2 < 0. \tag{41}$$

¿From Maxwell's equations and the boundary conditions, we can obtain general forms of wavenumber for TE and TM mode as

$$k_{2\mathrm{DTE}} = \pm\frac{\omega}{c}\sqrt{\frac{\frac{\epsilon_1}{\mu_1}-\frac{\epsilon_2}{\mu_2}}{(\frac{1}{\mu_1}-\frac{1}{\mu_2})(\frac{1}{\mu_1}+\frac{1}{\mu_2})}} \quad \text{(TE mode)}, \tag{42}$$

$$k_{2\mathrm{DTM}} = \pm\frac{\omega}{c}\sqrt{\frac{\frac{\mu_1}{\epsilon_1}-\frac{\mu_2}{\epsilon_2}}{(\frac{1}{\epsilon_1}-\frac{1}{\epsilon_2})(\frac{1}{\epsilon_1}+\frac{1}{\epsilon_2})}} \quad \text{(TM mode)}. \tag{43}$$

We can derive a propagation condition of 2D optical waves from Eqs.(42) and (43). Here we take signs of relative permittivity and permeability of D, ND, NP and NI as $\epsilon > 0$ $\mu > 0$, $\epsilon_{\mathrm{ND}} < 0$ $\mu_{\mathrm{ND}} > 0$, $\epsilon_{\mathrm{NP}} > 0$ $\mu_{\mathrm{NP}} < 0$ and $\epsilon_{\mathrm{NI}} < 0$ $\mu_{\mathrm{NI}} < 0$, respectively. For example, for the case of D/NP, 2D optical wave is TE mode and the propagation condition is given by $|\mu_{\mathrm{NP}}| > \mu$. We can summarize four conditions for the case of $\epsilon_1\epsilon_2 < 0$ or $\mu_1\mu_2 < 0$ as follows:

$$|\epsilon_{\mathrm{ND}}| > \epsilon \qquad \text{(D/ND, TM mode)}, \tag{44}$$

$$|\mu_{\mathrm{NP}}| > \mu \qquad \text{(D/NP, TE mode)}, \tag{45}$$

$$\epsilon_{\mathrm{NP}} > |\epsilon_{\mathrm{NI}}| \quad \text{(NI/NP, TM mode)}, \tag{46}$$

$$\mu_{\mathrm{ND}} > |\mu_{\mathrm{NI}}| \quad \text{(NI/ND, TE mode)}. \tag{47}$$

These conditions are trivial from duality of Maxwell's equations. Furthermore, D/NI and ND/NP are also interesting systems because these interfaces can support the propagation of both TE and TM modes. These are the case of $\epsilon_1\epsilon_2 < 0$ and $\mu_1\mu_2 < 0$. The propagation condition can be mapped in 2D parameter space in $|\epsilon_2/\epsilon_1|$ and $|\mu_2/\mu_1|$, which are more complicated than four interfaces above. We summarize these results in Fig. 23.

Nowadays NP at near-infrared(IR) frequency is realized as metamaterials. Metal nanowire composites have been proposed theoretically as NI

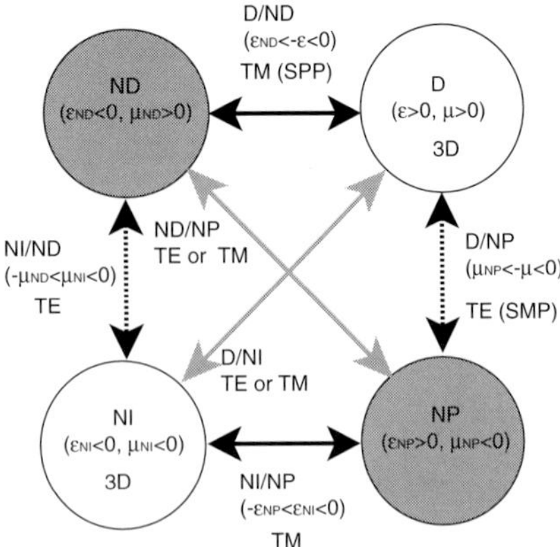

Fig. 23. Possible six interfaces supporting 2D optical waves in parameter space of ϵ and μ. Dielectric (D), negative dielectric (ND), materials with negative permeability (NP), negative-index materials (NI) are shown. 2D optical wave propagates as TM and/or TE mode.

materials at near-IR frequencies and recent progress in nanotechnology may help in their development.[32] Further investigations of these low-dimensional optical waves at exotic interfaces are required from the viewpoint of both basic science and nano-optics. The combination of NP, NI materials and low-dimensional optical waveguides has great potential for the future: in the 21st century, it may well lay the functions for the establishment of an entirely new branch of optics.

Acknowledgements

We would like to thank to Tetsuro Kobayashi (Professor Emeritus, Osaka University) for useful discussions and encouragements. We would like to thank to Dr. Fuminori Kusunoki, Suguru Yamagishi, Noriyuki Miyoshi, Yohei Matsui, Tadayoshi Yoneda and Kenichiro Ohmori for analytic and numerical calculations. We would like to thank to Dr. Akira Otomo, Dr. Kazuyoshi Kurihara and Dr. Kazuhiro Yamamoto for useful discussion for tapered waveguides. This research is supported in part by the 21st Century COE Program (G18) "Core Research and Advanced Education Center for Materials Science and Nano Engineering" of the Japan Society for the Promotion of Science.

References

1. W. L. Barnes, A. Dereux and T. W. Ebbesen, *Nature* **424**, 824 (2003) and references therein.
2. S. Kawata and H. Masuhara eds., Handai Nanophotonics Series Vol. 2, *Nanoplasmonics*, (Elsevier Sciences, 2006).
3. J. Takahara, S. Yamagishi, H. Taki, A. Morimoto and T. Kobayashi, *Opt. Lett.*, **22**, 475 (1997).
4. J. Takahara and T. Kobayashi, *Optics & Photonics News*, **15** (10), 54 (2004).
5. M. Born and E. Wolf, *Principles of Optics*, 7th (expanded) edition (Cambridge University Press, 1999).
6. V. Veselago, *Soviet Physics Uspekhi* **10**, 509 (1968).
7. D R Smith, W J Padilla, D C Vier, S C Nemat-Nasser, and S Schultz, *Phys. Rev. Lett.* **84**, 4184 (2000).
8. J.B. Pendry et al.: *Optics Express*, *11*, 639 (2003) .
9. C. Caloz and T. Itoh, *Electromagnetic Metamaterials*, (John Wiley & Sons, 2006).
10. J. Takahara and T. Kobayashi, in Handai Nanophotonics Series Vol. 2, S. Kawata and H. Masuhara eds., *Nanoplasmonics*, pp.55–77, (Elsevier Sciences, 2006).
11. J. Takahara and T. Kobayashi, *Proc. SPIE*, **5604**, 158 (2004).
12. J. Takahara, F. Kusunoki and T. Kobayashi, Proc. SPIE, **5928**, 59280D-1 (2005).
13. H. Raether, *Surface plasmons on smooth and rough surfaces and on gratings*, (Springer, Berlin, 1988).
14. E. N. Economou, *Phys. Rev.* **182**(2), 539 (1969).
15. J. J. Burke, G. I. Stegeman and T. Tamir, *Phys. Rev.* **B33**(8), 5186 (1986).
16. M. Fukui, V-C.Y. So and R. Normandin, *Phys. Status Solidi*, **B91**, K61 (1979).
17. F. Kusunoki, J. Takahara and T. Kobayashi, *Appl. Phys. Lett.*, **86**, 211101 (2005).
18. F. Kusunoki, T. Yotsuya and J. Takahara, *Optics Express* **14**, 5651 (2006).
19. J. Takahara and F. Kusunoki, *IEICE Trans. Electron.* E90-C(1), 87 (2007).
20. J.C. Ashley and L.C. Emerson, *Surface Science* **41**, 615 (1974).
21. C.A. Pfeiffer, E.N. Economou and K.L. Ngai, *Phys. Rev.* **B10**(8), 3038 (1974).
22. S. Ramo, J.R. Whinnery and T. Van Duzer, *Fields and Waves in Communication Electronics*, 3rd Edition (John Wiley & Sons, 1994).
23. J. Takahara, in *Springer Series in Optical Sciences* **84**, S. Kawata, M. Ohtsu and M. Irie, eds., *Nano-Optics*, pp.126–133, (Springer-Verlag, 2002).
24. E. D. Palik, *Handbook of optical constants of solids*, (Academic Press, New York, 1998).
25. M. I. Stockman, *Phys. Rev. Lett.*, **93**(13), 137404 (2005).
26. Kh. V. Nerkararyan, *Phys. Lett.* **A237**, 103 (1997).
27. A. J. Babadjanyan, N. L. Margaryan and Kh. V. Nerkararyan, *J. Appl. Phys.*, **87**(8), 3785 (2000).

28. N. Miyoshi, J. Takahara, T. Kobayashi and F. Kusunoki, *Proc. of SPIE*, **6050**, 60500T-1 (2005).
29. D. F. P. Pile and D.K. Gramotnev, *Appl. Phys. Lett.*, **89**, 041111 (2006).
30. K. Kurihara, A. Otomo, S. Syouji, J. Takahara, K. Suzuki, S. Yokoyama, *J. Phys. A: Math. Theor.* **40**, 12479 (2007).
31. J. Takahara, *In Abstracts of 9th International Conference on Near-Field Optics, Nanophotonics & Related Techniques (NFO-9)*, 243 , Lausanne, Switzerland (Sep., 2006).
32. A. V. Kildishev, W. Cai, U. K. Chettiar, H. Yuan, A. K. Sarychev, V. P. Drachev, and V. M. Shalaev: *J. Opt. Soc. Am. B* **23**(3), 423 (2006).

Chapter 3

NANOPARTICLE PLASMON WAVEGUIDES

Stefan A. Maier
Experimental Solid State Group, Department of Physics, Imperial College, London SW7 2AZ, UK
S.Maier@imperial.ac.uk

Metallic nanoparticles of a size significantly smaller than the wavelength of light can sustain strong dipolar excitations in the form of localized surface plasmon resonances. Electromagnetic coupling between particles tightly arranged in a one-dimensional array therefore leads to the propagation of electromagnetic energy. Such *metal nanoparticle plasmon waveguides* can show a high mode confinement well below the diffraction limit upon excitation in the vicinity of the particle resonances, however with associated large losses. An elementary treatment of this problem is presented, alongside initial experimental investigations of the coupling phenomenon. Many open question regarding the nature of the coupling and the length scale over which energy propagation occurs remain, making the further study of these systems highly topical. Possible applications lie in energy concentration and focusing over micron-scale distances in the visible part of the spectrum. In addition, a short glimpse at opportunities to employ metal nanoparticles as the building blocks for two-dimensional photonic-crystal-like plasmon waveguides is provided. While in this case the lateral mode confinement is of the order of the wavelength, low losses allow intruiging applications for surface waveguides and sensors at near-infrared frequencies, including efficient coupling to optical fibres.

1. Introduction to localized surface plasmons

This chapter discusses plasmon waveguides of a different sort than the rest of this book. As opposed to electromagnetic energy transport via propagating surface plasmon polaritons sustained at planar metal/dielectric interfaces, here we describe waveguides consisting of coupled metal nanocavities. While descriptions of *coupled resonator optical waveguides* usually assume dielectric cavities,[1] for example microspheres sustaining whispering gallery modes or point defects in photonic crystals, in our case the resonators are metal nanoparticles sustaining *localized surface plasmons*. Such *metal nanoparticle plasmon waveguides* exhibit important differences compared to their dielectric coupled-cavity counterparts. For once, mode confinement below the diffraction limit in the surrounding dielectric is possible due to energy concentration in the near-field zone. As we will see, this high confinement implies large losses, in line with the fundamental trade-off between localization and loss occuring in plasmonic guiding geometries. In most cases, these losses are almost entirely due to absorption processes in the particles themselves, as opposed to mostly radiation-loss limited dielectric coupled-resonator waveguides.

Localized surface plasmons are *non-propagating* excitations of the conduction electrons of metallic nanostructures coupled to the electromagnetic field. These modes arise naturally in the scattering problem of a small, sub-wavelength conductive nanoparticle placed in an oscillating electromagnetic field.[2] The curved surface of the particle excerts an effective restoring force on the driven electrons, so that resonances can arise, leading to field amplification both inside and in the near-field zone outside the particle. Another consequence of the curved surface is that plasmon resonances can be excited by direct light illumination without the employment of phase-matching techniques, in contrast to propagating surface plasmon polaritons.

For particles of sub-wavelength size, the localized plasmon resonance has the character of a dipole. In the case of gold and silver nanoparticles embedded in a low-permittivity dielectric, the dipolar resonance falls into the visible region of the electromagnetic spectrum. A striking consequence of this are the bright colors exhibited by metallic nanoparticles both in transmitted and reflected light, due to resonantly enhanced absorption and scattering. This effect has found applications for many hundreds of years, for example in the staining of glass for windows or ornamental cups.

Before embarking on a description of energy transport in metal nanoparticle arrays, we will discuss the basics of localized surface plasmon modes, in a description adapted from a recent book on plasmonics by the author.[3] We then present the basic physics of interparticle coupling, which can in a first approximation be understood via a point-dipolar model, neglecting finite-size effects (i.e., higher order modes). We will see that the simplest treatment possible, a quasi-static model exclusively built on near-field coupling, cannot describe the whole problem, both quantitatively and under certain conditions also not qualitatively. Crucially, the quasi-static model seems to underestimate the distance over which energy can be propagate. As the main experimental investigations of energy transport in particle chains relied on this modelling, new experiments on waveguides with parameters optimized using the full-field model are called for. Metal nanoparticle waveguides therefore remain an exciting field of study, almost a decade after their initial description.

2. Dipolar plasmon resonances in metal nanoparticles in the quasi-static approximation

The interaction of a particle of size $2a$ with the electromagnetic field can be analyzed using the simple *quasi-static approximation* provided that the particle is much smaller than the wavelength of light in the surrounding medium, i.e. $2a \ll \lambda$. The phase of the harmonically oscillating electromagnetic field is practically constant over the particle volume, so that one can solve the scattering problem by assuming the simplified problem of a particle in an electrostatic field. This lowest-order approximation of the full scattering expansion describes the optical properties of nanoparticles of dimensions sufficiently below 100 nm adequately for many purposes. For simplicity, we will in the following assume nanoparticles of a spherical shape.

An easy analytical treatment is possible for a homogeneous, isotropic sphere of radius a located at the origin in a uniform, static electric field $\mathbf{E} = E_0\hat{\mathbf{z}}$ (Fig. 1). We assume that the surrounding medium is isotropic and non-absorbing with dielectric constant ε_m, and that the field lines are parallel to the z-direction at sufficient distance from the sphere. The dielectric response of the sphere is further described by a dielectric function $\varepsilon(\omega)$, which we take for the moment as a simple complex number ε.

In the electrostatic approach, we are interested in a solution of the *Laplace equation* for the potential, $\nabla^2\Phi = 0$, from which we will be able to calculate the electric field $\mathbf{E} = -\nabla\Phi$. Due to the azimuthal symmetry of the problem, the general solution is of the form[4]

$$\Phi(r,\theta) = \sum_{l=0}^{\infty}\left[A_l r^l + B_l r^{-(l+1)}\right] P_l(\cos\theta), \tag{1}$$

where $P_l(\cos\theta)$ are the Legendre Polynomials of order l, and θ the angle between the position vector $\mathbf{r}$ at point P and the z-axis (Fig. 1). Due to

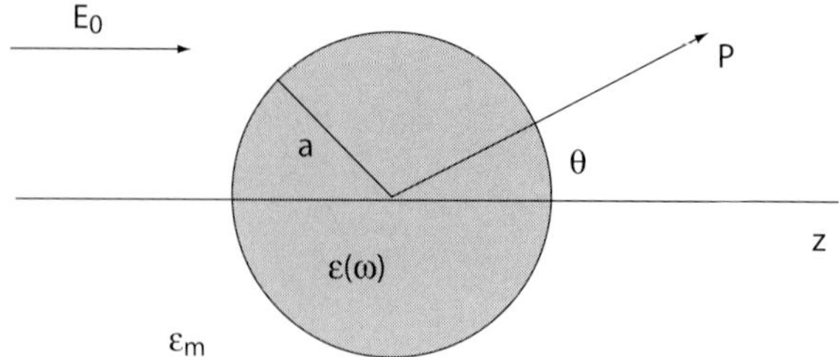

Fig. 1. Sketch of a homogeneous sphere placed into an electrostatic field.

the requirement that the potentials remain finite at the origin, the solution for the potentials Φ_{in} inside and Φ_{out} outside the sphere can be written as

$$\Phi_{\text{in}}(r,\theta) = \sum_{l=0}^{\infty} A_l r^l P_l(\cos\theta) \tag{2a}$$

$$\Phi_{\text{out}}(r,\theta) = \sum_{l=0}^{\infty} \left[B_l r^l + C_l r^{-(l+1)}\right] P_l(\cos\theta). \tag{2b}$$

The coefficients A_l, B_l and C_l can now be determined from the boundary conditions at $r \to \infty$ and at the sphere surface $r = a$. The requirement that $\Phi_{\text{out}} \to -E_0 z = -E_0 r\cos\theta$ as $r \to \infty$ demands that $B_1 = -E_0$ and $B_l = 0$ for $l \neq 1$. The remaining coefficients A_l and C_l are defined by the boundary conditions at $r = a$. Equality of the tangential components of the electric field demands that

$$-\frac{1}{a}\frac{\partial\Phi_{\text{in}}}{\partial\theta}\bigg|_{r=a} = -\frac{1}{a}\frac{\partial\Phi_{\text{out}}}{\partial\theta}\bigg|_{r=a}, \tag{3}$$

and the equality of the normal components of the displacement field

$$-\varepsilon_0\varepsilon\frac{\partial\Phi_{\text{in}}}{\partial r}\bigg|_{r=a} = -\varepsilon_0\varepsilon_{\text{m}}\frac{\partial\Phi_{\text{out}}}{\partial r}\bigg|_{r=a}. \tag{4}$$

Application of these boundary conditions leads to $A_l = C_l = 0$ for $l \neq 1$, and via the calculation of the remaining coefficients A_1 and C_1 the potentials evaluate to[4]

$$\Phi_{\text{in}} = -\frac{3\varepsilon_m}{\varepsilon + 2\varepsilon_m} E_0 r\cos\theta \tag{5a}$$

$$\Phi_{\text{out}} = -E_0 r\cos\theta + \frac{\varepsilon - \varepsilon_m}{\varepsilon + 2\varepsilon_m} E_0 a^3 \frac{\cos\theta}{r^2}. \tag{5b}$$

It is interesting to interpret equation (5b) physically: Φ_{out} describes the superposition of the applied field and that of a dipole located at the particle center. We can rewrite Φ_{out} by introducing the dipole moment $\mathbf{p}$ as

$$\Phi_{\text{out}} = -E_0 r \cos\theta + \frac{\mathbf{p}\cdot\mathbf{r}}{4\pi\varepsilon_0\varepsilon_m r^3} \tag{6a}$$

$$\mathbf{p} = 4\pi\varepsilon_0\varepsilon_m a^3 \frac{\varepsilon - \varepsilon_m}{\varepsilon + 2\varepsilon_m}\mathbf{E}_0. \tag{6b}$$

We therefore see that the applied field induces a dipole moment inside the sphere of magnitude proportional to $|\mathbf{E}_0|$. If we introduce the polarizability α, defined via $\mathbf{p} = \varepsilon_0\varepsilon_m\alpha\mathbf{E}_0$, we arrive at the central result

$$\alpha = 4\pi a^3 \frac{\varepsilon - \varepsilon_m}{\varepsilon + 2\varepsilon_m}. \tag{7}$$

Equation (7) represents the (complex) polarizability of a small sphere of sub-wavelength diameter in the electrostatic approximation.

Fig. 2 shows the absolute value and phase of α with respect to frequency ω (in energy units) for a dielectric constant varying as $\varepsilon(\omega)$ of the Drude form. It is apparent that the polarizability experiences a resonant enhancement under the condition that $|\varepsilon + 2\varepsilon_m|$ is a minimum, which for the case of small or slowly-varying Im $[\varepsilon]$ around the resonance simplifies to

$$\text{Re}\left[\varepsilon\left(\omega\right)\right] = -2\varepsilon_m. \tag{8}$$

This relationship is called the Fröhlich condition and the associated mode (in an oscillating field) the *dipole surface plasmon* of the metal nanoparticle.

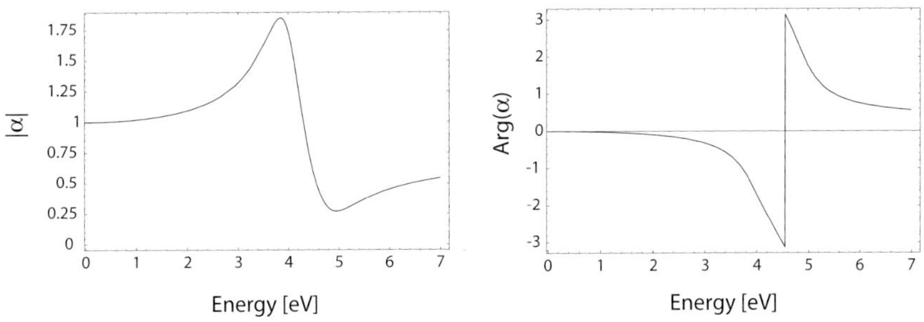

Fig. 2. Absolute value and phase of the polarizability α (7) of a sub-wavelength metal nanoparticle with respect to the frequency of the driving field (expressed in eV units). Here, $\varepsilon(\omega)$ is taken as a Drude fit to the dielectric function of silver.[5]

For a sphere consisting of a Drude metal with a dielectric function $\varepsilon(\omega)$ located in air, the Fröhlich criterion is met at the frequency $\omega_0 = \omega_{\mathrm{p}}/\sqrt{3}$. (8) further expresses the strong dependence of the resonance frequency on the dielectric environment: The resonance red-shifts as ε_m is increased. We note that the magnitude of α at resonance is limited by the incomplete vanishing of its denominator, due to $\mathrm{Im}\left[\varepsilon(\omega)\right] \neq 0$.

The distribution of the electric field $\mathbf{E} = -\nabla\Phi$ can be evaluated from the potentials (5) to

$$\mathbf{E_{in}} = \frac{3\varepsilon_m}{\varepsilon + 2\varepsilon_m}\mathbf{E}_0 \tag{9a}$$

$$\mathbf{E_{out}} = \mathbf{E}_0 + \frac{3\mathbf{n}\left(\mathbf{n}\cdot\mathbf{p}\right) - \mathbf{p}}{4\pi\varepsilon_0\varepsilon_m}\frac{1}{r^3}. \tag{9b}$$

As expected, the resonance in α also implies a resonant enhancement of both the internal and dipolar fields. It is this field-enhancement at the plasmon resonance on which many of the prominent applications of metal nanoparticles in optical devices and sensors rely.

We will now leave the electrostatic description and turn our attention to the electromagnetic fields radiated by a small particle excited at its plasmon resonance. The dipolar approximation will be upheld, but we will allow for time-varying fields, neglecting spatial retardation effects over the particle volume. Under plane-wave illumination with $\mathbf{E}(\mathbf{r},t) = \mathbf{E}_0 e^{-i\omega t}$, the fields induce an oscillating dipole moment $\mathbf{p}\left(t\right) = \varepsilon_0\varepsilon_m\alpha\mathbf{E}_0 e^{-i\omega t}$, with α given by the electrostatic result (7). The radiation of this dipole leads to *scattering* of the plane wave by the sphere, which can be represented as radiation by a point dipole.

In the context of coupling between metallic nanoparticles for applications in waveguiding, it is useful to briefly review the basics of the electromagnetic fields associated with an oscillating electric dipole. The total fields $\mathbf{H}(t) = \mathbf{H}e^{-i\omega t}$ and $\mathbf{E}(t) = \mathbf{E}e^{-i\omega t}$ in the near, intermediate and radiation zones of a dipole can be written as[4]

$$\mathbf{H} = \frac{ck^2}{4\pi}\left(\mathbf{n}\times\mathbf{p}\right)\frac{e^{ikr}}{r}\left(1 - \frac{1}{ikr}\right) \tag{10a}$$

$$\mathbf{E} = \frac{1}{4\pi\varepsilon_0\varepsilon_m}\left\{k^2\left(\mathbf{n}\times\mathbf{p}\right)\times\mathbf{n}\frac{e^{ikr}}{r} + \left[3\mathbf{n}\left(\mathbf{n}\cdot\mathbf{p}\right) - \mathbf{p}\right]\left(\frac{1}{r^3} - \frac{ik}{r^2}\right)e^{ikr}\right\}, \tag{10b}$$

with $k = 2\pi/\lambda$ and $\mathbf{n}$ the unit vector in the direction of the point P of interest. In the near zone ($kr \ll 1$), the electrostatic result (9b) for the electric field is recovered,

$$\mathbf{E} = \frac{3\mathbf{n}\,(\mathbf{n}\cdot\mathbf{p}) - \mathbf{p}}{4\pi\varepsilon_0\varepsilon_m}\frac{1}{r^3} \tag{11a}$$

and the accompanying magnetic field present for oscillating fields amounts to

$$\mathbf{H} = \frac{i\omega}{4\pi}\,(\mathbf{n}\times\mathbf{p})\,\frac{1}{r^2}. \tag{11b}$$

We can see that within the near field, the fields are predominantly electric in nature, since the magnitude of the magnetic field is about a factor $\sqrt{\varepsilon_0/\mu_0}\,(kr)$ smaller than that of the electric field. For static fields ($kr \to 0$), the magnetic field vanishes.

In the opposite limit of the radiation zone, defined by $kr \gg 1$, the dipole fields are of the well-known spherical-wave form

$$\mathbf{H} = \frac{ck^2}{4\pi}\,(\mathbf{n}\times\mathbf{p})\,\frac{e^{ikr}}{r} \tag{12a}$$

$$\mathbf{E} = \sqrt{\frac{\mu_0}{\varepsilon_0\varepsilon_m}}\mathbf{H}\times\mathbf{n}. \tag{12b}$$

We will see that the full description (10) is needed to adequately describe interparticle coupling in metal nanoparticle arrays in the later sections.

Another consequence of the resonantly enhanced polarization α is a concomitant enhancement in the efficiency with which a metal nanoparticle scatters and absorbs light. The corresponding cross sections for scattering and absorption C_{sca} and C_{abs} amount in the small-particle limit to[6]

$$C_{\text{sca}} = \frac{k^4}{6\pi}\,|\alpha|^2 = \frac{8\pi}{3}k^4a^6\left|\frac{\varepsilon-\varepsilon_m}{\varepsilon+2\varepsilon_m}\right|^2 \tag{13a}$$

$$C_{\text{abs}} = k\text{Im}\,[\alpha] = 4\pi ka^3\text{Im}\left[\frac{\varepsilon-\varepsilon_m}{\varepsilon+2\varepsilon_m}\right]. \tag{13b}$$

For small particles with $a \ll \lambda$, the efficiency of absorption, scaling with a^3, therefore dominates over the scattering efficiency, which scales with a^6. We point out that no explicit assumptions were made in our derivations so far that the sphere is metallic. The expressions for the cross sections (13) are thus valid also for dielectric scatterers. For metallic scatteres however,

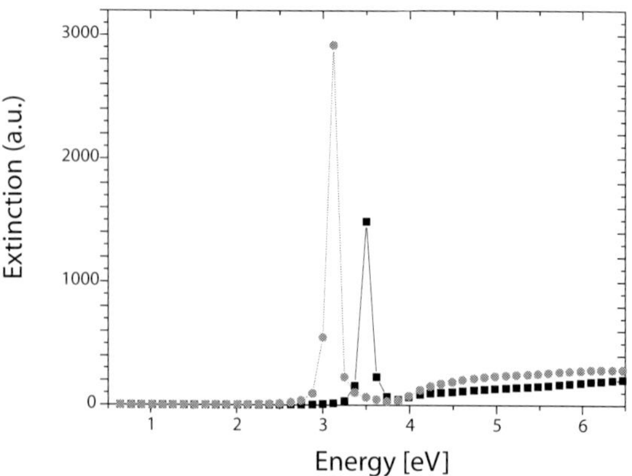

Fig. 3. Extinction cross section calculated using (14) for a silver sphere in air (black curve) and silica (gray curve), with the dielectric data taken from.[5]

equations (13) show that both absorption and scattering (and thus extinction) are resonantly enhanced at the dipole particle plasmon resonance, i.e. when the Frölich condition (8) is met.[2] For a sphere of volume V and dielectric function $\varepsilon = \varepsilon_1 + i\varepsilon_2$ in the quasi-static limit, the explicit expression for the extinction cross section $C_{\mathrm{ext}} = C_{\mathrm{abs}} + C_{\mathrm{sca}}$ is

$$C_{\mathrm{ext}} = 9\frac{\omega}{c}\varepsilon_m^{3/2} V \frac{\varepsilon_2}{\left[\varepsilon_1 + 2\varepsilon_m\right]^2 + \varepsilon_2^2}. \tag{14}$$

Fig. 3 shows the extinction cross section of a silver sphere calculated using this formula for immersion in two different media.

We now relax the assumption of a spherical nanoparticle shape. A slightly more general geometry amenable to analytical treatment in the electrostatic approximation is that of an ellipsoid with semiaxes $a_1 \leq a_2 \leq a_3$, specified by $\frac{x^2}{a_1^2} + \frac{y^2}{a_2^2} + \frac{z^2}{a_3^2} = 1$. A treatment of the scattering problem in ellipsoidal coordinates[6] leads to the following expression for the polarizabilities α_i along the principal axes ($i = 1, 2, 3$):

$$\alpha_i = 4\pi a_1 a_2 a_3 \frac{\varepsilon(\omega) - \varepsilon_m}{3\varepsilon_m + 3L_i\left(\varepsilon(\omega) - \varepsilon_m\right)} \tag{15}$$

L_i is a geometrical factor given by

$$L_i = \frac{a_1 a_2 a_3}{2} \int_0^\infty \frac{dq}{(a_i^2 + q)\, f(q)}, \tag{16}$$

where $f(q) = \sqrt{(q + a_1^2)\,(q + a_2^2)\,(q + a_3^2)}$. The geometrical factors satisfy $\sum L_i = 1$, and for a sphere $L_1 = L_2 = L_3 = \frac{1}{3}$. As an alternative, the polarizability of ellipsoids is also often expressed in terms of the *depolarization factors* $\tilde{L}_i$, defined via $E_{1i} = E_{0i} - \tilde{L}_i P_{1i}$, where E_{1i} and P_{1i} are the electric field and polarization induced inside the particle by the applied field E_{0i} along a principal axis i, respectively. $\tilde{L}$ is linked to L via

$$\tilde{L}_i = \frac{\varepsilon - \varepsilon_m}{\varepsilon - 1} \frac{L_i}{\varepsilon_0 \varepsilon_m}. \tag{17}$$

An important special class of ellipsoids are *spheroids*. For *prolate* spheroids, the two minor axes are equal ($a_2 = a_3$), while for *oblate* spheroids, the two major axes are of same size ($a_1 = a_2$). An examination of (15) reveals that a spheroidal metal nanoparticle exhibits two spectrally separated plasmon resonances, corresponding to oscillations of its conduction electrons along the major or minor axis, respectively. The resonance due to oscillations along the major axis can show a significant spectral red-shift compared to the plasmon resonance of a sphere of the same volume. Thus, plasmon resonances can be lowered in frequency into the near-infrared region of the spectrum using metallic nanoparticles with large aspect ratio. For a quantitative treatment, we note however that (15) is only strictly valid as long as the major axis is significantly smaller than the excitation wavelength.

3. Beyond the quasi-static approximation

For particles of larger dimensions, where the quasi-static approximation is not justified due to significant phase changes of the driving field over the particle volume, a rigorous *electrodynamic* approach is required. In a seminal paper, Mie in 1908 developed a complete theory of the scattering and absorption of electromagnetic radiation by a sphere, in order to understand the colors of colloidal gold particles in solution.[7] The approach of what is now know as *Mie theory* is to expand the internal and scattered fields into a set of *normal modes* described by vector harmonics. The quasi-static results valid for sub-wavelength spheres are then recovered by a power series expansion of the absorption and scattering coefficients and retaining only the first term. For a full presentation of Mie theory we refer the reader to a variety of books.[2,6]

We will now analyze changes to the spectral position and width of the plasmon resonance with particle size not captured by the quasi-static description. Two regimes will be considered: Firstly, that of larger particles where this approximation breaks down due to retardation effects, and secondly the regime of very small metal particles of radius $a < 10$ nm, where the particle dimensions are appreciably smaller than the mean free path of its oscillating electrons.

Starting with larger particles, a straight-forward expansion of the first TM mode of Mie theory yields for the polarizability of a sphere of volume V the expression[8,9]

$$\alpha_{\text{Sphere}} = \frac{1 - \left(\frac{1}{10}\right)(\varepsilon + \varepsilon_m)\, x^2 + O\left(x^4\right)}{\left(\frac{1}{3} + \frac{\varepsilon_m}{\varepsilon - \varepsilon_m}\right) - \frac{1}{30}\left(\varepsilon + 10\varepsilon_m\right) x^2 - i\frac{4\pi^2 \varepsilon_m^{3/2}}{3}\frac{V}{\lambda_0^3} + O\left(x^4\right)} V, \quad (18)$$

where $x = \frac{\pi a}{\lambda_0}$ is the so called *size parameter*, relating the radius to the free-space wavelength. Compared to the simple quasi-static solution (7), a number of additional terms appear in the numerator and denominator of (18), each having a distinct physical significance. The term quadratic in x in the numerator includes the effect of retardation of the exciting field over the volume of the sphere, leading to a shift in the plasmon resonance. The quadratic term in the denominator also causes an energy shift of the resonance, due to the retardation of the *depolarization field*[8] inside the particle. For Drude and the noble metals, the overall shift is towards lower energies: the spectral position of the dipole resonance red-shifts with increasing particle size. Intuitively, this can be understood by recognizing that the distance between the charges at opposite interfaces of the particle increases with its size, thus leading to a smaller restoring force and therefore a lowering of the resonance frequency. For the noble metals, this red-shift also implies that effects of interband transitions (described by an increase in $\text{Im}\left[\varepsilon_2\right]$) not captured by the Drude theory decrease as the plasmon resonance moves away from the interband transition edge.

The quadratic term in the denominator also increases the magnitude of the polarization, and thus inherently lessens the influence of the absorption due to the imaginary part of ε. However, this increase in strength is counteracted by the third, imaginary term in the denominator, which accounts for *radiation damping*. An inclusion of terms of higher order in expression (18) will lead to the occurance of higher-order resonances.

Radiation damping is caused by a direct radiative decay route of the coherent electron oscillation into photons,[10] and is the main cause of the weakening of the strength of the dipole plasmon resonance as the particle volume increases.[11] Thus, despite the fact that an increase in particle volume decreases the strength of the non-radiative decay pathway (namely absorption), a significant broadening of the plasmon resonance sets in.

We can summarize that the dipole plasmon resonance of particles beyond the quasi-static regime is damped by two competing processes: a radiative decay process into photons, dominating for larger particles, and a non-radiative process due to absorption. The non-radiative decay is caused by the creation of electron-hole pairs via either intraband excitations within the conduction band or interband transitions from lower-lying d-bands to the sp conduction band (for noble metal particles). More details on the physics of the damping can be found in a variety of studies.[12,13]

In order to arrive at a quantitative description, these two damping processes can be incorporated into a simple two-level model of the plasmon resonance, developed by Heilweil and Hochstrasser.[14] Using it, the homogeneous linewidth Γ of the plasmon resonance, which can be determined using for example extinction spectroscopy, can be related to the internal damping processes via the introduction of a *dephasing time* T_2. In energy units, the relation between Γ and T_2 is

$$\Gamma = \frac{2\hbar}{T_2}. \tag{19}$$

We note that in analogy to dielectric resonators, the strength of a plasmon resonance can also be expressed using the notion of a *quality factor* Q, given by $Q = E_{\mathrm{res}}/\Gamma$, where E_{res} is the resonant energy.

In this theory, dephasing of the coherent excitation is either due to energy decay, or scattering events that do not change the electron energy but its momentum. This can be expressed by relating T_2 to a population relaxation or decay time T_1, describing both radiative and non-radiative energy loss processes, and a *pure dephasing time* T_2^* resulting from elastic collisions:

$$\frac{1}{T_2} = \frac{1}{2T_1} + \frac{1}{T_2^*}. \tag{20}$$

Via an examination of the details of plasmon decay, for example with pump-probe experiments,[12] it can be shown that $T_2^* \gg T_1$,[12] so that $T_2 = 2T_1$.

For small gold and silver nanoparticles, in general 5 fs $\leq T_2 \leq$ 10 fs, depending on size and the surrounding host material.

We now turn the attention to the regime of very small metallic particles. For gold and silver particles of radius $a < 10$ nm, an additional damping process, loosely termed *chemical interface damping*, must be considered. Here, the rate of dephasing of the coherent oscillation is increased due to elastic scattering at the particle surface, since the size of the particle is substantially smaller than the electron mean free path (of the order of 30-50 nm). Empirically, the associated broadening of the experimentally observed plasmon linewidth Γ_{obs} can be modeled via[2]

$$\Gamma_{\text{obs}}(R) = \Gamma_0 + \frac{Av_F}{R}. \tag{21}$$

Here, Γ_0 describes the plasmon linewidth of particles that are outside the regimes where interface damping or radiation damping dominate, i.e. where Γ is defined by $\text{Im}\left[\varepsilon(\omega)\right]$ alone. v_F the Fermi velocity of the electrons, and $A \approx 1$ a factor incorporating details of the scattering process.[15] In addition to the broadening of the resonance, shifts in resonance energy have also been reported for particles of dimensions below 10 nm. For particles with a radius of the order of or below 1 nm, quantum effects begin to set in.

We want to finish this section by taking a brief look at the interaction of localized plasmon resonances with gain media. The motivation for this is twofold: the field enhancement sustained by the metallic nanostructures upon resonant excitation can lead to a reduction in the threshold for achieving inversion in the optically active surrounding medium, and the presence of gain can counteract the inherent absorption losses in the metal. The strengthening of plasmon resonances in gain media has up to this point only been confirmed indirectly via the observation of enhancement of Rayleigh scattering in a mixture of silver nanoparticles with an optical dye.[16] Amplification of fluorescence due to field enhancement in mixtures of laser dyes with metal nanoparticles has also recently been reported.[17]

In its simplest form, the problem of a gain-induced increase in the strength of the plasmon resonance can be treated by analyzing the case of a sub-wavelength metal nanosphere embedded in a homogeneous medium exhibiting optical gain. The quasi-static approach presented at the beginning of this chapter can be followed, and the presence of gain incorporated by replacing the real dielectric constant ε_m of the insulator surrounding the sphere with a complex dielectric function $\varepsilon_2(\omega)$.

Using this straightforward analytical model, Lawandy has shown that the presence of gain, expressed by $\mathrm{Im}[\varepsilon_2] < 0$, can lead to a significant strengthening of the plasmon resonance.[18] This is due to the fact that in addition to the cancellation of the real part of the denominator of the polarizability α (7), the positive imaginary part of ε_2 can in principle lead to a complete cancellation of the terms in the denominator and thus to an infinite magnitude of the resonant polarizability. Taking as a starting point the expressions for the electric fields (9), the depolarization field $\mathbf{E}_{\mathrm{pol}} = \mathbf{E}_{\mathrm{in}} - \mathbf{E}_0$ inside the particle is given by

$$\mathbf{E}_{\mathrm{pol}} = \frac{\varepsilon_2 - \varepsilon}{\varepsilon + 2\varepsilon_2}\mathbf{E}_0. \tag{22}$$

For a Drude metal with electron scattering rate $\gamma \ll \omega$, the incomplete vanishing of the denominator in (22) upon resonance can be now overcome by optical gain. Ignoring gain saturation, it can be shown that the critical gain value α_{c} at the plasmon resonance ω_0 for the singularity to occur can be approximated as

$$\alpha_{\mathrm{c}} = \frac{\gamma\,(2\mathrm{Re}\,[\varepsilon(\omega_0)] + 1)}{2c\sqrt{\mathrm{Re}\,[\varepsilon(\omega_0)]}}. \tag{23}$$

For silver and gold particles, this results in $\alpha_{\mathrm{c}} \approx 10^3\ \mathrm{cm}^{-1}$. Of course, in real examples the divergence in field amplification will be suppressed due to saturation effects.

4. Intuitive description of coupling between localized plasmons

In particle ensembles, frequency shifts of the plasmon resonances are expected to occur due to electromagnetic interactions between localized modes. For small particles, these interactions are essentially of a dipolar nature, and the particle ensemble can in a first approximation be treated as an ensemble of interacting dipoles.

We will now describe the effects of such interactions in ordered metal nanoparticle arrays. Here, we assume that the particles of size $2a$ are arranged within an ordered one-dimensional array with interparticle spacing d. We further assume that $2a \ll d$, so that the dipolar approximation is justified, and the particles can be treated as point dipoles.

Two regimes have to be distinguished, depending on the magnitude of the interparticle distance d. For closely spaced particles, $d \ll \lambda$, near-field interactions with a distance dependence of d^{-3} dominate, and the particle array can in a first approximation be described as an array of point dipoles interacting via their near field (see (11)). In this case, strong field localization in nano-sized gaps between adjacent particles has been observed for regular one-dimensional particle chains.[20] The field localization is due to a suppression of scattering into the far-field via excitation of plasmon modes in particles along the chain axis, mediated by near-field coupling. Fig. 4 illustrates this fact by showing the experimentally observed (a, c) and simulated (b, d) distribution of the electric field above single gold nanoparticles and a particle chain. In this study by Krenn and co-workers, the structures were excited using prism coupling from the substrate side and the optical near field was probed by near-field microscopy in collection mode. From the images, it can clearly be seen that scattering is drastically suppressed for closely spaced particles, and that the fields are instead highly localized at interstitial sites.

One can intuitively see that interparticle coupling will lead to shifts in the spectral position of the plasmon resonance compared to the case of an isolated particle. Using the simple approximation of an array of interacting

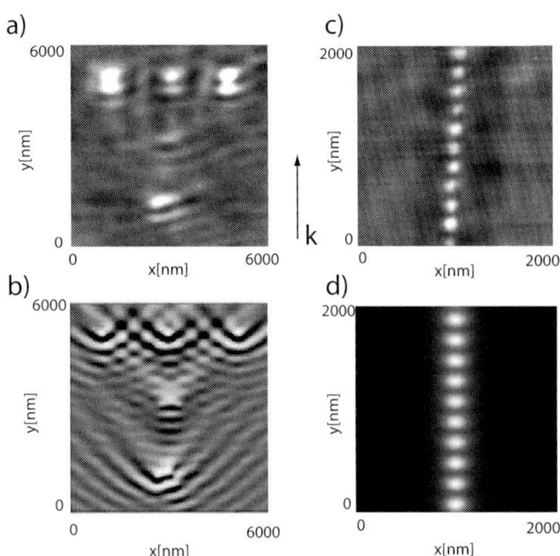

Fig. 4. Experimentally observed (a, c) and simulated (b, d) intensity distribution of the optical near field above an ensemble of well-separated gold particles (a, b) and a chain of closely spaced gold nanoparticles (c, d). While for separated particles interference effects of the scattered fields are visible, in the particle chain the fields are closely confined in gaps between adjacent particles. Plasmon resonances were excited using prism coupling with the direction of the in-plane moment component as outlined in the pictures. Reprinted with permission from.[19] Copyright 2001 by Blackwell Publishing.

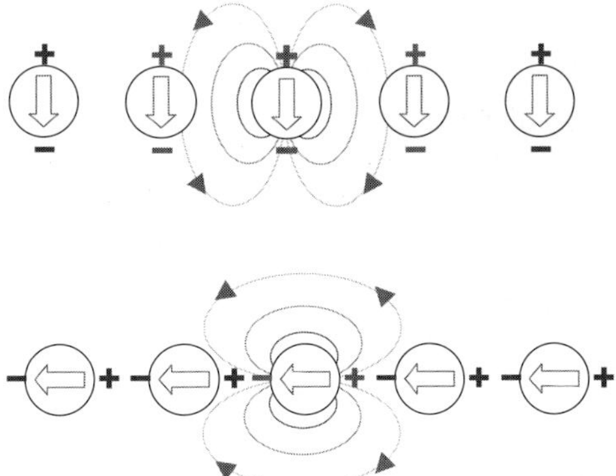

Fig. 5. Schematic of near-field coupling between metallic nanoparticles for the two different polarizations.

point dipoles, the direction of the resonance shifts for in-phase illumination can be determined by considering the Coulomb forces associated with the polarization of the particles. As sketched in Fig. 5, the restoring force acting on the oscillating electrons of each particle in the chain is either increased or decreased by the charge distribution of neighboring particles. Depending on the polarization direction of the exciting light, this leads to a blue-shift of the plasmon resonance for the excitation of transverse modes, and a red-shift for longitudinal modes.

Using one-dimensional arrays of 50 nm gold particles with varying interparticle distance (Fig. 6a), these shifts in resonance energy were experimentally demonstrated using far-field extinction spectroscopy.[21] The dependence of the spectral position of the plasmon resonance on interparticle distance is shown in Fig. 6b both for longitudinal and transverse polarization. Due to the strong scaling of the interaction strength with d^{-3} (see (11)), particle separations in excess of 150 nm are sufficient to recover the behavior of essentially isolated particles.

The spatial extent of near-field interactions can further be quantified by analyzing the dependence of the resonance shifts on the length of the particle arrays.[22] Fig. 7 shows results from finite-difference time-domain (FDTD) calculations and comparisons to experimental shifts obtained for chains of gold nanoparticles with fixed interparticle distance and varying chain lengths. In the FDTD simulations, the time-dependence of the electric field was monitored at the center of a particle of a chain consisting of seven 50 nm gold spheres separated by 75 nm in air (left panel). The upper

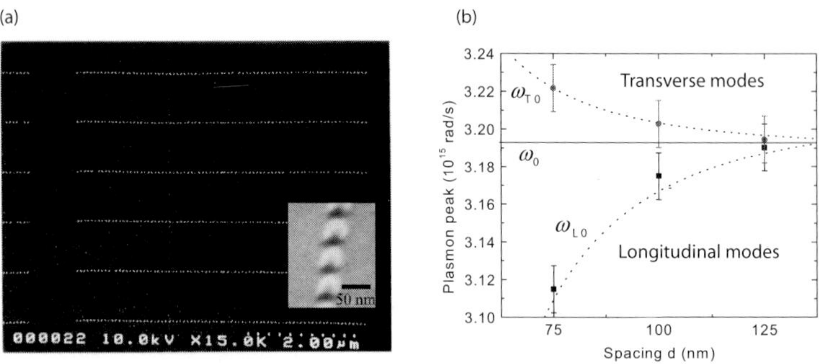

Fig. 6. SEM image of arrays of closely spaced gold nanoparticles (a) and dependence of the spectral position of the dipole plasmon resonance on interparticle spacing (b). The dotted lines show a fit to the d^{-3} dependence of the coupling expected from a quasi-static point-dipole model. Reprinted with permission from.[21] Copyright 2002 by the American Physical Society.

inset shows the distribution of the initial electric field around the structure upon in-phase excitation with longitudinal polarization, and the lower inset the Fourier transform of the time-domain data, peaking at the longitudinal resonance frequency E_L. A comparison with chains fabricated on a silica

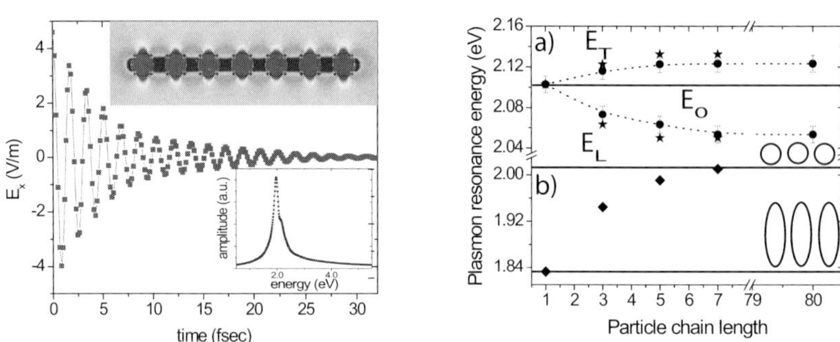

Fig. 7. Left: Time-dependence of the electric field monitored at the center of a particle of a chain consisting of seven 50 nm gold spheres separated by 75 nm in air obtained using FDTD. For details see text. Right: a) Collective plasmon resonance energies for both longitudinal (E_L) and transverse (E_T) excitations for gold nanoparticle arrays (of the same geometry as in the left panel) obtained via far-field spectroscopy on fabricated arrays (circles) and FDTD simulations (stars). b) Simulation results for the collective plasmon resonance energies for transverse excitation of gold spheroids with aspect ratios 3:1 (diamonds). Reproduced with permission from.[22] Copyright 2002, American Institute of Physics.

Fig. 8. Sketch of the nanoparticle waveguide geometry.

substrate using electron beam lithography is shown in the right panel (a). As apparent, the collective plasmon resonance energies for both longitudinal (E_L) and transverse (E_T) excitations for gold nanoparticle arrays of different lengths asymptote already for a chain length of about 5 particles due to the near-field nature of the coupling. The coupling strength between adjacent particles can be increased by changing the geometry to spheroidal particles (Fig. 7b). We point out that phase retardation effects lead to the breakdown of monotonic plasmon resonance shifts with particle chain length for particle sizes and interparticle distances not significantly in the sub-wavelength regime, as has been experimentally confirmed.[23]

After these initial investigations, a number of different studies using both near- and far-field detection techniques have investigated the distance-dependence of near-field interactions in particle arrays[24] as well as particle pairs.[25,26] Of huge significance for applications in waveguiding, it has been found that even for interparticle spacings $d << \lambda$, the coupling is not adequately described by near-field terms only,[27,28] as we will see in the next section. For a detailed analysis of near-field interactions in particle ensembles of various lengths and shapes using Mie theory, we refer to the treatment by Quinten and Kreibig.[29]

5. Metal nanoparticle waveguides

Equipped with a basic understanding of localized plasmon resonances in metallic nanoparticles as well as an intuitive picture of interparticle coupling phenomena, we can now attempt an elementary treatment of electromagnetic energy transport in a one-dimensional particle chain. The geometry is sketched in Fig. 8, and consists of an array of metallic nanospheres of size $2a$ regularly spaced with centre-to-centre distance d and embedded in a homogeneous dielectric. If a particle (or a region of multiple particles) in such a chain is excited by an external field, propagating polarization waves can ensue, polarized either parallel to the nanoparticle axis (the longitudinal mode) or perpendicular to it (two transverse modes).

The transport of energy along such a chain has been analyzed in a number of approximations, starting with the initial proposal of this guiding geometry by Quinten and co-workers.[30] The study was motivated by the problem of finding optical waveguide geometries allowing transverse mode sizes considerably smaller than the wavelength of light. Previously, Takahara and co-workers had pointed out in a seminal paper[31] that contrary to dielectric waveguides, metallic guiding modalities would allow for sub-diffration-limit mode sizes and the absence of mode cut-off with reduction in physical waveguide size for certain geometries sustaining surface plasmon polaritons. Quinten and co-workers realized that similiar localization effects should occur in coupled resonator waveguides if the mode size of the fundamental resonances can be made arbitrarily small (within the limits of classical electrodynamics), which is the case for localized plasmon resonances in metal nanoparticles. Not only can these modes show energy localization below the diffraction limit, due to the sharp fall-off of the near field outside the particle surface, but the enhanced absorption and scattering cross sections upon resonance should also allow for reasonable in-coupling efficiencies, despite a large mismatch between particle size and wavelength. In this initial description, energy propagation along a particle array was studied using generalized Mie theory, i.e. via a scattering formalism including the effects of higher multipoles. However, only the distribution of the electric field in the near-field zone was analyzed, i.e. in essence a quasi-static approach used. The first particle in a 50-particle long chain was excited at its dipole plasmon resonance frequency with a plane wave polarized either in longitudinal or transverse direction, and the electromagnetic near field around the array directly computed. Strong evidence for energy transport with typical $1/e$ energy attenuation lengths on the order of 1 μm was obtained for longitudinal polarization only. An optimal ratio for maximizing the energy propagation length of about $d/a \approx 3$ was further predicted. It is unclear why no evidence for energy propagation for transverse polarization was found; this could possibly have been due to the excitation conditions employed.

While this study hinted at the possibility of energy transfer and arrived at estimates of sub-micron energy propagation lengths, subsequent studies concentrated on the dispersion properties, starting with the most simplified analytical treatment possible: the representation of the chain as an array of point-dipoles, and using the quasi-static approximation.[32] It is important to realize that the point-dipole approximation is a different constraint than the quasi-static approximation. While the later neglects retardation effects via limiting interactions to the near-field terms of the field expansion, valid in the regime $kd \ll 1$, the restriction to point-dipolar excitation precludes multipoles and is therefore only valid for small particle size. As we will see below, a satisfying description of energy transport is only possible by relaxing one or even both assumptions, depending on the circumstances.

Nevertheless, the quasi-static point-dipole model is still of value for didactic purposes, and as we will see approximately valid under certain restrictive conditions for the description of energy propogation. In order to arrive at the dispersion relation in this model, the equation of motion for the dipole moment $p_{i,m}$ of particle m with polarization in the i direction can be written as

$$\ddot{p}_{i,m} = -\omega_0^2 p_{i,m} - \Gamma \dot{p}_{i,m} - \gamma_i \omega_1^2 \left(p_{i,m-1} + p_{i,m+1}\right). \tag{24}$$

Here, ω_0 is the single-particle dipole plasmon resonance frequency, Γ an absorptive damping term describing relaxation processes, γ_i a polarization-dependent constant, and $\omega_1 \propto d^{-3}$ the electromagnetic coupling term in the near-field approximation. Note that in an additional simplification, (24) takes only nearest-neighbor interactions into account, motivated by the strong distance dependence of the near-field coupling. Substitution of propagating wave solutions $p = P \exp\left[-\alpha m d + i\left(\omega t \pm kmd\right)\right]$ into this equation results in the dispersion relation

$$\omega^2 = \omega_0^2 + 2\gamma_i \omega_1^2 \cos kd \cosh \alpha d, \tag{25}$$

which in the small-damping limit $\alpha d \ll 1$ simplifies to

$$\omega^2 = \omega_0^2 + 2\gamma_i \omega_1^2 \cos kd. \tag{26}$$

This quasi-static, point-dipolar dispersion relation (26) is shown as solid curves in Fig. 9 for both longitudinal and transverse polarisation. As apparent, the *group velocity* for energy transport, given by the slope of the dispersion curves, is highest for excitation at the single particle plasmon frequency, occurring at the center of the first Brillouin zone. Realistic geometric and materials parameters give group velocities on the order of $0.1c$. This model[32] predicts energy attenuation lengths of similar magnitude than those obtained in the Mie-theory treatment of the problem.[30] It is also noteworthy that transverse polarization waves propagate with negative phase velocity for excitation near the single-particle resonance, as

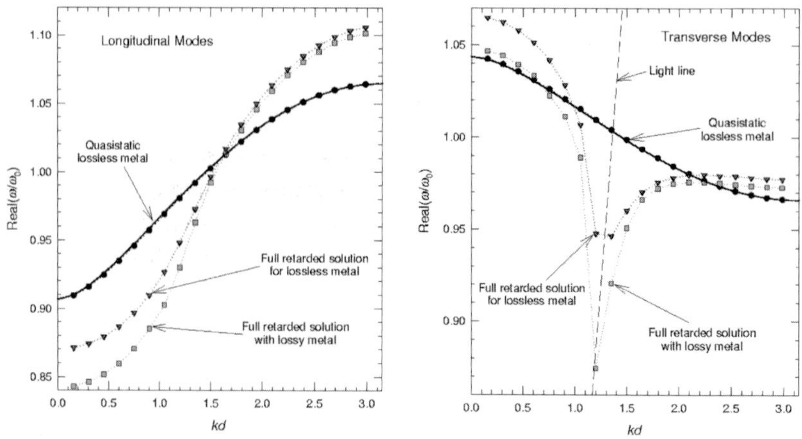

Fig. 9. Dispersion of longitudinal (left panel) and transverse (right panel) modes sustained by an infinite chain of spherical particles in the quasi-static approximation (solid lines,[32]), for a finite 20-sphere chain in the quasi-static approximation (full circles), and for the fully retarded solution with a lossy metal (squares) and for a lossless metal (triangles). Differences between the models are pronounced for transverse polarization. Reprinted with permission from.[33] Copyright 2004 by the American Physical Society.

has been confirmed via full-field finite-difference time-domain simulations[34] (Fig. 10).

Corrections to this solution by considering higher order multipoles - albeit still in the quasi-static approximation $kd \ll 1$ - have been obtained by Park and Stroud using a tight-binding approach.[35] As expected, deviations from the point-dipole model occur when the particles are closely spaced, as now multipolar fields influence the polarization of neighboring particles. However, for $a/d \leq 0.3$, a good agreement with the dipolar model was found. If we take the splitting between the longitudinal and transverse modes at $k = 0$ as a measure of strength of the particle interactions, then in the dipole limit this splitting increases with increasing a/d. In the higher-order model however the splitting reaches a maximum just before percolation sets in.

More significant insight into the full dynamics of energy propogation is obtained when the quasi-static assumption is relaxed and coupling to the radiation continuum included. The full expansion for the electric dipole field is used, and the polarizability changed from its quasistatic form α_{qs} to $1/\alpha \to 1/\alpha_{qs} - \frac{2i}{3}\frac{\omega^3}{c^3}$ to ensure self-consistency via the inclusion of radi-

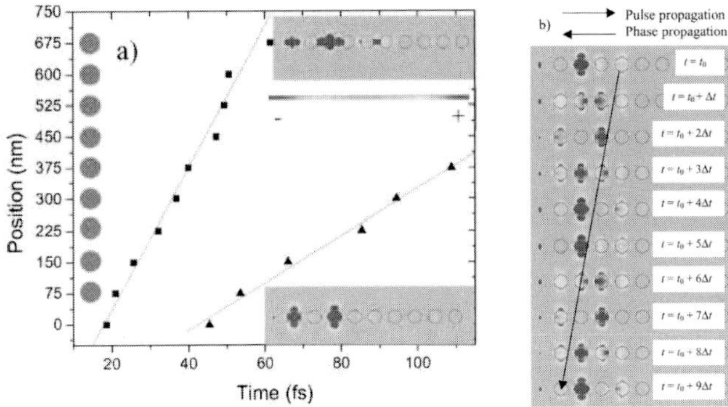

Fig. 10. Finite-difference time-domain simulation of a pulse propagating through a chain of 50 nm gold spheres with a 75 nm center-to-center distance. (a) Position of the peak of a pulse centered around the single particle resonance frequency with time as it propagates through the chain for longitudinal (squares) and transverse (triangles) polarization. The insets show snapshots of the electric field distribution. (b) Snapshots of the electric field distribution of a transverse mode traveling with negative phase velocity. The arrow denotes the movement of a particular phase front. Reprinted with permission from.[34] Copyright 2003 by the American Physical Society.

ation damping[33,36,37a]. The effects of these changes are shown in Fig. 9, which shows the dispersion calculation for a chain of point dipoles under these condition.[33] A significant modification of the dispersion relation for the transverse mode near the light line is revealed, due to phase-matching between the transverse dipolar mode to photons propagating along the waveguide at the same frequency,[33] resulting in two distinct branches due to polariton-like anticrossing.[37]

A more rigorous calculation based on a combination of an eigendecomposition method and Mie theory yields similar results (albeit with a small red-shift of the dispersion bands).[38] We note that negative phase velocity is now only predicted for modes showing strong radiation damping, and that excitation at the single-particle plasmon resonance frequency results in the excitation of a mode in this regime.

For longidutinal modes, coupling to the radiation continuum cannot take place, and the obtained curves are similar to the quasistatic result. However, the bandwidth of the longitudinal mode is more than doubled compared to the quasi-static result, resulting in an almost doubled group

[a]An inclusion of dynamic depolarization effects in addition to radiation damping results in only a small correction.[28]

velocity. We note that the point-dipole approximation reduces the validity of these results to the regime $a/d \leq 0.3$. Regarding energy propagation lengths, contrary to the results by Quinten also strong propagation for transverse modes was found, with energy attenuation lengths that can reach multiple μm for tranverse excitation just below the light line.[37] Naturally, for $k < \omega/d$, strong radiation damping sets in. For modes outside the light cone, these predicted modifications of the dispersion relation not captured by the quasi-static model have been experimentally confirmed using angle-resolved extinction spectrocopy.[27,28]

On a side note, another consequence of retardation effects is a strong frequency dependence of the spatial distribution of light localization in finite particle chains when illuminated by a plane wave in the direction of the array axis.[39,40] Depending on the frequency of the illuminating wave, electromagnetic energy can be preferentially localized at the beginning or the end of an array. This effect can be understood intuitively when considering the particle structure as an array of phased scatterers, leading to constructive or destructive interference along the chain.

Examples of the electric field distribution of the guided modes are shown in Fig. 10, which depicts results from finite-difference time-domain simulations of pulse propagation through a chain of 50 nm gold spheres separated by a center-to-center distance of 75 nm in air. These simulations have also confirmed the negative phase velocity of transverse modes[34] when excited at their single particle resonance frequency.

We now turn our attention to a direct experimental study of energy transport. The excitation of traveling waves requires a local excitation scheme, since far-field excitation only excites modes around the $\mathbf{k} = 0$ point in the dispersion diagram. By analysing the shift of the plasmon resonance compared to that of a single particle (or an array of sufficiently separated particles), due to interparticle coupling upon in-phase excitation, the strength of the coupling can be judged. Fig. 11 shows as an example a waveguide consisting of silver rods of aspect ratio $90 \times 30 \times 30$ nm^3 separated by a gap of 50 nm, and far-field exctinction spectra of the chain as well as of well-separated particles. A significant blue-shift shift due to particle coupling is apparent for the chain.

In order to locally excite a traveling wave on this structure, the tip of a near-field optical microscope was used as a local illumination source, and the energy transport along the particle array detected via fluorescent polymer beads (Color Plate 1a).[41] In this study, the tip of the near-field microscope was scanned over an ensemble of waveguides (Color Plate 1b),

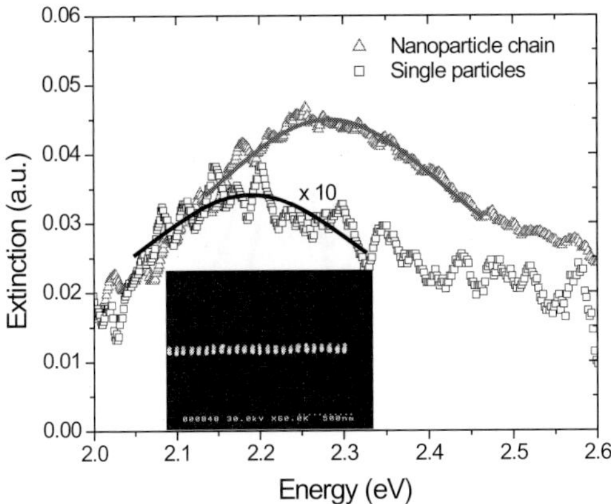

Fig. 11. Experimentally observed plasmon resonance for single silver rods and a chain of closely-spaced rods under transverse illumination (along the long axis of the rods). The blue-shift between the two spectra is due to near-field interactions between particles in the chain. Reprinted by permission of Macmillan Publishers Ltd: Nature Materials,[41] copyright 2003.

and the recorded fluorescent spots in the obtained near-field images compared between beads situated at a distance from the waveguides (Color Plate 1c) and those deposited on top of them (Color Plate 1d). The latter showed an elongation of the spot profile along the direction of the waveguide due to excitation at a distance via the particle waveguide: energy is transfered from the tip to the waveguide, and channeled to the fluorescent particle (see scheme in Color Plate 1a). Representative cross cuts through the fluorescent spots are shown in Fig. 12, suggesting energy transport along the particle chain over a distance of 500 nm. A numerical analysis has confirmed the major aspects of this coupling scheme.[42]

While this study presented the first direct hints at the possibility of energy transport experimentally, it needs to be pointed out that the nanoparticle waveguide geometry employed was based on mainly quasi-static considerations. Crucially, the waveguide was excited at the single particle plasmon frequency, which according to full-field studies would result in significant radiation damping in addition to absorptive losses. This implies that the waveguide was not excited at the frequency of lowest attenuation loss; guiding over multiple μm might indeed be possible according to full-field studies.[27,37] New experimental investigations are therefore clearly

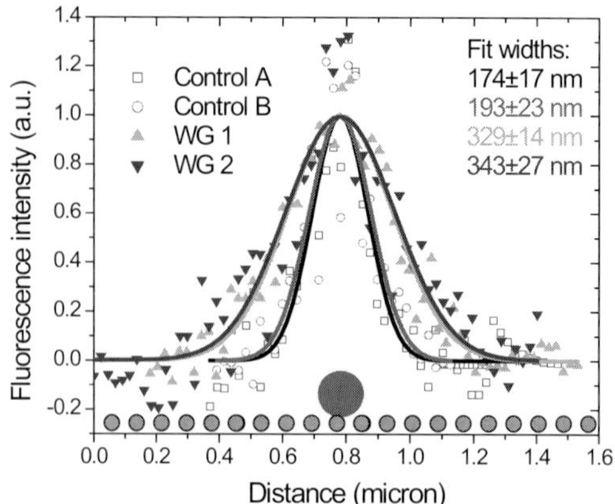

Fig. 12. Intensity of the fluorescence signal along the cross-cuts indicated in Fig. 6c,d for control fluorescent spheres located away from waveguides (squares) and spheres located on top of particle waveguides (triangles). The increase in width of the fluorescence peaks for the latter is indicative of excitation at a distance via the particle waveguides (see sketch in Fig. 6a). Reprinted by permission of Macmillan Publishers Ltd: Nature Materials,[41] copyright 2003.

called for, both in order to clarify the real potential of metal nanoparticle arrays for waveguiding, and also for possible applications in energy focusing[43] or sub-wavelength imaging.[44]

6. Two-dimensional photonic metal nanoparticle plasmon waveguides

We want to finish this chapter with a description of a nanoparticle plasmon waveguide of a different flavour, combining plasmonic excitations with photonic-crystal-based paradigms. As we have seen, excitation of metal nanoparticles in the vicinity of their dipole plasmon resonance implies large absorption losses. Significantly longer propagation lengths can be achieved by using non-resonant particle excitation at lower frequencies. However, while the absorptive losses are lowered, now radiative losses begin to overwhelm the guiding, and a different approach than one-dimensional chains is needed to keep the energy confined to the waveguide. A promising approach to achieve this was demonstrated in the form of a nanoparticle plasmon waveguide operating in the telecommunications window at

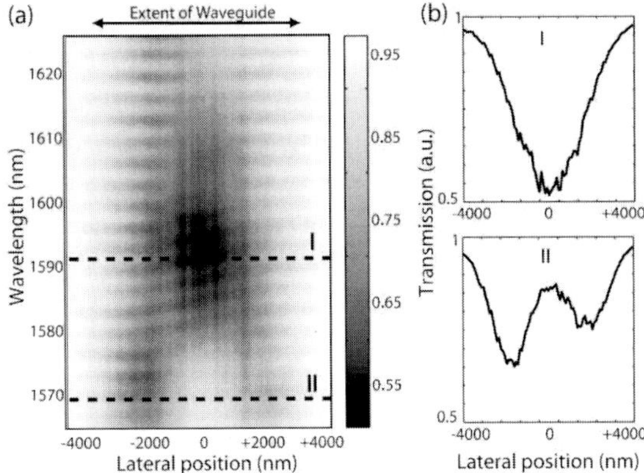

Fig. 13. (a) Power transferred past the coupling region vs. wavelength and transverse taper position. Both the fundamental and the first-order mode are shown in data cuts (b). Reprinted with permission from.[46] Copyright 2005, American Institute of Physics.

$\lambda_0 = 1.5\ \mu$m.[45,46] The new design exhibited a confinement superior to long-ranging stripe waveguides, while still exhibiting propagation lengths of the order of 100 μm. The waveguide is based on a two-dimensional lattice of metal nanoparticles on a thin, undercut silicon membrane (Color Plate 2d). Vertical confinement is achieved by a hybrid plasmon/membrane-waveguide mode, while transverse confinement can be achieved by using a lateral grading of nanoparticle size, thus in a sense creating a higher effective refractive index in the waveguide center. This way, the mode is confined to the higher-index region, leading to wavelength-scale transverse confinement and sub-wavelength scale vertical confinement (Color Plates 2b,c).

Due to the periodicity in the propagation direction, the plasmon mode is zone-folded back into the first Brillouin zone (Color Plate 2a). This suggests a convenient scheme for excitation using fiber tapers placed on top of the waveguide: contra-directional phase-matched evanescent coupling between the fiber taper and the plasmon mode can take place (see Color Plate 3).

The fiber taper is also a convenient means to investigate both the spatial and the dispersive properties of the nanoparticle waveguide. For a spatial mapping of the guided modes, the fiber simply has to be moved over the waveguide in the transverse direction, and the wavelength-dependent power transferred past the coupling region monitored. As an example, Fig. 13a shows the power transmitted past the coupling region vs. wavelength and transverse location of the taper over the waveguide. Both the fundamental and the first higher-order mode of the plasmon waveguide manifest themselves via power drops at 1590 nm and 1570 nm (Fig. 13b,c), depending on whether the taper is located over the waveguide center or at its edges. The spatial resolution is of course limited by the diameter of the taper, which in this case was about 1.5 μm.

Translation of the taper in the direction of the waveguide moves the point of phase-matching via a change in taper diameter. This can be used to map out the dispersion relation, and confirm the contra-directional nature of the coupling (Fig. 14a): As the diameter of the taper is increased (and thus its dispersion curve moves closer to the silica light line), the point of phase-matching shows a red-shift. A look at the dispersion diagram of Fig. 6a confirms that this is only the case for coupling to the zone-folded upper band. The maximum power transfer efficiency demonstrated experimentally using this geometry is about 75% (Fig. 14b).

These low-loss metal nanoparticle waveguides could be employed in applications for coupling radiation transmitted through optical fibers into two-dimensional SPP modes with high efficiency. After the coupling region, guiding to desired structures on a chip for sensing is possible, perhaps after interfacing with higher-confinement waveguides for field focusing.

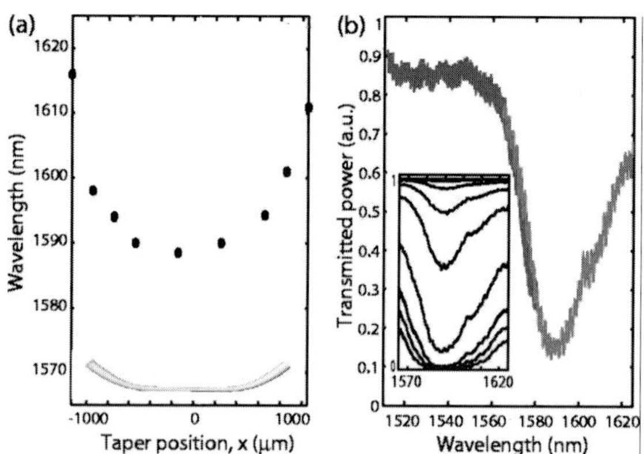

Fig. 14. (a) Spectral position of the phase matching point vs. taper position as the taper is moved along the waveguide axis, demonstrating the contra-directional nature of the coupling. (b) Power transmitted past the coupling region for the condition of optimum coupling, demonstrating transfer efficiencies of about 75%. The inset shows the evolution of the coupling profile as the gap between the taper and the waveguide is descreased. Reprinted with permission from.[46] Copyright 2005, American Institute of Physics.

References

1. A. Yariv, Y. Xu, R. K. Lee, and A. Scherer, Coupled-resonator optical waveguide: a proposal and analysis, *Opt. Lett.* **24**(11), 711–713 (June, 1999).
2. U. Kreibig and M. Vollmer, *Optical properties of metal clusters.* (Springer, Berlin, 1995).
3. S. A. Maier, *Plasmonics - Fundamentals and Applications.* (Springer, New York, NY, 2007), 1st edition.
4. J. D. Jackson, *Classical Electrodynamics.* (John Wiley & Sons, Inc., New York, NY, 1999), 3rd edition.
5. P. B. Johnson and R. W. Christy, Optical constants of the noble metals, *Phys. Rev. B.* **6**(12), 4370–4379 (Dec., 1972).
6. C. F. Bohren and D. R. Huffman, *Absorption and scattering of light by small particles.* (John Wiley & Sons, Inc., New York, NY, 1983), first edition.
7. G. Mie, Beiträge zur Optik trüber Medien, speaiell kolloidaler Metallösungen" , *Ann. Phys.* **25**, 377, (1908).
8. M. Meier and A. Wokaun, Enhanced fields on large metal particles: dynamic depolarization, *Opt. Lett.* **8**(11), 581–583 (nov, 1983).
9. H. Kuwata, H. Tamaru, K. Esumi, and K. Miyano, Resonant light scattering from metal nanoparticles: Practical analysis beyond Rayleigh approximation, *Appl. Phys. Lett.* **83**(22), 4625–2627 (Dec., 2003).
10. T. Kokkinakis and K. Alexopoulos, Observation of radiative decay of surface plasmons in small silver particles, *Phys. Rev. Lett.* **28**(25), 1632–1634 (June, 1972).
11. A. Wokaun, J. P. Gordon, and P. F. Liao, Radiation damping in surface-enhanced Raman scattering, *Phys. Rev. Lett.* **48**(14), 957–960 (Apr., 1982).
12. S. Link and M. A. El-Sayed, Shape and size dependence of radiative, non-radiative and photothermal properties of gold nanocrystals, *Int. Reviews in Physical Chemistry.* **19**(3), 409–453, (2000).
13. C. Sonnichsen, T. Franzl, T. Wilk, G. von Plessen, J. Feldmann, O. Wilson, and P. Mulvaney, Drastic reduction of plasmon damping in gold nanorods, *Phys. Rev. Lett.* **88**(7), 077402 (Feb., 2002).
14. E. J. Heilweil and R. M. Hochstrasser, Nonlinear spectroscopy and picosecond transient grating study of colloidal gold, *J. Chem. Phys.* **82**(9), 4762–4770, (1985).
15. H. Hovel, S. Fritz, A. Hilger, U. Kreibig, and M. Vollmer, Width of cluster plasmon resonances: Bulk dielectric functions and chemical interface damping, *Phys. Rev. B.* **48**(24), 18178–18188 (Dec., 1993).
16. M. A. Noginov, G. Zhu, M. Bahoura, J. Adegoke, C. E. Small, B. A. Ritzo, V. P. Drachev, and V. M. Shalaev, Enhancement of surface plasmons in an Ag aggregate by optical gain in a dielectric medium, *Opt. Lett.* **31**(20), 3022–3024 (Oct., 2006).
17. G. D. Dice, S. Mujumdar, and A. Y. Elezzabi, Plasmonically enhanced diffusive and subdiffusive metal nanoparticle-dye random laser, *Appl. Phys. Lett.* **86**, 131105, (2005).

18. L. M. Lawandy, Localized surface plasmon singularities in amplifying media, *Appl. Phys. Lett.* **85**(21), 5040–5042, (2004).
19. J. R. Krenn, M. Salerno, N. Felidj, B. Lamprecht, G. Schider, A. Leitner, F. R. Aussenegg, J. C. Weeber, A. Dereux, and J. P. Goudonnet, Light field propagation by metal micro- and nanostructures, *J. Microscopy.* **202**, 122–128 (Apr., 2001).
20. J. R. Krenn, A. Dereux, J. C. Weeber, E. Bourillot, Y. Lacroute, J. P. Goudonnet, G. Schider, W. Gotschy, A. Leitner, F. R. Aussenegg, and C. Girard, Squeezing the optical near-field zone by plasmon coupling of metallic nanoparticles, *Phys. Rev. Lett.* **82**(12), 2590–2593 (Mar., 1999).
21. S. A. Maier, M. L. Brongersma, P. G. Kik, and H. A. Atwater, Observation of near-field coupling in metal nanoparticle chains using far-field polarization spectroscopy, *Phys. Rev. B.* **65**, 193408, (2002).
22. S. A. Maier, P. G. Kik, and H. A. Atwater, Observation of coupled plasmon-polariton modes in Au nanoparticle chain waveguides of different lengths: Estimation of waveguide loss, *Appl. Phys. Lett.* **81**, 1714–1716 (Aug., 2002).
23. Q.-H. Wei, K.-H. Su, S. Durant, and X. Zhang, Plasmon resonance of finite one-dimensional Au nanoparticle chains, *Nano Letters.* **4**(6), 1067, (2004).
24. G. A. Wurtz, J. S. Im, S. K. Gray, and G. P. Wiederrecht, Optical scattering from isolated metal nanoparticles and arrays, *J. Phys. Chem. B.* **107**(51), 14191–14198, (2003).
25. K. Su, Q.-H. Wei, X. Zhang, J. J. Mock, D. R. Smith, and S. Schultz, Interparticle coupling effects on plasmon resonances of nanogold particles, *Nano Letters.* **3**(8), 1087–1090 (Aug., 2003).
26. A. Sundaramurthy, K. B. Crozier, G. S. Kino, D. P. Fromm, P. J. Schuck, and W. E. Moerner, Field enhancement and gap-dependent resonance in a system of two opposing tip-to-tip Au nanotriangles, *Phys. Rev. B.* **72**, 165409, (2005).
27. A. F. Koenderink, R. de Waele, J. C. Prangsma, and A. Polman, Experimental evidence for large dynamic effrects on the plasmon dispersion of subwavelength metal nanoparticle waveguides, *Phys. Rev. B.* **76**, 201403(R), (2007).
28. K. B. Crozier, E. Togan, E. Simsek, and T. Yang, Experimental measurement of the dispersion relations of the surface plasmon modes of metal nanoparticle chains, *Opt. Express.* **15**(26), 17482 (Dec., 2007).
29. M. Quinten and U. Kreibig, Absorption and elastic scattering of light by particle aggregates, *Appl. Opt.* **32**(30), 6173–6182 (Oct., 1993).
30. M. Quinten, A. Leitner, J. R. Krenn, and F. R. Aussenegg, Electromagnetic energy transport via linear chains of silver nanoparticles, *Opt. Lett.* **23**(17), 1331–1333 (Sept., 1998).
31. J. Takahara, S. Yamagishi, H. Taki, A. Morimoto, and T. Kobayashi, Guiding of a one-dimensional optical beam with nanometer diameter. **22**(7), 475–477 (Apr., 1997).
32. M. L. Brongersma, J. W. Hartman, and H. A. Atwater, Electromagnetic energy transfer and switching in nanoparticle chain arrays below the diffraction limit, *Phys. Rev. B.* **62**(24), R16356–R16359 (Dec., 2000).

33. W. H. Weber and G. W. Ford, Propagation of optical excitations by dipolar interactions in metal nanoparticle chains, *Phys. Rev. B.* **70**, 125429, (2004).
34. S. A. Maier, P. G. Kik, and H. A. Atwater, Optical pulse propagation in metal nanoparticle chain waveguides, *Phys. Rev. B.* **67**, 205402, (2003).
35. S. Y. Park and D. Stroud, Surface-plasmon dispersion relations in chains of metallic nanoparticles: An exact quasistatic solution, *Phys. Rev. B.* **69**, 125418, (2004).
36. D. S. Citrin, Coherent excitation transport in metal-nanoparticle chains, *Nano Letters.* **4**(9), 1561–1565 (Sept., 2004).
37. A. F. Koenderink and A. Polman, Complex response and polariton-like splitting in periodic metal nanoparticle chains, *Phys. Rev. B.* **74**, 033402, (2006).
38. K. H. Fung and C. T. Chan, Plasmonic modes in periodic metal nanoparticle chains: a direct dynamic eigenmode analysis, *Opt. Lett.* **32**(8), 973 (Apr., 2007).
39. J. V. Hernandez, L. D. Noordam, and F. Robicheaux, Asymmetric response in a line of optically driven metallic nanospheres, *J. Phys. Chem. B.* **109**, 15808, (2005).
40. R. de Waele, A. F. Koenderink, and A. Polman, Tunable nanoscale localization of energy on plasmon particle arrays, *Nano Letters.* **7**(7), 2004, (2007).
41. S. A. Maier, P. G. Kik, H. A. Atwater, S. Meltzer, E. Harel, B. E. Koel, and A. A. G. Requicha, Local detection of electromagnetic energy transport below the diffraction limit in metal nanoparticle plasmon waveguides, *Nat. Mat.* **2**(4), 229–232 (Apr., 2003).
42. C. Girard and R. Quidant, Near-field optical transmittance of metal particle chain waveguides, *Opt. Express.* **12**(25), 6141–6146 (Dec., 2004).
43. K. Li, M. I. Stockman, and D. J. Bergman, Self-similar chain of metal nanospheres as an efficient nanolens, *Phys. Rev. Lett.* **91**(22), 227402 (Nov., 2003).
44. C. R. Simvoski, A. J. Viitanen, and S. A. Tretyakov, Resonator mode in chains of silver spheres and its possible application, *Phys. Rev. E.* **72**, 06606, (2005).
45. S. A. Maier, P. E. Barclay, T. J. Johnson, M. D. Friedman, and O. Painter, Low-loss fiber accessible plasmon waveguides for planar energy guiding and sensing, *Appl. Phys. Lett.* **84**(20), 3990–3992 (May, 2004).
46. S. A. Maier, M. D. Friedman, P. E. Barclay, and O. Painter, Experimental demonstration of fiber-accessible metal nanoparticle plasmon waveguides for planar energy guiding and sensing, *Appl. Phys. Lett.* **86**, 071103, (2005).

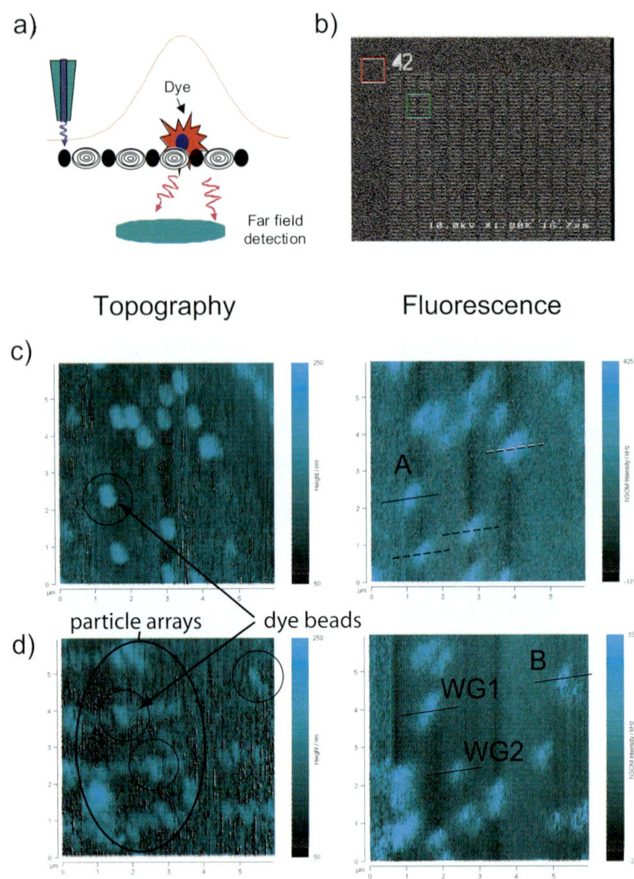

Color Plate 1. Local excitation and detection of energy transport in metal nanoparticle plasmon waveguides. Schematic of the experiment (a), SEM images of plasmon waveguides (b), and images of the topography and fluorescence (c, d). The images presented in (c) show fluorescent spheres deposited in a region without waveguides, while (d) shows spheres deposited on top of the ends of four nanoparticle chains. The circles and lines mark the fluorescent spots analyzed in Fig. 12. Reprinted by permission of Macmillan Publishers Ltd: Nature Materials,[41] copyright 2003.

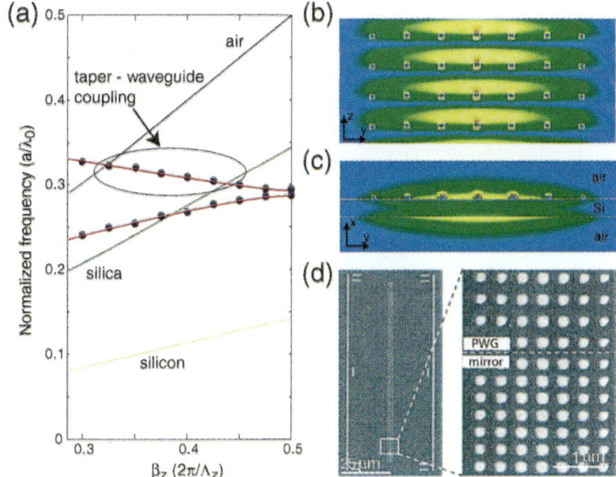

Color Plate 2. Dispersion relation (a) and mode profiles in top (b) and side (c) view of a metal nanoparticle plasmon waveguide on a thin Si membrane operating in the near-infrared. (d) SEM picture of a fabricated device. Reprinted with permission from.[45] Copyright 2004, American Institute of Physics.

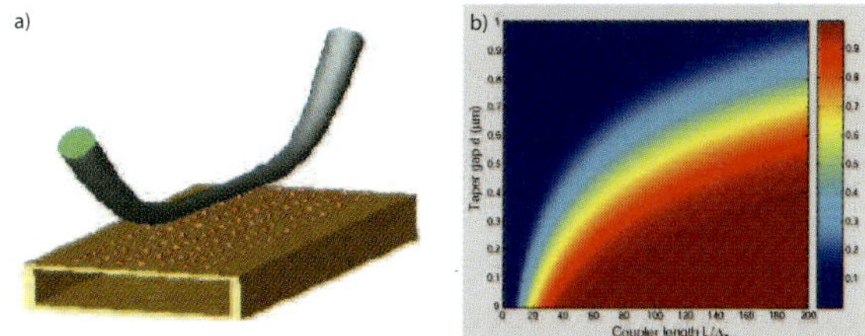

Color Plate 3. (a) Schematic of evanescent coupling to metal nanoparticle plasmon waveguides via optical fibre tapers. (b) Transferred power vs. taper/waveguide gap and length of coupling region.

Chapter 4

SURFACE PLASMON-POLARITON GAP WAVEGUIDE AND ITS APPLICATIONS

Kazuo Tanaka

Department of Electronics and Computer Engineering, Gifu University
Yanagido 1-1 Gifu City, Japan 501-1193
E-mail: tanaka@tnk.info.gifu-u.ac.jp

The phase velocities of surface plasmon polaritons (SPPs) that propagate in the nanosize metallic gap strongly depend on the gap width. This means that it is possible to control the phase velocities of SPPs by varying the gap. Nanosize optical waveguide circuits and devices can be realized by using this characteristic. These nanometer optical waveguides can be called surface plasmon gap waveguides (SPGWs)[1]. In this chapter, basic characteristics of SPGWs are demonstrated by three-dimensional (3D) numerical simulations based on the volume integral equation method. Applications of SPGWs to nanosize planar optical circuits and optical probes for near-field scanning optical microscopes (NSOMs) are also demonstrated.

1. The structure of SPGWs

Surface plasmon polaritons (SPPs) are waves that result from an intimate interaction between electromagnetic waves and metallic surface electrons. When using conventional dielectric optical waveguides, sizes and densities of integration are limited by diffraction limit of light. By contrast, a SPP-based optical waveguide can overcome this limit. Optical circuits using SPPs are thus promising candidates for future nanoscale circuitry [1-5]. In this chapter, the basic characteristics of interesting SPP waveguides are demonstrated by 3D numerical simulations. Throughout this chapter, we assume that the wavelength of the light is $\lambda = 573$ nm and that the material supporting the SPP is silver (Ag) whose relative permittivity is $\varepsilon_1/\varepsilon_0 = -124 - j0.85$ assuming time dependence $\exp(j\omega t)$, where ε_1 and ε_0 are the permittivities of the metal and of vacuum, respectively. In this chapter, all spaces outside the metallic materials are assumed to be vacuum in order to reduce the parameters in the numerical simulations. In many cases, these spaces will be filled with dielectric media in future practical applications; however, this will not drastically change the fundamental waveguide characteristics. All lengths used are normalized by the free-space wavenumber $k_0 = 2\pi/\lambda = \omega/c$ (where c is the speed of light in vacuum) throughout this chapter. Many important lengths are also expressed by approximate real lengths in nm in parentheses for the convenience of readers.

We start with a simple two-dimensional (2D) problem. We first calculate the propagation constants of a SPP in a nanosize gap (permittivity ε_0) between two parallel semi-infinite metallic plates. The

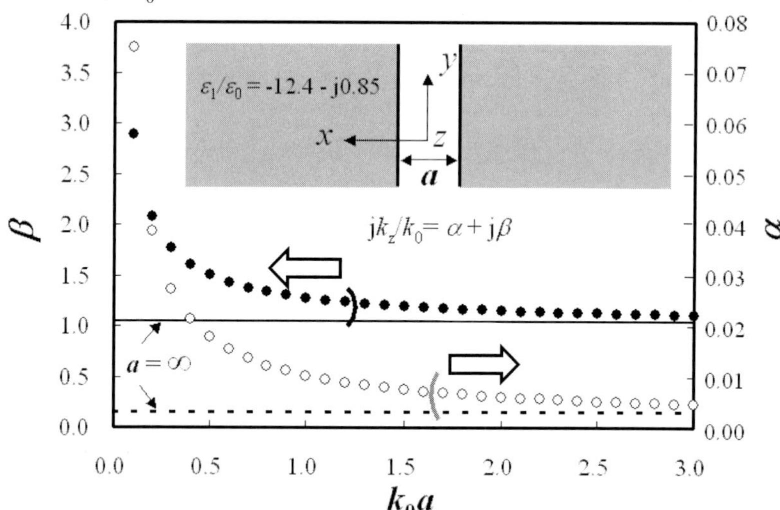

Fig. 1. Complex propagation constant in the gap between two parallel semi-infinite metallic plates. The solid and open circles represent the phase (β) and attenuation (α) constants normalized by the free-space wavenumber k_0, respectively. The solid and dotted lines show the results of SPPs in the semi-infinite plate.

geometry of the problem is shown in the inset in Fig. 1. Since variation along the y-axis can be neglected in the geometry shown in the inset, the problem is 2D and a dispersion equation for the SPP in this structure can be readily obtained and solved to yield the complex propagation constant in the z-direction in the inset. We assume that the dependence of propagating SPP on the z-coordinate is expressed as $\exp(\pm jk_z z)$. The dependences of the propagation constant k_z normalized by the free space wavenumber $jk_z / k_0 = \alpha + j\beta$ in the z-direction on the normalized gap width $k_0 a$ are shown in Fig. 1, where α and β are the (normalized) attenuation and phase constants, respectively. It can be seen that the narrow gap width gives a large phase constant β (low phase velocity) and large attenuation constant α for SPPs in the gap between two parallel metallic plates. It is also found that the attenuation constant α remains at an acceptable value from the viewpoint of nanosize optical circuits. For example, $\alpha \approx 0.0153$ and $\beta \approx 1.4323$ for the case of $k_0 a = 0.6$ (55 nm). In this case, the propagation length, that is the distance of $1/e$ field amplitude reduction, is 5934 nm inside the gap. This propagation length is adequate for future nanosize optical integrated circuits. With this behavior of the SPP in mind, we consider the waveguide whose cross-

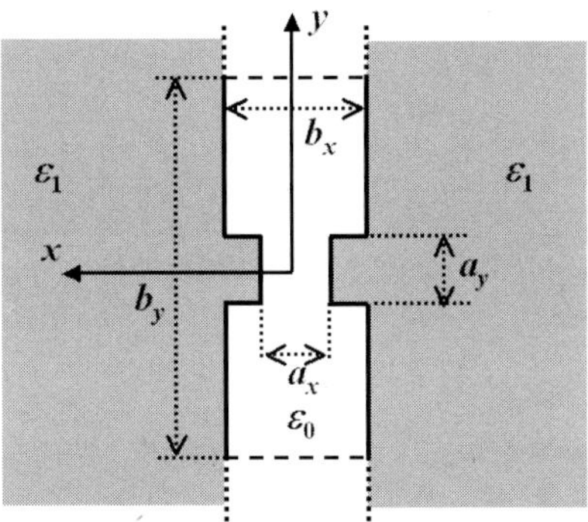

Fig. 2. The basic structure of a surface plasmon gap waveguide (SPGW).

section is shown in Fig. 2. A narrow-gap region, whose size is $a_x \times a_y$, is sandwiched by two wide-gap regions, each of size $b_x \times (b_y - a_y)/2$, between two metallic plates. Since the phase velocity of a SPP in the narrow-gap region is smaller than that in the wide-gap region (based on an analogy with dielectric waveguides) it is expected that the optical field will be confined near the narrow-gap region and guided along it. One can imagine many modified structures based on the principles and structures of Figs. 1 and 2. We call optical waveguides based on this structure surface plasmon gap waveguides (SPGW) [1] throughout this chapter.

2. Simulation of excitation and propagation of SPPs in a straight SPGW

In order to understand the basic characteristics of a SPGW, first consider the simulated excitation of guided SPPs in a straight SPGW through a small aperture shown in Fig. 3. We investigate the propagation characteristics of this SPGW numerically[6]. The geometry of the problem is shown in Fig. 3. Two identical thick metallic slabs whose size is given by $D \times d \times l$ are placed parallel to the *x*-*z* plane with a nanosize gap of *w*. This structure constitutes a straight SPGW, where the sizes of the narrow- and wide-gap regions are given by $a_x = w$, $a_y = d$, $b_x \approx \infty$, and $b_y \approx \infty$ in Fig. 2. Throughout this chapter, we refer to *w* and *d* as the gap

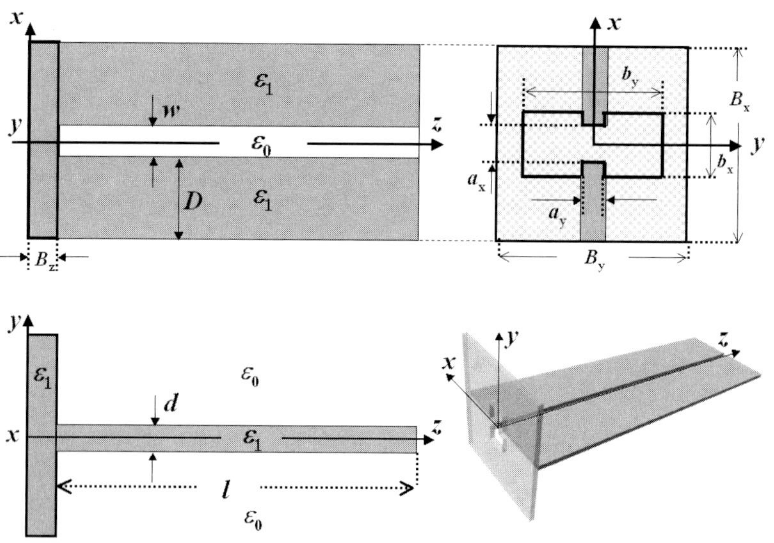

Fig. 3. The excitation of guided modes in a straight SPGW through an I-shaped aperture in a metallic screen.

width and gap depth, respectively. Optically guided SPPs in the straight SPGW can be excited through an entrance aperture created in a rectangular thick metallic screen shown in Fig. 3. The entrance aperture is also a SPGW whose structure is identical to that shown in Fig. 2, i.e., a rectangular narrow-gap region sandwiched between two rectangular wide-gap regions. We call this entrance the I-shaped aperture[7]. The screen is made of the same material (silver) as the metallic slabs and is positioned parallel to the x-y plane shown in Fig. 3. Its center coincides with the central axis of the straight SPGW. The size of the rectangular metallic screen is given by $B_x \times B_y \times B_z$ and is assumed to be $B_x = (2 \times D + w)$ for convenience in the numerical simulations. A Gaussian beam is incident normal to the screen from the negative z direction in Fig. 3. The electric-field polarization at $z = 0$ is parallel to the x axis and the amplitude on the beam axis at $z = 0$ is unity. It is also assumed that the beam axis coincides with the center of the SPGW. The spot size of the incident Gaussian beam is sufficiently smaller than the cross-section of the metallic screen that negligibly small guided waves are excited by the screen into the straight SPGW by the incident field. The guided fields are therefore excited in the straight SPGW by the Gaussian beam only through the entrance I-shaped aperture in Fig. 3.

2.1 *Simulation by a volume integral equation*

We solve the scattering problem for the metallic structure shown in Fig. 3 by using a volume integral equation given by

$$\mathbf{E}_i(\mathbf{x}) = \mathbf{D}(\mathbf{x}) / \varepsilon_r(\mathbf{x}) - k_0^2 \mathbf{A}(\mathbf{x}) - \nabla \left[\nabla \bullet \mathbf{A}(\mathbf{x}) \right], \quad (1)$$

where $\mathbf{D}(\mathbf{x})$ is the total electric flux, $\mathbf{E}_i(\mathbf{x})$ is the electric field of the incident beam, and $\mathbf{A}(\mathbf{x})$ is the vector potential, which is expressed by the following volume integral [8, 9],

$$A(x) = (1/\varepsilon_0) \iiint_V \{[\varepsilon_r(x') - \varepsilon_0] / \varepsilon_r(x')\} G(x \mid x') D(x') dv'. \quad (2)$$

Here, $G(\mathbf{x} \mid \mathbf{x}')$ is a free-space Green's function given by

$$G(\mathbf{x}\,|\,\mathbf{x}') = \exp(-jk_0\,|\,\mathbf{x}-\mathbf{x}'|)/(4\pi\,|\,\mathbf{x}-\mathbf{x}'|)\,. \tag{3}$$

The volume integral region V in (2) represents the entire space, and $\varepsilon_r(\mathbf{x})$ is the permittivity, where $\varepsilon_r(\mathbf{x})=\varepsilon_1$ in the metal of the straight SPGW and the screen and $\varepsilon_r(\mathbf{x})=\varepsilon_0$ in the surrounding free space. An expression for the incident Gaussian beam can be found in the literature [10]. In the numerical technique employed in this chapter, the entire region must be a rectangular parallelepiped of size $B_x \times B_y \times (l+B_z)$ including the free-space regions shown in Fig. 3 and this region is divided into small discretized cubes of size $\delta\times\delta\times\delta$. Under these conditions, the volume integral equation (1) with (2) is discretized by the method of moments using roof-top functions as basis and testing functions. The resultant system of linear equations is then solved by iteration using the generalized minimized residual method (GMRES) [11] with a fast Fourier transform (FFT). The numerical expressions are omitted here since they can be found in the literature [8-11].

2.2 *Results by numerical simulations*

The following parameters are used in the numerical simulations for Fig. 3: size of the metallic screen: $k_0B_x = k_0B_y = 112$ (1021 nm), $k_0B_z = 20$ (182 nm); cross-section of the entrance I-shaped aperture: $k_0b_z = 3.0$ (274 nm), $k_0b_z = 4.0$(365 nm), $k_0a_x = k_0a_x = 1.0$ (91 nm); length of the SPGW: $k_0l = 59.2$(5399 nm); gap width: $k_0w = k_0a_x = 1.0$(91 nm); gap depth: $k_0d = k_0a_y = 1.0$(91 nm); size of the discretized cube used in the discretization: $k_0\delta = 0.1$ (9 nm); diameter of the spot size of the incident Gaussian beam at $z=0$: λ (573 nm). The simulation results of optical intensities $|\mathbf{E}(\mathbf{x})|^2$ are shown in Figs. 4(a) and (b). They plot the distribution of total optical intensity on the x-z plane and on the y-z plane, respectively. Since the SPGW is not terminated by the matched boundary, standing waves can be clearly seen to arise along the straight SPGW in Fig. 4. These results show that the guided wave is efficiently excited through the entrance I-shaped aperture. Optical intensities of the field components $|E_x(\mathbf{x})|^2$, $|E_y(\mathbf{x})|^2$, and $|E_z(\mathbf{x})|^2$ on a plane parallel to the x-y plane placed at $k_0z = 1.0$ inside the entrance I-shaped aperture are shown in Figs. 5(a), (b) and (c), respectively. Notice that the intensity range

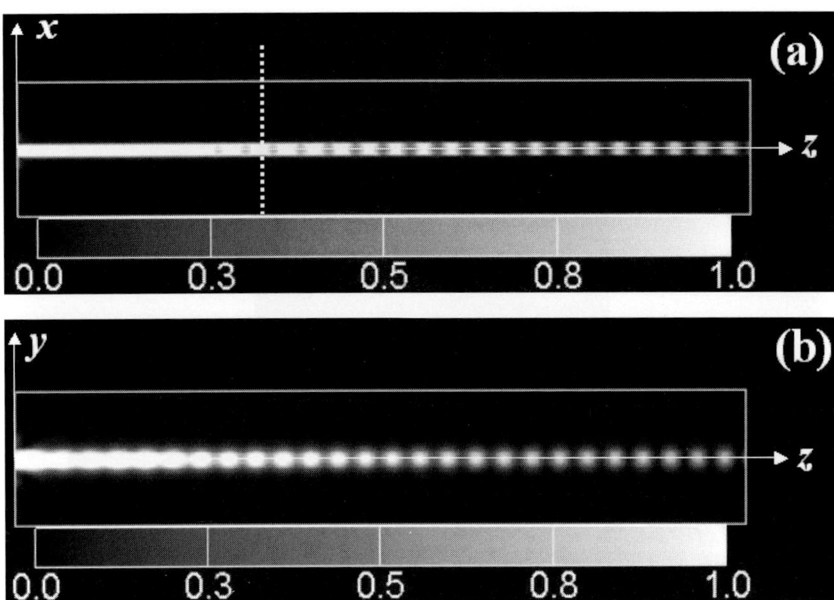

Fig. 4. Total optical intensity $|\mathbf{E}(\mathbf{x})|^2$ (a) on the x-z plane and (b) on the y-z plane in the straight SPGW for the case of $k_0 w = k_0 d = 1.0$ (91 nm). The intensity range is 0.0 to 1.0.

normalized to the incident intensity is 0.0 to 4.0 in Fig. 5(a), 0.0 to 1.0 in (b), and 0.0 to 0.4 in (c). We can see that the optical field is enhanced in the narrow-gap region inside the I-shape aperture shown in Fig. 4(a). Optical intensities of the field components $|E_x(\mathbf{x})|^2$, $|E_y(\mathbf{x})|^2$, and $|E_z(\mathbf{x})|^2$ on a plane parallel to the x-y plane placed at $k_0 z = 20.4$ indicated by the white dotted line in Fig. 4(a) are also shown in Figs. 6(a), (b), and (c), respectively, with a range of 0.0 to 2.0. The electric field of the guided wave excited in the SPGW principally has an x-component. In Figs. 5 and 6, the guided wave is confined in the gap region and is guided along it in the SPGW. In order to determine the vector characteristics of the guided waves in the SPGW in detail, the optical fields $Re[E_x(\mathbf{x})]$, $Re[E_y(\mathbf{x})]$, and $Re[E_z(\mathbf{x})]$ on planes parallel to the x-y plane placed at the proper points (not the same for each component) along the straight SPGW are shown in Figs. 7(a), (b), and (c), respectively. Notice the field distribution near the edges of the two metallic slabs that constitute the straight SPGW in Fig. 7. The y-component of the electric field is anti-symmetric about the x-axis, while

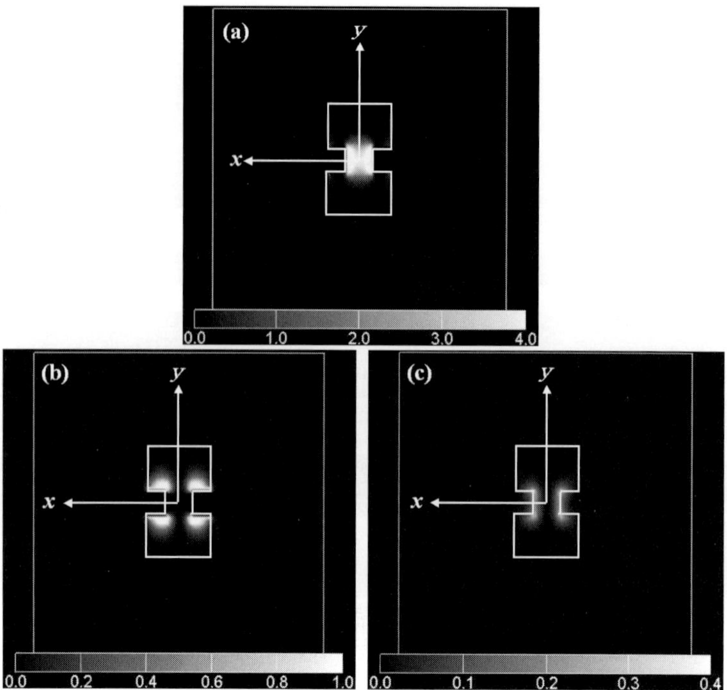

Fig. 5. Optical intensities of electric field components (*a*) $|E_x(\mathbf{x})|^2$, (b) $|E_y(\mathbf{x})|^2$, and (c) $|E_z(\mathbf{x})|^2$ on a plane parallel to the *x*-*y* plane placed at $k_0z = 1.0$ inside the entrance I-shaped aperture in Fig. 2. The intensity ranges are (a) 0.0 to 4.0, (b) 0.0 to 1.0, and (c) 0.0 to 0.4.

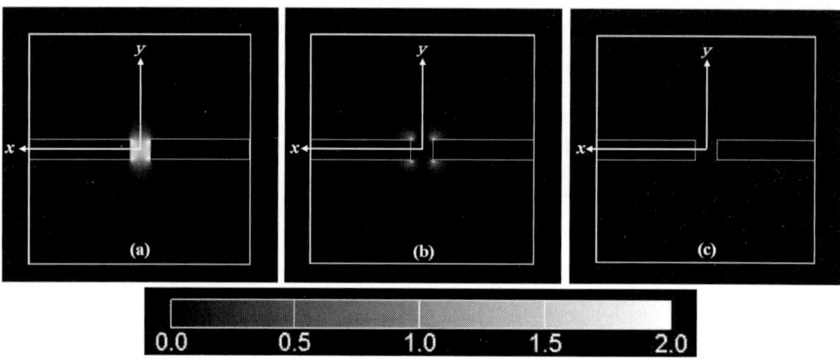

Fig. 6. Optical intensities (*a*) $|E_x(\mathbf{x})|^2$, (b) $|E_y(\mathbf{x})|^2$, and (c) $|E_z(\mathbf{x})|^2$ on a plane parallel to the *x*-*y* plane placed at $k_0z = 20.4$ of the SPGW shown in Fig. 2. The intensity range is 0.0 to 2.0.

the z-component is symmetric near the edges as shown in Figs. 7(b) and (c). These characteristics imply that the guided mode in the SPGW is coupled to the short-range SPP (SR-SPP) mode that propagates along the metallic slabs [12]. This result is important for understanding the dependence of the attenuation on the gap depth.

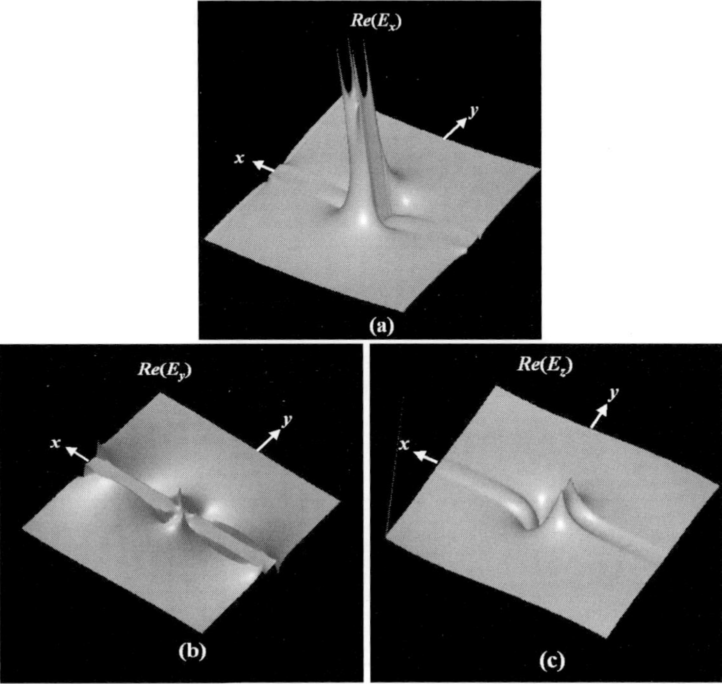

Fig. 7. Three-dimensional illustrations of electric field components of (a) $Re[E_x(\mathbf{x})]$, (b) $Re[E_y(\mathbf{x})]$ and (c) $Re[E_z(\mathbf{x})]$ on planes parallel to the x-y plane placed at the proper points in the SPGW in Fig. 2. The vertical range is arbitrary.

2.3 *Field distributions along the SPGW*

Typical one-dimensional distributions of the optical intensity $|E_x(\mathbf{x})|^2$ along the z-axis inside the straight SWPG are shown in Figs. 8 and 9. The results with a gap width w for the case of $k_0 d = 1.0$ (91 nm) and with a gap depth d for the case of $k_0 w = 1.0$ (91 nm) are shown in Figs. 8 and 9, respectively. Apparently, we can understand that the smaller

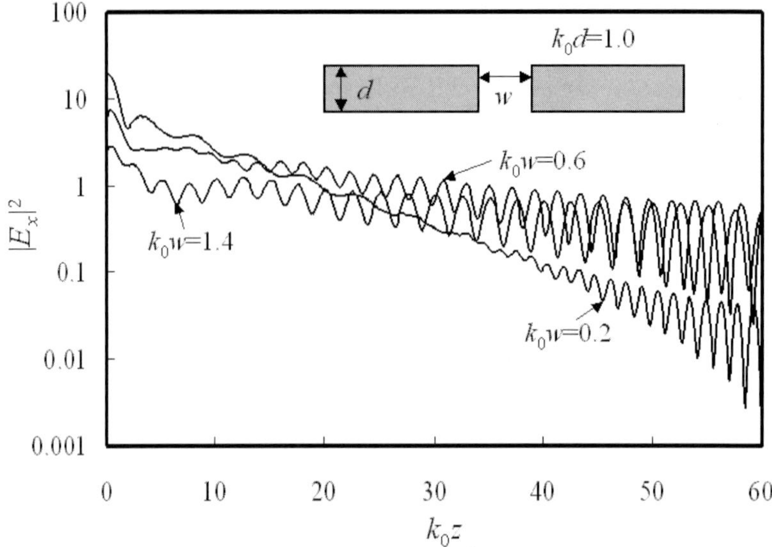

Fig. 8. Dependence of the optical intensity $|E_x(\mathbf{x})|^2$ on the propagation distance k_0z in the straight SPGW as a function of the normalized gap depth k_0w for the case of $k_0d = 1.0$.

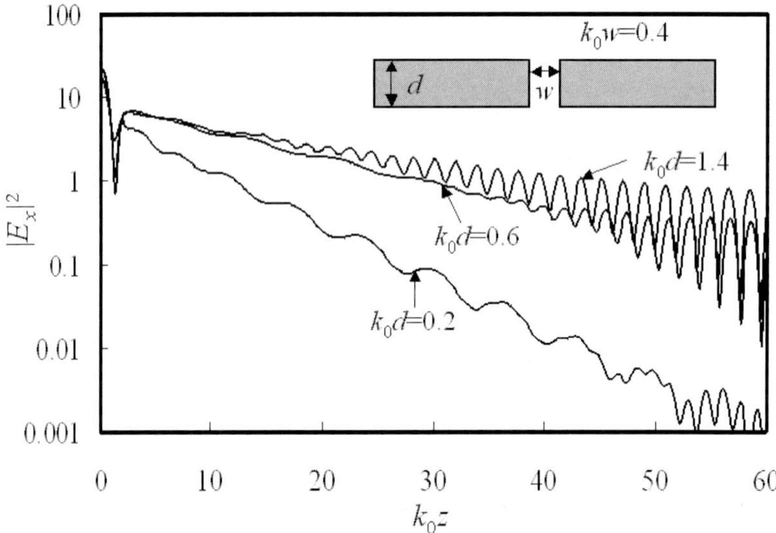

Fig. 9. Dependence of the optical intensity $|E_x(\mathbf{x})|^2$ on the propagation distance k_0z in the straight SPGW as a function of the normalized gap depth k_0d for the case of $k_0w = 0.4$.

values of the gap width and the gap depth increase the attenuation in the SPGW. The optical fields are enhanced inside the entrance I-shaped aperture, i.e., in the range $k_0z < 20$. Smaller values of the gap width give a larger enhancement in Fig. 8. Figures 8 and 9 show a mostly monotonic decay of the envelope of the standing waves together with an oscillation along the propagation distance z for $k_0z > 20$. These characteristics show that two guided waves that travel in opposite directions have one propagation constant k_z in this range, i.e., single-mode propagation. So it is possible to calculate a single complex propagation constant k_z from the simulated optical fields shown in Figs. 8 and 9.

From the above results of the numerical simulations, we determine the following interesting characteristics of the SPGW:

***Confinement*:** The optical fields are confined in the nanosize gap region of the SPGW. The size of the confined optical field distribution can be reduced to a size much smaller than the wavelength.

***Guidance*:** The optical fields can be guided along the nanosize gap region of the SPGW.

***Enhancement*:** The optical fields are enhanced in the nanosize gap region of the SPGW.

2.4 *Calculation of complex propagation constants by least-squares fitting*

The most fundamental parameter of a SPGW is the propagation constant k_z in the application of SPGWs to practical optical circuits or devices. The permittivity of the metallic material that supports SPPs in a SPGW has a complex value in the optical frequency. Therefore, the propagation constant has a complex value and the calculation of k_z is tedious. A rigorous solution of the guided mode for the geometry shown in Fig. 2 is not known. Approximate values of the propagation constant can usually be obtained by the effective-index modeling (EIM) [13]. However, EIM is not valid in the present case. We calculate the propagation constant using the simulation results in Figs. 8 and 9, i.e., we calculate the normalized attenuation constant α and phase constant β of the guided waves from the electric field distributions by least-squares fitting. When the attenuation constant is not large and sufficiently long SPGWs cannot be used in the simulation, waves reflected by the waveguide end cannot be neglected. In the regions close to the entrance aperture and to the end of the SPGW shown in Fig. 3, the optical fields do not consist of simple single-mode propagation, based on the numerical examples shown in Figs. 8 and 9. The evanescent near-fields resulting from excitation through the entrance aperture and reflection by the waveguide end must be included. However, in the limited range $Z_{min} \le z \le Z_{max}$ between the entrance ($z = B_z$) and end ($z = B_z + l$) of the SPGW it can be assumed that the x-component of the electric fields $e_x(z)$ along the z-axis can be approximated by the following expression,

$$e_x(z) = A\exp[-(\alpha + j\beta)k_0 z] + B\exp[+(\alpha + j\beta)k_0 z] \qquad (4)$$

where A and B are complex constants. The first term in Eq. (4) represents a wave propagating in the positive z-direction while the second term represents a wave propagating in the negative z-direction in Fig. 1. The unknown constants A, B, α, and β can be approximately calculated by minimizing the following least-squares error at discrete points z_j along the proper line parallel to the z-axis inside the SPGW,

$$J(A,B,\alpha,\beta) = \sum_j | E_x(z_j) - e_x(z_j) |^2 \qquad Z_{min} \le z_j \le Z_{max} \qquad (5)$$

where $E_x(z_j)$ represents discrete values of the x-component of the electric field at discrete points z_j in the numerical simulation. The parameters Z_{min} and Z_{max} are minimum and maximum values of the range over which the least-squares fitting is employed. The values of $k_0 Z_{min} = 20$ and $k_0 Z_{max} = 50$ are employed in this chapter based on consideration of the numerical results in Figs. 8 and 9. The propagation constants obtained in this way can be confirmed by comparing the electric field from the simulation with the fields reconstructed from Eq. (4). We used only the x-component of the electric field in the calculation of Eq. (5), because E_y and E_z are much smaller than E_x along the central axis of the straight SPGW as shown in Fig. 6.

The calculated dependences of the normalized attenuation constant α and phase constant β on the normalized gap width $k_0 w$ for the case of $k_0 d = 1.0$(91 nm) are shown in Fig. 10. Both constants α and β decrease

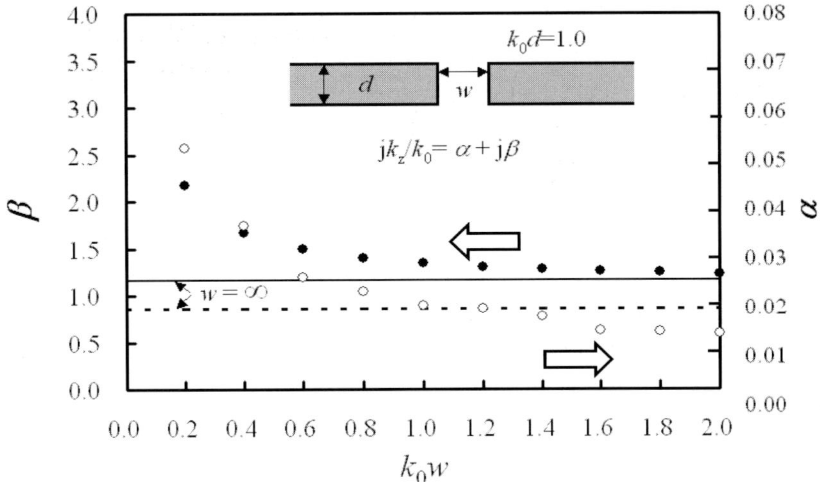

Fig. 10. Dependence of the normalized attenuation constant α (open circles) and phase constant β (solid circles) on the gap width k_0w for the case of $k_0d = 1.0$ (91 nm). The solid and broken lines show the phase and attenuation constants, respectively, for the case of $k_0w = \infty$ (edge mode).

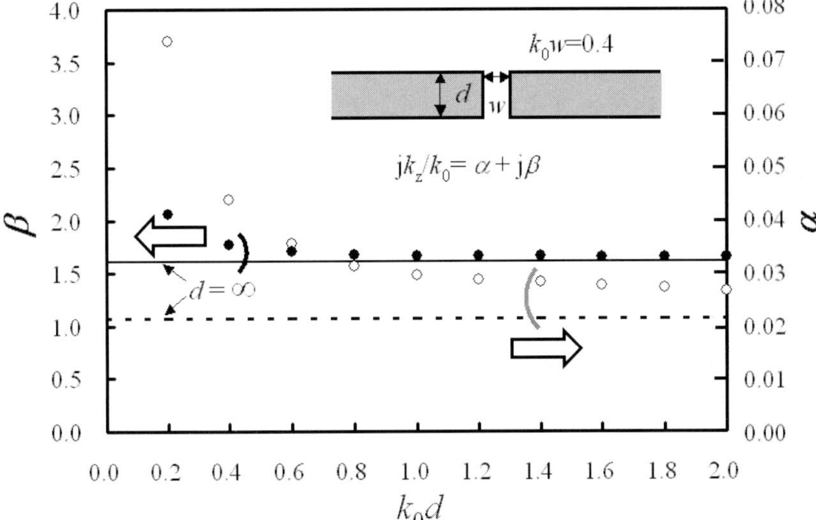

Fig. 11. Dependence of the normalized attenuation constant α (open circles) and phase constant β (solid circles) on the gap depth k_0d for the case of $k_0w = 0.4$ (36 nm). The solid and broken lines show the phase and attenuation constants, respectively, for the case of $k_0d = \infty$ from Fig. 1.

with increasing gap width w and asymptotically approach the results for $w=\infty$ shown in Fig. 10. Results for the case $w=\infty$ can be obtained by removing one of the slabs that constitute the straight SPGW in Fig. 2. It is known that SPPs can propagate along the edge of the slab. We call this guided wave the edge mode [14, 15]. Under the conditions of this chapter, the dependence of β on the gap width w is weak for $k_0 w > 1.0$ in Fig. 10. It is found that the attenuation constant α is smaller than that of the edge mode ($w=\infty$) in the range of $k_0 w > 1.0$ in Fig. 10. For example, the phase constant β of the SPGW is larger than that of the edge mode for the case of $k_0 w = 1.2$ and $k_0 d = 1.0$. However, the attenuation constant α of the SPGW is similar to that of edge mode in this same case. This result implies the existence of an optimum gap width that gives the smallest attenuation constant for guided waves in the SPGW. Similar phenomena are also found for the 2D calculation of the SPP gap mode [13]. The dependence of the normalized attenuation constant α and phase constant β on the gap depth d for the case of $k_0 w = 0.4$ (36 nm) are shown in Fig. 11. Both constants α and β decrease with increasing gap depth d and asymptotically approach the results for $d=\infty$ (Fig. 1) shown in Fig. 11. It is found that their dependence on the gap depth d is weak for the case of $k_0 d > 0.6$ in Fig 11. The reason that α and β increase with decreasing gap depth d is that the guided wave shown in Fig. 7 is coupled with the SR-SSP mode excited in the metallic slabs. So in general strongly confined fields induce large attenuations of the guided wave in a SPGW.

2.5 *Other structures of SPGWs*

We can consider many variations in the structure of the SPGW. It is interesting to investigate possible reductions of the attenuation constant from the viewpoint of applications. Based on the numerical results in Figs. 10 and 11, we find that the attenuation constants α can be reduced by increasing the gap width w and/or the gap depth d. When $w \to \infty$ in the SPGW, the waveguide structure only supports an edge mode. The dependence of the propagation constants of the edge mode on the slab depth d is shown in Fig. 12. Comparing Figs. 11 and 12, the phase and attenuation constants of the edge mode for the case $k_0 w$=0.4 are smaller than those for all values of $k_0 d$. However, as shown in Fig. 10, the attenuation constant of the edge mode is not always smaller than that of the SPGW. The intensity distributions of the field components $|E_x(\mathbf{x})|^2$, $|E_y(\mathbf{x})|^2$, and $|E_z(\mathbf{x})|^2$ of the edge mode are shown in Fig. 13. It is found that the intensities are also confined near the edge. When we compare Figs. 6 and 13, we see that the intensity range of $|E_y(\mathbf{x})|^2$ is similar to that of $|E_x(\mathbf{x})|^2$. This means that the degree of confinement of the edge mode is weaker than that of the SPGW.

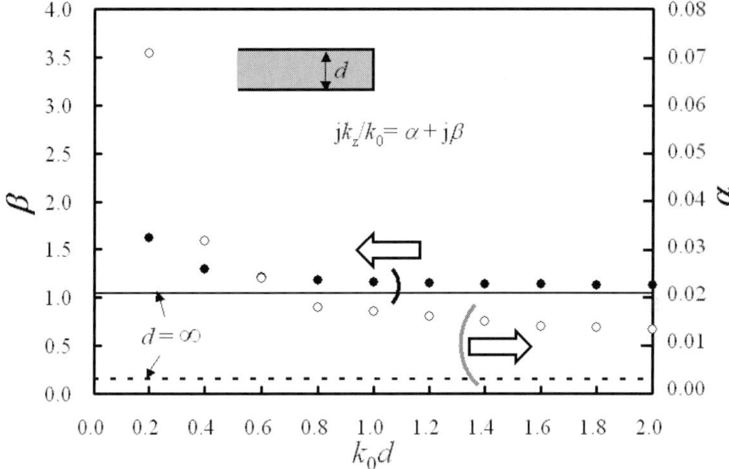

Fig. 12. Dependence of the normalized attenuation constant α (open circles) and phase constant β (solid circles) of the edge mode on the slab depth. The solid and broken lines show the phase and attenuation constants, respectively, for the case of $k_0 d = \infty$.

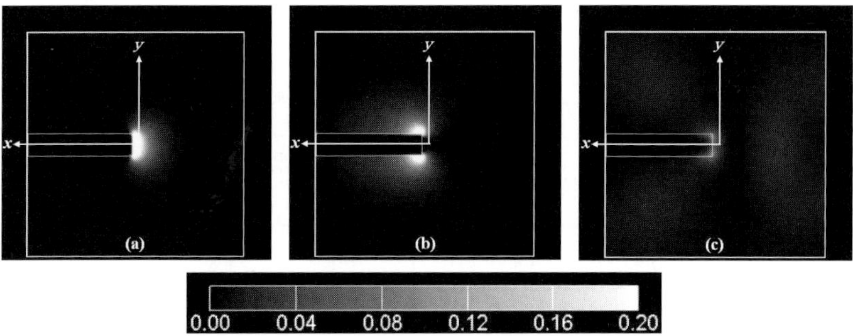

Fig. 13. Typical optical intensities of the field components (a) $|E_x(\mathbf{x})|^2$, (b) $|E_y(\mathbf{x})|^2$, and (c) $|E_z(\mathbf{x})|^2$ in a plane parallel to the x-y plane placed at $k_0 z$=5.0 of the SPGW (edge mode) in Fig. 2. The intensity range is 0.0 to 0.2.

Consider the case where the slab depth of the SPGW is increased. The guided mode in the SPGW is coupled with the SR-SPP mode of the slab as shown in Fig. 7. The attenuation of the SR-SPP mode increases with decreasing slab width. We expect a reduction of the attenuation constant when one of the slabs is replaced by a thick metallic plate, i.e., $d \to \infty$. The results in this case are shown in Fig. 14 for the case of $k_0 d = 1.0$ and in Fig. 15 for the case of $k_0 d = 0.4$. The attenuation and phase constants in Fig. 14 are smaller than those of the original SPGW shown in Fig. 10, as expected. We can find an optimum gap width *w* that gives the smallest attenuation constant in the SPGWs of Figs. 14 and 15. These smallest attenuation constants are about 20 - 30% smaller than that of the edge mode. Typical intensity distributions of the field components for the case $k_0 w = k_0 d = 1.0$ are shown in Fig. 16. The degree of confinement is in Fig. 16 is smaller than that of the original SPGW shown in Fig. 6. Other ideas can be considered in order to further reduce the attenuation constants of the SPGW. It would be useful to find the structure that gives an attenuation constant close to that of a SPP on an infinite plate, i. e., $\mathrm{Im}[\varepsilon_0 \varepsilon_1 /(\varepsilon_0 + \varepsilon_1)]$.

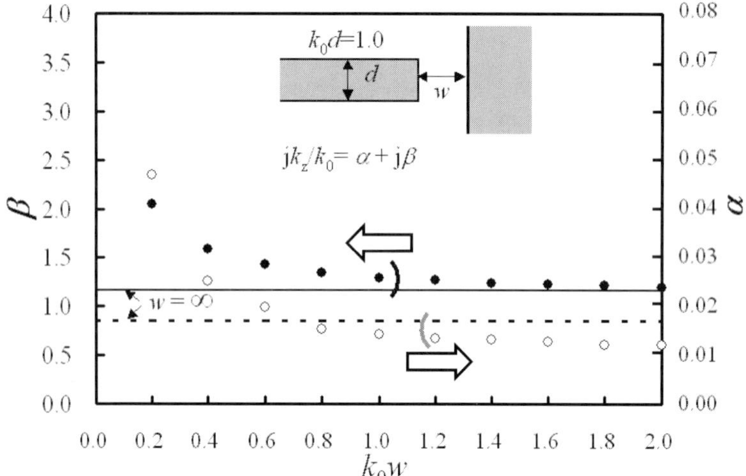

Fig. 14. Dependence of the normalized attenuation constant α (open circles) and phase constant β (solid circles) of SPPs on the gap width for a slab-plate structure in the case $k_0 d = 1.0$. The solid and broken lines show the phase and attenuation constants, respectively, for the case of $k_0 w = \infty$ (edge mode).

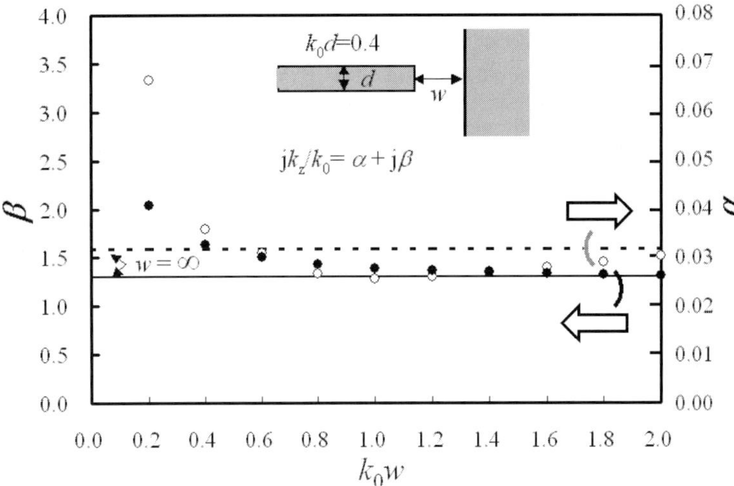

Fig. 15. Dependence of the normalized attenuation constant α (open circles) and phase constant β (solid circles) of SPPs for a slab-plate structure in the case $k_0d = 0.4$. The solid and broken lines show the phase and attenuation constants, respectively, for the case of $k_0w = \infty$ (edge mode).

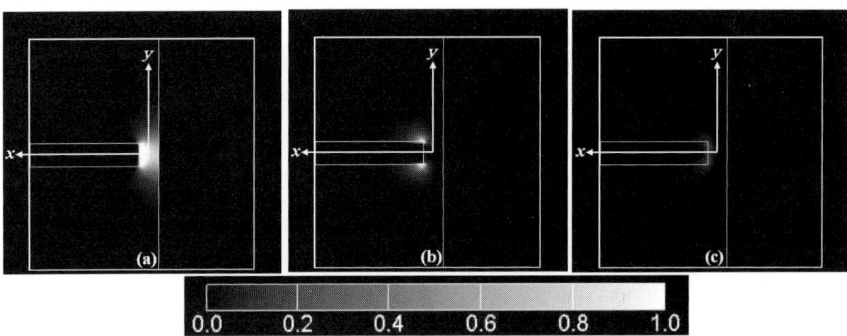

Fig. 16. Typical optical intensities (*a*) $|E_x(\mathbf{x})|^2$, (b) $|E_y(\mathbf{x})|^2$, and (c) $|E_z(\mathbf{x})|^2$ in a plane parallel to the *x*-*y* plane placed at k_0z=11.2 for the SPGW (slab-plate structure) in Fig. 2 for the case $k_0w=k_0d$=1.0. The intensity range is 0.0 to 1.0.

3. Application of SPGWs to nanosize planar optical circuits

Consider the applications of SPGWs to practical nanosize optical circuits. Since the electric field is mainly composed of the *x*-component in the SPGW shown in Fig. 3, we can consider two kinds of planar optical circuits by analogy to microwave circuits. One is the E-plane planar circuit [1, 16, 17] where the main electric field component is parallel to the circuit plane. The other is the H-plane circuit [1, 18] where the principal electric field component is perpendicular to the circuit plane.

3.1 *E-plane optical circuits*

When we construct a gap-structure parallel to the *x*-*z* plane, rather than the straight SPGW of Fig. 3, we obtain an E-plane planar optical circuit. An example of an E-plane circuit is shown in Fig. 17. An asymmetric four-port branching structure is constructed by the SPGWs whose gap size and gap depth are *w* and *d*, respectively, in Fig. 17. The output ports of the branching SPGWs are open to free space. Like the geometry shown in Fig. 3, guided waves in the SPGWs are excited by a Gaussian beam through an entrance aperture created in a square thick metallic screen parallel to the *x*-*y* plane in Fig. 17. The electric field polarization at $z = 0$ is parallel to the *x* axis and the amplitude on the beam axis at z=0 is unity. The size of the screen is given by $B_x \times B_y \times B_z$ and B_x is equal to the width of the circuit plane in the *x* direction in Fig. 17. The screen is made of the same material as that of the SPGW circuit. A single small rectangular aperture of size $b_x \times b_y$ is used as the entrance aperture in this case. A Gaussian beam is incident normal to the screen from the negative *z* axis and the amplitude on the beam axis at z=0 is unity. The beam axis coincides with the *z*-axis. The parameters used in the simulation are as follows: Size of the metallic screen: $k_0 B_x = k_0 B_y = 10.7$ (976 nm), $k_0 B_z = 1.0$ (91 nm). Cross-section of entrance rectangular aperture:

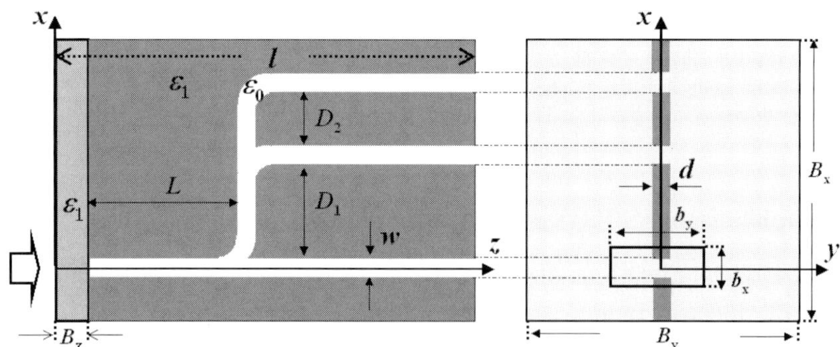

Fig. 17. Schematic of a nanosize E-plane planar optical circuit with SPGWs.

$k_0b_x = 0.4$ (36 nm), $k_0b_y = 1.8$ (164 nm). Size of the circuit: $k_0l = 8.8$ (803 nm), $k_0L = 2.5$ (228 nm), $k_0D_1 = 2.05$ (189 nm), and $k_0D_2 = 1.35$ (123 nm). Average radius of the curved sections: $k_0r = 1.0$ (91 nm). Size of the discretized cube: $k_0\delta = 0.05$ (4.5 nm). Diameter of the spot size of the incident Gaussian beam at $z = 0$ is λ (573 nm). The illustration shown in Fig. 17 does not represent the actual geometry used in the numerical simulation. The center of the entrance aperture must coincide with the center of the square metallic screen in the geometry used in the simulations. Because the size of the entrance rectangular aperture b_y perpendicular to the polarization is smaller than π, the TE_{01} dominant mode in the rectangular waveguide (made of a perfect conductor) is cut off. The size of the waveguide cross-section that gives rise to this cutoff condition can be significantly decreased in the case of a SPP rectangular waveguide. This is an interesting characteristic of a SPP waveguide [19]. In the examples here, the SPP is efficiently excited in the SPGW through this size of rectangular aperture.

The results of the numerical simulations are shown in Color Plates 1 and 2 (color figures at end of chapter). Each of the six panels in these figures consist of two parts. The main illustration shows the distribution of total optical intensity $|\mathbf{E}(\mathbf{x})|^2$ in the x-z plane. The narrow illustrations on their right sides show the total optical intensity $|\mathbf{E}(\mathbf{x})|^2$ in the plane parallel to the x-y plane in the region indicated by the white dotted line in the main illustration. The results of the simple S-shaped circuit and asymmetric three-port branching circuit are shown in Color Plates 1(a) and (b), respectively. The structures of these circuits are made by simplifying the original four-port branching circuit of Fig. 17. In Color Plate 2, the results for four-port branching circuits are shown for the four combinations of $k_0w = 0.2, 0.4$ (18 nm, 36 nm) and $k_0d = 0.2, 0.4$. Note that the size of these circuits in the x-z plane is approximately $1.7\lambda \times 1.6\lambda$. Standing waves can be seen to arise along the branching SPGW structures in the circuit shown in Color Plates 1 and 2. It is possible to calculate approximate phase constants β by using the distance between adjacent peaks of these standing waves to obtain (a) $\beta \approx 2.51$ $(k_0w = 0.2, k_0d = 0.2)$, (b) $\beta \approx 2.17$ $(k_0w = 0.2, k_0d = 0.4)$, (c) $\beta \approx 1.80$ $(k_0w = 0.4, k_0d = 0.4)$, and (d) $\beta \approx 1.75$ $(k_0w = 0.4, k_0d = 0.2)$. We see that the phase constant decreases with increasing gap width w

and/or gap depth d. Since these values are larger than the speed of light in a dielectric medium whose index of refraction is $n \approx 1.5$, it is possible to fabricate the E-plane circuit of Fig. 17 on a glass substrate. The optical fields are confined in the nanometer region and are guided along the branching structure as designed. These results show that it is possible to excite optical waves in a complicated nanosize optical SPGW circuit using a conventional light source. The optical fields inside the circuits are strongly enhanced in these figures and the smaller gap width gives a larger enhancement of the guided fields. This is the one of the interesting characteristics of the nanosize optical circuits made of SPGW. Since the total energy of the nanosize optical fields are small, due to the field confinement in nanosize space regions, the enhancement of the optical fields will give rise to interesting practical applications.

3.2 *H-plane optical SPGW circuits*

We next consider H-plane planar optical SPGW circuits. An example is shown in Fig. 18. First, the metallic substrate is shaved while preserving the ridge structure of height h which forms a four-port asymmetric branching circuit in the y-z plane shown in Fig. 18. Then a thick metallic cover plate is placed over the ridge structure, leaving a small gap between them. The substrate and cover plate are thicker than the optical skin depth of the metal. The space between the substrate and cover plate is connected to the open space in Fig. 18. The width of the ridge (gap depth) is d and the gap between the ridge and the cover plate (gap width)

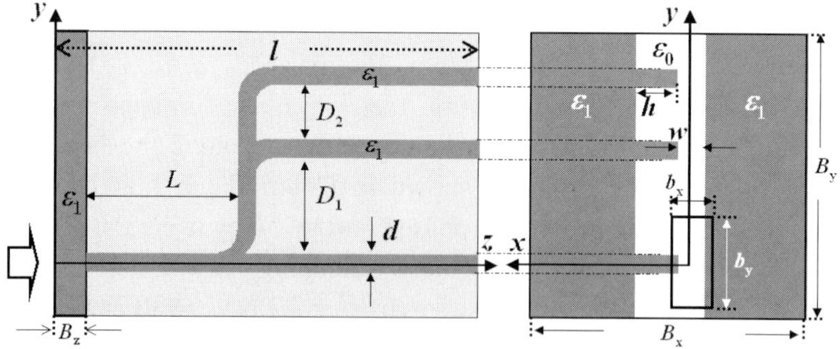

Fig. 18. Schematic of a nanosize H-plane planar optical circuit with SPGWs.

is w. We use the same (slab-plate) structure of SPGW as shown in Figs. 14 and 15 for the gap between the metallic slab and the plate. The entrance rectangular aperture is also the same as that used in the E-plane circuit in Fig. 17. A Gaussian beam is incident normal to the screen from the negative z axis and the amplitude on the beam axis at z=0 is unity. The electric field polarization at $z = 0$ is parallel to the x axis. As for Fig. 17, Fig. 18 does not represent the real geometry used in the numerical simulation. Most of the parameters used in the numerical simulations are identical to those used in Fig. 17. The only new parameter is the ridge height, given by $k_0 h = 2.0$ (182 nm). Total optical intensities obtained from the numerical simulations are shown in Color Plate 3 for the four possible combinations of $k_0 w = 0.2, 0.4$ (18 nm, 36 nm) and $k_0 d = 0.2, 0.4$. The main illustration shows the distribution of optical intensity $|\mathbf{E}(\mathbf{x})|^2$ in the y-z plane. The narrow illustrations on their right sides show the total optical intensity $|\mathbf{E}(\mathbf{x})|^2$ in the plane parallel to the x-y plane indicated by the white dotted line in the main illustration. Approximate values of phase constants calculated from the standing waves in Color Plate 3 are (a) $\beta \approx 2.17$ $(k_0 w = 0.2, k_0 d = 0.2)$, (b) $\beta \approx 2.02$ $(k_0 w = 0.2, k_0 d = 0.4)$, (c) $\beta \approx 1.75$ $(k_0 w = 0.4, k_0 d = 0.2)$, and (d) $\beta \approx 1.65$ $(k_0 w = 0.4, k_0 d = 0.4)$. We also see that phase constant decreases with increasing gap width w and/or gap depth d. It makes sense that all of the values of β are smaller than those in Color Plate 1, because a slab-plate structure is used instead of slab-slab structure. The optical fields are confined in the nanometer region and are guided along the branching structure as designed. The smaller gap width creates a large enhancement of the optical intensity in the SPGW similar to results in Color Plate 2.

It is interesting to compare the results shown in Color Plate 3 with those of the edge mode waveguide in Fig. 12. If the cover plate is removed in Fig. 18, the H-plane planar circuit becomes an edge-mode waveguide circuit. The circuit characteristics can then be simulated using similar methods as before. The results are shown in Color Plate 4. Optical waves are guided along the branching ridge structure without a cover plate. Approximate values of the phase constant β calculated from the standing waves in Color Plate 4 are (a) $\beta \approx 1.57 (k_0 d = 0.2)$ and (b) $\beta \approx 1.16 (k_0 d = 0.4)$. The large difference between the results in Color

Plates 3 and 4 is due to the degree of enhancement in Color Plate 4. The degree of enhancement in Color Plate 4 is much smaller than that in Color Plate 3. It depends on many circuit parameters and further investigation would be required to fully describe it. It should be noteced that the attenuation constant of the edge mode is not always smaller than that of the SPGW shown in Figs.14 and 15.

4. Creation of strongly enhanced and strongly confined optical fields by a SPGW

Creation of a strongly confined and strongly enhanced optical field from a conventional light source is a fundamental problem in nano-photonics and plasmonics [20-22]. Such optical fields can be applied to various practical devices. One obvious application is the optical probe of a near-field scanning optical microscope (NSOM). A SPGW can simultaneously provide both a strongly enhanced optical field and a small spot size. An example of the technique by which a confined and enhanced optical near-field on a tapered metallic probe tip can be formed using a SPGW is described in this section [23, 24].

4.1 *Geometry of the problem*

A schematic structure of the metallic probe tip used to create strongly confined and strongly enhanced optical fields is shown in Fig. 19. A uniform metallic screen $B_x \times B_y$ with thickness B_z and complex-valued permittivity ε_1 is placed on the x-y plane. A pyramidal structure of the same metal is fabricated on top of this screen, and a SPGW (an I-shaped aperture) is formed. The cross-section of this SPGW is I-shaped and consists of a rectangular narrow-gap region measuring $a_x \times a_y$ and two wide-gap regions of size $b_x \times (b_y - a_y)/2$. The pyramidal structure is divided into two sections by the SPGW. The height from the screen to the tip of the left pyramidal section is assumed to be larger than that of the right section in Fig. 19. A Gaussian beam is incident normal to the screen from the negative z-direction below the metallic screen, and the electric field polarization at z=0 is parallel to the x-axis. The SSPs exiting the I-shaped aperture are confined and enhanced in the narrow–gap

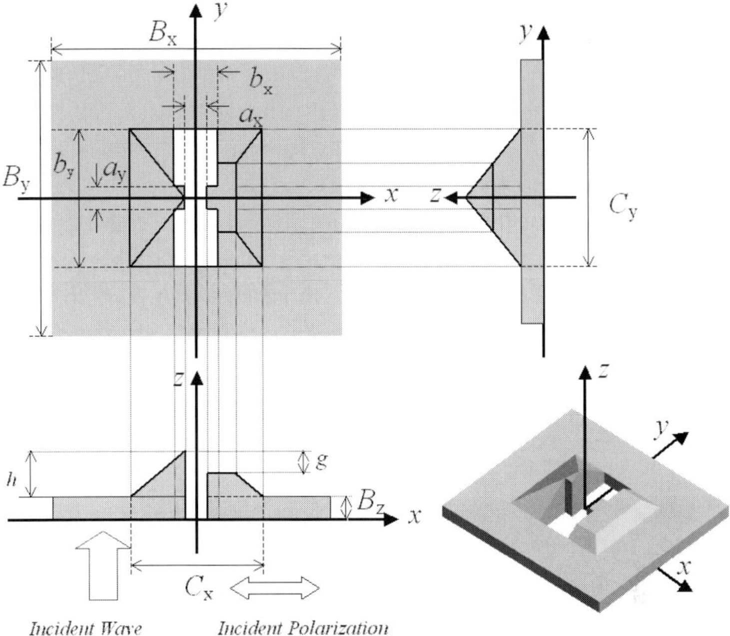

Fig. 19. Optical probe to create a strongly confined and strongly enhanced near-field on a probe tip. The probe consists of a SPGW (an I-shaped aperture) in a pyramidal structure.

region through the pyramidal structure and propagates along the left section. Finally the SPPs reach the tip of the left section shown in Fig. 19 and are focused at it with a small radius. The pyramidal tip is the tapered end of a plasmon waveguide. The optical field is focused by decreasing the cross-section of the waveguide along the propagation direction of the SPP [25]. Therefore we expect that a near-field with high intensity and small spot size will be created around the probe tip. In this probe, the tip is illuminated by SPPs that are confined and enhanced by the SPGW. In the simulation, the amplitude of the incident Gaussian beam is unity at the center axis at $z=0$. The parameters used in the simulation are as follows: Size of the metallic screen: $k_0 B_x = k_0 B_y = 21.2$ (1933 nm), $k_0 B_x = 0.7$ (64 nm). I-shaped aperture: $k_0 a_x \times k_0 a_y = 0.4 \times 0.4$ (36 nm× 36 nm), $k_0 b_x \times k_0 b_y = 1.0 \times 4.0$ (91 nm × 365 nm). Pyramidal probe: $k_0 h = 1.2$ (109 nm), $k_0 g = 0.4$ (36 nm), $C_x = C_y = b_y$. Size of the discretized cube: $k_0 \delta = 0.05$ (4.5 nm). The diameter of the spot size of the beam is 2λ.

4.2 *Enhanced and localized optical intensity distribution*

We plot typical numerical simulation results for the probe in Fig. 20. Total optical intensities $|\mathbf{E}(0,k_0y,k_0z)|^2$ and $|\mathbf{E}(k_0x,0,k_0z)|^2$ inside the SPGW and pyramidal structure, i.e., the intensity distributions on the y-z and x-z planes, are shown in Figs. 20(a) and (b), respectively. The shapes of the cross-section of the probe tips are shown in the right-hand portion of each panel in Fig. 20. These figures show that the probe tip is composed of a single discretized cube whose size is $k_0\delta = 0.05$ (4.6 nm). The standing wave (interference between incident and reflected guided waves) and the enhancement of the guided optical waves can be observed along the narrow-gap region inside the SPGW and pyramidal structure in Fig. 20. The optical intensity is apparently confined and

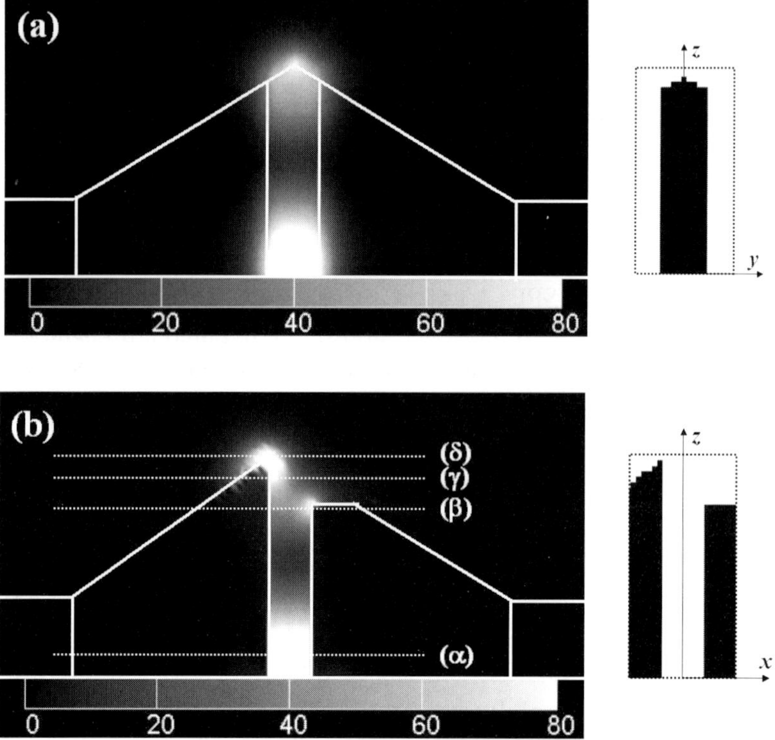

Fig. 20. Total optical intensity distributions on the (a) y-z and (b) x-z planes. Notice that the normalized intensity range is 0 to 80. The right-hand portion of each panel shows the cross-section that contains the probe tip of the structure used in the simulation.

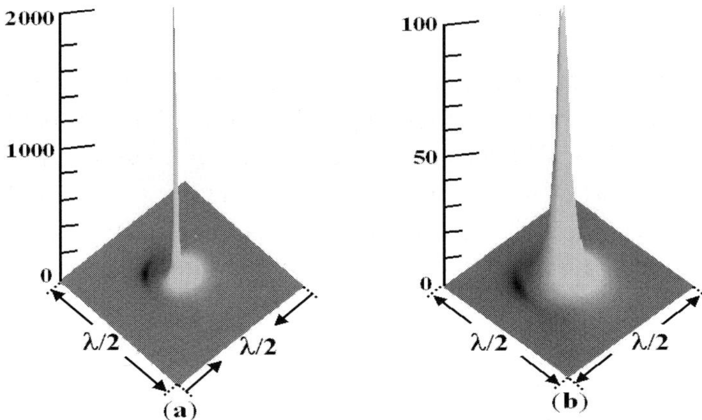

Fig. 21. Two-dimensional graph of the total optical intensity distributions on the probe tip. (a)$|\mathbf{E}(k_0x, k_0y, k_0w + k_0h + 0.025)|^2$ and (b)$|\mathbf{E}(k_0x, k_0y, k_0w + k_0h + 0.075)|^2$. The normalized intensity range is 0 to 2000 in (a) and 0 to 100 in (b).

enhanced near the tip of the left pyramidal section, i.e., the tapered probe tip shown in Fig. 20. Figures 21(a) and (b) show two-dimensional distributions of the confined and enhanced total optical intensities on planes parallel to the x-y plane above the probe tip, i.e., $|\mathbf{E}(k_0x, k_0y, k_0w+k_0h+0.025)|^2$ (distance between tip and observation plane ≈ 2.3 nm) and $|\mathbf{E}(k_0x, k_0y, k_0w+k_0h+0.075)|^2$ (distance between tip and observation plane ≈ 6.8 nm), respectively. It is found that the enhanced intensity close to the tip in Figs. 21(a) and (b) is greater than 1000× and 100× the incident intensity, respectively. The approximate full-width-at-half-maximum (FWHM) value normalized by the wavenumber k_0 of the optical intensity shown is $k_0 \times FWHM \approx 0.08$ (7 nm) in Fig. 21(a) and $k_0 \times FWHM \approx 0.25$ (23 nm) in Fig. 21(b), respectively, in both the x- and y-directions. Since the maximum value in Fig. 21(b) is about 100× the incident intensity, the decay of the localized and enhanced intensity in the z-direction is faster than those in the x- and y-directions. Figure 22 shows the total optical intensities on the cross-sections indicated by the dotted lines (α)-(δ) parallel to the x-y plane of Fig. 19(b). First, the excited optical field is confined and enhanced in the narrow-gap region of the SPGW shown in (α). Next, the optical field is

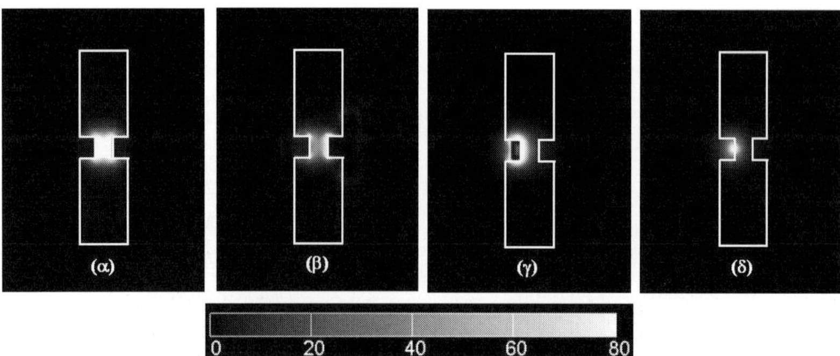

Fig. 22. Two-dimensional total optical intensity distributions on the cross-sections indicated by (α)-(δ) in Fig. 20(b) parallel to the x-y plane. Normalized intensity range is 0 to 80.

guided along the SPGW shown in (β) and transferred to the guided mode along the tapered SPP waveguide shown in (γ). Finally, the optical field is enhanced by the tapered structure of the waveguide and a strongly enhanced optical field is created on the tip shown in (δ). Figures 23(a), (b), and (c) show the x-, y-, and z-components of the electric field, i.e., $Im[E_x(k_0x, k_0y, k_0w + k_0h + 0.025)]$, $Im[E_y(k_0x, k_0y, k_0w + k_0h + 0.025)]$ and $Im[E_z(k_0x, k_0y, k_0w + k_0h + 0.025)]$, respectively. From the results shown in Fig. 23, we find that the confined and enhanced optical near-field above the tip principally consists of the z-component. The x-directed polarization due to the incident wave vanishes in the plane shown in Fig. 23. We also see that the x-component is anti-symmetric about the y-axis and the y-component is anti-symmetric about the x-axis, as shown in Figs. 23(b) and (c). These results show that the excited mode along the tapered probe (waveguide) is a SR-SPP mode. In such a SR-SPP mode, strong confinement induces large attenuation or dissipation in the propagation. Therefore the structure of the probe tip must be carefully designed [24].

It is possible to simulate the imaging of a SNOM with a probe using the SPGW shown in Fig. 19 to confirm operation of the proposed probe[26]. The simulation results demonstrate that this probe works illumination, collection-reflection, and collection modes, and that is it not necessary to vibrate the probe tip in order to remove background noise. They also show that the resolution of the scanned image is approximately equal to the diameter of the probe tip. Furthermore the throughput of the probe is much larger than that of conventional aperture probes of similar

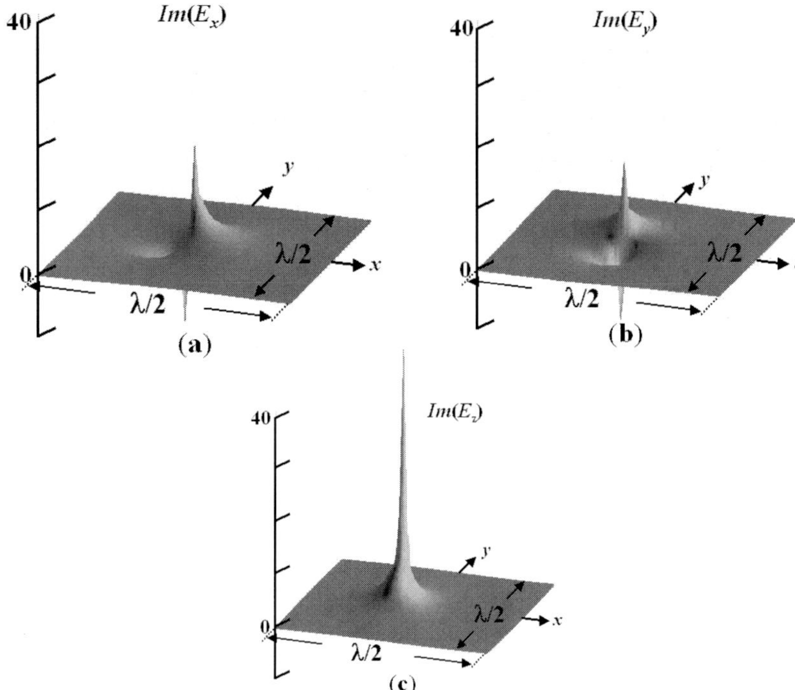

Fig. 23. Two-dimensional distributions of the imaginary part of the electric field components on the probe tip. (a) $Im[E_x(k_0x, k_0y, k_0w+k_0h+0.025)]$, (b) $Im[E_y(k_0x, k_0y, k_0w+k_0h+0.025)]$, and (c) $Im[E_z(k_0x, k_0y, k_0w+k_0h+0.025)]$.

resolution. These results imply that the probe in Fig. 19 combines advantages of both aperture probes and scattering probes and is expected to have excellent characteristics for use as a scanning probe for NSOM. Of course, the application of this probe to the optical head for the optical storage system is also interesting.

5. Conclusions

The SPGW can create optical fields that are strongly confined and strongly enhanced in the nanosize gap region and can guide these fields along the nanosize gap structures. The phase velocity of the SPP propagating along the waveguide can be controlled by the gap width of the SPGW. These characteristics are useful for realizing nanosize optical circuits and devices. In this chapter, only a limited number of SPGW structures and a limited range of parameters have been considered. Many other SPGW structures can be considered leading to other interesting characteristics [15, 27-29]. The application of these ideas to practical nanosize optical devices will be important in future work in nano-photonics and plasmonics.

Acknowledgments

This work was partially supported by a Grant-in-Aid for Scientific Research from the Ministry of Education, Science, Sports, and Culture of Japan. The author gratefully thanks Prof. M. Tanaka and Mr. T. Sugiyama of Gifu University and Dr. K. Katayama of Hamamatsu University for their help with the numerical simulations.

References

1. K. Tanaka and M. Tanaka, *Appl. Phys. Lett.* **82**, 1158 (2003).
2. J. Takahara, S. Yamagishi, H. Taki, A. Morimoto and T. Kobayashi, *Opt. Lett.* **22**, 4765 (1997).
3. S. A. Maier, M. L. Brongersma, P. G. Kik, S. Meltzer, A. A. G. Requicha and H. A. Atwater, *Adv. Mat.* **13**, 1501 (2001).
4. H. Ditlbacher, J. R. Krenn, G. Schider, A. Leitner and F. R. Aussenegg, Appl. Phys. Lett. **81**,1753 (2002).
5. S. I. Bozhevolnyi, V. S. Volkov, E. Devaux, J. –Y. Laluet and T. W. Ebbesen, *Nature* **440**, 508 (2006).
6. K. Tanaka, M. Tanaka, K. Katayama and D. Miyahara, *Comp. Rend. Physique*, to be published.
7. K. Tanaka and M. Tanaka, J. Microscopy **210**, 294 (2003).
8. E. K. Miller, L. Medgyesi-Mitschnag and E. H. Newsman, Eds., Computational Electromagnetics Frequency-Domain Method of Moments (Institute of Electrical and Electronics Engineers, Inc., 1992).
9. P. Zwamborn and P. M. van den Berg, *IEEE Trans. on MTT* **40**, 1757 (1992).
10. G. S. Smith, An introduction to classical electromagnetic radiation (Cambridge University, New York, 1997).
11. R. Barrett, M. Berry, T. F. Chan, J. Demmel, J. Donato, J. Dongarra, V. Eijkhout, R. Pozo, C. Romine and H. van der Vorst, *Templates for the Solution of Linear Systems: Building Blocks for Iterative Methods*, (Society for Industrial and Applied Mathematics, 1994).

12. J. J. Burk, G. L. Stegeman and T. Tamir, Phys. Rev. B, **33**, 5186 (1986).
13. S. I. Bozhevolnyi, *Opt. Express* **14**, 9467 (2006).
14. A. D. Boardman, G. C. Aers and R. Teshima, Phys. Rev. B, **24**, 5703 (1981).
15. D. F. P. Pile, T. Ogawa, D. K. Gramotnev, Y. Matsuzaki, K. C. Vernon, K. Yamaguchi, T. Okamoto, M. Haraguchi and M. Fukui, *Appl. Phys. Lett.* **87**, 261114 (2005).
16. K. Tanaka and M. Tanaka, *Jpn. J. Appl. Phys.* **42**, L585 (2003).
17. K. Tanaka, M. Tanaka and T. Sugiyama, *IEEE J. Light. Tech.* **23**, 3857 (2005).
18. K. Tanaka, M. Tanaka and T. Sugiyama, *Opt. Express* **13**, 256 (2005).
19. R. Gordon and A. G. Brolo, *Opt. Express* **13**, 1933 (2005).
20. M. Ohtsu, Eds., *Near-field Nano/Atom Optics and Technology* (Springer-Verlag, Tokyo, 1998).
21. M. Ohtsu and H. Hori, *Near-Field Nano-Optics* (Kluwer Academic/Plenum Publishers, New York, 1999).
22. S. Kawata, M. Ohtsu and M. Irie, Eds., *Nano-Optics* (Springer-Verlag, Tokyo, 2001).
23. K. Tanaka, M. Tanaka and T. Sugiyama, *Appl. Phys. Lett* **87**, 151116 (2005)
24. K. Tanaka, M. Tanaka and T. Sugiyama, *Opt. Express* **14**, 832 (2006).
25. M. I. Stockman, *Phys. Rev. Lett.* **93**, 137404 (2004).
26. K. Tanaka, M. Tanaka and K. Katayama, *Opt. Express* **14**, 10603 (2006).
27. B. Wang and G. P. Wang, *Appl. Phys. Lett.* **85**, 3599 (2005).
28. L. Liu, Z. Han and S. He, *Opt. Express* **13**, 6645 (2005).
29. B. Wang and G. P. Wang, *Opt. Express* **13**, 10558(2005).

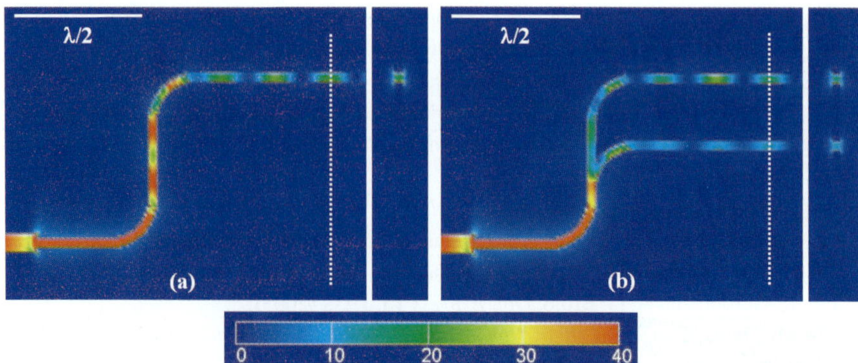

Color Plate 1. Total optical intensities for E-plane (a) S-shaped circuit and (b) asymmetric three-port branching circuit in the *x*-*z* plane for the case of $k_0 w = 0.2$ and $k_0 d = 0.2$. Normalized intensity range is 0 to 40.

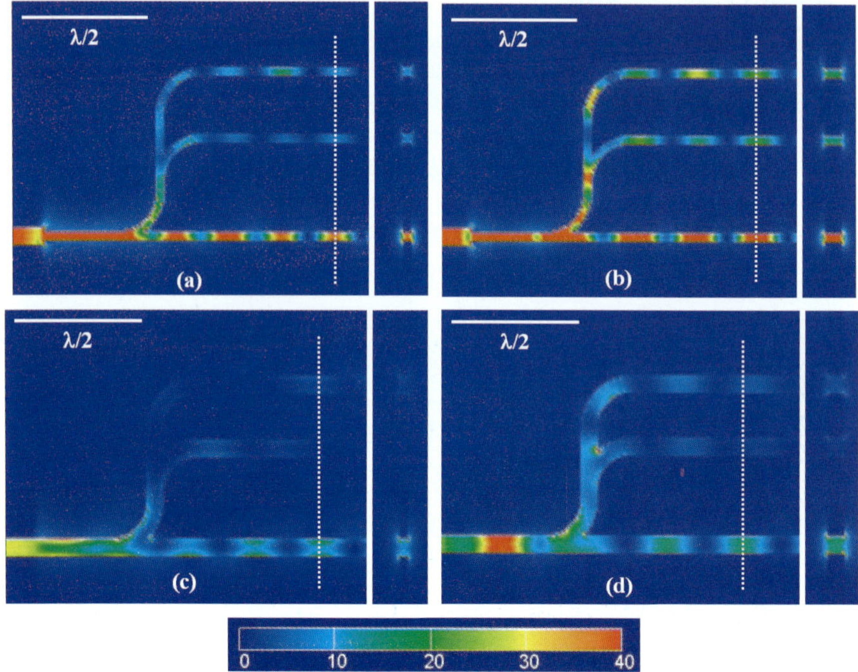

Color Plate 2. Total optical intensities of asymmetric four-port branching circuits in the *x*-*z* plane for (a) $k_0 w = 0.2$, $k_0 d = 0.2$; (b) $k_0 w = 0.2$, $k_0 d = 0.4$; (c) $k_0 w = 0.4$, $k_0 d = 0.2$; and (d) $k_0 w = 0.4$, $k_0 d = 0.4$. Normalized intensity range is 0 to 40.

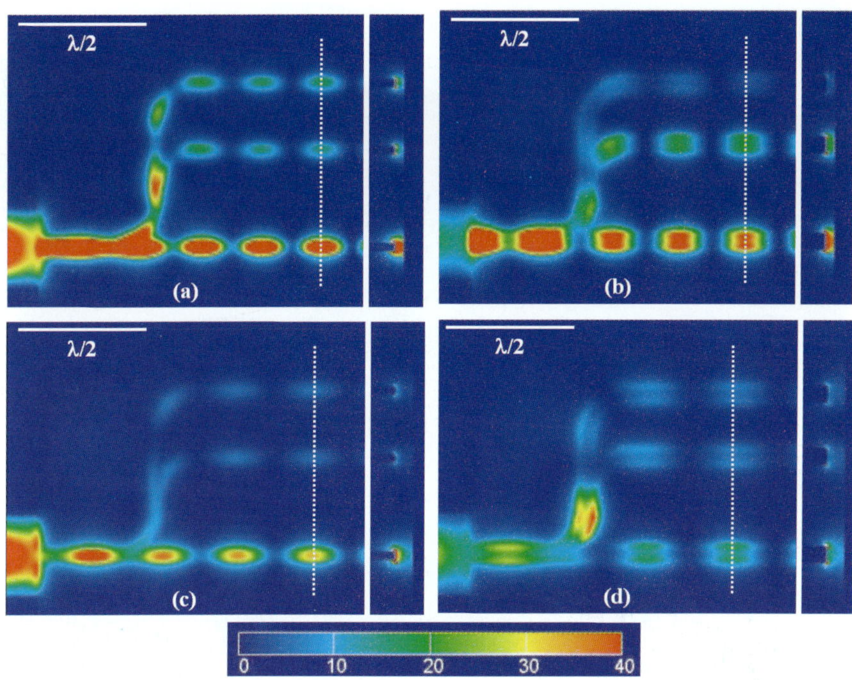

Color Plate 3. Total optical intensities of asymmetric four-port branching circuits for (a) $k_0 w = 0.2$, $k_0 d = 0.2$; (b) $k_0 w = 0.2$, $k_0 d = 0.4$; (c) $k_0 w = 0.4$, $k_0 d = 0.2$;

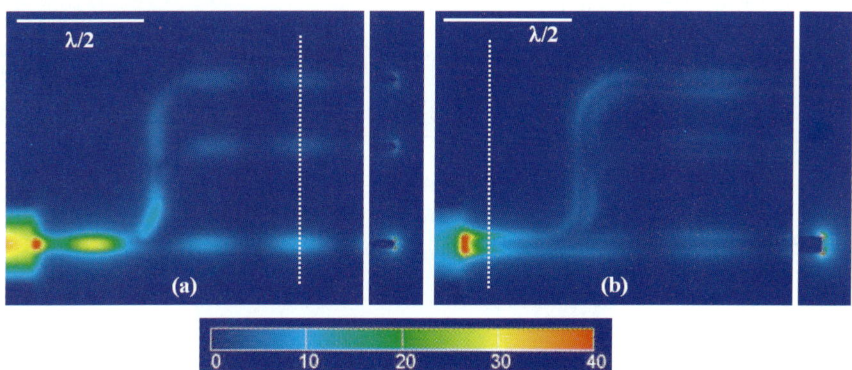

Color Plate 4. Total optical intensities in the asymmetric four-port branching circuit using an edge mode for (a) $k_0 d = 0.2$ and (b) $k_0 d = 0.4$. Normalized intensity range is 0 to 40.

Chapter 5

METAL HETEROSTRUCTURES

Guo Ping Wang

Key Laboratory of Acoustic and Photonic Materials and Devices, Ministry of Education and Department of Physics
Wuhan University, Wuhan 430072, P. R. China
E-mail: kp_wang@hotmail.com

Recent progresses in the design of metal heterostructures (metal gap heterowaveguides, planar metal heterostructures, and ridged metal heterostructures) for nanofocusing, nanoguiding, directional beaming of light, enhancement of local electromagnetic field, as well as the numerical demonstration of the above nanophotonic properties by using finite-difference time-domain simulation are reviewed.

1. Effective index of refraction of metal heterostructures

1.1. *Metal gap waveguides*

Considering a metal gap waveguide (MGW) composed of two parallel metal plates [inset of Fig. 1(a)], when gap width w of the MGW is reduced below the diffraction limit, optical modes of a TM-polarized incident light (magnetic field parallel to the y direction) in the waveguide transforms to surface plasmon polariton modes (SPPs) on metallic surfaces. The dispersion equation of SPPs in the waveguide can be read as [1]

$$\frac{\varepsilon_1 p}{\varepsilon_m k} = \frac{1-\exp(kw)}{1+\exp(kw)} \ ,$$

with

$$k = k_0\sqrt{(\frac{\beta}{k_0})^2 - \varepsilon_1} \; ; \; p = k_0\sqrt{(\frac{\beta}{k_0})^2 - \varepsilon_m} \, . \tag{1}$$

Here k and p are the wave numbers of SPPs in the dielectric of the guide region and the metal, respectively, $\beta = \beta_R + j\beta_I$ is the propagation constant of SPPs, and k_0 (= $2\pi/\lambda$, λ is the wavelength of incident light) is the wave number of light in air, while ε_l and ε_m are the dielectric functions of dielectric in guide region and metal plate, respectively. The term ε_m is from the Drude formula [2]: $\varepsilon_m = 1 - \omega_p^2 / [\omega(\omega + i\gamma)]$. Here ω_p and γ are the plasma frequency and loss component of metal, respectively, and ω (= ck_0, c is the light velocity in air) is the frequency of incident light. The effective index of refraction of the waveguide for SPPs can be defined as

$$n_{eff} = \frac{\beta}{k_0} = \frac{\beta_R + j\beta_I}{k_0} = n_{eff}' + jn_{eff}'' \tag{2}$$

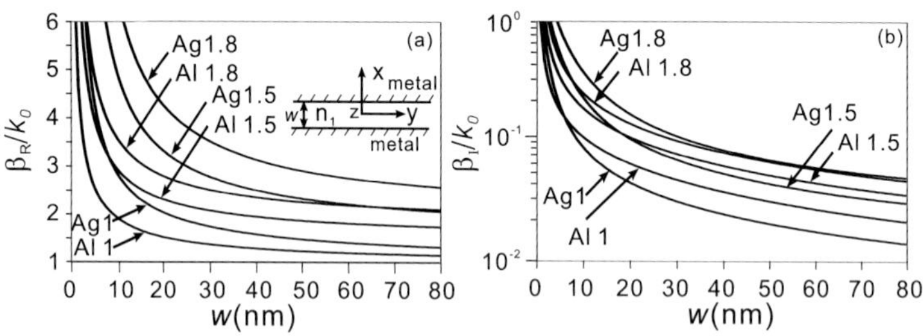

Fig.1. Dependence of (a) real part and (b) imaginary part of β of SPPs on gap width w and the refractive index n_l (= 1.0, 1.5, and 1.8, respectively) of the dielectric in the guide region. Inset of (a), scheme of the gap waveguides (adapted from Ref. 3).

From Eqs. (1) and (2), one can see that n_{eff} is simultaneously determined by gap width w and the dielectric constants of metals (ε_m) and dielectrics filled in the guide region (ε_l).

Figure 1 shows the dependence of n_{eff} of SPPs in Al ($\varepsilon_m = \varepsilon_2$) and Ag ($\varepsilon_m = \varepsilon_3$) gap waveguides, respectively, on width w of the gap guide as a function of the refractive index of dielectric ($n_l = \varepsilon_l^{1/2}$) in the guide region [3]. In the calculations, $\lambda = 539.1$ nm, $\varepsilon_2 = -42.13 + j11.96$, and $\varepsilon_3 =$

- 10.55 + j0.84, respectively [2], are used (in this chapter, the metal materials and exciting light are fixed except of additional notation).

1.2. *Metal-dielectric-air structures*

Regarding a metal-dielectric-air structureas shown in the inset of Fig. 2, only the fundamental electromagnetic mode (TM_0) is allowable on the interface between the metal and dielectric layer when the thickness of the dielectric layer is far below the incident wavelength [1]. The dispersion relationship of TM electromagnetic modes on the interface of metals and dielectrics can be written as [1]

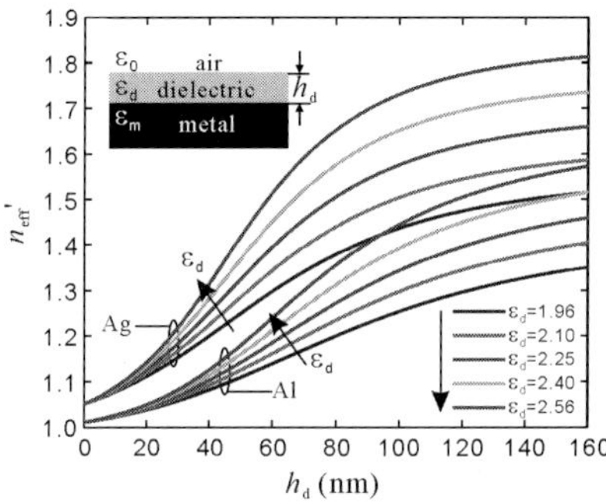

Fig. 2. Dependence of n_{eff}' on h_d and ε_d of the dielectric layer covered on metals Al and Ag. Inset: scheme of the metal-dielectric-air structures. Arrows denote the increase of ε_d (adapted from Ref. 4).

$$k_d h_d = \operatorname{atan}(\frac{\varepsilon_d k_a}{\varepsilon_0 k_d}) + \operatorname{atan}(\frac{\varepsilon_d k_m}{\varepsilon_m k_d}),$$

with

$$k_m = \sqrt{\beta^2 - \varepsilon_m k_0^2}\ ;\ k_d = \sqrt{\varepsilon_d k_0^2 - \beta^2}\ ;\ k_a = \sqrt{\beta^2 - \varepsilon_0 k_0^2}\ . \qquad (3)$$

Where ε_m, ε_d, and ε_0 are the dielectric constants of metal, dielectric, and air, respectively. The effective index of refraction of the electromagnetic mode in the dielectric layer is from Eq. (2).

Figure 2 shows the dependence of the calculated real part of the effective index of refraction n_{eff}' on the thickness h_d and dielectric constant ε_d of the dielectric layer covered on the surface of metals Ag and Al. We can see that the effective index of refraction of Ag guide is bigger than that of Al guide, i.e., the phase velocity v_p of SPPs in the former is lower than that in the latter [4].

1.3. *Metal films*

Considering an air-metal-air system (free standing metal films) as shown in the inset of Fig. 3, where d is the thickness of metal film, ε_l and ε_m represent the relative permittivities of air and metal material, respectively, we can write the amplitude distribution of electric field $\boldsymbol{E}_z$ of SPP modes in the interfaces of metal and air as

$$E_z = \begin{cases} A\exp(-px), & 0 \le x < \infty \\ B\exp(-kx) + C\exp(kx), & -d \le x \le 0 \\ D\exp(px), & -\infty < x \le -d \end{cases} \tag{4}$$

where k and p are the transverse propagation constants (along the x axis) of SPPs in dielectric (ε_l) and metal (ε_m), and A, B, C, and D are the mode amplitudes of SPPs in the metal and air, respectively. A similar magnetic field distribution $\boldsymbol{H}_y$ can be obtained by applying Maxwell's equations to Eq. (4). By employing the boundary condition that $\boldsymbol{E}_z$ and $\boldsymbol{H}_y$ must be continuous in the interfaces and solving the $\boldsymbol{E}_z$ and $\boldsymbol{H}_y$ equations, we can get the dispersion relation of SPPs in the system as Eq. (5) [5]. For a certain thickness d of metal film, we can get the propagation constant β of SPPs from Eq. (5), and then the effective index of refraction of SPPs on the interfaces between the metal and air from Eq. (2).

$$kd = \ln\left|\frac{1-\dfrac{p\varepsilon_m}{k\varepsilon_1}}{1+\dfrac{p\varepsilon_m}{k\varepsilon_1}}\right| \tag{5}$$

Figure 3 shows the dependence of n_{eff}' on the thickness d of the Ag and Al films, respectively. From the figure, one can see that, with the same thickness d, n_{eff}' of SPPs on the Ag surface is always greater than that on the Al surface, and the difference of n_{eff}' is enhanced as d reduces in a certain range. Furthermore, n_{eff}' decreases and asymptotically tends to be infinite as d is approaching zero for each case [6].

From the principle of integrated optics, we know that light waves trend to propagate in higher refractive index media or, in other words, light waves prefer to be with lower phase velocity v_p (= c/n, c and n are the speed of light in air and the refractive index of dielectrics, respectively) [7]. This is the base of conventional dielectric waveguides and optical fibers. However, the size of the light beam in such common waveguides and fibers is limited by the cut-off frequency and is generally on the order of one light wavelength. As an evanescent wave excited on the interface between metals and dielectrics, SPPs can localize light energy in a nanoscale domain and circumvent the limitation of the cut-off

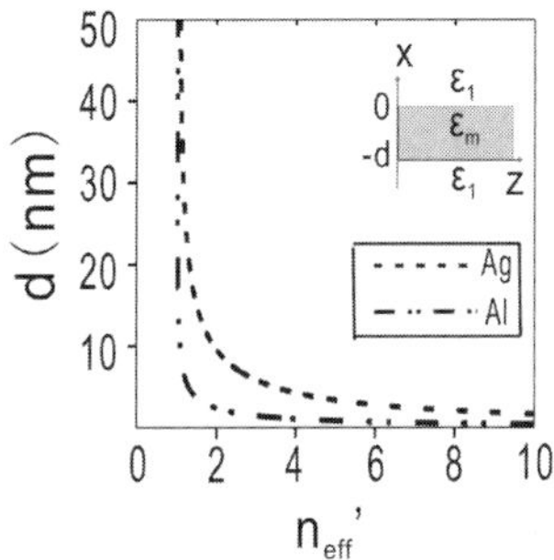

Fig. 3. Dependence of n_{eff}' on the thickness d of Ag and Al films. Inset: scheme of the structures (adapted from Ref. 6).

frequency [8], and have propagation behavior similar to light waves of trending to be with lower phase velocity [9]. From Figures 1 to 3, we can see that β_R/k_0 of SPPs on the surface of Ag film is greater than that of Al. As a result, v_p of SPPs on the former is smaller than that on the latter. Therefore, if a structure is constructed with metals Ag and Al, then SPPs tend to be confined to the surface of Ag. In this chapter, we will review the recent progresses in the design of a class of metal heterostructures constructed with two different metal materials for nanofocusing, nanoguiding, directional beaming of light, the enhancement of local electromagnetic field and the numerical demonstration of the above nanophotonic properties by using finite-difference time-domain (FDTD) simulation [10].

2. Metal heterostructures for nanofocusing of light

2.1. *Rectangular metal gap heterowaveguides*

It is well known that the spatial resolution of an optical device is limited by the diffraction limit. By using an aperture probe with nanometeric size as either light sources or detectors, scanning near-field optical microscopy (SNOM) has reached a resolution much beyond the diffraction limit [11]. Although the smaller the aperture, the better the resolution of the SNOM, a small aperture usually bears low light efficiency. To overcome this shortcoming, Ebbesen *et al.* demonstrated a focus structure for enhancing the transmission of light through the nanoscale holes in metal films [12]. From Fig. 1, we know that, in the case where the waveguides are made up of a single metal (Ag or Al) with different width, β_R/k_0 of SPPs in the narrower guide is always greater than that in the wider one. As a result, $v_p = c/n_{eff}' = \omega/\beta_R$ of SPPs in the former is smaller than that in the latter. Based upon this property, Tanaka *et al.* have numerically confirmed that a MGW with narrower gap width can focus and guide light in nanoscale cross section [9].

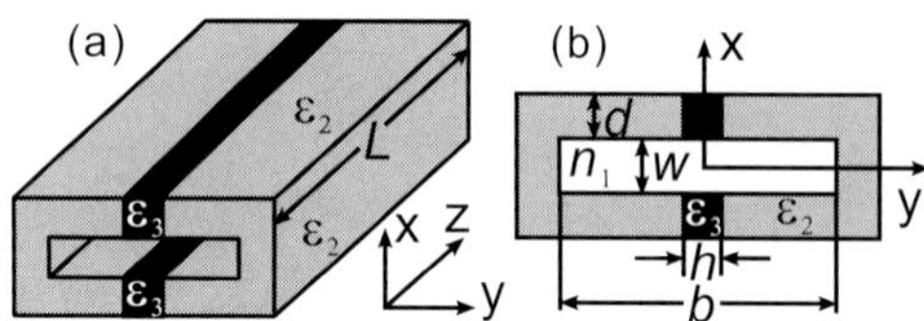

Fig. 4. (a) Scheme of the MGHWs and (b) their cross section in the x-y plane.

Instead of changing the geometry of the MGWs, we construct a three-dimensional (3D) rectangular metal gap heterowaveguide (MGHWs) by using metals Ag and Al as shown in Fig. 4 and search for its nanofocusing property to light by numerical simulation of 3D FDTD method [3]. Of which, the Ag waveguide is inlaid in the centric part of the Al waveguide. We choose metals Ag and Al to construct the MGHWs because both metals have higher dielectric contrast in real parts but with relative smaller values in imaginary parts of the dielectric constants in our interesting spectrum range. Geometric parameters of the waveguide such as the guide length L (= 400 nm, fixed), the thickness of metallic wall d (= 85 nm, fixed), the width w and height h, b of Ag and Al guides are shown in Fig. 4.

The light used to excite the SPPs of the waveguide is a TM-polarized wave normally incident to the entrance face of the waveguide. The intensity distribution of the incident light is a Gaussian profile as shown in Fig. 5(a) with a full-width at half-maximum (FWHM) of 150 nm.

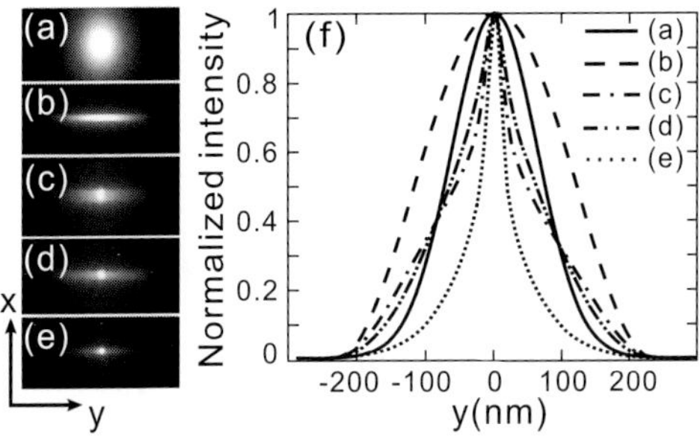

Fig. 5. Gray distributions of $|\boldsymbol{E}|^2$ of (a) incident and (b)-(e) output light beams in the x-y plane and (f) their normalized profiles on a line (x = 0) parallel to the y axis as gap width w = (c) 35, (d) 25, and (e) 15 nm, respectively (adapted from Ref. 3).

Figures 5(b)-5(f) show, respectively, the $|\boldsymbol{E}|^2$ distributions of output light in a plane 15 nm away from the end of the waveguides (same in the following Figs. 6 and 7) and their normalized profiles on a line (x = 0) parallel to the y axis in different cases, where h = 25 nm, b = 435 nm, d = 85 nm, and n_l = 1.5 are used. Figure 5(b) shown is the case where the waveguide is constructed only with metal Ag. We see that the FWHM of the output light beam in the y direction extends to 235 nm. Obviously, the conventional rectangular waveguides show no focusing effect. However, when the light is incident into the Ag-Al formed MGHWs, one can see from Figs. 5(c)-5(e) that the output light tends to be confined to the centric part of the guide (the Ag guide region $h \times w$, see Fig.4). As w decreases, the size of the output spot will become to be smaller. For instance, as w changes from w = 35 nm to 25 and 15 nm, FWHM of the spots in the y direction decreases from 130 nm to 105 and 40 nm, respectively. We attribute it to the propagation property of SPPs as discussed in part 1.1. The difference of β_R of SPPs in both Ag and Al waveguides increases as the width w of the guide is narrowed [Fig. 1(a)]. As a result, the phase velocity of SPPs in Ag and Al guide regions show bigger difference, which will enhance the confinement of light.

It should be noted that our discussion is based on the propagation characteristics of SPPs in two-dimensional (2D) MGWs like the inset of Fig. 1(a), but we believe that the conclusion deduced from the 2D cases is qualitatively reasonable for understanding the focusing effect of our 3D structures in the case of $b >> w$ (Fig. 4).

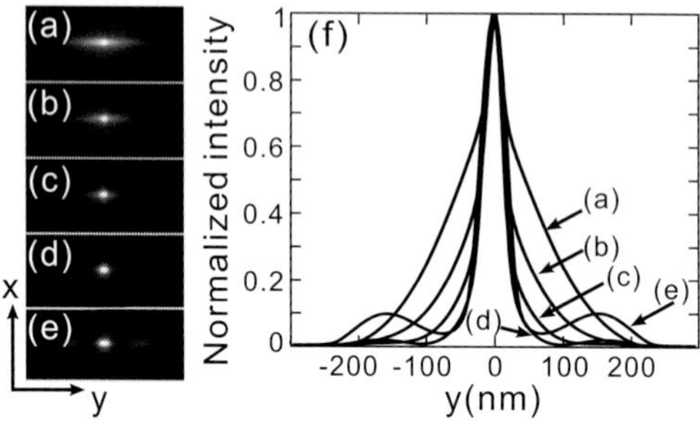

Fig. 6. (a)-(e) Gray distributions of $|\boldsymbol{E}|^2$ of the output light beams in the x-y plane and (f) their normalized profiles on a line (x = 0) parallel to the y axis as Ag film width h = (a) 15, (b) 25, (c) 35, (d) 45, and (e) 65 nm, respectively (adapted from Ref. 3).

Figures 6(a)-6(f) present the dependence of $|\boldsymbol{E}|^2$ distributions of the output light and their normalized profiles on the height h of Ag waveguide (y direction) as w = 15 nm, b = 435 nm, and n_l = 1.5, respectively. When h is chosen as (a) 15, (b) 25, (c) 35, (d) 45, and (e) 65 nm, FWHM of the output light spot is 95, 45, 35, 35, and 45 nm, respectively [Fig. 6(f)], which means that, to achieve a relative small light spot, an optimal h is required. This is due to the fact that propagation constant β_R of SPPs on a metallic surface is also dependent on the width of metal film [13]. To obtaining an optimal β_R of SPPs on the surface of a metal film, one has to select a proper film width.

On the other hand, from Fig. 1(a) we see that $n_{eff}' = \beta_R/k_0$ and its difference of SPPs in Ag and Al waveguides are also dependent on the refractive index n_l of the dielectric filled in the guide region. The difference of β_R/k_0 of SPPs in both guides is increased with n_l. This means that the difference of v_p of SPPs in Ag and Al guide regions becomes large and hence is helpful for the focusing of light.

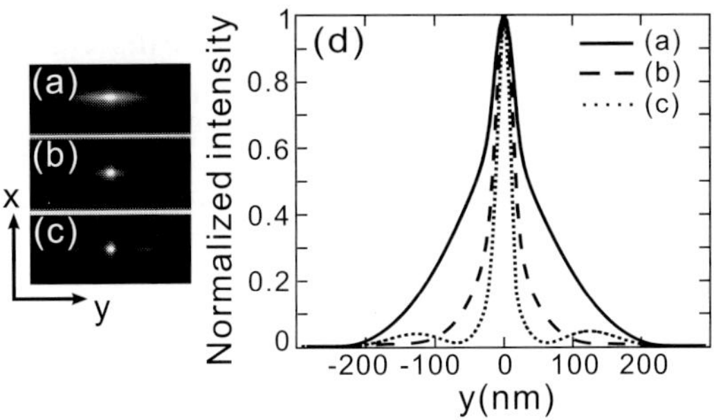

Fig. 7. Gray distributions of $|E|^2$ of the output light beams in the x-y plane and (f) their normalized profiles on a line ($x = 0$) parallel to the y axis as refractive index n_l = (a) 1.0, (b) 1.5, and (c) 1.8, respectively (adapted from Ref. 3).

Figures 7(a)-7(c) show the $|E|^2$ profiles of output light from the MGHW as n_l = 1.0, 1.5, and 1.8, respectively, while other parameters are the same as the case of Fig. 6(c). The results show that the FWHM of the emission spots in the y direction decreases from 75 nm to 35 to 30 nm, respectively, as n_l is increased [Fig. 7(d)], confirming that the focusing spot of light can be shrunk by raising the refractive index of the filled dielectric other than by varying the cross section of Ag waveguide. In the case of n_l = 1.8, a light spot with the size (FWHM of $|E|^2$ in the x-y plane) of about 20 × 30 nm^2 ($\lambda/27 \times \lambda/18 \approx 0.002\ \lambda^2$), significantly smaller than the diffraction limit is obtained [Fig. 7(d)]. Defining the focusing efficiency of the output spot as the $\Sigma|E|^2$ of light injected into the Ag guide region (with the cross section $h \times w$, 10 nm away from the entrance inside the guide, Fig. 4) divided by that of illuminating light on the entrance plane of the guide within the cross section of the whole waveguide ($b \times w$), we can get 69.8 %, 24.1 %, and 15.5 % focusing efficiencies as n_l = 1.0, 1.5, and 1.8, respectively. From the results, one can see that, although a higher n_l produces a smaller focusing spot due to the enhanced contrast of β_R and v_p of SPPs in Ag and Al guide regions, it accompanies a lower focusing efficiency. This can be attributed to both stronger propagation loss of SPPs in the guide [Fig. 1(b)] and the reflection of light on the input surface of the waveguide as the guide region is filled with a higher refractive index dielectric.

Most interesting, when the focusing efficiency is defined as $\Sigma|\boldsymbol{E}|^2$ of light injected into the Ag guide region divided by that of illuminating light on the entrance plane of the guide within the cross section of Ag waveguide (Fig. 4, $h \times w$), then an efficiency higher than 1 can be reached. For example, in the case of $n_I = 1.5$, a 188.8 % focusing efficiency is obtained, implying that the light outside the cross section of the Ag guide region is also coupled into the guide.

2.2. *Trapezoid metal gap heterowaveguides*

To further improve the focusing efficiency of MGHWs, we here consider the inlaid Ag guide to be a trapezoid structure as shown in Fig. 8(a) [14]. The geometric parameters of the Ag guide are denoted with the width w, the heights h_1 and h_2 at the input and output planes of the MGHWs, respectively. The FWHM of the Gaussian beam used to excite SPPs in the MGHWs is about 214 nm. Other parameters such as the length of MGHWs L (350 nm, fixed), the thickness of the Al film (85 nm, fixed), and the cross section of the Al guide (b_x = 85 nm and b_y = 405 nm > $\lambda/2$, fixed) are denoted in Fig. 8(a).

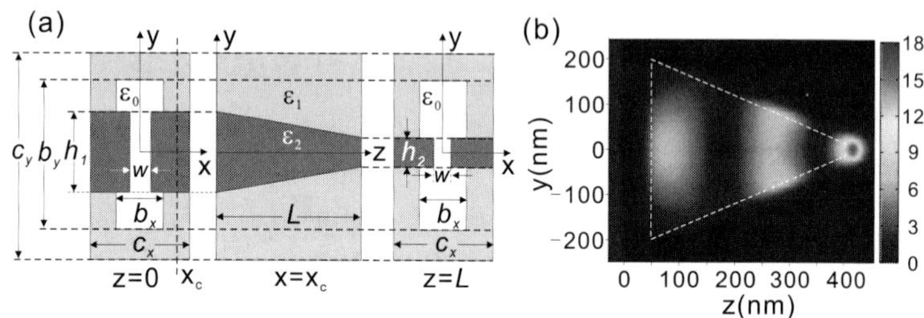

Fig. 8. (a) Schemes of the cross sections of the trapezoid MGHWs (Ag, black and Al, gray) in the x-y and y-z planes, respectively. (b) A typical gray distribution of $|\boldsymbol{E}|^2$ in the y-z plane at $x = 0$ as SPPs pass through a trapezoid MGHW with h_1 = 405 nm, $w = h_2$ = 35 nm (adapted from Ref. 14).

Figure 8(b) displays a typical $|\boldsymbol{E}|^2$ distribution of SPPs passing through a MGHW in the y-z plane at $x = 0$. We can see that SPP energy is confined mainly in Ag guide, and a strong focusing effect appears at the output of MGHW. To understand the dependence of the focusing effect on the geometry of the MGHWs, we calculate the $|E|^2$ distributions of SPPs in a x-y plane 15 nm away from the output of the MGHWs as functions of w, h_2 (Fig. 9), and h_1 (Fig. 10), respectively. We see that, with a fixed h_1 = 405 nm, when w = 35 nm but h_2 decreases from 35 nm to 25 and 15 nm, respectively, the size of focusing spot changes from 60 × 45 nm^2 to 60 × 40 and 60 × 35 nm^2, respectively. The corresponding focusing efficiencies are 98.6 %, 90.3 %, and 77.8 %, respectively [Figs. 9(a)-9(c)]. As w also decreases from 35 nm to 25 and 15 nm, respectively, the focusing spot shrinks further to 50 × 25 and 35 × 20 nm^2, and the efficiency decreases to 96.8 % and 26.2 % [Figs. 9(d) and 10(a)]. One can see that the size of focusing spot seems to be determined by that of the output of Ag waveguide: larger w and h_2 produce a bigger focusing spot, and vice versa.

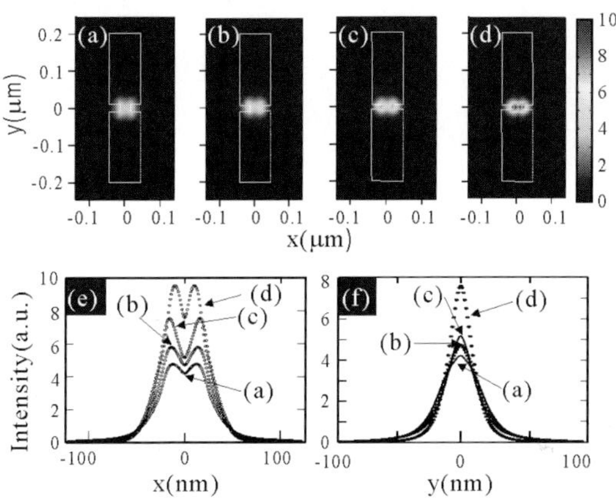

Fig. 9. (a)-(d) Gray distributions of $|E|^2$ in a plane (x-y plane) 15 nm away from the output end of the MGHWs and (e)-(f) their profiles in the x and y directions, respectively, as (a) $w = h_2$ = 35 nm, (b) w = 35 nm, h_2 = 25 nm, (c) w = 35 nm, h_2 = 15 nm, and (d) w = 25 nm, h_2 = 15 nm (adapted from Ref. 14).

However, when h_1 decreases from 405 nm to 305, 205, and 105 nm, while $w \times h_2$ keeps constant (15 nm × 15 nm), the focusing spot changes from 35 × 20 nm^2 to 40 × 20, 40 × 20, and 35 × 20 nm^2, respectively [Figs. 10(a)-10(d)], and the corresponding focusing efficiency falls from

26.2 % to 6.7 %, 3.3 %, and 1.4 %. This is to say that, in the cases of fixing the output size of the Ag guide, the focusing efficiency of the MGHWs is increased with the height h_1 of the Ag guide in the input plane, whereas the size of focusing spot shows a complex dependence on h_1.

From Fig. 1(a) we know that, in the cases in which the width w of Ag guides is narrowed but that of Al ones is fixed, the difference of phase velocity (Δv_p) of SPPs in both guides is increased, which will produce a smaller focusing spot [Figs. 9(c), 9(d), and 10(a)], because SPPs trend to propagate with slower v_p (Ag guides). This means that the larger the Δv_p of SPPs, the stronger the focusing effect [3]. However, a narrower w also produces a bigger β_I [Fig. 1(b)], which corresponds to stronger energy damping. Hence, a smaller focusing spot with lower focusing efficiency appears as w is decreased [Figs. 9(c) and 10(a)].

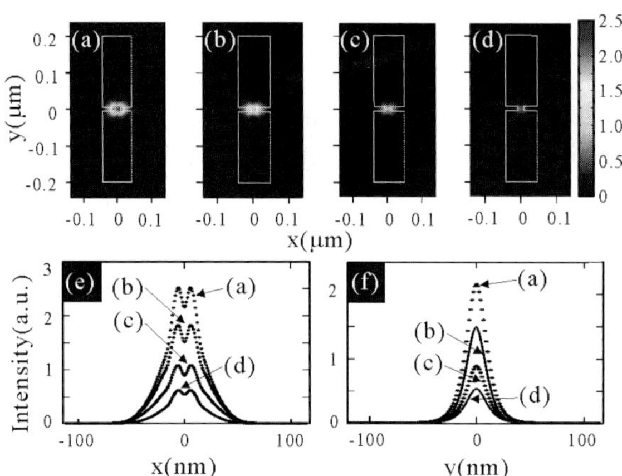

Fig. 10. (a)-(d) Gray distributions of $|E|^2$ in a plane (x-y plane) 15 nm away from the output end of the MGHWs and (e)-(f) their profiles in the x and y directions, respectively, as h_1 = (a) 405 nm, (b) 305 nm, (c) 205 nm, and (d) 105 nm (adapted from Ref. 14).

Additionally, a larger h_2 is helpful for SPPs propagation [15]. Therefore the energy efficiency of light spot is gradually increased with h_2, and vice versa [Figs. 9(a)-9(c)]. Nevertheless, in the case of w = 25 nm while other parameters are fixed, the MGHW produces not only a smaller focusing spot (50×25 nm$^2 \approx 0.004\ \lambda^2$) but also a higher energy efficiency (96.8 %) [Fig. 9(d)]. This can be attributed to the fact that, to get larger Δv_p (for smaller light spot) but with a smaller β_1 (for higher efficiency), an optimized geometry of MGHWs may be required (Fig. 1). In the cases of fixing the output size of Ag guides, we see that the focusing efficiency of MGHWs is gradually increased with h_1 (Fig. 10), because a Ag guide with higher h_l makes use of a larger part of incident light energy to excite SPPs of the Ag guide at the input of MGHWs. The size fluctuation of focusing spot with h_l further confirms that careful selection of w and h of MGHWs is beneficial for SPPs propagation [15].

3. Metal heterostructures for nanoguiding of light

3.1. *Metal gap heterowaveguides*

The simultaneous realization of a strongly confined light beam with sufficiently low propagation loss for practical applications is of central importance in integrated photonics. SPPs-based structures are widely regarded as feasible candidates in constructing novel waveguides with nanometric cross section. Of which, typical examples include metal nanowires [16], linear chain of closely packed metal nanoparticles [17], triangular groove [18-22], index-guided 2D waveguides [23,24], 3D gap waveguide [9, 25], etc.

In this section, we will consider the nanoguiding characteristics of Ag-Al constructed MGHWs [14], in which a rectangular Ag guide is considered [$h_l = h_2 = h$, see Fig. 8(a)]. Because the refractive index n of dielectrics in the guide region also plays an important role in phase velocity of SPPs [3], we present in Figs. 11(a)-11(c) the calculated $|\boldsymbol{E}|^2$ distributions of SPPs along MGHWs in the y-z plane at $x = 0$ as $n = 1.0$, 1.2, and 1.5, respectively, where geometric parameters $w = b_x = 15$ nm, $h_1 = 35$ nm, and $b_y = 405$ nm are used. The figures show that SPPs is mainly transported through the centric part of the MGHWs (Ag guide). The appearance of standing waves is due to the reflection of SPPs from the output end of the waveguides. Figure 11(d) shows the $|\boldsymbol{E}|^2$ profiles of SPPs at the crests of standing waves. Since the size of $|\boldsymbol{E}|^2$ in the x direction is mainly determined by the guide width w, we consider only the FWHM of $|\boldsymbol{E}|^2$ in the y direction to evaluate the beam size of SPPs in all cases. From Fig. 11(d), we see that, as n increases from 1.0 to 1.2 and 1.5, FWHMs of $|\boldsymbol{E}|^2$ are decreased gradually from 120 nm to 68 and 35 nm, respectively. The higher the refractive index n of the medium in guide region is, the smaller the beam size of SPPs is. This is because the difference $\Delta\beta_R$ or Δv_P of SPPs in Ag and Al waveguides is increased with the index [3].

For propagation loss (η) of SPPs in MGHWs, we use a method introduced by Ref. 26. From the FDTD simulations, we can get the electric field distribution of SPPs along the guide direction (z axis) as

$$E(x, y, z) = E(x, y, 0)e^{j(\beta z-\omega t)}. \tag{6}$$

By fitting the electric field distribution, we can get

$$\beta_I = \frac{1}{Z}\ln\left(\left|\frac{E(0,0,0)}{E(0,0,z)}\right|\right). \tag{7}$$

Then the propagation loss η of SPPs along the MGHWs can be read as

$$\eta = \frac{20}{Z}\log\left(\left|\frac{E(0,0,0)}{E(0,0,z)}\right|\right)=8.69\beta_1(dB/m). \tag{8}$$

From this formula, we get the values of η of SPPs as 6.4, 10.1, and 18.5 dB/μm, with n equal to 1.0, 1.2 and 1.5, respectively. This is to say that the higher the refractive index of dielectric in the guiding region is, the stronger the energy damping of SPPs is. This is because that β_I of SPPs in both Ag and Al waveguides is also increased with the index [Fig. 1(b)]. Consequently, a smaller beam size of SPPs will accompany with an enhanced energy loss [3].

Figure 11(e) presents the dependence of FWHMs of SPPs in MGHWs on the height h_I and width w of Ag guide as n = 1.5. One sees that, when h_I changes from 15 nm to 55 nm while $w = b_x$ = 15 nm are fixed, SPPs can achieve a minimal FWHM: as h_I = 35 nm, a FWHM of 35 nm is obtained. This is partly because of the dependence of propagation constant β_R of SPPs on the height h of metal films [13]. When w decreases while h_I is fixed, the FWHM of $|\boldsymbol{E}|^2$ of SPPs becomes narrower [inset of Fig. 11(e)]. This is due to the increase of $\Delta\beta_R$ of SPPs in both Ag and Al waveguides as w is reduced in a certain range of width (Fig. 1), which is beneficial for compressing the spread of SPPs.

From the above discussions, we see that, to reduce β_I or the propagation loss η of SPPs, the dielectrics in the guiding region should be with smaller refractive index, whereas the guide width should be as wide as possible. However, in this case, $\Delta\beta_R$ of SPPs in Al and Ag guides will be reduced, and hence the beam size of SPPs will be larger. Therefore, to simultaneously reduce the beam size and energy damping of SPPs, we can fill the guiding region with air (n = 1.0) and further enhance the width difference of Al and Ag guides. In this case, Ag and Al guides in MGHWs will have different widths [see Fig. 8 (a)].

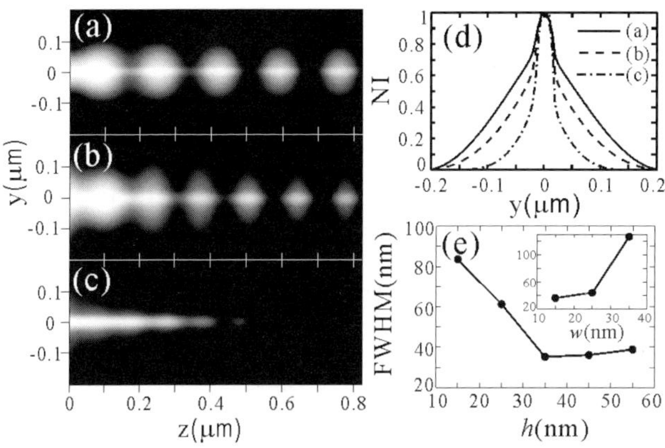

Fig. 11. (a)-(c) Gray distributions of $|\boldsymbol{E}|^2$ of SPPs at $x = 0$ of the y-z plane of the MGHWs with rectangular Ag guide ($h_1 = h_2 = h$) and (d) their normalized intensity (NI) profiles in the y direction at the crests of standing waves ($x = 0$) as n = (a) 1.0, (b) 1.2, and (c) 1.5, respectively. (e) Dependence of the FWHM of $|\boldsymbol{E}|^2$ in the y direction on h_1 as w = 15 nm and n = 1.5. Inset: dependence of the FWHM of $|\boldsymbol{E}|^2$ on Ag and Al guide widths $w = b_x$ as h = 35 nm (adapted from Ref. 14).

Figure 12 shows the $|\boldsymbol{E}|^2$ distributions [parts (a)-(c)], the cross-section profiles [part (d)], and the intensity damping [part (e)] of SPPs passing through L = 800 nm long MGHWs with different parameters. In the cases of fixing the width of Al guide at b_x = 85 nm and the height of Ag guide at h_1 = 35 nm, when the width of Ag guide is w = 15 nm, we get a 40 nm wide SPPs beam (y direction) with 18.8 dB/μm propagation loss [Fig. 12(a)]. Widening w to 55 nm, a beam width of 65 nm with 2.0 dB/μm energy damping is obtained [Fig. 12(b)]. As w is set at 35 nm, one can get a 50 nm wide beam with 5.1 dB/μm energy attenuation [Fig.

12 (c)]. This is to say that a narrower Ag guide will produce a smaller beam size of SPPs but a stronger propagation loss as well. This can be understood that, in the cases of fixing the size of Al guide, narrower Ag waveguides will not only enhance the Δv_p of SPPs in Ag and Al waveguides but also lead to larger β_l of SPPs in Ag guides. When w is fixed at 35 nm while the height of Ag guides h_l is changed, similar results can also be obtained. For example, when h_l varies from 15 nm to 35 and 55 nm, we can get 30, 50, and 70 nm wide beams with 9.7, 5.1, and 3.5 dB/μm propagation losses, respectively. This is because a wider metallic surface is helpful for the simultaneous propagation and spread of SPPs [13]. When the width of Al guide b_x is also narrowed, Δv_p of SPPs in Al and Ag waveguides will be reduced [Fig. 1(a)]. As a result, the beam size of SPPs in MGHWs will become wider. For instance, as b_x changes from 85 nm to 55 nm while $w \times h_l$ = 35 nm × 35 nm is fixed, the width of SPPs field increases from 50 nm to 55 nm, but the propagation loss of SPPs reduces from 5.1 dB/μm to 2.0 dB/μm.

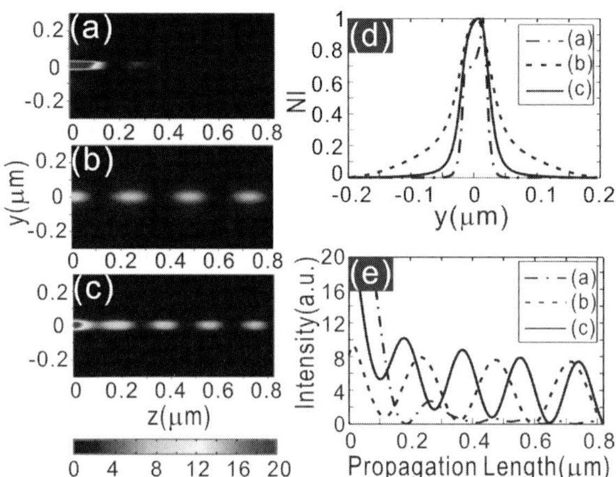

Fig. 12. (a)-(c) Gray distributions of $|E|^2$ of SPPs at $x = 0$ of the y-z plane of MGHWs and (d) their normalized intensity (NI) profiles in the y direction. (e) Dependence of $|E|^2$ on L as SPPs pass through L = 800 nm long MGHWs as w = (a) 15 nm, (b) 55 nm, and (c) 35 nm, respectively (adapted from Ref. 14).

3.2. *Planar metal heterowaveguides*

In terms of the results obtained in part 1.2, we demonstrate here a planar metal heterowaveguide (PMHW) for confinement and guiding of SPPs [27]. Figure 13 shows the PMHWs constructed by inlaying a striped Ag film into the centric part of Al film while coating them with a uniform dielectric layer. The thickness and refractive index of the dielectric layer on metals Al and Ag are labeled with t and n, respectively, and w stands for the width of Ag film.

Since the difference of effective index between Ag and Al waveguides remains small, we can use the effective index method to characterize the propagation properties of SPPs in the waveguides [19]. The propagation length L_{sp} is deduced from the imaginary part β_I of the propagation constant $\beta = \beta_R + j\beta_I$ by $L_{sp} = 1/2\beta_I$ [28]. The beamwidth W of SPPs is obtained by calculating the FWHM of SPP intensity in the waveguide. The effective index n_{eff} of the waveguide is obtained from Eq. (2). In the calculations, only TM-polarized modes (electric field perpendicular to the metal surface) in the waveguide is considered.

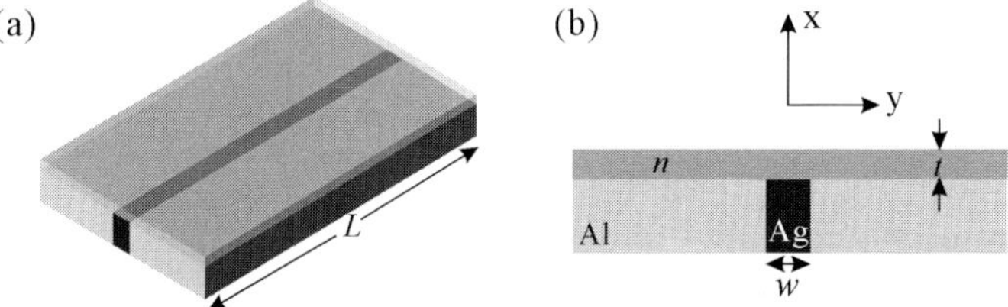

Fig. 13. (a) Scheme of Al-Ag constructed PMHWs with a uniformly covered dielectric layer. (b) Cross section of the PMHWs (adapted from Ref. 27).

Figures 14(a) and 14(b) show the dependence of real part of the effective index n_{eff}', propagation length L_{sp}, and beamwidth W of SPPs in the waveguide on thickness of the dielectric layer. Here the width of Ag film and the refractive index of dielectric layer are fixed at w = 150 nm and n = 1.5, respectively. From the figures, one sees that n_{eff}' of the waveguide increases with the thickness t of dielectric layer. But L_{sp} of SPPs dramatically decreases with the increase of t at first and reaches a minimum value at $t \approx 90$ nm. The W of SPPs shows a similar dependence on t but appears the minimum of 208 nm at $t \approx 100$ nm. Figures 14(c) and 14(d) show the dependence of n_{eff}', L_{sp}, and W of SPPs on Ag film width w. Here the thickness of dielectric layer is fixed at t = 100 nm. From the figures, one sees that, except the increase of n_{eff}' with w, both L_{sp} and W of SPPs undergo their owner decrease with increased w. When w approaches to 150 nm, W reaches its minimum of 208 nm and L_{sp} is about 1.76 μm. This means that, to balance the losses of SPPs in the waveguide derived from scattering and absorption, one has to optimize the thickness of dielectric layer and the width of Ag film simultaneously for improving the propagation length of SPPs. In addition, by covering the metal films instead with a gain medium [29, 30], it is expected to effectively compensate the effect of losses and enhance the propagation efficiency of SPPs [31].

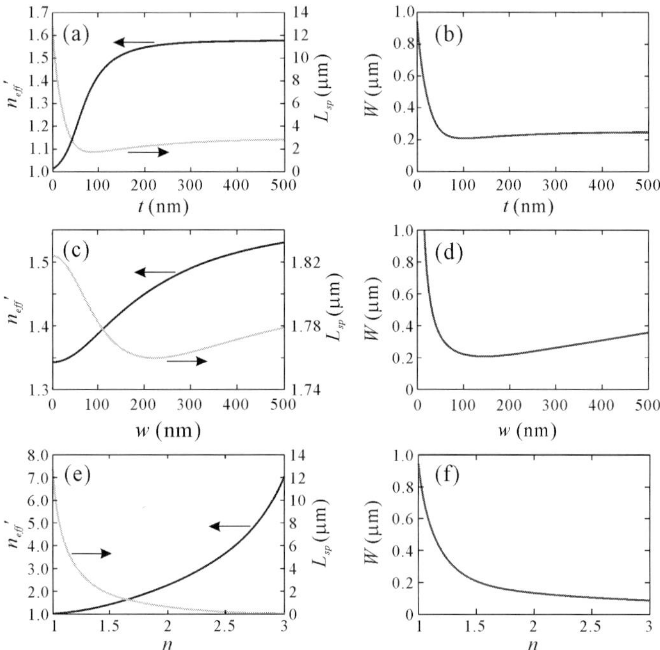

Fig. 14. Dependence of n_{eff}', L_{sp}, and W of SPPs on (a)-(b) thickness t of the dielectric layer, (c)-(d) width w of the Ag film, and (e)-(f) refractive index n of the dielectric layer (adapted from Ref. 27).

Additionally, n_{eff}', L_{sp}, and W of SPPs are also dependent on the refractive index n of the dielectric layer. As shown in Figs. 14(e) and 14(f), both L_{sp} and W of SPPs decrease monotonously with n. The reason is that both the imaginary part β_I of propagation constant and the difference of n_{eff}' of SPPs in Ag or Al constructed metal waveguides increase with n simultaneously [5].

Though the effective index method predicts that strongly confined SPPs can be achieved through optimizing the parameters of the waveguides, the field distribution in the waveguides is still not explicit. Here we employ FDTD simulation to depict the electric field distribution of SPPs in the PMHWs. The whole length of the waveguides in the calculation space is set to be 3 μm, and other parameters are given as t = 100 nm, w = 150 nm, and n = 1.5, respectively.

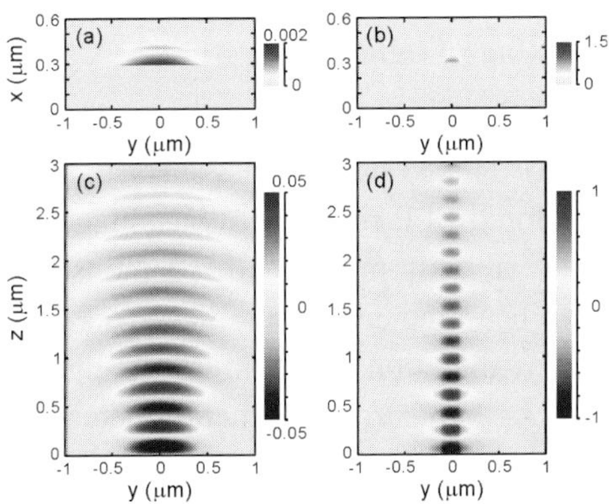

Fig. 15. Gray distributions of (a)-(b) $|\boldsymbol{E}_y|^2$ in the x-y plane and (c)-(d) $\boldsymbol{E}_y$ in the y-z plane of an Al waveguide [(a), (c)] and a PMHW [(b), (d)] (adapted from Ref. 27).

Figures 15(a) and 15(b) show the $|\boldsymbol{E}_y|^2$ distributions of SPPs in the x-y plane of a dielectric layer-coated planar Al waveguide and a PMHW, respectively. A TM-polarized Gaussian light with a FWHM of 600 nm is incident from one end of the guides. One sees from the figures that SPPs in Al waveguide are widely spread, while in the PMHW, SPPs show a strong confinement. The cross section of SPPs in the x-y plane of the PMHW is about 171 × 33 nm^2 (0.32 × 0.061 λ^2) [Fig. 15 (b)], which is slightly different from the beamwidth obtained from the effective index method (208 nm in the x direction). This can be attributed to the fact that effective index method neglects the stronger localization of electromagnetic field on Ag surface than that on Al surface [28], and hence neglects the forceful field intensity of SPPs on the center of Ag film in the heterostructures.

Figures 15(c) and 15(d) show the electric field distributions of SPPs in the *y*-*z* plane at a distance 20 nm above the metal surface in Al waveguide and the PMHW, respectively. One sees clearly that SPPs spread out along the *y* direction in Al waveguide while that are well confined in the PMHW. From the period of the field distribution we get a n_{eff}' of 1.42 of SPPs in the PMHW, agreeing well with the value of 1.42 obtained from the effective index method [see Fig. 14(a)].

4. Metal heterostructures for other nanophotonic functions

4.1. *Directional beaming of light*

It was reported that the directional beaming of light were observed in a nanohole (or nanoslit) flanked by periodic surface corrugations [32-34], and surface geometry-tailored SPP resonance was suggested to result in the effect. In this part, we will numerically demonstrate that such SPP resonance can also be engineered by dielectric properties of metal materials and highly directional beaming of light from a nanoslit surrounded by Al-Ag constructed metal heterostructures can be achieved[5].

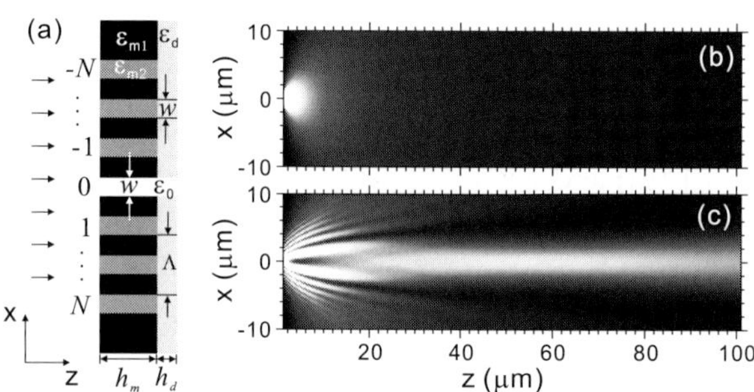

Fig. 16. (a) Metal heterostructures with a nanoslit. h_m and h_d: thicknesses of the metal films and dielectric layer; ε_{m1}, ε_{m2}, and ε_d : dielectric constants of Al, Ag, and dielectric layer; *w*: widths of the Ag film and nanoslit; Λ and N : period and number of the Ag films. (b)-(c) Gray distributions of $|H_y|^2$ for $N = 0$ and 16, respectively (adapted from Ref. 4).

Figure 16(a) schematically illustrates the proposed metal heterostructures, where a nanoslit with a width of w is introduced in the center of the heterostructures, which is constructed by alternatively stacking Al and Ag films with a period of Λ. The width of Ag films is w, and N denotes the number of the Ag films. Both Al and Ag films have the same thickness of h_m. The upper surface of the heterostructures (except the slit) is covered by a dielectric layer with a thickness of h_d. When a TM-polarized plane electromagnetic wave is incident from the bottom of the slit (without a dielectric layer), it couples to SPPs in the metal nanoslit and then is diffracted to the output surface (with a dielectric layer) as well as the forward region in the air. The diffracted electromagnetic waves on the output surface will propagate along the metal-dielectric-air structures formed by Al or Ag films and their clad dielectric layer (see the inset of Fig. 2).

From Fig. 2, one sees that n_{eff}' of Al and Ag guides are different for fixed h_d and ε_d. For example, n_{eff}' of Al and Ag guides for h_d = 100 nm and ε_d = 2.25 are n_{eff1}' = 1.34 and n_{eff2}' = 1.58, respectively. The fluctuation of the refractive index on the output surface of the metal heterostructures [Fig. 16(a)] will scatter electromagnetic waves which would have been propagating in the metal-dielectric-air structures. The leaky electromagnetic waves from the structures will emit into the air and interfere with the diffracted waves from the nanoslit, resulting in the redistribution of electromagnetic field in the air.

Figures 16(b)-16(c) show the gray distributions of $|\boldsymbol{H}_y|^2$ for a different number of Ag films. The other parameters of the heterostructures are given as follows: Λ = 400 nm, h_d = 80 nm, ε_d = 2.25, h_m = 360 nm (fixed), and w = 100 nm. For N = 0, which means the nanoslit is flanked by only one metal (Al) film, the electromagnetic wave is scattered uniformly to the air and no leaky wave is generated [Fig. 16(b)]. When N is increased to 16, which means that the nanoslit is flanked by Al-Ag constructed metal heterostructures, however, a strong directional beaming of light occurs [Fig. 16(c)]. Notice that a very elongated light beam appears in the far-field region. The appearance of the fringes is due to the interference between the leaky waves from the exit surface and diffracted waves from the nanoslit, respectively.

Figure 17(a) shows the normalized $|\boldsymbol{H}_y|^2$ profiles of light along the x axis on the line z = 100 μm for different values of N. It appears clearly that, as N is increased, the beam intensity increases while the width of the beam decreases simultaneously. The FWHM of the beam is 5.7 μm for N = 16, which means a small divergence angle of 3.27 deg in the far-field region. Figure 17(b) depicts the distributions of the Poynting vector $\boldsymbol{S}_x$ along the x axis on the line z = 0 for N = 0 and 16, respectively. Due to the absorption and radiation loss, the energy flux density is decreased as the electromagnetic waves propagate along the metallic surface. The periodic fluctuation resulted from the damping of $\boldsymbol{S}_x$ field implies the emission of the leaky waves from the output surface of the nanoslit. A larger N is not necessary for increasing the beam intensity [for example, larger than 20, see Fig. 17(a)] on account of weaker surface wave emission and coupling.

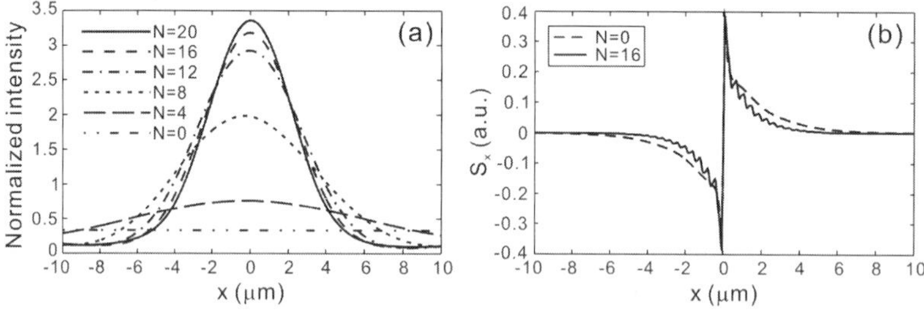

Fig. 17. (a) Normalized intensity profiles of $|\boldsymbol{H}_y|^2$ along the x axis on the line z = 100 μm for N = 0 (bottom) to 4, 8, 12, 16, 20 (top). (b) Poynting vector $\boldsymbol{S}_x$ along the x axis on the line z = 0 for N = 0 and 16. The negative values indicate the energy transports along the $-x$ direction (adapted from Ref. 4).

Figures 18(a)-18(d) illustrate the gray distributions of $|\boldsymbol{H}_y|^2$ of light for different h_d and the corresponding normalized intensity profiles along the x axis on the line z =100 μm, respectively. Here we choose Λ = 380 nm while other parameters are the same as that of Fig. 16(c). For a thinner dielectric layer (h_d = 60 nm), the emitted light is focused at a distance from slit exit and then spreads with a large divergence angle. When h_d is increased to a proper value (h_d = 100 nm), the divergence angle decreases and there is a strong directional beam with a FWHM of

5.7 μm on the line z = 100 μm [see Figs. 18(b) and 8(d)]. For larger h_d, the electromagnetic field separates from the output surface of the slit and there is no focus observed in the forward region. The process can actually be predicted by the Huygens's principle. The Ag films (where leaky waves are generated) can be regarded as scattering centers on the output surface of the slit. Because beaming occurs as the in-phase leaky waves reemit from the scattering centers [34], the following resonance condition should be satisfied

$$k_0[n_{eff1}'(\Lambda - w) + n_{eff2}'w] = 2m\pi \text{ , } m = 1, 2, 3 \ldots \tag{9}$$

where the left two items present the phase variation after electromagnetic wave propagating though a single Al and Ag film, respectively. The equation indicates that the phase accumulated by the optical length of adjacent scattering centers (the center of Ag films) should be integer times of 2π. If the summary of the left two items of Eq. (9) is slightly less than $2m\pi$, the emitted light will be focused so that the phase difference of two adjacent scattering centers can be compensated by the optical path in the air [Fig. 18(a)], and, contrariwise, it will spread with a large divergence angle from the beginning and no focus in the forward region will be found [Fig. 18(c)].

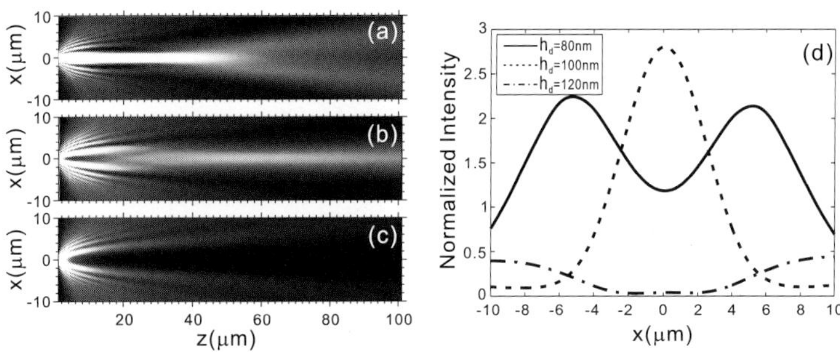

Fig. 18. Gray distributions of $|\boldsymbol{H}_y|^2$ for h_d = (a) 80 nm, (b) 100 nm, (c) 120 nm and Λ = 380 nm and (d) their profiles along the x axis on the line z =100 μm. Other parameters are the same as that of Fig. 16(c) (adapted from Ref. 4).

With the parameters used in Fig. 18, the predicted h_d according to Eq. (9) at which directional beaming occurs is 104.8 nm, which agrees well with the value (100 nm) we have used in FDTD simulation. When Λ and w remain at fixed values, the resonance condition is dominated by the real parts of effective index of refraction of the Al (n_{eff1}') and Ag (n_{eff2}') guides. Figure 19(a) shows the normalized $|\boldsymbol{H}_y|^2$ profiles of light on the line z = 100 μm for different values of ε_d. As ε_d is increased from 1.96 to 2.56, the light beam reaches to a maximum intensity and a minimum width at ε_d = 2.25. For lager ε_d, however, the intensity reduces and the width is increased. The process is similar to that shown in Fig. 18. Since $n_{\rm eff}$' increases with h_d and ε_d (see Fig. 2), the modulation to n_{eff}' by changing h_d or ε_d is equivalent. On the other hand, because the resonance condition is decided by Λ for fixed n_{eff}' and w [Eq. (9)], when h_d = 100 nm and ε_d = 2.25 are fixed but Λ is changed from 340 to 420 nm, the process shown in Fig. 18 will recur as well and a highly directional light beam appears for Λ = 380 nm.

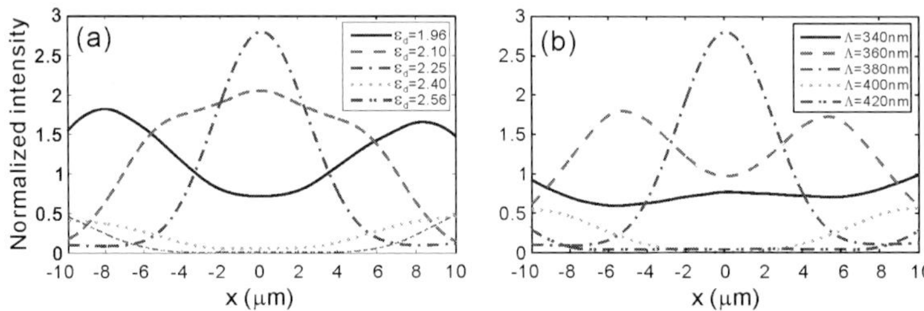

Fig. 19. Normalized intensity profiles of $|\boldsymbol{H}_y|^2$ along the x axis on the line z = 100 μm as a function of (a) ε_d (Λ = 380 nm) and (b) Λ (ε_d = 2.25). Other parameters are the same as that of Fig. 16(c) (adapted from Ref. 4).

Considering the proposed metal heterostructures here, if w is much smaller than Λ and the difference of n_{eff}' is not large, Eq. (9) can be simplified as $k_0 n_{eff1}'\Lambda = 2\pi$ for m = 1. Note that the wavelength of the surface modes can be given as $\lambda_{sm} = \lambda / n_{eff1}' \equiv 2\pi/(k_0 n_{eff1}')$. It is equal to Λ according to the above simplified equation. In this case, the physics behind the directional beaming effect of the metal heterostructures with flat surfaces presented here is accorded with that of the metal homostructures with corrugated surface where beaming occurs as the corrugation period (Λ) coincides with the surface mode wavelength λ_{sm}[34].

It should be pointed out that the proposed metal nanoslit heterostructures can be extended to 2D cases where the beaming effect can be realized through a hole surrounded by annular metal heterostructures constructed with metals Al and Ag.

4.2. *Nanoantennas*

It was demonstrated that the local optical intensity close to a sharp metal Au tip can be enhanced to 1000 times higher than that of the incident light [35-37], while the intensity enhancement produced by the "bowtie" antennas [38-41] at the so-called "hot spot" is in excess of 1500 [41]. Such enhanced local field intensity is crucial in highly-sensitive biological and chemical detection such as surface enhanced Raman scattering (SERS) [42-44] etc. Although some triangular metal nanocrystals constructed with homogeneous metal materials have also been synthesized for the purpose of local enhancement [45-47], the lack of controllable fabrication and the localization of strong local field intensity around the corners of these structures [50] are not convenient in some cases for realistic applications. In terms of the propagation property of SPPs on the surface of metal films discussed in part 1.3, we designed a type of ridged metal heterostructures (RMHs), as is schematically shown in Fig. 20(a), for enhancing the local field of light, where a triangular Ag ridge is sandwiched by two Al ridges. The cross section of the ridges is an isosceles triangle [6].

Figures 20(b) and 20(c) show the calculated gray distributions of $|\boldsymbol{E}|^2$ at the middle part of the RMHs in the x–z plane and along the y axis as the structure is with the lengths of Al at $h_1 = h_3 = 250$ nm and Ag at $h_2 = 65$ nm, and vertex angle at $\theta = 30°$, respectively, while the height of the structures along the z direction is fixed at 375 nm. In the calculations, to get a symmetrical electric field distribution, we assume two plane waves (in the y–z plane) with unity amplitude to symmetrically illuminate the RMHs from both sides of the structures [Fig. 20(a)]. The wave vector $\boldsymbol{k}$ and electric field vector $\boldsymbol{E}$ are set in the x and z directions, respectively. To quantitatively estimate the enhancement effect of the RMHs on the local electric field, we define the normalized electric field intensity $|\boldsymbol{E}|^2 = |\boldsymbol{E}_1|^2/|\boldsymbol{E}_2|^2$ to represent the field distributions in all the situations, where $|\boldsymbol{E}_1|^2$ and $|\boldsymbol{E}_2|^2$ represent the steady-state electric field intensities of both without and with the RMHs, respectively.

From the figure, we can see that not only the electric field is tightly confined around the ridge of the RMHs [Fig. 20(b)], but is mainly localized in the part of Ag ridge [Fig. 20(c)]. Figure 20(d) shows the corresponding profile of the $|\boldsymbol{E}|^2$ distribution along the y direction (parallel to the ridge of the RMHs). It can be seen that the normalized electric field intensity reaches about 1200 at the middle part of the RMHs, while in the side parts (Al ridges), a much weaker electric field intensity appears.

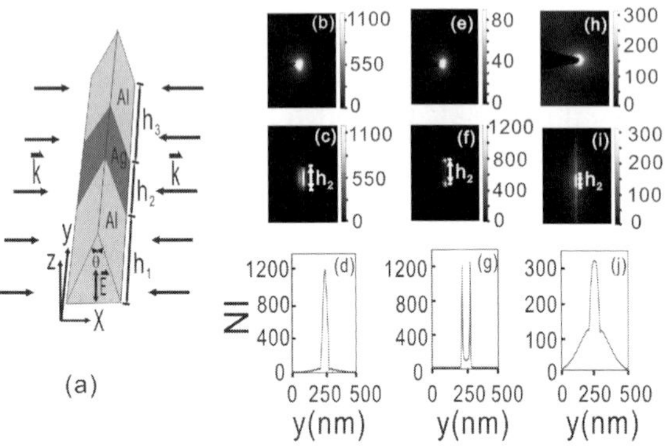

Fig. 20. (a) Schematic for the designed RMHs. Gray distributions of $|\boldsymbol{E}|^2$ at the middle part of Ag guide in [(b), (e), (h)] x–z plane and [(c), (f), (i)] along the y axis. (d), (g), and (j) Normalized intensity (NI) profiles along the y direction as the RMHs are with [(b)–(d), (h)–(j)] and without Al ridges [(e)–(g)], respectively. In the calculations, $h_1 = h_3 = 250$ nm, $h_2 = 65$ nm, and $\theta = 30°$are used. In (b)–(g), the structures are with a sharp tip while in (h)–(j) the RMH is with a round tip with a 15 nm radius (adapted from Ref. 6).

When the Al parts in the RMHs are removed, the structure consists of Ag only and is similar to the conventional triangular metal nanocrystals [45-47]. Figures 20(e)–20(g) shows the calculated electric field intensity $|\boldsymbol{E}|^2$ at the middle part of the Ag ridge in the x–z plane, along the y axis, and the corresponding profile of $|\boldsymbol{E}|^2$ along the y direction of the structure, respectively, where the geometric parameters of the Ag ridge (375 nm high and 65 nm long) and illuminating conditions are the same as that of Figs. 20(b)–20(d). As it is expected that, though as strong as about 1240 normalized electric field intensity can be reached, the giant local electric field is localized around both tips of the Ag ridge, while at its middle part, only about a 90 normalized electric field intensity is observed [Figs. 20(f) and 20(g)], which is similar to the cases of homogeneous metal nanocrystals produced [47].

Making the electric field vector ***E*** of illuminating light perpendicular to the *z* direction but keeping other geometric parameters fixed, the calculated result reveals that the normalized electric field intensity shows no enhancement effect. This suggests that it is crucial to make the polarization direction of the illuminating light perpendicular to the ridge of RMHs (along the *z* axis) for obtaining an enhanced local electric field [35, 37].

The physics behind the giant enhancement effect of the RMHs can be attributed to following reasons: firstly, when the polarization direction of the incident light is perpendicular to the ridge direction (*z* axis), there is a large surface charge accumulation on the ridge, which will induce a giant local field intensity in the vicinity of the ridge [35, 37]. Secondly, because electromagnetic waves tend to propagate with lower phase velocity [8], we know from Fig. 3 that SPPs will concentrate into the Ag region of the RMHs. Furthermore, because n_{eff}' reduces and asymptotically tends to be infinite as the thickness d of the Ag film is approaching zero, the phase velocity of SPPs on the surface of the Ag film tends to zero, which leads to the slowdown and asymptotically stop of SPPs at the vertex of the ridge. As a result, the accumulation of SPPs on the ridge of the Ag region occurs.

However, the realization of the sharp features of a nanostructure is limited in actual situation, which in turn may reduce the field enhancement of RMHs. Figures 20(h)–20(j) shown are the calculated electric field intensity $|\boldsymbol{E}|^2$ at the middle part of a RMH in the *x*–*z* plane, along *y* axis, and the corresponding profile of $|\boldsymbol{E}|^2$ along the *y* direction, respectively, as the RMH is with a round tip with 15 nm radius while all the other conditions are the same as that of Figs. 20(b)–20(d). It can be seen that the normalized electric field decreases from 1200 [Fig. 20(d)] to about 315 [Fig. 20(j)]. As the tip radius is set at 10, 20, and 25 nm, respectively, the normalized electric intensity changes to 380, 280, and 250, respectively, meaning that the electric intensity is decreased with enlarged tip radius. This can be attributed to the reduction of charge density and less accumulation of SPPs at the tip as its radius is increased.

In addition, it should be remarkde that the enhanced electric field intensity $|\boldsymbol{E}|^2$ of the RMHs is also dependent on the length h_2 of Ag ridge and the vertex angle θ of the structures [6]. To achieve stronger electric field intensity, in additional to an optimal h_2, a smaller θ is beneficial for the enhancement of the electric field intensity [6]. This is due to the enhancement of charge density and slowdown of SPPs propagation on the ridge of RMHs as the vertex angle θ is decreased.

In terms of the RMHs, we construct a nanoscale antenna consisting of two RMHs with 20 nm gap as shown in the inset of Fig. 21. Figure 21 illustrates the gray distribution of $|\boldsymbol{E}|^2$ of light at the middle part (x–z plane) of the nanoscale antenna, of which all the parameters are the same as that of Figs. 20(b)–20(d). The highest normalized field intensity can reach as high as 1605 and the light spot is confined in a size of 5×20 nm^2 in the x–z plane.

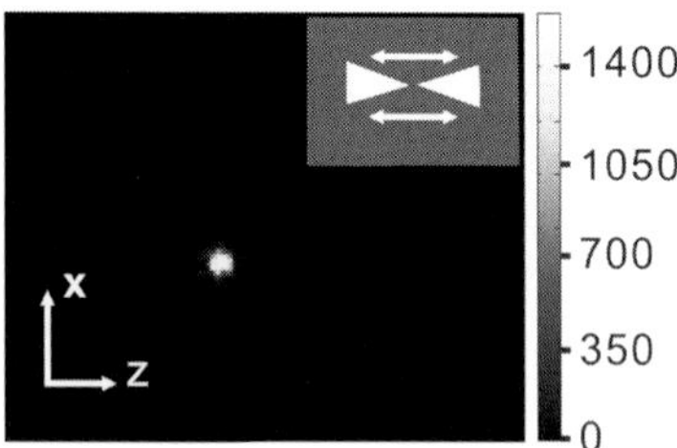

Fig. 21. Gray distribution of $|\boldsymbol{E}|^2$ in the x-z plane of a nanoscale antenna. Inset, scheme of the antenna. The parameters used for the calculation are the same as that of Figs. 20(b)–20(d) (adapted from Ref. 6).

From the realistic point of view, the welding of two nanosized materials may result in the formation of a blur boundary due to the inter-diffusion or boundary stress, which may limit the resultant geometry of the structures. By carefully controlling and optimizing the fabrication techniques of the metal heterostructures, it is possible to get a sharper boundary between two different metals [48].

5. Summary

I have reviewed the recent progresses in the design and simulation of three classes of metal heterostructures: metal gap heterowaveguides, planar metal heterostructures, and ridged metal heterostructures for the realizations of nanofocusing, nanoguiding, directional beaming of light, and enhancement of local electromagnetic field. I would like to mention that the metal heterostructures can also be extended to construct other interesting nanophotonic devices such as high efficiency bending nanowaveguides [49], SPP splitters and interferometers [27, 49, 50], SPP Bragg gratings and cavities [51], SPP refraction components [48], as well as array waveguides for array nanofocusing and beam controlling [50], metal superlattices for observing plasmonic Bloch oscillations [52] and resonant Zener tunneling [53], etc. Our results imply that the metal heterostructures may provide a novel platform for the investigations of both technology and physics potentials of SPPs in optical communications, near-field optics, biophotonics, and nanophotonic integration etc.

Acknowledgments

I am grateful to Dr. B. Wang and my Ph D students W. Lin and L. Chen for help in preparing this chapter. This work is supported by the National Basic Research Program (Grant 2007CB935300), the National Natural Science Foundation (Grants 60736041, 10774116, 10574101), and the Program for New Century Excellent Talents in University of China (Grant 04-0678).

References

1. I. P. Kaminow, W. L. Mammel and H. P. Weber, *Appl. Opt.* **13**, 396 (1974).
2. E. D. Palik, *Handbook of optical constants of solids* (Academic, New York, 1985).
3. B. Wang and G. P. Wang, *Appl. Phys. Lett.* **85**, 3599 (2004).
4. B. Wang and G. P. Wang, *Appl. Phys. Lett.* **88**, 013114 (2006).
5. E.N. Economow, *Phys. Rev.* B **182**, 539 (1969).
6. L. Chen and G. P. Wang, *Appl. Phys.* B **89**, 573 (2007).
7. P. K. Tien, *Rev. Mod. Phys.* **49**, 361 (1977).
8. W. L. Barnes, A. Dereux and T. W. Ebbesen, *Nature* **424,** 824 (2003); and references therein.
9. K. Tanaka and M. Tanaka, *Appl. Phys. Lett.* **82**, 1158 (2003).

10. K. S. Yee, *IEEE Trans. Antennas Propagat.* AP-**14**, 302 (1966).
11. E. Betzig and R. J. Chichester, *Science* **262**, 1422 (1993).
12. T. W. Ebbesen, H. J. Lecec, H. J. Lezec, H. F. Ghaemi, T. Thio and P. A. Wolff, *Nature* **391**, 667 (1998).
13. P. Berini, *Phys. Rev.* B **61**, 10484 (2000).
14. G. P. Wang and B. Wang, *J. Opt. Soc. Am.* B **23**, 1660 (2006).
15. J. C. Weeber, Y. Lacroute and A. Dereux, *Phys. Rev.* B **68**, 115401 (2003).
16. J. Takahara, S. Yamagishi, H. Taki, A. Morimoto and T. Kobayashi, *Opt. Lett* **22**, 475 (1997).
17. S. A. Maier, P. G. Kik, H. A. Atwater, S. Meltzer, E. Harel, B. E. Koel and A. A. G. Requicha, *Nature Mater.* **2**, 229 (2003).
18. D. F. P. Pile and D. K. Gramotnev, *Opt. Lett.* **29**, 1069 (2004).
19. D. K. Gramotnev and D. F. P. Pile, *Appl. Phys. Lett.* **85**, 6323 (2004).
20. S. I. Bozhevolnyi, V. S. Volkov, E. Devaux and T. W. Ebbesen, *Phys. Rev. Lett.* **95**, 046802 (2005).
21. S. I. Bozhevolnyi, V. S. Volkov, E. Devaux, J.-Y. Laluet and T. W. Ebbesen, *Nature* **440**, 508 (2006).
22. D. F. P. Pile, T. Ogawa, D. K. Gramotnev, Y. Matsuzaki, K. C. Vernon, K. Yamaguchi, T. Okamoto, M. Haraguchi and M. Fukui, *Appl. Phys. Lett.* **87**, 261114 (2005).
23. F. Kusunoki, T. Yotsuya, J. Takahara and T. Kobayashi, *Appl. Phys. Lett.* **86**, 211101(2005).
24. G. Veronis and S. Fan, *Appl. Phys. Lett.* **87**, 131102 (2005).
25. K. Tanaka, M. Tanaka and T. Sugiyama, *Opt. Express* **13**, 256 (2005).
26. Z. Y. Li and K. M. Ho, *Phys. Rev. Lett.* **92**, 063904 (2004).
27. B. Wang and G. P. Wang, *Appl. Phys. Lett.* **90**, 013114((2007).
28. H. Raether, *Surface Plasmons on Smooth and Rough Surfaces and on Gratings*, (Springer, Berlin, 1988).
29. M. P. Nezhad, K. Tetz, Y. Fainman, *Opt. Express* **12**, 4072(2004).
30. S. A. Maier, *Opt. Commun.* **258**, 295 (2006)
31. W. Lin and G. P. Wang (unpublished data)
32. H. J. Lezec, A. Degiron, E. Devaux, R. A. Linke, L. Martín-Moreno, F. J. García-Vidal and T. W. Ebbesen, *Science* **297**, 820 (2002).
33. F. J. García-Vidal, H. J. Lezec, T. W. Ebbesen and L. Martín-Moreno, *Phys. Rev. Lett.* **90**, 213901 (2003).
34. L. Martín-Moreno, F. J. García-Vidal, H. J. Lezec, A. Degiron and T. W. Ebbesen, *Phys. Rev. Lett.* **90**, 167401 (2003).
35. L. Novotny, R.X. Bian, X.S. Xie, *Phys. Rev. Lett.* **79**, 645 (1997).
36. M.I. Stockman, *Phys. Rev. Lett.* **93**, 137 404 (2004).
37. E.J. Sanchez, L. Novotny and X.S. Xie, *Phys. Rev. Lett.* **82**, 4014 (1999).
38. D.P. Fromm, A. Sundaramurthy, P.J. Schuck, G. Kino and W.E. Moerner, *Nano Lett.* **4**, 957 (2004).

39. P.Muhlschlegel, H.J. Eisler, O.J.F. Martin, B. Hecht and D.W. Pohl, *Science* **308**, 1607 (2005).
40. E. Cubukcu, E.A. Kort, K.B. Crozier and F. Capasso, *Appl. Phys. Lett.* **89**, 093 120 (2006).
41. P.J. Schuck, D.P. Fromm, A. Sundaramurthy, G.S. Kino and W.E. Moerner, *Phys. Rev. Lett.* **94**, 017 402 (2005).
42. S. Nie and S.R. Emory, *Science* **275**, 1102 (1997).
43. K. Kneipp, Y. Wang, H. Kneipp, L.T. Perelman, I. Itzkan, R. Dasari and M.S. Feld, *Phys. Rev. Lett.* **78**, 1667 (1997).
44. K. Kneipp, H. Kneipp, I. Itzkan, R.R. Dasari and M.S. Feld, *J. Phys. Condens. Matter* **14**, R597 (2002).
45. R.C. Jin, Y.W. Cao, C.A. Mirkin, K.L. Kelly, G.C. Schatz and J.G. Zheng, *Science* **294**, 1901 (2001).
46. R.C. Jin, Y.C. Cao, E. Hao, G.S.M. Traux, G.C. Schatz and C.A. Mirkin, *Nature* **425**, 487 (2003).
47. E. Hao, R.C. Bailey, G.C. Schatz, J.T. Hupp and S. Li, *Nano Lett.* **4**, 327(2004).
48. K. Tazuko, M. Shintaro, M. Yoshiteru, K. Kunio and T. Akira, *Chin. Phys. Lett.* **24**, 2827 (2007).
49. L. Chen, B. Wang and G. P. Wang, *Appl. Phys. Lett.* **89**, 243120(2006).
50. B. Wang and G. P. Wang, *Opt. Express*, **13**, 10558 (2005)
51. B. Wang and G. P. Wang, *Appl. Phys. Lett.* **87**, 013107(2005).
52. W. Lin and G. P. Wang, *Appl. Phys. Lett.* **91**, 143121 (2007).
53. W. Lin and G. P. Wang, (unpublished data)

Chapter 6

PLASMONIC SLOT WAVEGUIDES

Georgios Veronis

Department of Electrical and Computer Engineering,
Louisiana State University,
Baton Rouge, LA 70806, USA
Email: gveronis@lsu.edu

Shanhui Fan

Department of Electrical Engineering,
Stanford University,
Stanford, CA 94305, USA
Email: shanhui@stanford.edu

We investigate the properties of the modes supported by three-dimensional subwavelength plasmonic slot waveguides. We show that the fundamental mode supported by a symmetric plasmonic slot waveguide, composed of a subwavelength slot in a thin metallic film embedded in an infinite homogeneous dielectric, is always a bound mode. Its modal fields are highly confined over a wavelength range extending from zero frequency to the ultraviolet. We then show that for an asymmetric plasmonic slot waveguide, in which the surrounding dielectric media above and below the metal film are different, there always exists a cutoff wavelength above which the mode becomes leaky. In addition, we investigate the properties of compact couplers between high-index contrast dielectric slab waveguides and two-dimensional metal-insulator-metal subwavelength plasmonic waveguides. We show that a coupler created by simply placing a dielectric waveguide terminated flat at the exit end of a plasmonic waveguide can be designed to have a transmission efficiency of $\sim$70% at the optical communication wavelength. We also show that the transmission efficiency of the couplers can be further increased by using optimized multisection tapers. In both cases the transmission response is broadband. In addition, we investigate the properties of a Fabry-Perot structure in which light is coupled in and out of a plasmonic waveguide sandwiched between dielectric waveguides.

1. Modes of three-dimensional subwavelength plasmonic slot waveguides

Guiding electromagnetic waves with a mode at deep subwavelength scale is of great recent interest. In the visible wavelength range, the typical way to create a subwavelength waveguide involves the use of the so-called surface plasmon polaritons (SPPs). SPPs are bound non-radiative surface modes which propagate at metal-dielectric interfaces with field components decaying exponentially with distance away from the interface.[1] The decay length of the fields can be much smaller than the wavelength near the surface plasmon frequency. Several different plasmonic waveguiding structures have been proposed,[2–9] such as metallic nanowires,[3,4] metallic nanoparticle arrays,[5,6] and V-shaped grooves.[7,8] However, all these geometries[3–8] are fundamentally limited by the fact that they support a highly-confined mode *only* near the surface plasmon frequency. In this regime, the optical mode typically has low group velocity and short propagation length.

At microwave frequencies, where metals do not have a plasmonic response, two-conductor waveguides are used to guide subwavelength modes. Such waveguiding structures always support a fundamental TEM or quasi-TEM mode which can have deep subwavelength size and broad guiding bandwidth.[10] In addition, it has been shown that the guiding wavelength range of subwavelength modes by such structures extends into the infrared and visible wavelengths.[11] In the optical wavelength range, however, metals have a plasmonic response.[12] The properties of the optical modes supported by plasmonic two-conductor waveguides are therefore quite different from those of their counterparts at microwave frequencies, where metals behave almost as perfect electric conductors.[10] Several different plasmonic two-conductor waveguide structures have been proposed to guide light.[13–18] Because of the predicted attractive properties of plasmonic two-conductor waveguides, people have also started to explore such structures experimentally.[19,20]

Among all plasmonic two-conductor waveguides, a three-dimensional (3D) plasmonic waveguide based on a deep subwavelength slot in a thin metallic film was recently investigated.[16–18,21,22] The geometry of such a plasmonic slot waveguide is shown in Fig. 1(a). It consists of a slot in a thin metal film. The thin metal film is embedded in dielectric. The supported optical mode is highly localized in the slot, and its direction of propagation is parallel to the slot. Fig. 1(b) shows a cross-sectional view of the waveguide geometry. It was shown that such a plasmonic slot waveguide supports a single fundamental mode over a wide range of frequencies.[16] The size of this mode is completely dominated by the near field of the slot, and can be far smaller than the wavelength even when the effective index of the mode approaches that of the substrate. In addition, the group velocity of the mode is close to the speed of light in the substrate, and its propagation length is tens of microns at the optical communication wavelength (λ_0 =1.55 μm). Thus, such a plasmonic slot waveguide could be potentially important in providing an interface between conventional optics and subwavelength electronic and optoelectronic devices. The characteristics of the modes supported by plasmonic slot waveguides at visible wavelengths (λ_0 =632.8 nm) have also been investigated.[17,18]

In this section we investigate in detail the characteristics of the modes supported by 3D plasmonic slot waveguides. In particular, we illustrate the physics of such a waveguide by comparing it to a number of simplified geometries shown in Fig. 1 (c)-(e). We first consider a reference symmetric plasmonic slot waveguide structure, comprised of a slot in a thin metallic film embedded in an infinite homogeneous dielectric. We show that the fundamental mode supported by this symmetric plasmonic slot waveguide is always a bound mode for any combination of operating wavelength and waveguide parameters. We then consider an asymmetric plasmonic slot waveguide structure in which the surrounding dielectric media above and below the metal film are different. Unlike the symmetric case, in the asymmetric case the fundamental propagating mode is not always bound. We show that for a specific asymmetric plasmonic slot waveguide there may exist a cutoff slot width and/or a cutoff metal film thickness above which the mode becomes leaky, and there always exists a cutoff wavelength above which the mode becomes leaky.

1.1. *Simulation method*

We calculate the eigenmodes of plasmonic waveguides at a given wavelength λ_0 using a full-vectorial finite-difference frequency-domain (FDFD) mode solver.[16,23] For waveguiding structures which are uniform in the z direction, if an $\exp(-\gamma z)$ dependence is assumed for all field components, Maxwell's equations reduce to two coupled equations for the transverse magnetic field components H_x and H_y[23]

$$-\epsilon_r k_0^2 h_x + \epsilon_r \frac{\partial}{\partial y}[\epsilon_r^{-1}(\frac{\partial h_y}{\partial x} - \frac{\partial h_x}{\partial y})] - \frac{\partial}{\partial x}(\frac{\partial h_x}{\partial x} + \frac{\partial h_y}{\partial y}) = \gamma^2 h_x$$

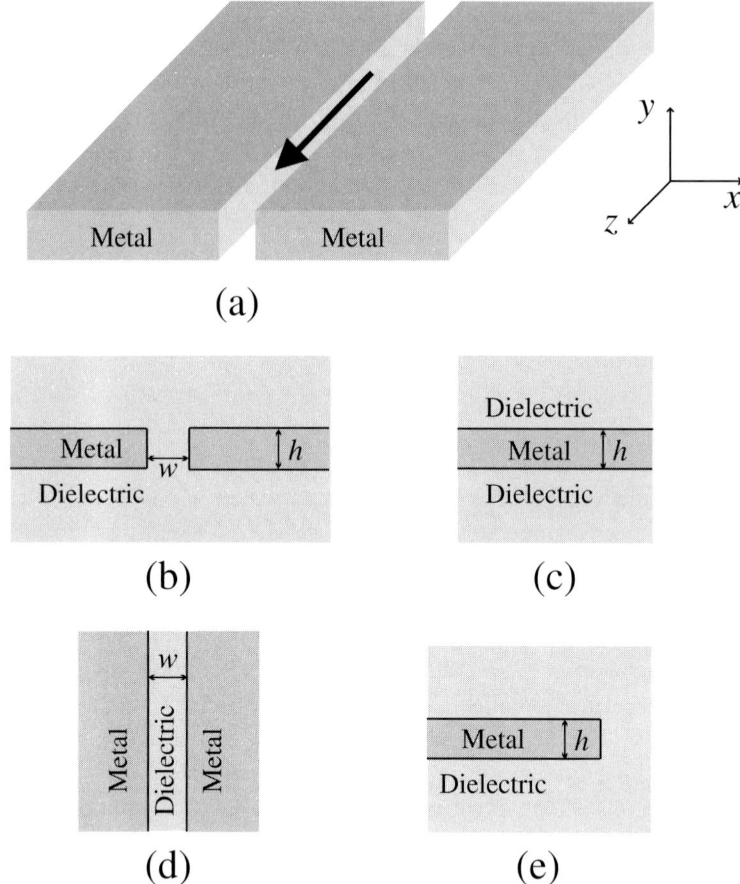

Fig. 1. (a) Geometry of a 3D plasmonic slot waveguide. The arrow shows the direction of propagation of the optical mode. (b) Cross-sectional view of the 3D plasmonic slot waveguide geometry. (c)-(e) Corresponding IMI, MIM, and truncated metal film structures for the 3D plasmonic slot waveguide of Fig. 1(b).

$$-\epsilon_r k_0^2 h_y - \epsilon_r \frac{\partial}{\partial x}[\epsilon_r^{-1}(\frac{\partial h_y}{\partial x} - \frac{\partial h_x}{\partial y})] - \frac{\partial}{\partial y}(\frac{\partial h_x}{\partial x} + \frac{\partial h_y}{\partial y}) = \gamma^2 h_y$$

where $\mathbf{H}(x,y,z) = \mathbf{h}(x,y)\exp(-\gamma z)$, $\epsilon_r = \epsilon_r(x,y)$ is the dielectric function, and $k_0^2 = \omega^2 \epsilon_0 \mu_0$. Solving these eigenvalue equations allows one to define the propagation length L_p and the effective index n_eff of a propagating mode through the equation $\gamma \equiv L_\mathrm{p}^{-1} + i\beta = L_\mathrm{p}^{-1} + i2\pi n_\mathrm{eff}\lambda_0^{-1}$. Also, the dispersion relation is defined as $\omega = \omega(\beta)$. These equations are discretized on a non-uniform orthogonal grid resulting in a sparse matrix eigenvalue problem of the form $\mathbf{A}\mathbf{h} = \gamma^2\mathbf{h}$, which is solved using iterative sparse eigenvalue techniques.[24] The discretization scheme is based on Yee's lattice.[23] To calculate the bound eigenmodes of the waveguide, we ensure that the size of the computational domain is large enough so that the fields are negligibly small at its boundaries,[25] while for leaky modes we use perfectly matched layer (PML) absorbing boundary conditions.[24] An important feature of this formulation is the absence of spurious modes.[25] In addition, metals have complicated dispersion properties in the optical wavelength range. As an example, Fig. 2 shows the real and imaginary part of the dielectric constant of silver at optical frequencies.[12,26] The frequency-domain mode solver allows us to directly use experimental data for the frequency-dependent dielectric constant of metals, including both the real and imaginary parts, with no approximation.

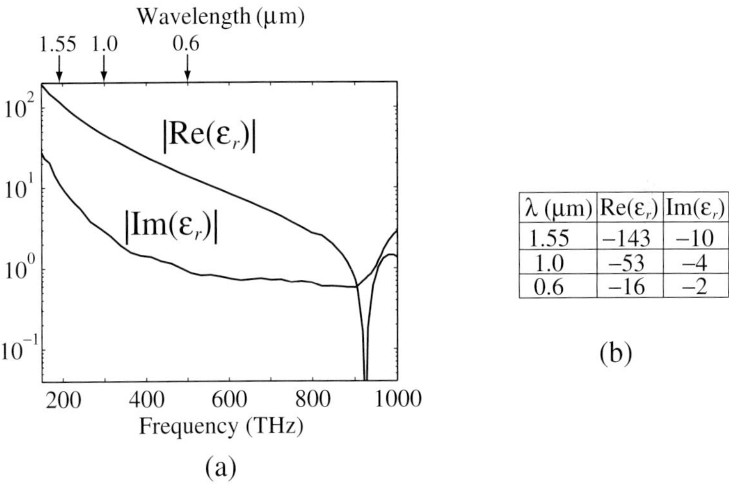

λ (μm)	$\mathrm{Re}(\varepsilon_r)$	$\mathrm{Im}(\varepsilon_r)$
1.55	−143	−10
1.0	−53	−4
0.6	−16	−2

Fig. 2. (a) Absolute values of the real ($\mathrm{Re}(\epsilon_r)$) and imaginary ($\mathrm{Im}(\epsilon_r)$) part of the dielectric constant of silver at optical frequencies. $\mathrm{Re}(\epsilon_r) < 0$ for frequencies below ~910 THz. (b) The dielectric constant of silver at selected wavelengths considered in this chapter.

1.2. *Simplified plasmonic structures*

In order to understand the modal structure of the slot waveguide shown in Fig. 1(a) and (b), we first consider the modes in corresponding simplified geometries that include two-dimensional (2D) insulator-metal-insulator (IMI)

(Fig. 1(c)), and metal-insulator-metal (MIM) (Fig. 1(d)) plasmonic waveguiding structures, as well as the edge on a truncated metal film (Fig. 1(e)). The asymptotic behavior of many properties of 3D plasmonic slot waveguides can be explained in terms of the properties of these simpler plasmonic structures. Below we focus only on those modes in these simplified structures which are directly related to the fundamental modes of the slot.

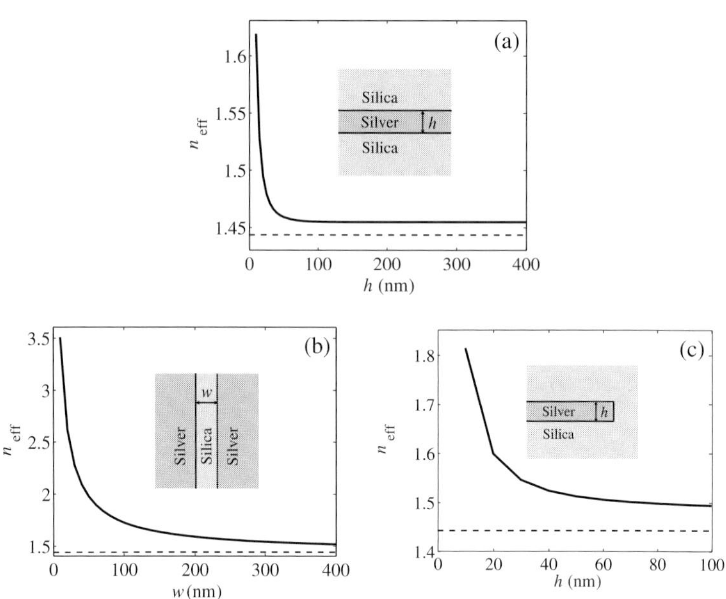

Fig. 3. (a) The effective index n_{eff} of the SR-SPP of a silica-silver-silica IMI structure (see inset) as a function of the width of the central silver region h (solid line). We also show the refractive index of silica (dashed line). (b) The effective index n_{eff} of the G-SPP of a silver-silica-silver MIM structure (see inset) as a function of the width of the central silica region w (solid line). We also show the refractive index of silica (dashed line). (c) The effective index n_{eff} of the fundamental edge mode of a truncated silver film embedded in silica as a function of the metal film thickness h (solid line). We also show the refractive index of silica (dashed line).

The corresponding IMI plasmonic structure (Fig. 1(c)) has the same metal film thickness as the film in the slot waveguide (Fig. 1(b)). Such a structure supports two modes below the surface plasmon frequency.[11] Of relevance here is the higher-index mode, which is conventionally called short-range SPP (SR-SPP), that has symmetric charge distribution. In Fig. 3(a) we show the effective index n_{eff} of the SR-SPP of a silica-silver-silica IMI structure as a function of the width of the central silver region

h. As a somewhat unusual feature, when h decreases, the fraction of the modal power in the metal increases, and n_{eff} therefore increases. In the opposite limit, as $h \to \infty$, the coupling between the surface plasmon modes of the two metal-dielectric interfaces vanishes, and n_{eff} therefore approaches asymptotically the effective index of the surface plasmon mode of a single metal-dielectric interface.

The corresponding MIM plasmonic structure (Fig. 1(d)) has a dielectric film thickness equal to the slot width (Fig. 1(b)). Such a structure supports a fundamental mode, the so called gap SPP (G-SPP), below the surface plasmon frequency with antisymmetric charge distribution.[11] In Fig. 3(b) we show the effective index n_{eff} of the G-SPP of a silver-silica-silver MIM structure as a function of the width of the central silica region w. As w decreases, the fraction of the modal power in the metal increases, and n_{eff} therefore increases. In the opposite limit, as $w \to \infty$, n_{eff} approaches asymptotically the effective index of the surface plasmon mode of a single metal-dielectric interface, as in the IMI structure. We also note that for a given width of the central region the coupling of the single-interface surface plasmon modes is substantially stronger in the MIM case compared to the IMI case, since in the former case the coupling occurs through dielectric, while in the latter it occurs through metal.

The corresponding truncated metal film structure (Fig. 1(e)) has the same metal film thickness as the film in the slot waveguide (Fig. 1(b)). Such a structure supports a fundamental edge mode. In Fig. 2.4 we show the power density ($S_z = \frac{1}{2}\text{Re}[\mathbf{E} \times \mathbf{H}^* \cdot \hat{z}]$) profile of the edge mode in a truncated silver film embedded in silica. Large modal intensity is observed at the silver-silica interface at the edge of the metal film. More specifically, the modal power density is maximum at the two edge corners, due to the singular behavior of the electric field near sharp edges.[27] In Fig. 3(c) we show the effective index n_{eff} of the fundamental edge mode of a truncated silver film embedded in silica as a function of the metal film thickness h (solid line). Similar to the IMI structure, n_{eff} increases, as h decreases. In the opposite limit, as $h \to \infty$, n_{eff} approaches asymptotically the effective index of the mode of a single 90° corner, which is higher than the effective index of a single-interface surface plasmon mode.[28]

1.3. *Symmetric plasmonic slot waveguide*

We now consider a symmetric plasmonic slot waveguide structure comprised of a slot in a thin metallic film embedded in an infinite homogeneous dielec-

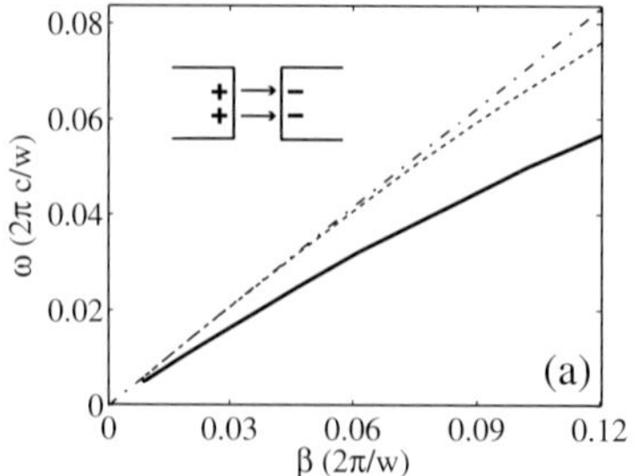

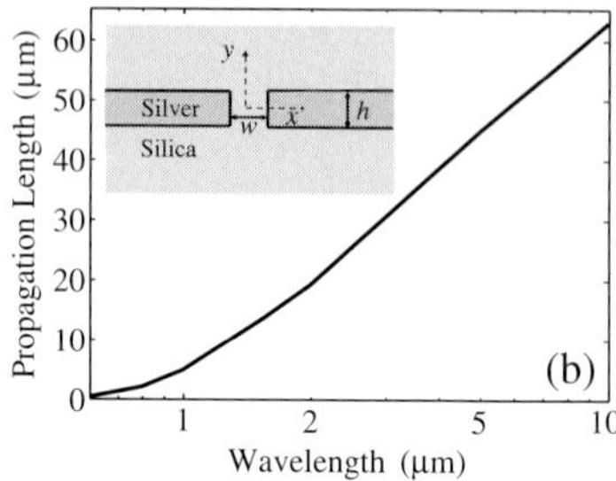

Fig. 4. (a) Dispersion relation of the fundamental mode of the symmetric plasmonic slot waveguide (shown with solid line) for w, h =50 nm (see inset of Fig. 4(b)). The inset shows a schematic of the charge and vector field distribution of the mode. The dash-dotted line is the light line of silica, while the dashed curve is the dispersion relation of the SR-SPP in the silica-silver-silica IMI thin film structure. (b) Propagation length of the fundamental mode of the symmetric plasmonic slot waveguide as a function of wavelength for w, h =50 nm.

tric. We are particularly interested in the regime where the dimensions of the slot are much smaller than the wavelength of light. Thus, our reference structure consists of a slot of width w =50 nm in a silver film of thickness h =50 nm embedded in silica (n_{s} =1.44) (inset of Fig. 4(b)).

In Fig. 4(a) we show the dispersion relation of the fundamental mode supported by our reference structure. We observe that the fundamental mode of the plasmonic slot waveguide has a wavevector larger than all radiation modes in silica, as well as all propagating modes in the silica-silver-silica IMI thin film structure over the entire frequency range. The fundamental mode supported by our reference plasmonic slot waveguide is therefore a bound mode. Since the slot dimensions are much smaller than the wavelength in the frequency range of interest, this waveguide does not support any higher order propagating bound modes.

In Fig. 4(b) we show the propagation length L_{p} of the fundamental mode of our reference plasmonic slot waveguide as a function of wavelength. The propagation length decreases as the wavelength decreases. This is consistent with the behavior of plasmonic structures in general,[1] since the fraction of the modal power in the metal increases at shorter wavelengths.

In Color Plates 2(a)–(d) we show the power density profile of the fundamental mode supported by our reference structure at different wavelengths. The fundamental mode is quasi-TEM with dominant field components E_x and H_y.[16] The charge distribution is therefore odd for the mirror plane normal to x, and even for the mirror plane normal to y (inset of Fig. 4(a)). We observe that over a wide wavelength range the modal fields are highly confined in the slot region and only slightly extend in the adjacent silica regions above and below the slot. Thus, the modal size is almost completely dominated by the near field of the slot.[16] In addition, the size of this mode is far smaller than the wavelength even when its dispersion relation approaches the light line of the surrounding dielectric media.[16] This behavior is fundamentally different from that of conventional dielectric waveguides, in which the mode extends significantly into the low-index cladding, as the dispersion relation of the optical mode approaches the cladding light line.[29,30] It is also fundamentally different from that of single-metal plasmonic waveguides (e.g. V-shaped grooves) in which a deep subwavelength mode is obtained only in a limited wavelength range, where the modal dispersion relation is far from the light line of the surrounding dielectric.[31]

Large modal intensity is observed at the silver-silica interfaces in the slot. Similar to the edge mode (Color Plate 1), the modal power density of a plasmonic slot waveguide reaches maximum at the corners of the slot, due to the singular behavior of the electric field near sharp edges.[27] However, the characteristics of the mode do not depend strongly on the sharpness of the slot corners.[21] At visible wavelengths (λ_0 =600 nm) the modal power is concentrated at the edges of the two semi-infinite metal film regions forming the slot, and the power density in the middle of the slot is small (Color Plate 2(a)). At this wavelength, the dielectric constant of silver is $\epsilon_r \simeq -16 - 2i$, and has a relatively small magnitude (Fig. 2). Hence the field penetrates strongly into the metal region. Also, since the fraction of the modal power in the metal is high enough, the edge mode of a truncated metal film (Fig. 1(e)) is tightly confined with a decay length in dielectric smaller than the slot width. At visible wavelengths the fundamental mode of the plasmonic slot waveguide therefore comprises of weakly-coupled edge modes of the two semi-infinite metal film regions forming the slot.[18] Since the coupling is weak, the effective index of the fundamental mode is very close to the effective index of a single edge mode. Due to the large fraction of the modal power in the metal at λ_0 =600 nm (ω =0.083 $2\pi c/w$), the effective index of the mode n_{eff} =2.45 is much larger than the index of silica n_{s} (Fig. 4(a)), and its propagation length L_{p} =0.6 μm is very short (Fig. 4(b)).

As the wavelength increases, the dielectric constant of metal increases in magnitude, the fraction of the modal power in the metal decreases, and therefore the effective index of the mode decreases, and its propagation length increases. For λ_0 =1, 1.55, 10 μm (ω =0.05, 0.032, 0.005 $2\pi c/w$) the modal effective indices are $n_{\rm eff}$ =2.05, 1.91, 1.72 (Fig. 4(a)) and the modal propagation length is $L_{\rm p}$ =5, 14, 63 μm (Fig. 4(b)), respectively. In addition, at infrared wavelengths the modal power density in the middle of the slot is significant (Color Plates 2(a)–(d)). Thus, at longer wavelengths, the mode can no longer be viewed as being comprised of edge modes. In particular, the effective index of the fundamental mode of the slot is much larger than the effective index of the edge mode.

In the remainder of this section, we investigate the effect of variations of the parameters of our reference plasmonic slot waveguide structure on the modal characteristics. We focus on the optical communication wavelength (λ_0 =1.55 μm) where subwavelength plasmonic slot waveguides have propagation lengths of tens of micrometers.[16]

In Fig. 5(a), (b) we show, respectively, the effective index $n_{\rm eff}$, and propagation length $L_{\rm p}$ of the fundamental mode of the plasmonic slot waveguide as a function of the slot width w (Fig. 4(b)). All other parameters are as in our reference structure (Fig. 4). We observe that as w decreases, $n_{\rm eff}$ increases, and $L_{\rm p}$ decreases. This is due to the fact that as w decreases, the fraction of the modal power in the metal increases. Similar behavior is observed in MIM plasmonic waveguides.[7,13,14] In fact for $w \to 0$ we observe that $n_{\rm eff}(w)$ is very close to $n_{\rm eff}(w)|_{\rm MIM}$ (Fig. 5(a)). In the limit of $w \to \infty$, $n_{\rm eff}(w)$ approaches asymptotically the effective index of the edge mode (shown with the dash-dotted line in Fig. 5(a)), which is slightly higher than the effective index of the SR-SPP of the corresponding IMI structure[11] (shown with the dashed line in Fig. 5(a)). The same asymptotic behaviors are also observed in the propagation length $L_{\rm p}(w)$ for $w \to 0$ and $w \to \infty$. In Fig. 5(c) we show the modal area (defined as the area in which the mode power density is larger than $1/e^2$ of its maximum value) of the fundamental mode of the plasmonic slot waveguide as a function of the slot area $w \times h$, as w is varied (Fig. 4(b)). As expected, for small w the modal area increases, as w increases. In addition, the coupling between the edge modes of the two semi-infinite metal film regions which form the slot decreases, as w increases. Therefore, when w is larger than the decay length of the edge mode in silica, the modal area decreases, as w increases. As $w \to \infty$, the modal area asymptotically approaches twice the area of the edge mode.

In Fig. 6(a), (b) we show, respectively, the effective index $n_{\rm eff}$, and propagation length $L_{\rm p}$ of the fundamental mode of the plasmonic slot waveguide as a function of the thickness h of the metallic film (Fig. 4(b)). All other

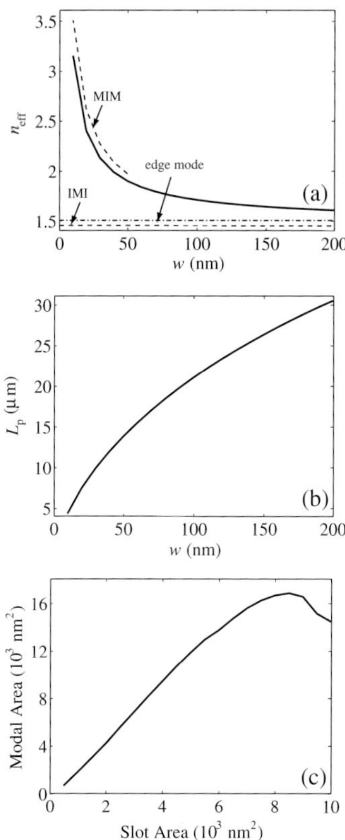

Fig. 5. (a) The effective index n_{eff} of the fundamental mode of the symmetric plasmonic slot waveguide (solid line) as a function of the slot width w (Fig. 4(b)). We also show the effective index of the G-SPP of the corresponding MIM structure (dashed line); the effective index of the SR-SPP of the corresponding IMI structure (dashed line); the effective index of the edge mode of the corresponding truncated metallic film (dash-dotted line). (b) The propagation length L_{p} of the fundamental mode of the symmetric plasmonic slot waveguide as a function of the slot width w (Fig. 4(b)). (c) The modal area of the fundamental mode of the symmetric plasmonic slot waveguide as a function of the slot area $w \times h$, as w is varied (Fig. 4(b)). All other parameters are as in our reference structure (Fig. 4).

parameters are as in our reference structure (Fig. 4). As $h \to \infty$, the effective index of the mode approaches asymptotically the effective index of the G-SPP of the corresponding MIM plasmonic waveguide ($\lim_{h\to\infty} n_{\text{eff}}(h) = n_{\text{eff}}|_{\text{MIM}}$), and the same asymptotic behavior is observed in the propagation length $L_{\text{p}}(h)$. In Fig. 6(c) we show the modal area of the fundamental mode of the plasmonic slot waveguide as a function of the slot area $w \times h$,

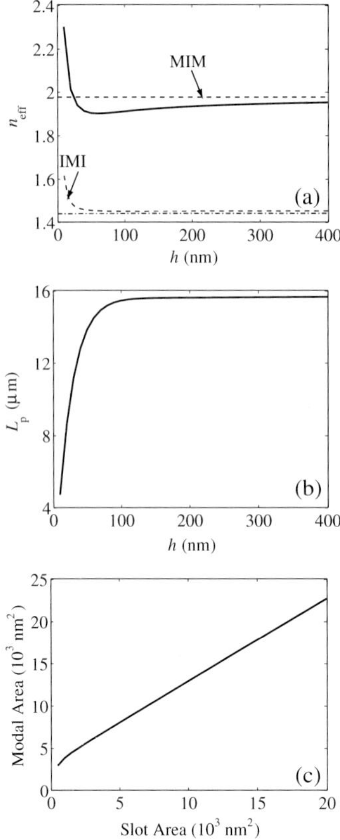

Fig. 6. (a) The effective index n_{eff} of the fundamental mode of the symmetric plasmonic slot waveguide (solid line) as a function of the thickness h of the metallic film (Fig. 4(b)). We also show the effective index of the G-SPP of the corresponding MIM structure (dashed line); the effective index of the SR-SPP of the corresponding IMI structure (dashed line); the refractive index of the substrate (dash-dotted line). (b) The propagation length L_{p} of the fundamental mode of the symmetric plasmonic slot waveguide as a function of the thickness h of the metallic film (Fig. 4(b)). (c) The modal area of the fundamental mode of the symmetric plasmonic slot waveguide as a function of the slot area $w \times h$, as h is varied (Fig. 4(b)). All other parameters are as in our reference structure (Fig. 4).

as h is varied (Fig. 4(b)). The modal area increases, as h increases, and for large h the increase is linear.

We also observe that as $h \to 0$, $n_{\text{eff}}(h)$ increases, and $L_{\text{p}}(h)$ decreases. We found that this behavior is related to the fringing fields of the mode rather than the fields in the slot region. The dominant components of the fringing fields are E_y and H_x, and they are maximum at the four lateral silver-silica interfaces at $y = \pm h/2$, $x > w/2$ or $x < -w/2$ (inset of Fig. 4(b)). As the film thickness h decreases, the fraction of modal power in the metal at these interfaces increases, and this results in increasing effective index n_{eff} and decreasing propagation length L_{p}. The same trend is also observed at visible wavelengths.[18] In Fig. 2.4(a), (b) we show the power density profile of the fundamental mode of the plasmonic slot waveguide for h =60 nm, h =10 nm, respectively. All other parameters are as in our reference structure (Fig. 4). We observe that, as the film thickness h decreases, the modal profile evolves into two weakly coupled edge modes. This is consistent with the increase of the effective index n_{eff}, which translates into smaller field decay length in dielectric, and also confirms that the effective index increase is related to the fringing fields of the mode at the semi-infinite metal film regions. We also observe that in the limit of $h \to 0$ the modal area approaches an asymptotic non-zero value (Fig. 6(c)), unlike the $w \to 0$ case. In other words, there is a minimum modal size as the film thickness approaches zero due to the fringing fields of the mode.

We also investigated the effect of varying the dielectric constant of the medium in which the metallic film with the slot is embedded. We found that as the permittivity of the dielectric is increased, the effective index n_{eff} of the fundamental mode increases, and its propagation length L_{p} decreases. As in the case of a single-interface surface-plasmon mode,[1] this is due in part to increased fraction of modal power in the metal as the permittivity of the dielectric increases, as well as decreased group velocity. Finally, we note that in the case of the symmetric plasmonic slot waveguide the wavevector of the fundamental mode is larger than all radiation modes in the surrounding dielectric, as well as all propagating modes in the corresponding IMI thin film structure (Fig. 1(c)), for any combination of operating wavelength and waveguide parameters. In other words, in the symmetric plasmonic slot waveguide there always exists a fundamental propagating bound mode, and there is no associated cutoff.

1.4. *Asymmetric plasmonic slot waveguide*

We now consider an asymmetric plasmonic slot waveguide structure in which the surrounding dielectric media above and below the metal film are different. More specifically, we consider a waveguide consisting of an air slot of width w in a metallic film of thickness h deposited on silica (inset of Fig. 7(a)).

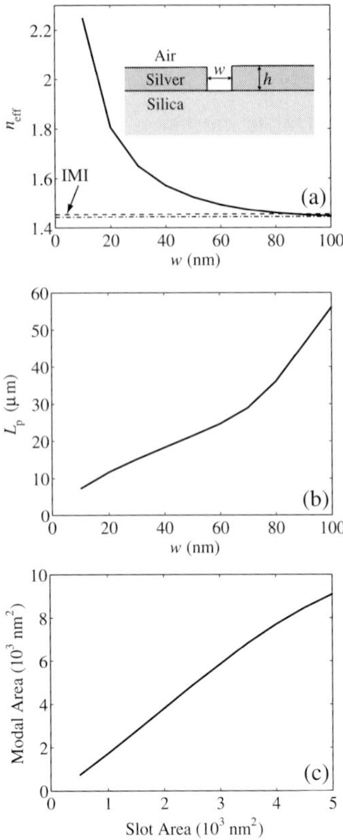

Fig. 7. (a) The effective index n_{eff} of the fundamental mode of the asymmetric plasmonic slot waveguide (solid line) as a function of the slot width w (see inset). We also show the effective index of the SR-SPP of the corresponding asymmetric IMI structure (dashed line); the refractive index of the substrate (dash-dotted line). (b) The propagation length L_{p} of the fundamental mode of the asymmetric plasmonic slot waveguide as a function of the slot width w (Fig. 7(a)). (c) The modal area of the fundamental mode of the asymmetric plasmonic slot waveguide as a function of the slot area $w \times h$, as w is varied (Fig. 7(a)). All other parameters are as in our reference structure (Fig. 4).

In Fig. 7(a), (b) we show, respectively, the effective index n_{eff}, and propagation length L_{p} of the fundamental mode of the asymmetric plasmonic slot waveguide as a function of the slot width w (inset of Fig. 7(a)). All other parameters are as in our reference structure (Fig. 4). In Fig. 7(c) we show the modal area of the fundamental mode of the asymmetric plasmonic slot waveguide as a function of the slot area $w \times h$, as w is varied. We observe that in general the effect of variation of the slot width w on the characteristics of the fundamental mode supported by the plasmonic slot waveguide is very similar in both the symmetric (Fig. 5) and asymmetric (Fig. 7) cases. However, there is an important difference between the two cases regarding the asymptotic behavior for $w \to \infty$. Unlike the symmetric case, in the asymmetric case the fundamental propagating mode is not always bound. If all other waveguide parameters are fixed, there is a slot width cutoff w_{cutoff}, and for $w > w_{\text{cutoff}}$ the mode is leaky. More specifically, we observe that if w is gradually increased to be w_{cutoff}, the modal power starts to leak into the propagating modes of the IMI air-silver-silica thin film structure, and if w is further increased so that $w > w_{\text{cutoff,substrate}}$, the modal power also leaks into the radiation modes in the silica substrate (Fig. 7(a)). In general as $w \to \infty$, $n_{\text{eff}}(w)$ approaches asymptotically the effective index of the edge mode of a truncated metal film deposited on the substrate. For the structure shown in the inset of Fig. 7(a) the effective index of the edge mode is smaller than the effective index of the SR-SPP of the asymmetric IMI (air-silver-silica) structure. Hence there exists a cutoff width w_{cutoff}. In addition, the effective index of the edge mode is also smaller than the refractive index of the substrate (silica in Fig. 7). Hence there also exists a cutoff width $w_{\text{cutoff,substrate}}$. We also observe that the propagation length L_{p} of the fundamental mode of the asymmetric plasmonic slot waveguide increases with w, even when the mode becomes leaky. This indicates that the dominant loss mechanism is the material loss in the metal. As w increases, the mode becomes less confined, and the power loss in the metal decreases. Thus, even though the radiation power loss increases with w, the overall power loss decreases. In fact, for $w > w_{\text{cutoff}}$ in the asymmetric case (Fig. 7(b)) the propagation length L_{p} increases more rapidly with w compared to the symmetric case (Fig. 5(b)), because leakage further reduces the mode confinement and therefore the power loss in the metal.

In Fig. 8(a), (b) we show, respectively, the effective index n_{eff}, and propagation length L_{p} of the fundamental mode of the asymmetric plasmonic slot waveguide as a function of the metallic film thickness h (inset of Fig. 7(a)). All other parameters are as in our reference structure (Fig. 4). In Fig. 8(c) we show the modal area of the fundamental mode of the asymmetric plasmonic slot waveguide as a function of the slot area $w \times h$, as h is varied. As in the case of slot width w variation (Fig. 5 and Fig. 7), we observe that in general the effect of metallic film thickness h variation on the characteristics of the fundamental mode supported by the plasmonic slot waveguide is very similar in both the symmetric (Fig. 6) and asymmetric (Fig. 8) cases. The most important difference between the two cases is

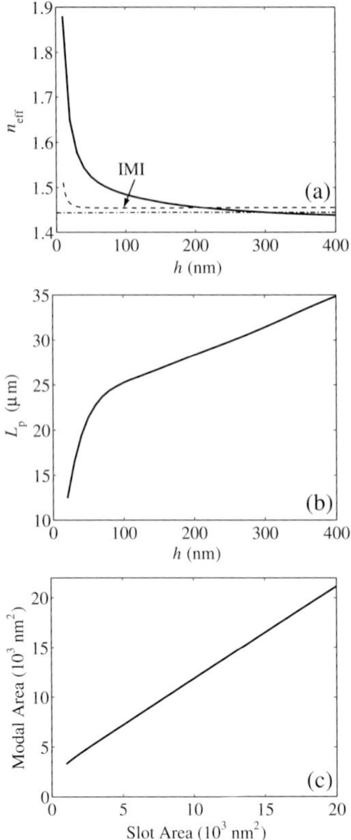

Fig. 8. (a) The effective index $n_{\rm eff}$ of the fundamental mode of the asymmetric plasmonic slot waveguide (solid line) as a function of the thickness h of the metallic film (Fig. 7(a)). We also show the effective index of the SR-SPP of the corresponding asymmetric IMI structure (dashed line); the refractive index of the substrate (dash-dotted line). (b) The propagation length $L_{\rm p}$ of the fundamental mode of the asymmetric plasmonic slot waveguide as a function of the thickness h of the metallic film (Fig. 7(a)). (c) The modal area of the fundamental mode of the asymmetric plasmonic slot waveguide as a function of the slot area $w \times h$, as h is varied (Fig. 7(a)). All other parameters are as in our reference structure (Fig. 4).

related to the asymptotic behavior for $h \to \infty$. In the asymmetric case, as $h \to \infty$, the effective index of the fundamental mode $n_{\rm eff}(h)$ approaches asymptotically the effective index of the G-SPP of the MIM plasmonic waveguide with dielectric core corresponding to the dielectric media in the slot. Since, for the structure shown in Fig. 7, the effective index of the G-SPP supported by this MIM structure (silver-air-silver in Fig. 7) is smaller than the effective index of the SR-SPP of the asymmetric IMI structure (air-silver-silica in Fig. 7), there exists a cutoff film thickness h_{cutoff}. For $h > h_{\text{cutoff}}$ there is no longer a fundamental guided mode. In addition, for this structure, since the G-SPP mode effective index of the MIM (silver-air-silver in Fig. 7) is smaller than the refractive index of the substrate (silica in Fig. 7), there also exists a cutoff film thickness due to the substrate $h_{\text{cutoff,substrate}}$ above which the fundamental mode leaks into the substrate. We also observe that the propagation length L_{p} of the fundamental mode of the asymmetric plasmonic slot waveguide increases with h, even when the mode becomes leaky (Fig. 8(b)), for similar reasons as those mentioned above for the case of increasing w.

As mentioned above, the cutoff slot width depends on the effective index of the corresponding edge mode. Similarly, the cutoff film thickness depends on the G-SPP mode effective index of the corresponding MIM plasmonic waveguide with dielectric core corresponding to the dielectric media in the slot. Therefore, the structure will not have a cutoff slot width or film thickness for the fundamental mode, if the slot is filled with the same dielectric as the substrate. We note however that for any asymmetric plasmonic slot waveguide there always exists a cutoff wavelength above which the mode becomes leaky. This is due to the fact that in the long wavelength limit metals behave as perfect electric conductors, so that the modal fields do not penetrate into the metal. Since some of the field lines are in the upper and some are in the lower dielectric media, the effective index of the mode always lies between the refractive indices of the upper and lower dielectric media, and the mode is therefore always leaky.[16] Finally, we note that as the structure becomes more asymmetric, the cutoff slot width, cutoff film thickness, and cutoff wavelength decrease.

2. Couplers between dielectric and plasmonic waveguides

In plasmonic two-conductor waveguides the propagation length of the fundamental subwavelength propagating mode is limited by material loss in the metal and is on the order of tens of micrometers at frequencies around the optical communication wavelength (λ_0 =1.55 μm).[13,16,21] Thus for longer distances it is expected that conventional dielectric waveguides with diffraction-limited optical mode confinement will be used to carry the optical signal. The propagation length in dielectric waveguides is primarily limited by fabrication related disorders and is orders of magnitude larger than the propagation length of plasmonic waveguides.[32] In short, it is expected that plasmonic waveguides will be used to address subwavelength optoelectronic devices, while conventional dielectric waveguides will be used to transfer the optical signal over distances longer than a few tens of microns. Couplers between plasmonic and dielectric waveguides will therefore be essential components for most applications involving the use of plasmonic waveguides such as optical interconnects.

As mentioned in the previous section, because of the predicted attractive properties of plasmonic two-conductor waveguides, people have started to explore such structures experimentally. In particular, Dionne *et al.*[19] have recently demonstrated waveguiding in a quasi-two-dimensional MIM geometry experimentally, showing clear evidence of a subwavelength guided mode with substantial propagation distances. With this as a background, it is important to explore the coupling of a dielectric slab waveguide into such a quasi-two-dimensional MIM geometry.

In this section we focus on 2D MIM plasmonic waveguides, and investigate the coupling of high-index contrast dielectric slab waveguides to such MIM waveguides.[33] We first show somewhat surprisingly that, despite the large difference between the modal sizes of dielectric and MIM plasmonic waveguides, a simple coupler created by placing the dielectric waveguide terminated flat at the exit end of the MIM waveguide without any tapering can be designed to have a transmission efficiency of ~70% at the optical communication wavelength. We then show that the transmission efficiency of the coupler can be increased to 93% by using a 400 nm long multisection taper. In both cases the transmission is found to be broadband. In addition, we investigate the properties of a Fabry-Perot structure in which light is coupled in and out of a MIM waveguide section sandwiched between dielectric waveguides.

In previous studies, the coupling between a dielectric slab waveguide consisting of a 1.25 μm silica core surrounded by air, and a 50 nm 2D gold-silica-gold MIM plasmonic waveguide has been theoretically investigated.[34] A coupling efficiency of ~70% was demonstrated by using a 2.5 μm long linearly tapered plasmon waveguide coupler. High coupling efficiency between silicon waveguides and metal slot waveguides using 4 μm long linear tapers has also been demonstrated experimentally.[20] In addition, efficient coupling between silicon-on-insulator dielectric waveguides and metallic stripe plasmonic waveguides has been demonstrated experimentally for a coupling length of 1.8 μm.[35] In contrast, our focus here is on coupling structures either without any transition region between the dielectric and plasmonic waveguides, or with submicron scale transition regions.

2.1. *Simulation method*

We use a 2D FDFD method[36,37] to investigate the properties of the couplers. For TM polarization we have $\mathbf{H} = H_z\hat{z}$ and

$$[\frac{\partial}{\partial x}(\frac{1}{\epsilon_r}\frac{\partial}{\partial x}) + \frac{\partial}{\partial y}(\frac{1}{\epsilon_r}\frac{\partial}{\partial y}) + k_0^2]H_z = j\omega\epsilon_0 M_z$$

where M_z is the magnetic source current. This equation is discretized on a nonuniform orthogonal grid[38] resulting in a sparse matrix problem of the form $\mathbf{Ax} = \mathbf{b}$, which is solved using direct sparse matrix methods. PML absorbing boundary conditions are used at all boundaries of the simulation domain.[24] Due to the rapid field variation at the metal-dielectric interfaces, a very fine grid resolution of ~1 nm is required at the metal-dielectric interfaces to adequately resolve the local fields. On the other hand, a grid resolution of $\sim \lambda/20$ is sufficient in the dielectric waveguide regions of the simulation domain. For example the required grid size in air at λ_0 =1.55 μm is ~77.5 nm which is almost two orders of magnitude larger than the required grid size at metal-dielectric interfaces.

In all cases the dielectric and plasmonic waveguides are aligned with their axes coinciding. To calculate the transmission and reflection coefficients of couplers between dielectric and MIM plasmonic waveguides, we excite the fundamental mode at the input waveguide and measure the power flux at both the input and output waveguides. These fluxes are then normalized with respect to the incident power flux in the input waveguide. We excite the fundamental TM mode of the dielectric waveguide by a line source with the appropriate mode profile calculated by solving the modal dispersion relation. The fundamental G-SPP mode of the MIM plasmonic waveguide can be excited by a single dipole point source, since all higher order modes are well below cutoff for deep subwavelength MIM waveguides.

We define the *transmission efficiency* of the coupler T_{ij} as the ratio of the transmitted power into the fundamental mode of the output waveguide j, and of the incident power of the fundamental mode of the input waveguide i. When the MIM waveguide is the output waveguide, the transmitted power into the MIM waveguide is measured right after the interface of the MIM waveguide with the coupler transition region or with the dielectric waveguide. Similarly, when the MIM waveguide is the input waveguide, the incident power is measured right before the interface of the MIM waveguide with the transition region or with the dielectric waveguide. This definition therefore provides an intrinsic measure of the property of the coupling region only. The presence of material loss, which is included in our simulations, obviously affects this coupling efficiency and is discussed in the next subsection. In addition, we note that both the dielectric and plasmonic waveguides considered here support a single guided mode. We always ensure that the output power in the dielectric waveguide side is measured far enough from the interface of the plasmonic and dielectric waveguides, so that the contribution of the radiation modes in the measured power flux is negligible.

We also note that due to reciprocity[10] the transmission efficiency from the dielectric to the plasmonic waveguide T_{dp} is equal to the transmission efficiency from the plasmonic to the dielectric waveguide T_{pd}. However the corresponding reflection coefficients R_{dp} and R_{pd} are not equal, since the power lost to radiation modes in the vacuum depends on whether the input waveguide is the dielectric or the plasmonic waveguide.

2.2. *Direct coupling*

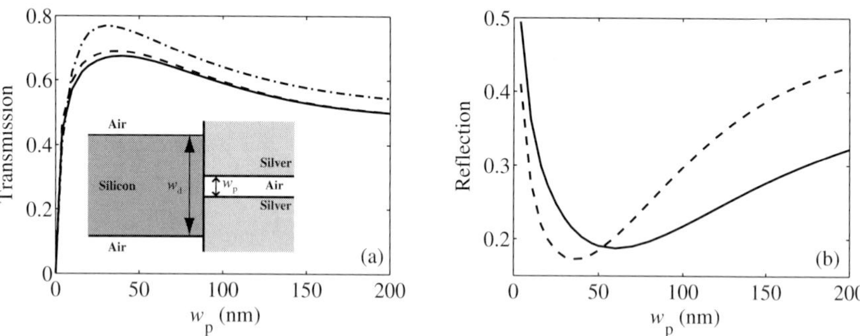

Fig. 9. (a) Power transmission efficiency (solid line) of a coupler between a dielectric and a MIM waveguide as a function of the width of the plasmonic waveguide w_p at λ_0 =1.55 μm calculated using FDFD. The coupler, created by placing the dielectric waveguide terminated flat at the exit end of the MIM waveguide, is shown in the inset. Results are shown for w_d = 300 nm. Also shown is the transmission efficiency, if the metal in the MIM waveguide is lossless (dashed line), or perfect electric conductor (dash-dotted line). (b) Coupler reflection coefficients R_{dp} (solid line), and R_{pd} (dashed line) as a function of the width of the plasmonic waveguide w_p. All other parameters are as in Fig. 9(a). Experimental data are used for the dielectric constant of the metal, including both the real and imaginary parts.

We first consider a coupler created by simply placing an air-silicon-air dielectric slab waveguide terminated flat at the exit end of a 2D silver-air-silver MIM plasmonic waveguide (inset of Fig. 9(a)). In Fig. 9(a) we show the coupler transmission $T_{dp}(= T_{pd})$ as a function of the width of the plasmonic waveguide w_p at λ_0 =1.55 μm. The width of the dielectric waveguide is w_d = 300 nm, which approximately corresponds to the optimal width of a silicon slab waveguide surrounded by air that achieves the minimum TM modal size. We observe that the transmission efficiency in this coupler is high and the maximum transmission of 68% is obtained for $w_p \simeq$40 nm. The transmission is also weakly dependent on w_p for w_p >20 nm. At the limit $w_p \rightarrow$0 the transmission goes to zero as expected. In Fig. 9(a) we also show the transmission for a coupler in which the metal in the MIM waveguide is lossless ($\epsilon_{\mathrm{met}} = \mathrm{Re}(\epsilon_{\mathrm{met}})$, neglecting the imaginary part of the dielectric permittivity $\mathrm{Im}(\epsilon_{\mathrm{met}})$) or perfect electric conductor (PEC) ($\epsilon_{\mathrm{met}} = -\infty$). In the former case we observe that the material losses in the metal do not affect significantly the transmission efficiency of the coupler. In the latter case (PEC), the transmission of the coupler is slightly higher, and its dependence on the MIM waveguide width is very similar to the plasmonic case. At λ_0 =1.55 μm the penetration of the electric field into the metal region is weak, and thus a PEC model provides a reasonable qualitative approximation. In Fig. 9(b) we show the coupler reflection coefficients R_{dp} and R_{pd} as a function of the width of the plasmonic waveguide w_p. The difference between R_{dp} and R_{pd} is due to the presence of radiation loss at the dielectric waveguide side.

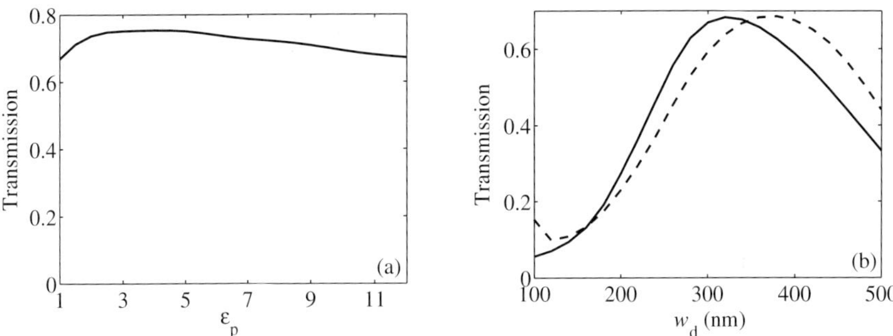

Fig. 10. (a) Transmission efficiency as a function of ϵ_p for w_p =50 nm. All other parameters are as in Fig. 9(a). Experimental data are used for the dielectric constant of the metal, including both the real and imaginary parts. (b) Transmission efficiency as a function of w_d for w_p =50 nm (solid line), w_p =100 nm (dashed line). All other parameters are as in Fig. 9(a). Experimental data are used for the dielectric constant of the metal, including both the real and imaginary parts.

In Fig. 10(a) we show the coupler transmission efficiency T_{dp} as a function of the dielectric permittivity ϵ_p of the gap region of the MIM waveguide. We observe that the transmission T_{dp} is weakly dependent on ϵ_p. In Fig. 10(b) we show the coupler transmission T_{dp} as a function of the width of the dielectric waveguide w_d for w_p =50 nm, and w_p =100 nm. In this case we observe that the transmission T_{dp} is quite sensitive to w_d. If w_d is decreased below ~300 nm, the modal energy spreads out of the dielectric region. As a result, the modal size of the dielectric waveguide increases rapidly, and the transmission T_{dp} decreases. Similarly for large w_d, the size of the fundamental mode of the dielectric waveguide becomes large and the transmission T_{dp} decreases. If w_p =50 nm (w_p =100 nm), the transmission is maximized for $w_d \simeq$320 nm ($w_d \simeq$380 nm).

In general we found that for a given width of the subwavelength MIM waveguide w_p, there is an optimum width of the dielectric waveguide w_d which maximizes the transmission efficiency and vice versa. We also found that for a given w_d the optimum w_p is significantly smaller than w_d. This is due to the fact that a subwavelength MIM waveguide collects light from an area significantly larger than its cross-sectional area.[39] More precisely, the transmission cross section of a MIM waveguide (in the unit of length in two dimensions), defined as the transmitted power into the waveguide normalized by the incident plane wave power flux, is significantly larger than its geometric cross-sectional area. As an example, we found that the transmission cross section of a MIM waveguide with w_p =50 nm is ~185 nm at λ_0 =1.55 μm. On the other hand, the transmission cross section of dielectric waveguides is approximately equal to their geometrical area. For example, for a waveguide consisting of a silicon slab surrounded by air with w_d =320 nm we found that the transmission cross section is ~340 nm at λ_0 =1.55 μm.

We also note that almost perfect transmittance has been previously reported for metallic grating structures, consisting of a periodic arrangement of subwavelength slits on a metal film.[40–42] It has also been shown that enhanced transmission is also present in single subwavelength slits,[43] and that the transmission can be further enhanced by corrugating the metal surface in the vicinity of the slit.[44] In all these previous studies the transmission was calculated for a plane wave excitation. In contrast, in our case the light funneling is provided by the dielectric waveguide.

In Color Plates 4(a), (b) we show the profiles of the magnetic and electric field respectively for a coupler with w_d =300 nm, w_p =50 nm (Fig. 9(a)). The power is incident from the left so that the incident and reflected waves result in an interference pattern in the dielectric waveguide. We note that the electric field intensity is significantly enhanced in the MIM waveguide with respect to the dielectric waveguide, while similar enhancement is not observed for the magnetic field intensity. This can be understood if we note that from Maxwell's equations we have $E_y = -\frac{1}{j\omega\epsilon}\frac{\partial H_z}{\partial x} = \frac{\gamma}{j\omega\epsilon}H_z$, so that for the Poynting vector we have $S = \frac{1}{2}\text{Re}(E_y H_z^*) \propto \frac{\beta}{\omega\epsilon}|H_z|^2 \propto \frac{\omega\epsilon}{\beta}|E_y|^2$, where $\gamma = \alpha + j\beta$ is the propagation constant of the mode. Thus, for the magnetic field enhancement we have $|\frac{H_{zp}}{H_{zd}}|^2 \sim \frac{w_d}{w_p}\frac{\beta_d/\epsilon_d}{\beta_p/\epsilon_p}$, while for the electric field enhancement we have $|\frac{E_{yp}}{E_{yd}}|^2 \sim \frac{w_d}{w_p}\frac{\epsilon_d/\beta_d}{\epsilon_p/\beta_p}$. The observed field enhancements (Color Plate 4) are consistent with these relations.

2.3. *Multisection taper*

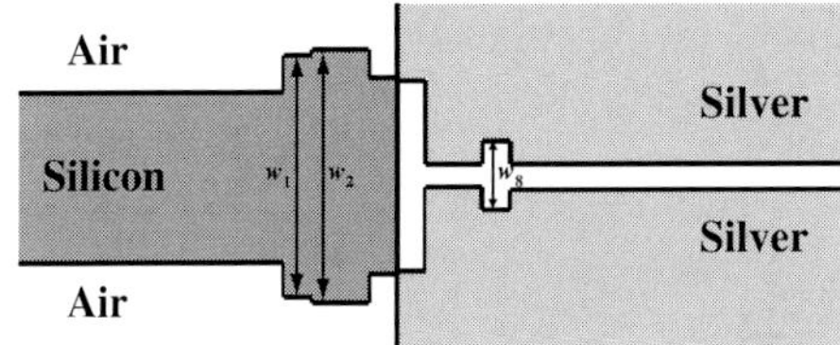

Fig. 11. Schematic of a coupler consisting of a multisection taper.

As mentioned above, a simple coupler created by placing a dielectric waveguide terminated flat at the exit end of a MIM waveguide (inset of Fig. 9(a)), can be designed to have transmission efficiency of 68%. To further increase the transmission, we design a coupler consisting of a multisection taper shown in Fig. 11. Such tapers, consisting of a number of waveguide sections, have been used as couplers between dielectric waveguides with highly different widths.[45–47] It has been shown theoretically and confirmed experimentally that they can be designed to have higher transmission efficiency than conventional tapers of the same length with linear or parabolic shapes.[46,47] The coupler design used here consists of a number of dielectric waveguide and MIM waveguide sections. The widths of these sections are optimized using a genetic global optimization algorithm in combination with FDFD. More specifically, we use a microgenetic algorithm which has been shown to reach the near-optimal region much faster than large population genetic algorithms.[48,49] Using this approach we designed a coupler with 93% transmission efficiency for w_d =300 nm, w_p =50 nm at λ_0 =1.55 μm. In this design we use 4 dielectric waveguide sections and 4 MIM waveguide sections. The lengths of all waveguide sections are

l_i =50 nm. Their widths $w_1, w_2, ..., w_8$ are optimized using the microgenetic algorithm, while the number of dielectric and MIM sections as well as their lengths are kept fixed during the optimization process. The designed coupler is extremely compact with a total length of 400 nm. The magnetic field profile for this optimized coupler design is shown in Color Plate 4(c).

Both the simple coupler of Fig. 9(a) and the multisection taper of Fig. 11 were optimized at a single wavelength of λ_0 =1.55 μm. In Fig. 12 we show the transmission efficiency of these couplers as a function of wavelength. We observe that in both cases the transmission efficiency is close to its maximum value in a broad range of wavelengths. This is due to the fact that in both cases the high transmission efficiency is not associated with any strong resonances. Similar broadband responses are observed in couplers between dielectric waveguides with highly different widths based on multisection tapers,[46,47] and in multisection impedance matching transformers used in microwave circuits.[10]

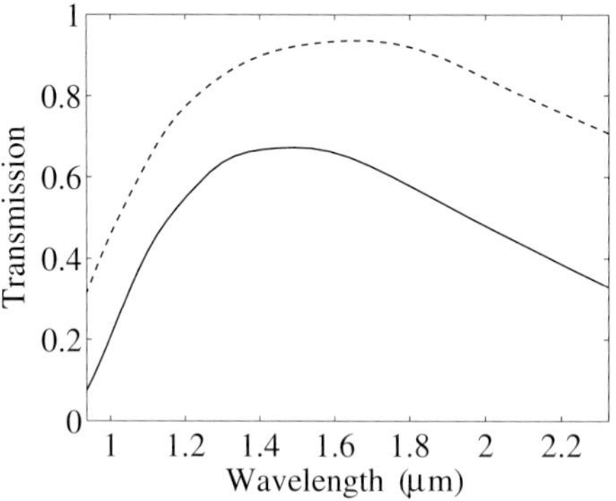

Fig. 12. Transmission efficiency as a function of wavelength for the couplers of Fig. 9(a) (solid line) and Fig. 11 (dashed line). In both cases the coupler parameters were optimized at a single wavelength of λ_0 =1.55 μm.

2.4. *Fabry-Perot structure*

In Fig. 13(a) we show a Fabry-Perot cavity structure consisting of a MIM waveguide of length l sandwiched between two dielectric waveguides. In this structure light is coupled from the wavelength-sized input dielectric waveguide to a subwavelength MIM waveguide, and then coupled back to the output dielectric waveguide. We excite the fundamental TM mode in the input dielectric waveguide and measure the power coupled into the fundamental mode of the output dielectric waveguide. The transmission efficiency of this structure as a function of the length l of the MIM waveguide section is shown in Fig. 13(b). We observe oscillations typical in the response of a Fabry-Perot structure. Such a device could be potentially used in nonlinear or sensing applications in which the large field enhancement achieved at metal-dielectric interfaces of plasmonic waveguides is desirable. Light could be transferred on chip with low loss dielectric waveguides and coupled efficiently in and out of the plasmonic waveguide-based device.

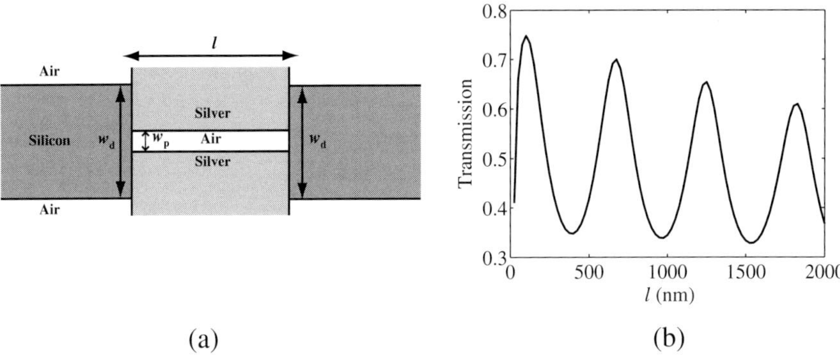

Fig. 13. (a) Schematic of a Fabry-Perot cavity structure consisting of a MIM waveguide sandwiched between two dielectric waveguides. (b) Transmission efficiency for the structure of Fig. 13(a) as a function of l at $\lambda_0 = 1.55$ μm. Results are shown for $w_d = 300$ nm, $w_p = 50$ nm.

References

1. W. L. Barnes, A. Dereux, and T. W. Ebbesen, *Nature.* **424**, 824–830, (2003).
2. J. Takahara, S. Yamagishi, H. Taki, A. Morimoto, and T. Kobayashi, *Opt. Lett.* **22**, 475–477, (1997).
3. J. C. Weeber, A. Dereux, C. Girard, J. R. Krenn, and J. P. Goudonnet, *Phys. Rev. B.* **60**, 9061–9068, (1999).
4. J. R. Krenn, B. Lamprecht, H. Ditlbacher, G. Schider, M. Salerno, A. Leitner, and F. R. Aussenegg, *Europhys. Lett.* **60**, 663–669, (2002).
5. M. L. Brongersma, J. W. Hartman, and H. A. Atwater, *Phys. Rev. B.* **62**, R16356–R16359, (2000).
6. S. A. Maier, P. G. Kik, H. A. Atwater, S. Meltzer, E. Harel, B. E. Koel, and A. A. G. Requicha, *Nat. Mater.* **2**, 229–232, (2003).
7. S. I. Bozhevolnyi, V. S. Volkov, E. Devaux, and T. W. Ebbesen, *Phys. Rev. Lett.* **95**, 046802, (2005).

8. S. I. Bozhevolnyi, V. S. Volkov, E. Devaux, J. Y. Laluet, and T. W. Ebbesen, *Nature.* **440**, 508–511, (2006).
9. E. Feigenbaum and M. Orenstein, *J. Lightwave Technol.* **25**, 2547–2562, (2007).
10. D. M. Pozar, *Microwave Engineering.* (Wiley, New York, 1998).
11. E. N. Economou, *Phys. Rev.* **182**, 539–554, (1969).
12. E. D. Palik, *Handbook of Optical Constants of Solids.* (Academic, New York, 1985).
13. R. Zia, M. D. Selker, P. B. Catrysse, and M. L. Brongersma, *J. Opt. Soc. Am. A.* **21**, 2442–2446, (2004).
14. K. Tanaka and M. Tanaka, *Appl. Phys. Lett.* **82**, 1158–1160, (2003).
15. F. Kusunoki, T. Yotsuya, J. Takahara, and T. Kobayashi, *Appl. Phys. Lett.* **86**, 211101, (2005).
16. G. Veronis and S. Fan, *Opt. Lett.* **30**, 3359–3361, (2005).
17. L. Liu, Z. Han, and S. He, *Opt. Express.* **13**, 6645–6650, (2005).
18. D. F. P. Pile, T. Ogawa, D. K. Gramotnev, Y. Matsuzaki, K. C. Vernon, K. Yamaguchi, T. Okamoto, M. Haraguchi, and M. Fukui, *Appl. Phys. Lett.* **87**, 261114, (2005).
19. J. A. Dionne, H. J. Lezec, and H. A. Atwater, *Nano Lett.* **6**, 1928–1932, (2006).
20. L. Chen, J. Shakya, and M. Lipson, *Opt. Lett.* **31**, 2133–2135, (2006).
21. G. Veronis and S. Fan, *J. Lightwave Technol.* **25**, 2511–2521, (2007).
22. N. N. Feng, M. L. Brongersma, and L. D. Negro, *IEEE J. Quantum Electron.* **43**, 479–485, (2007).
23. J. A. Pereda, A. Vegas, and A. Prieto, *Microwave Opt. Tech. Lett.* **38**, 331–335, (2003).
24. J. Jin, *The Finite Element Method in Electromagnetics.* (Wiley, New York, 2002).
25. S. J. Al-Bader, *IEEE J. Quantum Electron.* **40**, 325–329, (2004).
26. H. J. Hagemann, W. Gudat, and C. Kunz, *J. Opt. Soc. Am.* **65**, 742–744, (1975).
27. J. D. Jackson, *Classical Electrodynamics.* (John Wiley & Sons, New York, 1999).
28. D. F. P. Pile, T. Ogawa, D. K. Gramotnev, T. Okamoto, M. Haraguchi, M. Fukui, and S. Matsuo, *Appl. Phys. Lett.* **87**, 061106, (2005).
29. L. Vivien, S. Laval, E. Cassan, X. L. Roux, and D. Pascal, *J. Lightwave Technol.* **21**, 2429–2433, (2003).
30. S. L. Chuang, *Physics of Optoelectronic Devices.* (Wiley, New York, 1995).
31. E. Moreno, F. J. Garcia-Vidal, S. G. Rodrigo, L. Martin-Moreno, and S. I. Bozhevolnyi, *Opt. Lett.* **31**, 3447–3449, (2006).
32. M. Lipson, *J. Lightwave Technol.* **23**, 4222–4238, (2005).
33. G. Veronis and S. Fan, *Opt. Express.* **15**, 1211–1221, (2007).
34. P. Ginzburg, D. Arbel, and M. Orenstein, *Opt. Lett.* **31**, 3288–3290, (2006).
35. M. Hochberg, T. Baehr-Jones, C. Walker, and A. Scherer, *Opt. Express.* **12**, 5481–5486, (2004).
36. S. D. Wu and E. N. Glytsis, *J. Opt. Soc. Am. A.* **19**, 2018–2029, (2002).

37. G. Veronis, R. W. Dutton, and S. Fan, *Opt. Lett.* **29**, 2288–2290, (2004).
38. A. Taflove, *Computational Electrodynamics.* (Artech House, Boston, 1995).
39. H. Henke, H. Fruchting, and R. Winz, *Radio Sci.* **14**, 11–18, (1979).
40. J. A. Porto, F. J. Garcia-Vidal, and J. B. Pendry, *Phys. Rev. Lett.* **83**, 2845–2848, (1999).
41. P. Lalanne, J. P. Hugonin, S. Astilean, M. Palamaru, and K. D. Moller, *J. Opt. A.* **2**, 48–51, (2000).
42. S. Astilean, P. Lalanne, and M. Palamaru, *Opt. Commun.* **175**, 265–273, (2000).
43. Y. Takakura, *Phys. Rev. Lett.* **86**, 5601–5603, (2001).
44. F. J. Garcia-Vidal, H. J. Lezec, T. W. Ebbesen, and L. Martin-Moreno, *Phys. Rev. Lett.* **90**, 213901, (2003).
45. M. M. Spuhler, B. J. Offrein, G. L. Bona, R. Germann, I. Massarek, and D. Erni, *J. Lightwave Technol.* **16**, 1680–1685, (1998).
46. B. Luyssaert, P. Vandersteegen, D. Taillaert, P. Dumon, W. Bogaerts, P. Bienstman, D. V. Thourhout, V. Wiaux, S. Beckx, and R. Baets, *IEEE Photon. Technol. Lett.* **17**, 73–75, (2005).
47. B. Luyssaert, P. Bienstman, P. Vandersteegen, P. Dumon, and R. Baets, *J. Lightwave Technol.* **23**, 2462–2468, (2005).
48. K. Krishnakumar, *Proceedings of the SPIE.* **1196**, 289–296, (1989).
49. B. Wang, J. Jiang, and G. P. Nordin, *Opt. Express.* **12**, 3313–3326, (2004).

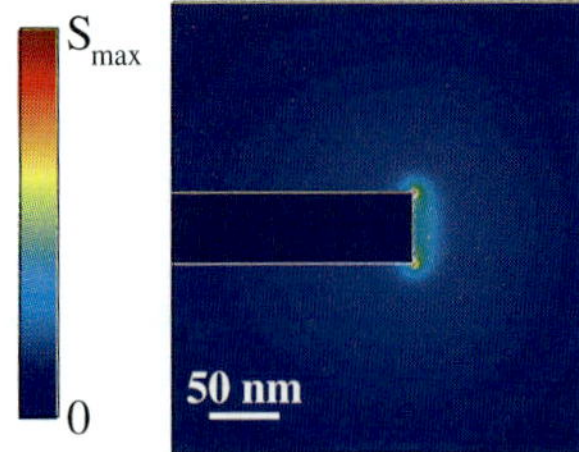

Color Plate 1. Power density profile of the edge mode supported by a truncated silver film embedded in silica for h =50 nm at λ_0 =1.55μm.

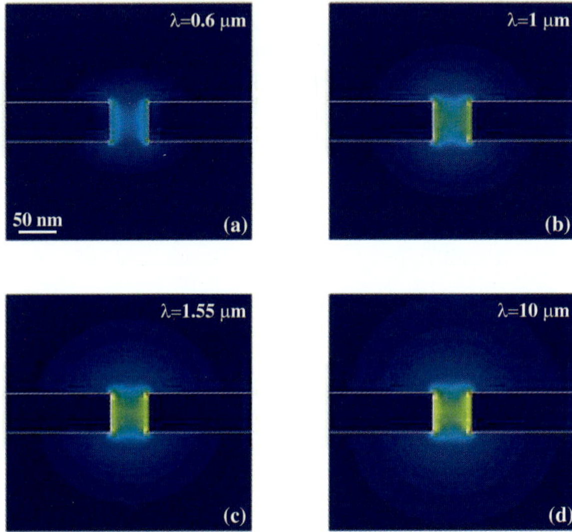

Color Plate 2. (a)-(d) Power density profile of the fundamental mode of the symmetric plasmonic slot waveguide (Fig. 4(b)) for w, h =50 nm at λ_0 = 0.6, 1, 1.55, 10 μm.

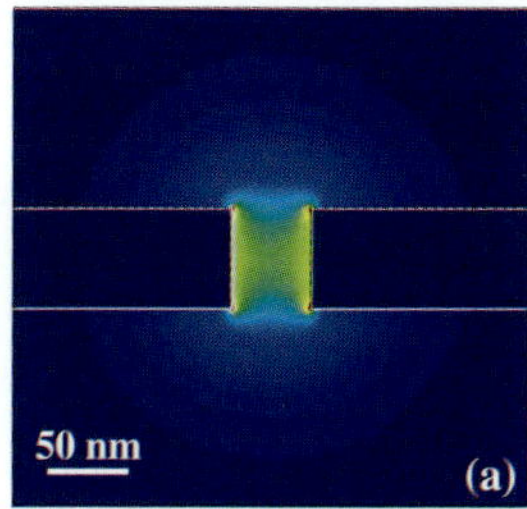

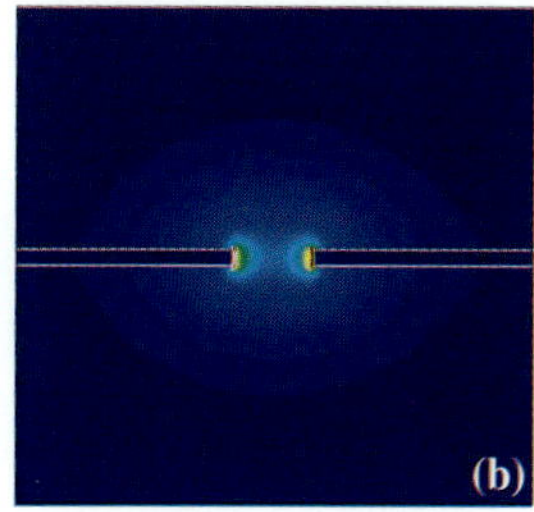

Color Plate 3. (a) Power density profile of the fundamental mode of the symmetric plasmonic slot waveguide (Fig. 4(b)) for w =50 nm, h =60 nm at λ_0 = 1.55 μm. (b) Same as in (a) except h =10 nm.

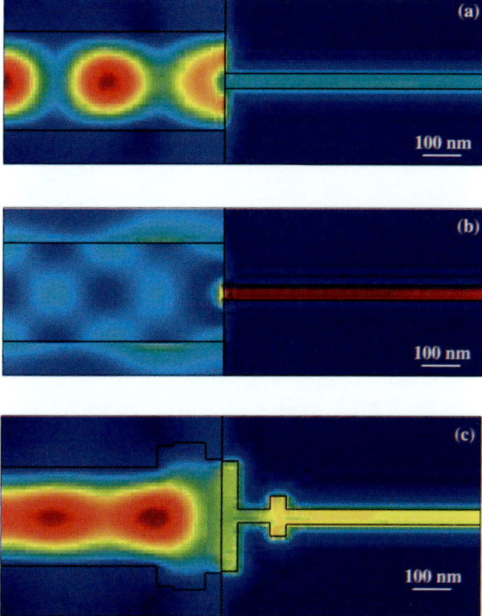

Color Plate 4. (a) Profile of the magnetic field amplitude $|H|$ for w_d =300 nm, w_p =50 nm (Fig. 9(a)). (b) Profile of the electric field amplitude $|E|$ for w_d =300 nm, w_p =50 nm (Fig. 9(a)). (c) Profile of the magnetic field amplitude $|H|$ of the optimized coupler design (Fig. 11) for w_d =300 nm, w_p =50 nm and 8 waveguide sections. The optimized widths of the dielectric waveguide sections are w_1 =420 nm, w_2 =440 nm, w_3 =440 nm, w_4 =340 nm, while the widths of the MIM waveguide sections are w_5 =330 nm, w_6 =40 nm, w_7 =40 nm, w_8 =120 nm.

Chapter 7

ALL-OPTICAL PLASMONIC MODULATORS AND INTERCONNECTS

Domenico Pacifici, Henri J. Lezec, Luke A. Sweatlock, Chris de Ruiter, Vivian Ferry, and Harry A. Atwater

Thomas J. Watson Laboratories of Applied Physics, California Institute of Technology, 1200 E California Blvd., Pasadena, CA 91125, USA

We present experimental results and design guidelines for optical modulators and interconnects based on metallo-dielectric plasmonic structures. The short wavelengths and high field confinements of surface plasmon polaritons at the metal/dielectric interfaces allow for the manipulation and guidance of optical signals in very compact and power-efficient devices, amenable to ultra-large scale integration of on-chip optical functionalities.

1. Introduction

Relative to electrons, photons are characterized by lower heat dissipation, better immunity to crosstalk, lower dispersion, and huge information capacity due to their large bandwidths at optical frequencies. Because of these attractive features, researchers have long pursued the use of photons as the new information carriers, for demanding computations and optical data processing,[1,2] and as a long-term solution to the interconnection bottleneck problem in microelectronics.[3-5] Essential building blocks for all-optical computation and networks, such as logic elements and modulators, have already been demonstrated.[6-9] However, despite the encouraging premises, all-optical integrated information processing has thus far remained elusive. Indeed, state-of-the-art optical modulators and interconnects typically rely upon weak, inefficient light-matter interactions, which call for large operating power densities[10] and long interaction distances (millimeters to centimeters),[11] thus limiting the opportunity for dense chip-based integration of all-optical functionalities. Moreover, due to diffraction, the wavelength of light typically sets a lower limit to further scaling down the feature size and increasing device density in conventional dielectric waveguide architectures.

Plasmonics, with its ability to confine and guide electromagnetic waves in subwavelength metallo-dielectric structures,[12,13] promises to be a valuable alternative to the implementation of compact, fast and power-efficient optical integrated networks.[14-16] In this chapter we investigate the use of subwavelength scatterers and corrugations in a metal as efficient, localized sources of surface plasmon polaritons (SPPs) propagating for several micrometers at the dielectric/metal interface. Several authors have reported, both theoretically and experimentally, that optical transmission through a subwavelength slit in metal films can be strongly influenced by the presence of nearby surface features, such as grooves or other slits.[17-24] We precisely measure and then modulate the propagation constants of the generated surface plasmon polaritons, using a set of surface wave interferometers consisting of a pair of parallel groove and slit, in which the distance between the slit and the groove is systematically varied up to several micrometers.[25]

Light directly incident on the slit interferes with the propagating SPP, which originates from the groove and reaches the slit location with a phase shift determined by a characteristic scattering phase lag[26,27] and a propagative phase depending on travelled distance. By varying the length of the interferometer arm, i.e. changing the distance between the slit and the groove, we determine the dispersion relation and propagation length of SPPs in a wide spectral range, i.e. from 400 to 1500 nm.

Active manipulation of the SPP is then achieved by coating the arm of the interferometer with an active medium, consisting, for instance, of a thin film of CdSe semiconductor quantum dots (QDs) fabricated by colloidal chemical synthesis in solution.[28] The QDs have been specifically designed to show an absorption bandgap at ~600 nm, therefore they can absorb high frequency surface plasmons (in the green-blue spectral range) while being transparent to surface plasmons at infrared wavelengths. Active modulation is achieved by means of two co-propagating SPPs launched by the groove. A control (pump) SPP having a free-space wavelength of 514.5 nm is absorbed in photogeneration of electron-hole pairs inside the QD volume while propagating along the metal-dielectric interface. A signal (probe) SPP at 1.5 μm copropagating with the control SPP is then absorbed by promoting an intraband transition in the quantum dots, that is, by re-exciting the electron to a higher energy level. As soon as the pump SPP is turned off, the probe SPP is no longer absorbed, as the QDs quickly (in less than 40 ns) return to their transparent ground state. The modulated amplitude of the probe SPP determines a quick change in the interference conditions at the slit, thus causing a change in the interferometer output state ("on" and "off"). Thanks to the highly confined nature of the SPP field and the large quantum-dot absorption cross-section, the fast all-optical plasmonic modulation can be achieved at very-low power densities, in micrometer-scale planar devices.[25]

We further design and fabricate a three-dimensional prototype of plasmonic modulator, employing a Mach-Zehnder interferometer in a multilevel metal-dielectric-metal stack, where the dielectric is made active by the presence of QDs. By opportunely varying the arms of the interferometers, higher modulation depths are achieved. Moreover, by exploiting the polarization properties of subwavelength slits in a metal film, we demonstrate optical isolation between the control and signal SPPs using cross-polarized incident beams incoupled into SPPs by perpendicularly-oriented slits. Finally, we present a design study for three-dimensional plasmonic vias and modulators, and investigate several schemes for coupling light into such devices, evaluating their power efficiency.

2. Surface plasmon interferometers

2.1. Fabrication and experimental methods

We fabricated several on-chip planar interferometers consisting of a single subwavelength aperture flanked by a subwavelength groove in a metal film. Identical subwavelength structures were fabricated by focused ion beam (FIB) milling into a 400 nm thick layer of Ag evaporated onto flat fused silica microscope slides. A low beam current (50 pA) was used to achieve surface features defined with a lateral precision of the order of 10 nm and characterized by near-vertical sidewalls and a minimal amount of edge rounding. The milled groove and slit were 10 μm long, 200 and 100 nm wide, and 100 nm and 400 nm deep, respectively. The slit-groove separation distance was systematically varied in the range 500-10400 nm, in step of 50 nm, with a precision of 1%. A series of single apertures is also milled for reference purposes. The same set of structures was milled in another silver film (co-evaporated with the previous sample) which was previously spin-coated with a thin film of densely-packed CdSe QDs obtained by colloidal chemical synthesis in solution.[28] The QDs were capped with trioctylphosphine oxide (TOPO) ligands. Multiple spin-coatings of the toluene solution containing CdSe QDs were performed at 1000 rpm for 1 minute, followed by a thermal treatment at 150°C for 5 minutes. A 24 nm-thick film of densely-packed CdSe QDs was formed, with a good uniformity over the entire silver surface. The QDs showed bright photoluminescence (PL) peaked at 625 nm.

Transmission measurements were performed on both QD-coated and uncoated interferometers by focusing a laser beam onto the patterned side (Ag-air side) of each sample. The film thickness is such that the devices are optically opaque at the groove position, and light is only transmitted through the slit aperture. This approach allows us to achieve high signal-to-noise ratios in our experiments. The light intensity transmitted through the slit of each interferometer was gathered by a 50X microscope objective, sent to a single grating monochromator and detected with a liquid-nitrogen-cooled CCD array detector (spectral range: 400–900 nm), with an ultra-sensitive liquid-nitrogen-cooled germanium detector (spectral range: 900–1700 nm), or with a fast photomultiplier tube with an extended near-infrared spectral response (range 300–1700 nm).

Colour plate 1(a) at the end of the chapter displays a schematic of a typical interferometer consisting of a single slit of subwavelength width etched completely through a metallic film on a transparent substrate flanked by a single parallel groove, etched partially through the metal.

The center-to-center separation distance between slit and groove is defined as *D*. The blue spheres represent CdSe QDs. Two beams (signal and control) uniformly illuminate the structure. Diffractive scattering by the groove excites SPPs at two different frequencies, which we call the control and signal SPP, respectively. Colour plate 1(b) reports an SEM micrograph of an uncoated interferometer, with visible groove and slit.

2.2. Theoretical framework

Let us consider a TM-polarized (*H* field parallel to the longer slit side) electromagnetic plane wave with free space wavelength λ (and free space wavevector $k = 2\pi/\lambda$) impinging on the entire surface at an angle θ with respect to the sample normal. Diffractive scattering by the groove couples a fraction β of the incident field amplitude *H* into a surface wave with amplitude H_s that reaches the slit with an overall phase shift:

$$\varphi = kD \sin\theta + k_s D + \varphi_g + i\,\alpha D/2, \qquad (1)$$

where the first term is the phase shift due to the free-space optical path length difference between the incident wavefronts reaching the aperture and the groove, the second term is the phase shift gained by the surface wave traveling with surface wavevector $k_s = n_s k = 2\pi n_s/\lambda$ (where n_s is the real part of the effective refractive index of the surface wave), and the third term is the phase shift introduced by scattering at the groove. The fourth imaginary term in Eq. (1) accounts for absorption or scattering loss experienced by the surface wave propagating along the surface (where $\alpha = 4\pi\kappa/\lambda$ is the overall absorption coefficient, κ being the imaginary part of the effective refractive index of the surface wave). At the slit entrance, both the directly incident light and the SPP arriving from the groove are coupled into guided slit modes through diffractive scattering, with respective amplitude fractions γ and δ. The respective field amplitudes emerging from the slit are then given by $H_{T,s} = t\gamma H$ and $H_{T,g} = t\delta\beta\, e^{i\varphi}$, respectively, where t is the complex transmission coefficient of the slit. The two contributions can interfere if spatial and temporal coherence is maintained between them, in general when the coherence length of the optical beam is greater than the slit-groove distance. The total transmitted amplitude through the aperture is then given by the following sum: $H_T = H_{T,s} + H_{T,g} = t(\gamma + \delta\beta\, e^{i\varphi})H$. Experimentally we observe the transmitted intensity:

$$I_T = |H_T|^2 = I_0 \left|1 + \eta e^{i\varphi}\right|^2, \tag{2}$$

where $I_0 = |t\gamma H|^2$ is the light intensity transmitted through an isolated slit (i.e. not flanked by a groove), and $\eta = \delta\beta/\gamma$ regroups the scattering efficiencies of the surface features. Eq. (2) can be written as:

$$T(P) \equiv \frac{I_T}{I_0} = 1 + \eta^2 e^{-\alpha(P)D} + 2\eta e^{-\frac{\alpha(P)D}{2}} \cos\left[kD(\sin\theta + n_s) + \varphi_g\right] \tag{3}$$

where $\alpha(P)$ is the absorption coefficient which we intend to modify with an external control beam of power P. Eq. (3) shows that maxima and minima in transmitted light intensity as a function of slit-groove distance result from constructive or destructive interference between the surface wave launched by the groove and the incident wave at the slit position.

There are several terms in Eq. (3) suggesting that such a slit-groove surface-wave interferometer could constitute an efficient all-optical modulator. Indeed, under the particular condition $kD \gg 1$, a small change in the real part of the refractive index n_s could produce a surface-wave phase shift of sufficiently high magnitude (relative to the fixed phase of the light directly illuminating the slit) to change a transmission maximum into a minimum or vice-versa, thus enabling the possibility of phase-modulating a probe signal. In another approach, for sufficiently long slit-groove distances D, a small change in the absorption coefficient α could lead to substantial attenuation of the surface wave during its trip from the groove to the slit. This would suppress interference with the light directly illuminating the slit, thus producing amplitude-modulation of the transmitted light. It is this second approach which we explore in the following sections.

2.3. *Uncoated plasmonic interferometers*

First we characterize the optical properties of uncoated interferometers. The top panel of Fig. 1 shows a schematic of the experiment; a single interferometric device is uniformly illuminated by an optical Gaussian beam. Diffractive scattering by the groove excites an SPP which propagates along the metal/air interface and interferes with the incident

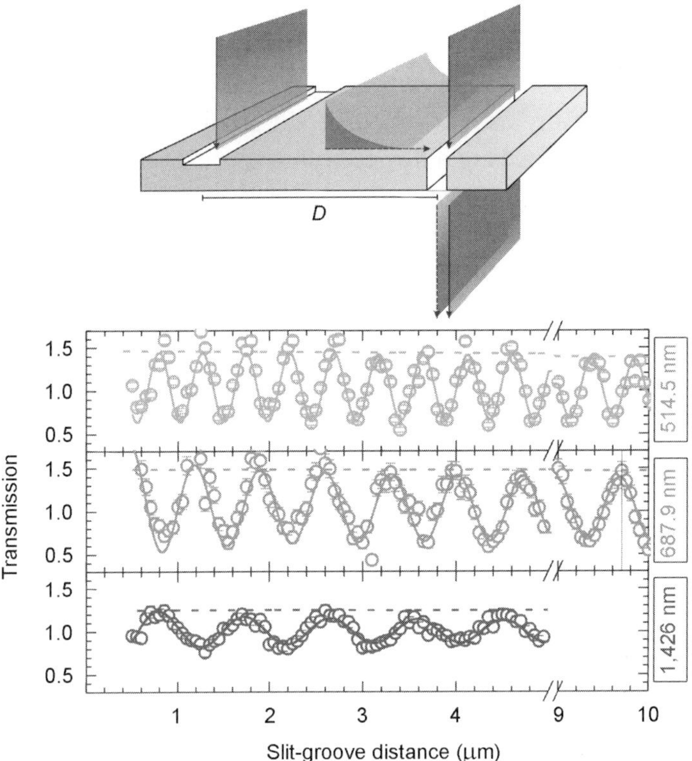

Fig. 1. Top panel: schematic of an uncoated interferometer consisting of a slit and a groove, uniformly illuminated by an optical beam normally incident on the metal surface. Bottom panels: measurements of the light intensity transmitted through the slit of a series of uncoated interferometers with different arm length (slit-groove distance), at various wavelengths.

beam at the slit position. Light transmitted through the slit carries information of the interference process between the reference beam (directly incident on the slit) and the propagating surface wave. The bottom panels of Fig. 1 report measurements of the light intensity transmitted through a set of interferometers with different arm lengths, i.e. slit-groove distance D, at various laser wavelengths (λ=514.5 nm, 687.9 nm, and 1426 nm, respectively). The angle of incidence is 0° in the first two cases and 32° in the third. The total power density incident on the sample surface is fixed at ~0.5 W/cm^2 in all three cases. The intensity of light transmitted by an isolated reference slit is used to normalize the transmission data through each device.

Compiling the normalized data yields a plot of normalized transmitted intensity as a function of slit-groove distance (where each open circle corresponds to a device with a specific slit-groove distance). Several features are apparent in Fig. 1: (1) at a given incident free-space wavelength λ, the normalized transmission through the subwavelength slit shows oscillatory behavior as a function of the slit-groove distance, as a result of constructive or destructive interference between the incident beam and the surface wave launched by the neighboring groove, (2) the spatial period of this oscillation increases with increasing λ, and (3) the envelope of the oscillation amplitude remains approximately constant as a function of interferometer arm length, suggesting that within the investigated spectral range, surface waves can propagate for several micrometers without significant loss.

The observed enhancement and suppression of the intensity transmitted through the slit can be explained within the theoretical framework developed in section 2.2. Using Eq. 3 with $k = 2\pi/\lambda$, D, and θ as known parameters, fits are performed to the data reported in Fig. 1, and shown as solid lines. These fits yield important physical parameters such as the surface wave launching efficiency η, which is about 20-25% in the investigated spectral range, the phase lag caused by diffraction at the slit φ_g, which depends on groove depth and it is approximately $\pi/2$, the absorption coefficient α and the effective refractive index n_s. In Fig. 2 we plot the derived value of the effective surface wave index n_s as a function of free-space wavelength λ (solid symbols) and compare it (continuous line) to the effective refractive index expected for an SPP at an Ag/air interface:[13]

$$n_{SPP} = \mathrm{Re}\left[\left(\frac{\varepsilon_1\varepsilon_2}{\varepsilon_1+\varepsilon_2}\right)^{1/2}\right], \quad (4)$$

where ε_1 is the complex dielectric constant of Ag, experimentally determined by ellipsometry, and $\varepsilon_2 = 1$ is the dielectric constant of air. The excellent agreement between the experimental data reported in Fig. 2 as solid symbols and Eq. (4), within the experimental errors, suggests

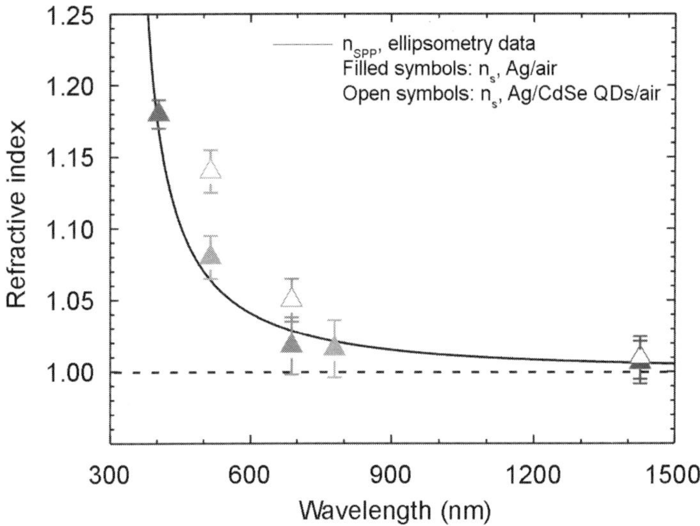

Fig. 2. Refractive index of the propagating surface wave modes as extracted by transmission measurements through uncoated and QD-coated interferometers.

that the long-range propagating surface wave is indeed an SPP. This demonstrates that diffractive scattering from a single subwavelength groove in a metal film can be an efficient source of guided surface plasmons at a metal-dielectric interface, indeed more compact than bulky prism- or grating-coupling approaches.

2.4. QD-coated plasmonic interferometers

In order to actively modify the propagation constants of the SPP launched by the groove, we coated the metal with a thin layer of semiconductor quantum dots, as explained in the previous experimental section.

As shown in Fig. 3, when excited by a laser beam at 404 nm the CdSe QDs show bright PL peaked at 625 nm with a FWHM of 32 nm, and red-shifted with respect to the 1S absorption edge which occurs at 595 nm, as evidenced by the absorption curve reported in Fig. 3. This emissive state is generally attributed to splitting of the lowest hole state as a result of crystal field and e-h exchange interaction,[29,30] as shown in the right inset

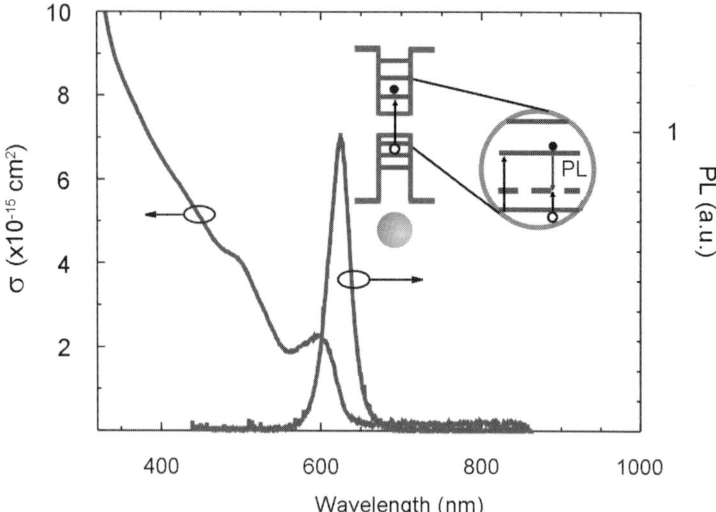

Fig. 3. Absorption cross section (σ) and photoluminescence spectrum (PL) for CdSe QDs. Insets: energy level scheme for a QD.

to Fig. 3. Both the PL and the absorption of the CdSe QD ensemble show broad features, indicating both homogeneous and inhomogeneous level broadening caused by the crystal-field induced removal of level degeneracy and by the size distribution of nanocrystals, respectively. It is interesting to note that the CdSe QDs have relatively high absorption cross section in the visible.[31] For example the measured absorption cross section at 514.5 nm is 3.5×10^{-15} cm^2. This value is 6 orders of magnitude higher than the typical absorption cross sections for atomic-like transitions in rare-earth ions, which are extensively used in telecommunication and as promising active materials in microphotonics.[32] To understand the role of the thin layer of CdSe QDs, we performed transmission experiments on similar slit-groove interferometers coated with QDs, and with varying arm distances. A schematic of the transmission experiment is reported in the top panel of Fig. 4. The normalized light intensities transmitted through the slit of various interferometers are reported in the bottom panels of Fig. 4, as a function of slit-groove separation distance, and at various free-space wavelengths λ. The first interesting observation is that the transmission maxima and minima occur at shorter slit-groove distances. This effect is

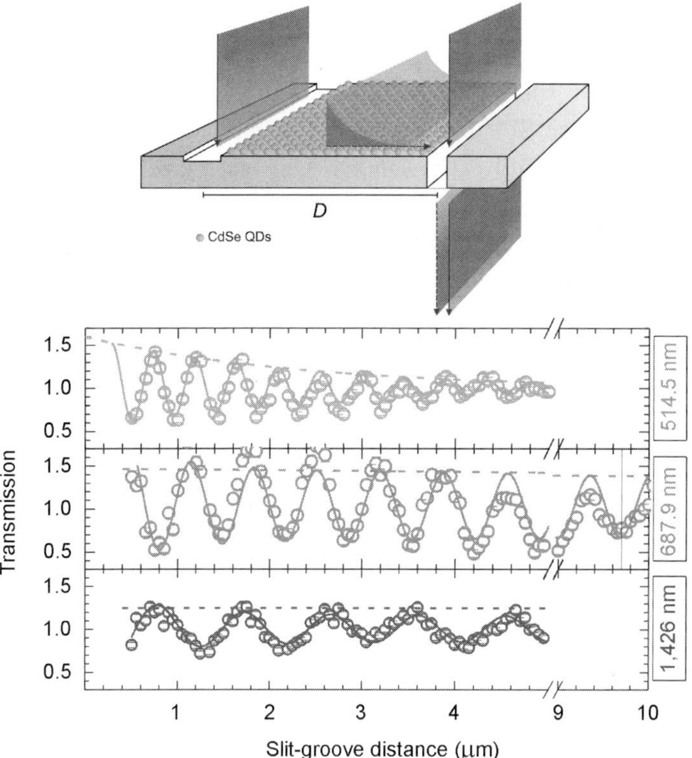

Fig. 4. Top panel: schematic of a QD-coated interferometer uniformly illuminated by an optical beam at normal incidence. Bottom panels: measurements of the light intensity transmitted through the slit of a series of QD-coated interferometers with different arm length (slit-groove distance), at various wavelengths.

particularly evident in interferometers with long arms. For example, the normalized light transmission through an interferometer with D=9.7 μm (indicated by vertical lines in Fig. 1 and Fig. 4, λ=687.9 nm) goes from a maximum (constructive interference) without QD-coating (Fig. 1) to a minimum (destructive interference) in the presence of QDs (Fig. 4). Such a remarkable change in the interference conditions can be achieved by a slight variation in the effective refractive index. Indeed, using Eq. (3) it is possible to fit the data in Fig. 4, and estimate the effective refractive index experienced by the surface wave n_s while propagating at the QD-coated metal interface. The fit values are reported in Fig. 2 as open symbols. The SPP refractive index systematically increases relative to the case of the uncoated interferometers, due to the presence of the high-refractive index film of CdSe QDs. The increase is consistent with calculations of the SPP mode profile and modal overlap with the CdSe QD layer, whose refractive index was measured by spectroscopic ellipsometry as a function of wavelength. We note that at λ=1426 nm the value of n_s is only marginally higher in the case of the coated sample compared to that of the uncoated sample. At λ = 514.5 nm, n_s is 6% higher in the QD-coated interferometer, consistent with an increased mode overlap with the QD layer due to the smaller skin depth of the SPP at shorter wavelengths.

Another striking result arises from a close comparison between the transmission data at λ=514.5 nm in Fig. 4 and Fig. 1, that is the envelope of the oscillating normalized transmission strongly decreases with slit-groove distance in the presence of QDs (Fig. 4, λ=514.5 nm), while it is approximately constant for the uncoated sample (Fig. 1, λ=514.5 nm). The extracted absorption coefficient α increases from 400 cm^{-1} without the QD layer to 8200 cm^{-1} with the QD layer. At first one may be tempted to explain the increased absorption by out-of-plane scattering of the SPP by the QDs. This hypothesis is easily ruled out by looking at the transmission data at longer wavelengths. Since Rayleigh scattering depends on the fourth power of the incident wavelength, one would expect stronger absorption at longer wavelengths. However, the oscillating transmission data at λ=687.9 nm and λ=1426 nm in Fig. 4 show a constant envelope amplitude as a function of slit-groove distance, similar to the uncoated case, suggesting that no substantial absorption is indeed occurring.

Instead, the increased absorption can be explained in terms of plasmonic excitation of electron-hole pairs in the QDs. For a nanocrystal radius of 2.4 nm, and a close-packed distribution of QDs, the estimated QD concentration in the film is $\varrho \sim 1 \times 10^{19}$ cm^{-3}. We estimate an SPP mode overlap with the 24-nm thick CdSe film $\Gamma \sim 25\%$. Therefore, the calculated effective absorption coefficient at λ=514.5 nm would be $\alpha = \Gamma\sigma\varrho =$ 8750 cm^{-1}. This value is in very good agreement with the value (8200 cm^{-1}) extracted from the fitting procedure, and strongly support the hypothesis that the QDs are indeed absorbing the propagating

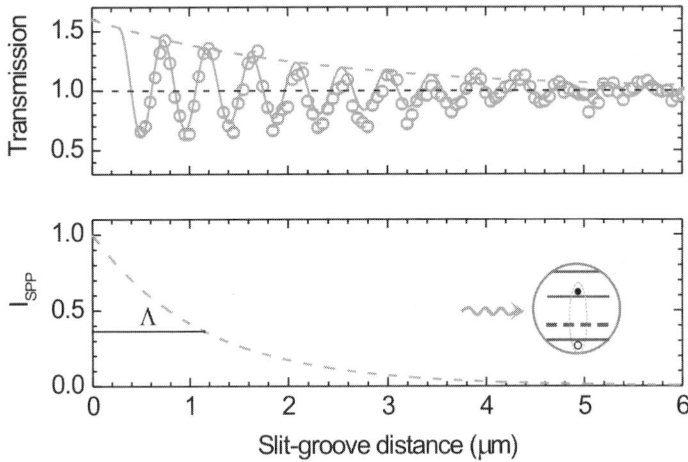

Fig. 5. Interferometry at λ=514.5 nm for QD-coated devices. Top panel: transmission measurements for a series of QD-coated interferometers with different arm length (slit-groove distance). Bottom panel: SPP intensity at the slit position as a function of slit-groove distance.

SPP at λ=514.5 nm. To stress this point, in Fig. 5 we report the transmission data at λ=514.5 nm for the QD-coated sample (top panel) together with the extrapolated values of the SPP intensity measured at the slit position (bottom panel), as a function of slit-groove distance. The SPP intensity decreases exponentially with an experimental decay constant (propagation length) Λ = 1.2 μm, again in good agreement with the value estimated by assuming bandgap absorption in the QDs (which gives α^{-1} = 1.14μm). The emerging picture is therefore the following: (1) SPPs with energy above the QD bandgap can be absorbed by CdSe QDs while propagating along the metal/dielectric interface, with subsequent photogeneration of bound electron-hole pairs (excitons) in the QD volume, as shown in the schematic of Fig. 5; (2) for sufficiently long interferometers, the SPP generated at the groove location can be completely absorbed by the QDs, and is no longer able to reach the slit; therefore, neither constructive nor destructive interference between the SPP and the incident beam at the slit position can occur, and the normalized transmission intensity tends to 1 (Fig. 4, λ=514.5 nm); (3) SPPs having energy smaller than the QD bandgap cannot be absorbed by the nanocrystals, thus they can propagate without attenuation, reach the slit and interfere with the incident beam, causing either constructive or destructive interference at the slit output mouth.

2.5. All-optical modulation in QD-coated plasmonic interferometers

So far we have studied the optical properties of plasmonic interferometers illuminated by a single light beam at a time. In all-optical modulators, two different light beams are generally employed: one of the two beams (control) is used to copy information onto the other (signal) at a different wavelength. In order to achieve modulation, the two beams, which are otherwise independent, need to be coupled by using some nonlinearity of the medium in which they propagate. Our intent is to use a control beam in the form of a propagating SPP to change the absorption coefficient α for another co-propagating signal SPP at a different frequency. The signal SPP interferes with the incident field at the slit, and the total light intensity transmitted through the slit is given by Eq. (3).

The exponential function of α provides for the necessary non-linear interaction between the two co-propagating SPPs. We performed continuous-wave pump (λ=514.5 nm, θ=0°) and probe (λ=1426 nm, θ=32°) transmission measurements on various quantum-dot coated interferometers. As illustrated in the schematic diagram of Colour plate 1(a), the control and signal beams simultaneously and uniformly impinge on the sample surface and are scattered by the subwavelength groove which launches two co-propagating SPPs with different energies (represented in green and red, respectively) towards the slit. The SPP at λ=514.5 nm is strongly absorbed by the CdSe QDs, while the SPP at λ=1426 nm propagates without absorption in the absence of the pump SPP. First we consider a structure with a slit-groove distance D=3.6 μm which displays a relative maximum in normalized transmission intensity at λ=1426 nm in absence of pump illumination (Fig. 4, λ=1426 nm). The top panel of Colour plate 2 displays (open symbols) the normalized transmission intensity of this device at λ=1426 nm as a function of the pump power at λ=514.5 nm. Both probe and pump beams are chosen to have TM polarization (H-field parallel to the slit and groove long axis).

The normalized transmitted probe intensity decreases as a function of pump power, reaching a minimum at a power density of ~600 W/cm^2, and then slightly increases as the pump power is increased further. Since the transmission enhancement for that particular device at zero pump power results from constructive interference between the SPP and the incident beam at the slit position, a decreased transmission can only be explained by attenuation of the surface wave due to an absorption process that is activated by the pump beam.

We propose that it is confined carrier absorption from excited CdSe QDs which attenuates the probe surface wave at λ=1426nm. In this model (schematic shown in the left inset of Colour plate 2) a confined exciton is first generated in a CdSe QD by photo-absorption of a control SPP at λ=514.5nm. The signal SPP at λ=1426 nm then re-excites the exciton by bringing the electron into a higher lying energy level of the QD, and is therefore absorbed. By increasing the pump power, a larger fraction of QDs is excited, and as a consequence more probe SPPs are absorbed. This phenomenon is analog to the free carrier absorption (FCA) of photons propagating in bulk semiconductors where free carriers are generated by thermal ionization of the dopants or by optical excitation of electron-hole pairs. The dynamics of this mechanism have been extensively studied in thick films of CdSe QDs by femtosecond transient absorption spectroscopy in the visible and near-infrared spectral ranges.[33,34]

If we denote with σ_{CCA} the cross section for the confined carrier absorption of a propagating signal SPP by an excited QD, the photoinduced absorption coefficient is then equal to:

$$\alpha = \Gamma_S \sigma_{CCA} \rho^*. \tag{5}$$

with Γ_S the mode overlap of the signal SPP with the active layer, and ρ^* the density of excited quantum dots in steady state. Approximating the QD with a two-level system, ρ^* is given by the following equation:

$$\rho^* = \frac{\sigma\phi\tau}{\sigma\phi\tau+1}\rho, \tag{6}$$

where ρ is the total concentration of QDs in the film, ϕ is the fraction of control SPP flux parallel to the metal/dielectric interface overlapping with the active film (in cm^{-2} s^{-1}), σ the QD absorption cross section at the pump wavelength, and τ the recombination lifetime of an excited QD. In the low pump power regime, such that $\sigma\phi\tau \ll 1$, Eq. (6) becomes:

$$\rho^* = \sigma\phi\tau\,\rho, \tag{7}$$

where ϕ can be calculated using the Poynting theorem and integrating the magnitude of the Poynting vector over the active region. Without loss of generality, ϕ is a linear function of the incident pump power density P:

$$\phi = \eta_C \frac{P}{\epsilon}, \tag{8}$$

with η_C an effective coefficient taking into account the incoupling efficiency, scattering cross section of the groove, and mode overlap of the control SPP with the active layer, and $\epsilon = h\nu$ the energy of an SPP of frequency $\nu = c/\lambda$, with h and c the Planck's constant and speed of light, respectively.

By combining the previous equations, Eq. (5) becomes:

$$\alpha(P) = \eta_c \Gamma_S \sigma_{CCA} \sigma\tau\,\rho \frac{\lambda}{hc} P, \tag{9}$$

which proves that $\alpha(P)$ is a linear function of the incident pump power density P.

To demonstrate that the observed all-optical modulation of light intensity at λ=1426 nm transmitted through the slit results solely from excitation of the CdSe QDs by the control SPP at λ=514.5 nm and not from direct illumination of the QDs by the incident pump beam, we changed the polarization of the pump beam at λ=514.5 nm from TM to TE (electric field parallel to the slit and groove long axis), a polarization for which no surface wave can be launched by the groove or the slit. When the two-colour pump-probe experiment is repeated under these conditions as a function of pump power (Colour plate 2, solid symbols), no significant modulation of transmitted intensity is observed as a function of pump power.

The lack of transmission modulation in absence of the surface wave shows that direct illumination of the surface is a negligible source of QD excitation over the explored range of pump powers. This is not surprising since the sum of the incident and reflected fields has, at normal incidence, a node at the metal-air interface, leading to a negligible contribution to optical power density in the CdSe-QD layer. On the other hand, the field amplitude of the control SPP launched by the groove has a maximum right at the interface, guaranteeing maximum mode overlap with the active QD layer. We conclude that the large amplitude transmission modulation observed in the case of a TM-polarized pump beam is the direct result of excitation of the CdSe QDs by a surface wave at that frequency launched by the groove and the slit.

For an interferometer with aperture-groove distance D=2.15 μm a transmission minimum occurs at λ=1426 nm, as shown in Fig. 4. The probe transmission variation with pump power through this particular device is given in the bottom panel of Colour plate 2. Unlike the variation observed for a distance corresponding to a transmission maximum, the intensity increases as a function of pump power density and then starts to decrease again with increasing pump power. In this case, photoinduced absorption of the probe SPP reduces its destructive interference with the incident beam at the slit, and thus increases the overall normalized transmission, making it closer to that of an isolated slit, i.e. equal to 1.

Figure 6 illustrates the temporal response of the transmitted signal to the pump excitation, for a 3.6-μm-long interferometer. At t =0 μs the pump beam, modulated at a frequency of 250 Hz, is turned on at a power density of 500 W/cm^2, and the probe transmission intensity shows an abrupt decrease. The switching time of the device is below the temporal resolution of our experimental set up (~40 ns). In order to better establish the timescale for the modulation effect, the frequency response of the probe signal to pump modulation was measured (Fig. 6, inset) from 1 to 1,000 Hz using a mechanical chopper, a Ge detector and a lock-in amplifier (solid triangles), and from 250 Hz to 50 MHz using an acousto–optic modulator, a photomultiplier-tube (PMT) detector extended in the near infrared and a multichannel scaler (solid circles). The change in probe signal (modulation depth) as a function of pump modulation frequency is constant up to 25 MHz, which again establishes a timescale of ~40 ns for the modulation effect (the roll-off at the end of the explored frequency range corresponding to the response time of our measurement system). The inferred response time of the device at ~40 ns is consistent with our suggested mechanism for device operation, consisting of probe SPP absorption by a fast intraband transition in the

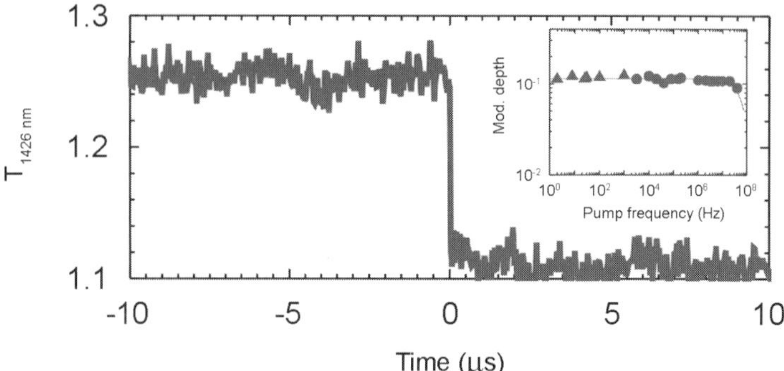

Fig. 6. Variation of the probe intensity transmitted through the slit of a 3.6-μm-long interferometer as a function of time; at t=0 μs the pump beam is turned on, and a fast (<40 ns) decrease in probe transmitted intensity is observed, due to photoinduced absorption in excited QDs. Inset: variation of the probe signal as a function of pump beam modulation frequency, at a power level of 500 W/cm^2.

CdSe QDs excited by the control SPP. The data reported in Fig. 6 clearly demonstrate the possibility of achieving modulation frequencies that are orders of magnitude faster than typical frequency responses of modulators relying on thermal effects, which are typically in the range of a few Hz. At power densities >600 W cm^2, both interferometers shown in Colour plate 2 experience a recovery of their original (that is, zero-pump) transmission values, owing to a reduction in absorption of the probe SPP from excited QDs.

In Fig. 7(a), the transmission intensity at λ=514.5 nm for QD-coated interferometer with a 2.6 μm slit–groove distance is shown to increase as a nonlinear function of pump power (open symbols). The transmission is at first equal to that of an isolated slit, due to QD-absorption of the SPP. At high pump powers, the transmission reaches values typical for an uncoated device (compare with Fig. 7(b), open symbols), in which no absorption from the QD occurs. The decrease in probe-SPP attenuation at high pump excitation rates is due to a two-photon process in which a pump SPP is absorbed by a previously excited QD, producing an electronic photo-ionization event that leads to charge trapping at nearby surface states. The QD is thereafter transparent to either incoming pump or probe SPPs (schematic inset of Fig. 7(a)).

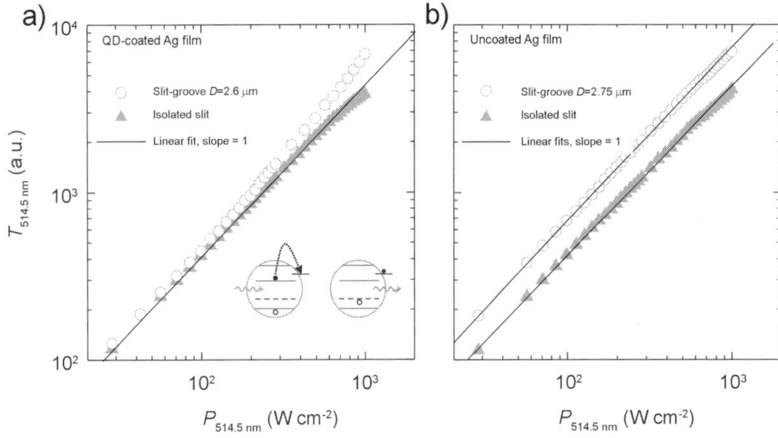

Fig. 7. Power dependence of the pump beam transmission at λ=514.5 nm. (a) Light intensity at λ=514.5 nm transmitted through the slit of a QD-coated interferometer with slit-groove distance D=2.6 μm (open symbols), and through an isolated slit (solid symbols); inset: schematic diagram of photoinduced bleaching in a QD. (b) Transmitted intensity at λ=514.5 nm as a function of power for an uncoated sample.

On the other hand, Fig. 7(b) indicates that pump transmission through an uncoated interferometer is instead linear at all powers, further confirming that the modulation effect observed in the QD-coated sample is not due to thermal or non-linear effects in the metal film, but to plasmon-induced absorption in CdSe QDs.

We note that a pump power density of ~100 Wcm^{-2} impinges on a groove area of $2\text{x}10^{-8}$ cm^2, so the total power incident on the groove is 2 mW. The experimentally determined power incoupling efficiency into an SPP is ~10%, so the pump SPP power is ~200 nW, or ~20 nW μm^{-1} per groove length. Thus, a 1-μm-long groove would enable modulator power operation at only 20 nW. In our scheme, based on fast intraband electronic transitions in QDs, the modulation switching time is ultimately limited by the QD exciton recombination lifetime (≤40 ns in our system), yielding an estimated minimum switching energy per operation of approximately 800 aJ. These results suggest the potential of our approach for chip-based all-optical modulator arrays whose switching energies may rival those of fully scaled Si-based CMOS electronic inverters.

3. Three-dimensional plasmonic modulators

3.1. New designs for improved figures of merit

In 2005, the ITRS set the requirements for a basic all-optical switching element, namely: it should (1) work at around 1.5 μm, (2) operate at low power, and (3) have a small component size. The planar interferometers presented so far are interesting because they fulfill all of these requirements. Nevertheless, the proof-of-concept design is affected by some important drawbacks that we would like to address and overcome.
First of all, even though the planar interferometers are characterized by a fast switching time and high modulation frequency, the typical modulation depth is relatively small, of the order of only ~10-20%. This is determined by the fact that even though the signal SPP is completely extinguished by photoinduced absorption in excited QDs, the light transmitted through the slit can never be zero, since the slit is always illuminated by the incident beam. A way to overcome this problem maintaining the same geometry would be to use localized excitation sources, such as the subwavelength aperture of a near-field optical microscope, to illuminate the groove and not the slit. In such a way, when the probe beam is on, the transmission through the slit is determined only by the propagating SPP term, i.e. $H_T = H_{T,g} = t(\delta\beta\, e^{i\varphi})H$, since the slit contribution is now $H_{T,s} = 0$. The normalized transmission in this case would be:

$$T(P) \equiv \frac{I_T(P)}{I_T(P=0)} = \frac{|H_T(P)|^2}{|H_T(P=0)|^2} = e^{-\alpha(P)D}, \tag{10}$$

where we have explicitly considered the power dependence of the absorption coefficient. In the absence of pump power, that is $P = 0$, $\alpha(P)D \ll 1$ and the normalized transmission through the slit is equal to "1," corresponding to the "on" state of the device. As soon as the pump beam is turned on, and the power set such that $\alpha(P)D \gg 1$, the signal SPP is completely absorbed by the excited QDs. Therefore no signal SPP can reach the slit, which in turn reduces the transmitted intensity to zero. Such a strategy would certainly enable orders of magnitude increase in modulation depth. However, this approach is not easy to implement in integrated dense optical chips.

Therefore, in order to solve the problem in a way that may lead to more realistic architectures, we designed and fabricated a novel plasmonic modulator, whose schematic and principle of operation are reported in Colour plate 3. The interferometer consists of a metal/dielectric/metal waveguide, and makes use of two slits milled in the top metal layer to incouple the incident beam into SPPs propagating in the dielectric waveguide. The dielectric can be a semiconductor, or even an insulating matrix codoped with CdSe quantum dots. By opportunely choosing its thickness, the dielectric layer can be made to only support the fundamental plasmonic mode while cutting-off all of the higher order photonic modes. A slit in the bottom metal layer collects the propagating SPPs; the interfered field is then guided in the slot cavity provided by the slit and scattered back in free space, at the bottom side of the device. It is interesting to note that this design could be merged with modern microelectronic circuits, by simply placing a CMOS-based detector right at the bottom output of the out-coupling slit. The device depicted in Colour plate 3 resembles a miniaturized Mach-Zehnder interferometer, with arms A and B, having lengths D_A and D_B, respectively. When the two input slits are coherently and uniformly illuminated with a signal beam at normal incidence, SPPs with field amplitudes $H_{SPP,A}$ and $H_{SPP,B}$ are launched and guided into the dielectric waveguide with the same amplitude βH and the same incoupling phase lag φ_0 (analogous to φ_g in Eq.(1)); the SPPs propagate with wavevector $k_{SPP} = n_{SPP}k$ and reach the outcoupler slit with a relative phase shift:

$$\Delta\varphi = \varphi_A - \varphi_B = \left(k_{SPP} + i\frac{\alpha(P)}{2}\right)(D_A - D_B), \tag{11}$$

which determines the interference condition between the two SPPs at the output slit position. For interferometers with relatively small arms, the material loss in the absence of pump power is negligible, and we can safely assume $\alpha(P=0) = 0$ in Eq. (11). The same arguments holds true if we choose the difference in optical path such that $\alpha(P=0)(D_A - D_B) \ll 1$. Either way, in the absence of a pump beam, by setting the distance of the two arms of the interferometer in such a way that the two SPPs are exactly out of phase, that is $\Delta\varphi = \pi$, the two SPPs perfectly cancel out at the slit position, thus determining an exact null in transmitted intensity through the bottom slit; this is the “off” state of the modulator. The condition for destructive interference can be altered by affecting the propagation of the SPP in one of the two arms using a pump SPP. For example, we could excite a control SPP in one of the two arms, say arm B, which would induce free-carrier absorption of the signal SPP propagating in arm B. This in turn would prevent the destructive interference with the other signal SPP propagating in arm A, which would then be transmitted through the slit, determining the "on" state of the modulator.

Another drawback of the simple planar proof-of-concept approach previously investigated in section 2 is that the control and signal SPPs are excited using the same subwavelength object (a groove), co-propagate in the modulator and are simultaneously transmitted through the slit. For many applications it would be desirable to maintain optical isolation between the two beams. Typically, this can be accomplished by using a filter in the far field to selectively transmit the frequency of interest. This approach is of course not convenient for large scale integration of the device. Therefore we recur to another trick. Subwavelength slits are very good polarizers, in that they can only transmit light with H-field component parallel to their long-axis (TM-polarized beam). At the same time, the launching efficiency for long-range propagating SPPs is maximum when the slit is illuminated with TM-polarized beams, and it goes to zero for TE-polarized beams. Therefore, to achieve the desired isolation, we use cross-polarized signal and control beams, and slits that are perpendicular to each other to independently incouple signal and control SPPs, respectively.

A practical implementation of the Mach-Zehnder plasmonic interferometer consists of two thermally evaporated 300-nm-thick silver films separated by a core layer of SiO_2 codoped with chemically synthesized CdSe colloidal quantum dots (QDs), capped with a different ligand (tris-hydroxylpropylphospine, THPP) to make them soluble in the spin-on-glass matrix. The size of these QDs is engineered such that in the ground state the QDs are transparent to the probe light, but once an exciton is created, the QDs can absorb the probe. Several devices have been fabricated by sequential evaporation of metal films, spin-coating of spin-on-glass codoped with QDs, and focused ion beam milling. Parallel slits are milled in the top and bottom silver films and aligned to previously milled slits in the bottom metal layer. To selectively create excitons only in one arm, slits parallel to the interferometer arm B are milled by FIB on the top metal layer. Pump light cross-polarized with respect to the probe beam is selectively coupled into this arm only.

An SEM micrograph of an array of 9 plasmonic Mach-Zehnder interferometers is shown in Colour plate 4(a). In this array, the length of arm A is varied from 2 μm (bottom device) to 2.4 μm (top device). A cross section of one of the devices is shown in Colour plate 4(b). Smooth core-cladding interfaces are clearly present, preventing SPPs from scattering loss in the waveguide. Colour plates 4(b) and (c) show, superimposed to the SEM micrographs, the working principle of the proposed modulator. In Colour plate 4(b) the device is uniformly illuminated with a probe beam, TM polarized, that is with magnetic field oscillating parallel to the long axis of the incoupling slits. Since the top metal layer is optically thick, light can only be transmitted through the two slits which act as the mirrors of the interferometer. Light transmitted through each slit is then coupled into SPPs propagating to the left and right of each slit.

We focus our attention on the SPPs that propagate toward the output slit milled in the bottom metal layer. By opportunely choosing the length of the two interferometric arms (designated with A and B in Colour plate 3) a destructive interference can be set such that the light intensity transmitted through the outcoupling slit is zero, or a minimum value. This scenario represents the "off" state of the modulator, which would correspond to a logic state "0." In Colour plate 4(c) the device is also illuminated with a pump beam which is cross-polarized with respect to the probe signal. Light can leak through the vertically aligned slits which act as sources of SPPs which propagate in the direction perpendicular to the right arm of the interferometer (arm B in Colour plate 3). These SPPs act as control beams in that they excite the QDs present in the dielectric and determine a photo-induced absorption of the probe SPP, as explained in section 2 for planar plasmonic interferometers. The signal SPP propagating in the right arm of the interferometer is absorbed along its path by free-carrier absorption in the excited QDs and is no longer able to reach the output slit. Therefore the destructive interference with the SPP propagating in the left arm is no longer possible. Light can now emerge through the slit, and the state of the modulator is set to "on,"

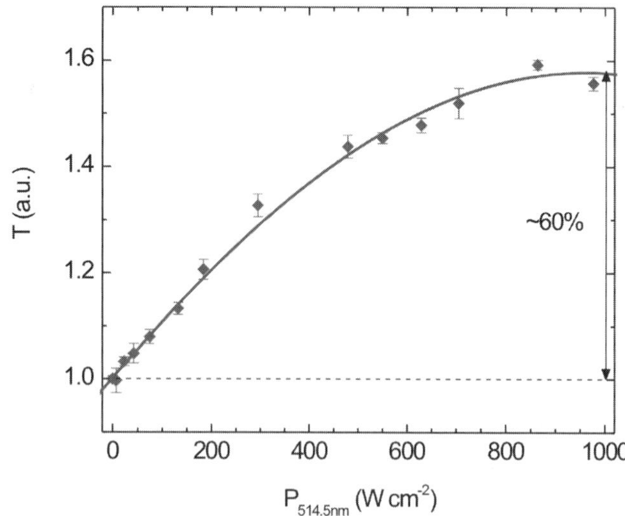

Fig. 8. Pump-power dependence of the transmitted probe intensity at 687.9 nm through a plasmonic modulator that shows a minimum in probe transmission at zero pump power. The solid line is a guide to the eye.

(logic state "1"). It is important to stress that the control SPPs are characterized by a longitudinal electric field oscillating in the direction parallel to the long axis of the output slit. For this polarization, no transmission of the pump beam is observed through the slit, which guarantees optical isolation between the control and the signal beams.

We experimentally investigate the operation of such interferometers by performing pump and probe experiments. First, the probe transmission trough the Mach-Zehnder plasmonic interferometers is studied as a function of the length of interferometer arm A. In practice, we scan the length of one of the two interferometer arms (arm A in this case) and look for a device that shows a minimum in transmission. When this condition is satisfied, destructive interference between the two SPPs is taking place at the slit. We find that a destructive interference state is obtained for a device with an interferometer arm length $D_A = 2.55\mu\text{m}$. Results of pump-probe experiments for such a device are presented in Fig. 8. Simultaneous illumination with both pump and probe light beams results in an increase of the transmitted probe intensity. By varying the power density, the probe signal transmission at 687.9 nm increases up to 60% with respect to the zero-pump value, as the result of increased absorption of the signal SPP propagating in the right arm of the interferometer.

4. Design of plasmonic modulators and interconnects

Inspired by the previous experimental results, we employ finite-difference time domain (FDTD) simulations to evaluate the performance of higher complexity plasmonic waveguide structures which are not readily tractable by analytic methods.

4.1. Multilevel plasmonic modulators and interconnects

In the following section, we study a multilayer metal/dielectric stack which defines two horizontal dielectric waveguides, separated from each other by a 150-nm-thick layer of Ag cladding. The cladding metal layer is thick enough to be optically opaque, thus suppressing crosstalk due to mode coupling between the top and bottom waveguides. Slits milled in the central metal cladding play the role of optical "vias" between the two dielectric waveguides. A subwavelength slit in the top metal layer acts as optical incoupler in the first waveguide, and three slits milled in the bottom metal layer serve as optical outcouplers. All the slits are 50 nm wide. A schematic cross section of the structure is reported in Fig. 9(a). When illuminated by an optical beam, the slit in the top metal layer is found to be effective at scattering light into guided modes of the first buried metal/dielectric/metal waveguide. These guided modes are launched with propagation direction transverse to the slit long axis. Two slits in the central metal layer partially scatter the guided modes into guided modes of the underlying waveguide. The waveguide dielectric cores are here modeled as a purely transparent dielectric with $n = (2.02+0i)$, corresponding to that of QD-doped silicon dioxide at the excitation wavelength $\lambda = 1.5$ μm. The relative fraction of light which is "tapped" off by the via can be adjusted by varying the via width, or by fabricating the via in a manner that provides dielectric contrast relative to the core material.

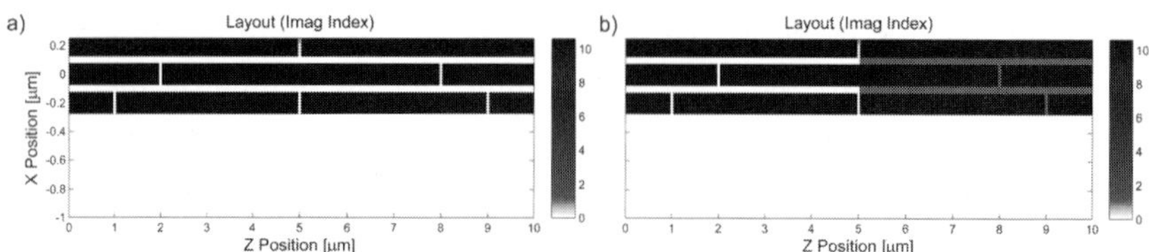

Fig. 9. Schematic cross sectional view of a simulated multilayer plasmonic interferometer. (a) Control device, in which the waveguide core is everywhere non-absorbing. (b) The waveguide core is given non-zero absorption at positions $z > 5\mu m$, simulating photoinduced absorption of the signal beam by means of a pump beam (not shown in the schematic).

From simulation performed on two stacked metal/dielectric/metal waveguides linked by a single slit as the optical via, we find that the via acts as a junction for modes incident from the waveguide, with ~18% of the power being reflected back in the waveguide, ~46% transmitted along the waveguide, and ~36% tapped by the via in the bottom waveguide. On the other hand, when the SPP is incident from the via side, the junction acts as a 50/50 splitter with negligible back-reflection.

We additionally consider filling the waveguide cores with a medium with non-zero absorption, $n = 2.02+i\kappa$ in the half-space $z > 5$ μm, as in Fig. 9(b). This investigation is inspired by the experimental results previously presented, and the values of imaginary index κ used here are chosen in order to achieve attenuation per unit length comparable to the experiment. However, in the current simulation we make two key simplifying assumptions. First, we do not explicitly simulate the propagation of a "pump" beam which activates the optical absorption in the buried layer, and second, we assume that the real part of the index stays constant even as κ varies. The simulated performance of a multilayer plasmonic interferometer is illustrated in Colour plate 5. The left column shows the magnetic field H_y magnitude, providing a clear view of propagation and standing wave patterns within the waveguide stack. In the right column we display the real part of the E_z field under the same conditions, which illustrate more clearly the "beams" of light which emanate from the output slits. One arm of the interferometer ($z < 5$ μm) is always filled with a non-absorbing medium ($n = 2.02$), while the other arm ($z > 5$ μm) is filled with an absorber having the following complex index of refractions: (a) $n = 2.02+0.025i$, (b) $n = 2.02+0.1i$, (c)

$n = 2.02+1.0i$. The device is illuminated by a plane wave inciden7t on the top input slit and produces three output beams through slits in the bottom metal surface. As absorption is increased, the right output beam is entirely suppressed, the center output beam is diminished, and the left output beam is unchanged.

4.2. Incoupling strategies

In order to consider integration of our plasmonic modulators or other subwavelength active devices as constituents of dense subwavelength photonic networks, we must also address strategies for efficiently coupling light in the plasmonic waveguides, and develop design strategies to achieve low overall loss. In this section, we employ FDTD simulations to survey alternative coupling schemes which are achievable with minimal incremental changes to the device structures we have already fabricated. In view of an increased compatibility with the existing Si-based microphotonics and mature CMOS technology, we investigated metal/dielectric/metal stack consisting of a core of 160-nm-thick Si layer, with 10-nm-thick silicon dioxide, which may be codoped with an active material, such as CdSe or Si QDs, showing photoinduced carrier absorption for the signal SPP. As cladding layer we consider 400-nm-thick Ag films on both sides of the thin waveguide. In all simulations the incident light is TM polarized, monochromatic, continuous wave excitation at $\lambda = 1.55$ μm. In the chosen coordinate system the waveguide propagation axis is z. In all cases the input port, whether slit or waveguide end facet, is centered at the center of the simulation volume, with coordinates $(z, x) = (2, 0)$ μm. At steady-state, the power which is considered "incoupled" in the plasmonic waveguide is that which passes through a monitor port consisting of a line segment which intersects the waveguide axis at $(z, x) = (2.5; -0.4{:}0.4)$ μm.

The following incoupling geometries are considered:

(1) Slit incoupling (Colour plate 6). The source is a gaussian beam with 1.5-μm waist in the plane $x = 1$ μm, about a half micrometer above the surface. The slit is an air-filled opening transverse to the waveguide, 400 nm wide, 490 nm deep through the metal cladding and halfway into the Si core. Note that the assumption that the focal radius equals λ is

somewhat arbitrary but intended as a "best case" estimate. The reported incoupling coefficient represents the energy coupled into the dielectric waveguide in the $+z$ direction; due to the symmetry of this coupling scheme, of course the same amount also couples into the $-z$ direction.

(2) Metal-clad taper incoupling (Colour plate 7). The source is the lowest order TM mode of an air-clad Si-core waveguide with a 970-nm thick core, same thickness as the entire stack. This waveguide is joined to the 160-nm-thick Si core of the metal/dielectric/metal stack by a 1 μm segment of concave parabolic taper. The tapered segment is clad in Ag.

(3) Thin (160 nm) waveguide incoupling (Colour plate 8). The source is the lowest order TM mode of a 160-nm-thick Si waveguide in air.

For comparison, we also consider a tapered waveguide incoupling geometry where the source is the lowest order TM mode of an air-clad Si-core waveguide with a 970-nm-thick core, joined to the 170-nm-thick Si core of the metal/dielectric/metal stack by a 1 μm segment of concave parabolic taper, clad in air; and to conclude, end-fire incoupling by using a gaussian beam with 1.5-μm waist, incident normal to the end facet of an abruptly truncated metal/dielectric/metal waveguide. The results of the FDTD data analysis are reported in Table 1.

We find that the insertion loss for the 160-nm-waveguide end-fire scheme, Table 1, is only -4.4 dB, an improvement of about 8 dB over the slit-coupling geometry. This type of analysis allows us to trade off the demand for increased performance with the desire to minimize added design complexity. Most importantly, the substantial improvement reported here can be achieved without varying the width of the Si core in the dielectric waveguide region relative to that of the core in the metal/dielectric/metal active region.

Table 1. Power incoupled to Ag/Si/Ag plasmonic waveguide.

Incoupling Geometry	Incoupled Power (%)	Incoupled Power (dB)
Slit	5.2	-12.8
Endfire	14.8	-8.3
Air-clad taper	20.4	-6.9
Metal-clad taper	24.0	-6.2
Thin (160 nm) waveguide	35.9	-4.4

Acknowledgements

We acknowledge from the Department of Energy, Basic Energy Sciences program under grant DOE DE-FG02-07ER46405.

References

1. A.A. Sawchuck and T.C. Strand, Digital optical computing, *Proc. IEEE* **72**, 758–779 (1984).
2. A. Huang, Architectural considerations involved in the design of an optical digital-computer, *Proc. IEEE* **72**, 780–786 (1984).
3. A.V. Krishnamoorthy and D.A.B. Miller, Scaling optoelectronic-VLSI circuits into the 21st century: a technology roadmap, *IEEE J. Sel. Top. Quantum Electron.* **2**, 55–76 (1996).
4. K.-N. Chen, M.J. Kobrinsky, B.C. Barnett and R. Reif, Comparisons of Conventional, 3-D, Optical, and RF Interconnects for On-Chip Clock Distribution, *IEEE Trans. Electron Devices* **51**, 233–239 (2004).
5. ITRS, International Technology Roadmap for Semiconductors, *Emerging Research Devices* (2005).
6. S.M. Jensen, The non-linear coherent coupler, *IEEE J. Quantum Electron.* **18**, 1580–1583(1982).
7. J.L. Jewell, Y.H. Lee, M. Warren, H.M. Gibas, N. Peyghambarian, A.C. Gossard and W. Wiegmann, 3-pJ, 82-MHz optical logic gates in a room-temperature GaAs-AlGaAs multiple-quantum-well etalon, *Appl. Phys. Lett.* **46**, 918–920 (1985).
8. S.R. Friberg, Y. Silberberg, M.K. Oliver, M.J. Andrejco, M.A. Saifi, and P.W. Smith, Ultrafast all-optical switching in a dual-core fiber nonlinear coupler. *Appl. Phys. Lett.* **51**, 1135–113 (1987).
9. S.D. Smith, A.C. Walker, F.A.P. Tooley and B.S. Wherrett, The demonstration of restoring digital optical logic, *Nature* **325**, 27–31 (1987).
10. Almeida, V. R., Barrios, C. A., Panepucci, R. R. and Lipson, M. (2004). All-optical control of light on a silicon chip, *Nature*, 431, 1081–1084.
11. A.S. Liu, R. Jones, L. Liao, D. Samara-Rubio, D. Rubin, O. Cohen, R. Nicolaescu and M.A. Paniccia, High-speed silicon optical modulator based on a metal-oxide-semiconductor capacitor, *Nature* **427**, 615–618 (2004).
12. E. Kretschmann, Determination of optical constants of metals by excitation of surface plasmons, *Z. Phys.* **241**, 313–324 (1971).
13. H. Raether, *Surface plasmons on smooth and rough surfaces and on gratings* (Springer-Verlag, Tracts in Mod. Phys., Vol. 111, New York, 1988).
14. R. Zia, J.A. Schuller, A. Chandran, and M.L. Brongersma, Plasmonics: the next chip-scale technology, *Mater. Today* **9**, 20–27 (2006).
15. S.A. Maier, *Plasmonics: fundamentals and applications* (Springer, New York, 2007).

16. H.A. Atwater, The promise of plasmonics, *Sci. Am.* **296,** 56–63 (2007).
17. F.J. Garcìa-Vidal, H.J. Lezec, T.W. Ebbesen, and L. Martin-Moreno, Multiple paths to enhance optical transmission through a single subwavelength slit, *Phys. Rev. Lett.* **90**, 213901 (2003).
18. H.J. Lezec, and T. Thio, Diffracted evanescent wave model for enhanced and suppressed optical transmission through subwavelength hole arrays, *Opt. Express* **12**, 3629–3651 (2004).
19. H.F. Schouten, N. Kuzmin, G. Dubois, T.D. Visser, G. Gbur, P.F.A. Alkemade, H. Blok, G.W. 't Hooft, D. Lenstra, E.R. Eliel, Plasmon-assisted two-slit transmission: Young's experiment revisited, *Phys. Rev. Lett.* **94**, 053901 (2005).
20. P. Lalanne, J.P. Hugonin and J.C. Rodier, Theory of surface plasmon generation at nanoslit apertures, *Phys. Rev. Lett.* **95**, 263902 (2005).
21. G. Gay, O. Alloschery, B. Viaris de Lesegno, C. O'Dwyer, J. Weiner and H.J. Lezec, The optical response of nanostructured surfaces and the composite diffracted evanescent wave model, *Nature Phys.* **2**, 262–267 (2006).
22. G. Gay, O. Alloschery, B. Viaris de Lesegno, J. Weiner, and H.J. Lezec, Surface wave generation and propagation on metallic subwavelength structures measured by far-field interferometry, *Phys. Rev. Lett.* **96**, 213901 (2006).
23. O.T.A. Janssen, H.P. Urbach, and G.W. 't Hooft, On the phase of plasmons excited by slits in a metal film, *Opt. Express* **14**, 11823–11832 (2006).
24. D. Pacifici, H.J. Lezec, H.A. Atwater and J. Weiner, Quantitative determination of optical transmission through subwavelength slit arrays in Ag films: Role of surface wave interference and local coupling between adjacent slits, *Phys. Rev. B* **77**, 115411 (2008).
25. D. Pacifici, H.J. Lezec and H.A. Atwater, All-optical modulation by plasmonic excitation of CdSe quantum dots, *Nature Photon.* **1**, 402–406 (2007).
26. G. Lévêque, O.J.F. Martin, J. Weiner, Transient behavior of surface plasmon polaritons scattered at a subwavelength groove, *Phys. Rev. B* **76**, 155418 (2007).
27. J. Weiner, Phase shifts and interference in surface plasmon polariton waves, *Opt. Express* **16**, 950–956 (2008).
28. C.B. Murray, D.J. Noms, and M.G. Bawendi, Synthesis and characterization of nearly monodisperse CdE (E = S, Se, Te) semiconductor nanocrystallites, *J. Am. Chem. Soc.* **115**, 8706–8715 (1993).
29. M. Nirmal, D.J. Norris, M. Kuno, M.G. Bawendi, A.L. Efros and M. Rosen, Observation of the dark exciton in CdSe quantum dots, *Phys. Rev. Lett.* **75**, 3728–3731 (1995).
30. V.I. Klimov, A.A. Mikhailovsky, S. Xu, A. Malko, J.A. Hollingsworth, C.A. Leatherdale, H.J. Eisler and M.G. Bawendi, Optical gain and stimulated emission in nanocrystal quantum dots, *Science* **290**, 314–317 (2000).
31. C.A. Leatherdale, W.-K. Woo, F.V. Mikulec and M.G. Bawendi, On the absorption cross section of CdSe nanocrystal quantum dots, *J. Phys. Chem. B* **106**, 7619–7621 (2002).

32. A. Polman, Erbium implanted thin film photonic materials, *J. Appl. Phys.* **82**, 1–39 (1997).
33. V.I. Klimov, C.J. Schwarz, D.W. McBranch, C.A. Leatherdale and M.G. Bawendi, Ultrafast dynamics of inter- and intraband transitions in semiconductor nanocrystals: Implications for quantum-dot lasers, *Phys. Rev. B* **60**, R2177–R2180 (1999).
34. V.I. Klimov, Optical nonlinearities and ultrafast carrier dynamics in semiconductor nanocrystals, *J. Phys. Chem. B* **104**, 6112–6123 (2000).

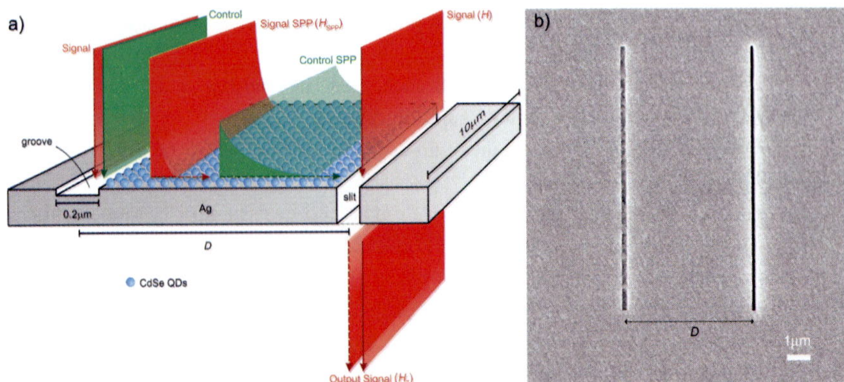

Colour Plate 1. (a) Schematic of a plasmonic interferometer consisting of a subwavelength slit flanked at a distance D by a subwavelength groove milled in a Ag film, coated with CdSe QDs. TM-polarized signal and control beams at different wavelengths uniformly illuminate the device. Diffractive scattering by the groove converts both beams into co-propagating surface plasmon polaritons (SPPs) at different frequencies that are made to interact via a thin layer of CdSe QDs. The signal beam (red) is transmitted through the slit with a magnetic field amplitude H_T which results from the interference between the incident field (H) and the propagating SPP (H_{SPP}) at the slit position. (b) SEM micrograph showing an uncoated interferometer.

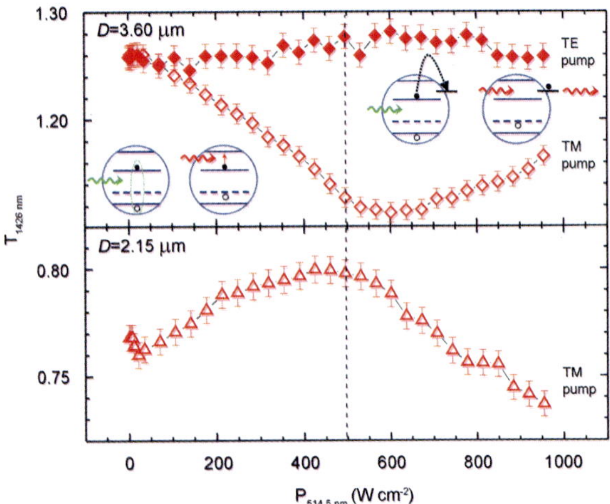

Colour Plate 2. All-optical modulation in QD-coated interferometers. Open symbols: probe beam transmitted intensity at λ=1426 nm versus pump power density for two QD-coated interferometers with slit-groove distances D=3.60 μm and D=2.15 μm; pump and probe beams are TM-polarized Solid symbols: probe transmission versus power density for a TE-polarized pump beam. Inset schematics: (left) attenuation of the probe SPP by an excited QD; (right) photobleaching of a QD.

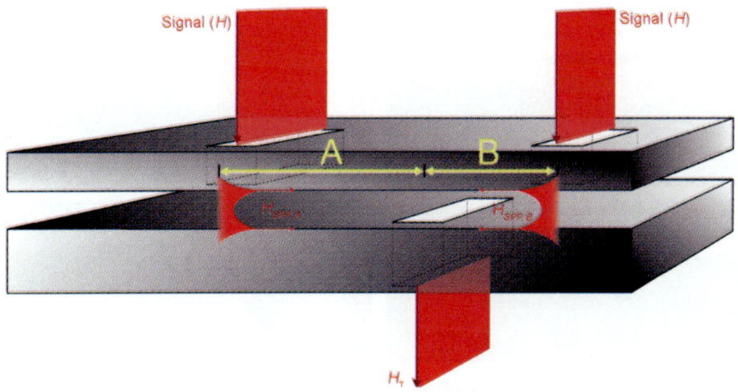

Colour Plate 3. 3D schematic of a metal/dielectic/metal Mach-Zehnder plasmonic interferometer. The top and bottom metal layers are optically-thick. Two subwavelength slits (top metal layer) incouple incident light into SPPs propagating in the dielectric waveguide. The SPPs in arms A and B interfere at the slit position, causing enhancement or suppression of light intensity transmitted through the output slit (bottom metal later).

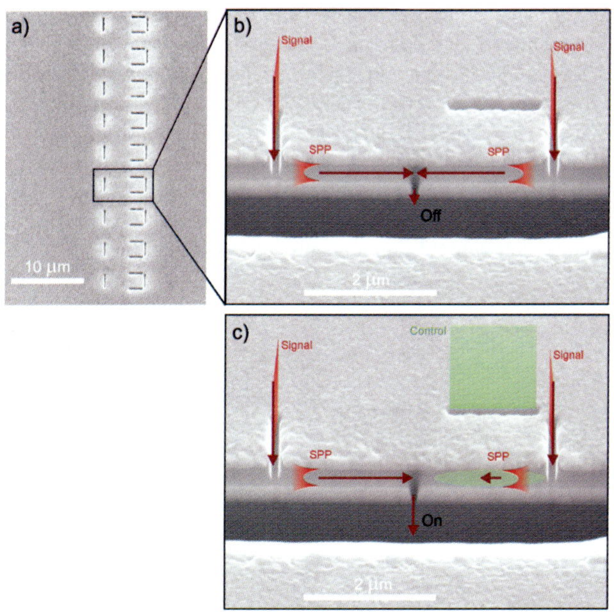

Colour Plate 4. (a) SEM micrograph of the top view of an array of 9 Mach-Zehnder plasmonic modulators, with fixed right-arm length, and varying left-arm length. (b) Cross-sectional SEM micrograph of a modulator; input slits in the top silver film are visible for both probe beam (vertical slits in image) and pump beam (horizontal slits); by opportunely varying the length of one interferometer arm (left arm or arm A in Colour plate 3), a destructive interference can be set between the two SPPs at the output slit position; the superimposed drawings schematically represent the "off" state of the plasmonic modulator. (c) "on" state of plasmonic modulator, achieved by selectively affecting the propagation length of the SPP in the right arm of the interferometer.

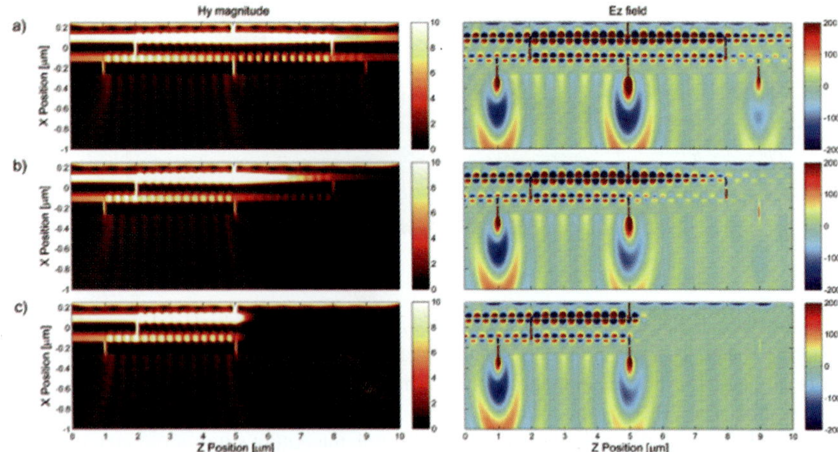

Colour Plate 5. Simulated performance of a multilayer plasmonic interferometer (left column: H_y magnitude; right column: E_z field amplitude). One arm of the interferometer ($z < 5$ µm) is always filled with a transparent medium ($n = 2.02$), while the other arm ($z >$ 5 µm) is filled with an medium with progressively increased absorption, in particular: (a) $n = 2.02{+}0.025i$, (b) $n = 2.02{+}0.1i$, and (c) $n = 2.02{+}1.0i$. The device is illuminated by a plane wave incident on the top input slit and produces three output beams through outcoupling slits in the bottom metal surface. As absorption is progressively increased from (a) to (c), the right output beam is entirely suppressed, the center output beam is diminished, and the left output beam transmission is left unchanged.

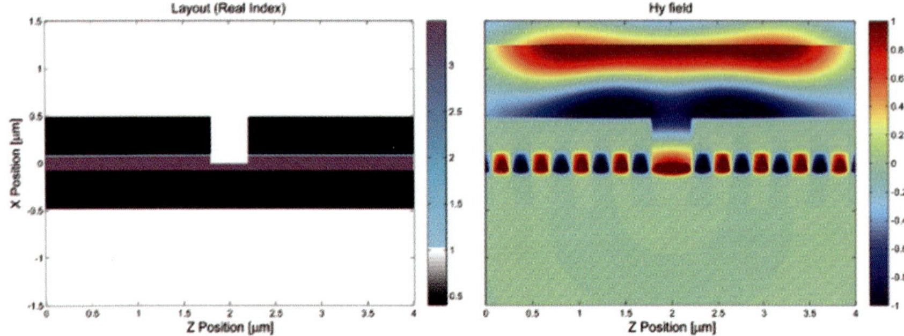

Colour Plate 6. Slit incoupling geometry. Simulated geometry for incoupling a tightly-focused, normally-incident, free-space Gaussian beam (wavelength $\lambda = 1.5$ µm, beam waist) into a metal/dielectric/metal plasmonic waveguide by means of a subwavelength slit. Left: colour scale represents the real optical index of the materials used in all space: air (white), Ag (black), Si (purple), and QD-doped silicon dioxide (blue) are visible. Right column: H_y field at simulated steady state.

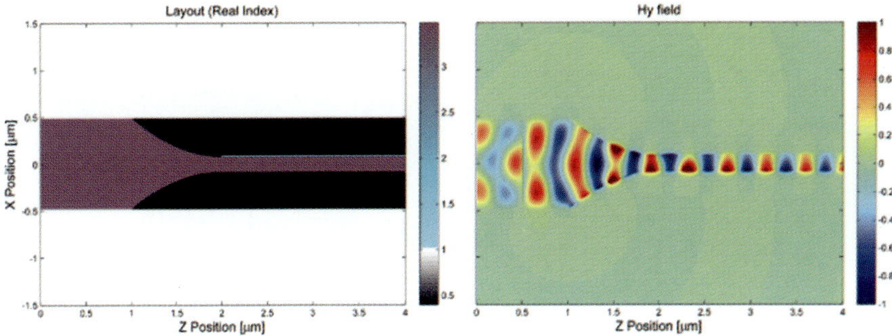

Colour Plate 7. Metal-clad taper incoupling geometry. Left: colour scale represents the real optical index of the materials used in all space: air (white), Ag (black), Si (purple), and QD-doped silicon dioxide (blue) are visible. Right column: H_y field at simulated steady state.

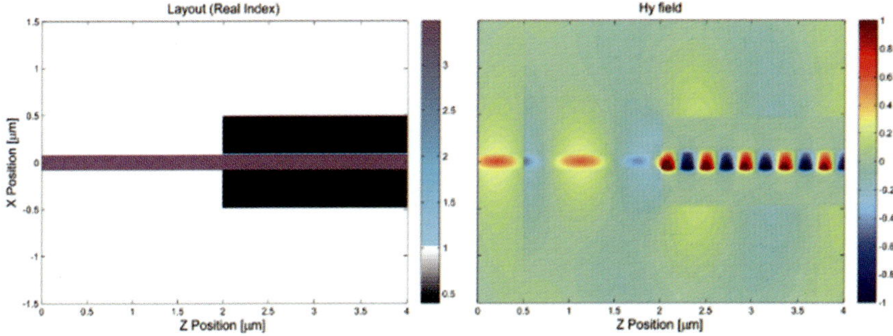

Colour Plate 8. Thin (160 nm) waveguide incoupling geometry. Left: colour scale represents the real optical index of the materials used in all space: air (white), Ag (black), Si (purple), and QD-doped silicon dioxide (blue) are visible. Right column: H_y field at simulated steady state.

Chapter 8

METAL TRENCH WAVEGUIDES: EXPERIMENTS AND ANALYSIS

Meir Orenstein

Department of Electrical Engineering, Technion
Haifa 3200, Israel
E-mail: meiro@ee.technion.ac.il

Propagation of plasmonic modes within dielectric voids embedded in metal layers is a favorable method of constructing plasmonic based nanocircuitry. In these structures – the metal cladding is providing the guiding mechanism and the tight confinement, while the dielectric core can be used for light-matter interactions for the realization of switches, modulators, nonlinear elements and for ultrasensitive probing. While the generic 1D plasmonic gap structure (MIM) can be solved in closed form to exhibit explicitly the merits of the plamonic modes – the 2D version of such waveguides – slots, trenches and channels, is by far more complex. This complexity stems from the fact that, in contrary to regular dielectric cladded waveguides, each metal-dielectric ingredient of the waveguide is a guide of surface plasmon polariton by itself. This includes flat interfaces as well as metallic wedges. A trench in a metal can be considered as a complex coupled assembly ("hybridization") of stripe and wedge plasmonic waveguides. Trench waveguide modes are studied here by mode-solving methods and the issue of structural symmetry and its significance for cutoff at nano dimensions is emphasized. Experimental measurements of modal field intensities of trench waveguides is presented and validating the theoretical considerations. As a conclusion we show that trench/slot waveguides are the favorable means to launch long propagating plamonic waves either millimeters long for micrometers wide trenches or tens of micrometers for nanometers wide trenches. The embedded mechanism in most cases is due to coupled edge (wedge) guiding – which has no parallel in regular dielectric waveguides.

1. Introduction

The structures studied in this chapter are trenches dissected in a finite thickness metal layer and filled with a dielectric material (Fig.1). The structures encompass via-trenches (going through the metal layer), channels (carved structures terminated within the metal) and shaped trenches and channels (V, W etc.). Many references to related structures can be found in previous chapters of this book, and not all of them will be repeated here.

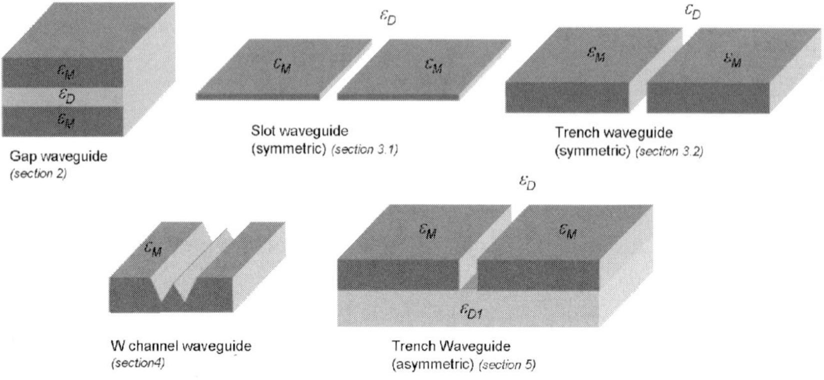

Fig. 1. The plasmonic waveguides studied in this chapter, starting from the 1D plasmonic gap waveguide. For each structure the relevant section in this chapter is indicated.

We present the basic guiding mechanisms and characteristics of such trench structures and convey in each section also the related experimental measurement results. The chapter is organized as follows: in Section 2 the 1D version of void plasmonic waveguide (the plasmonic gap waveguide) is briefly discussed – mainly to differentiate it from the 2D open trench waveguide configuration. In the consecutive sections – plasmonic trenches with progressively enhanced complexity are described. In Section 3 we start with the analysis of symmetrically embedded via-slots in a very thin (20 nm) metal layer, which exhibit modal characteristics determined by the coupling of the side metal "edge" modes. Subsequently, we examine symmetrical deep via-trenches in a thick (400 nm) metal layer, where the metallic side walls contribute significantly to the confined mode characteristics. Section 4, although deviating a little from the main track, is included in order exemplify the complex contribution of substructures to the modal characteristics, not only the flat metal segments but also corners (wedges) etc. In the final section, Sec. 5 – we examine the final frontier - the possibility of realizing nano-wide trench waveguides in very asymmetrical but practical scenarios (e.g. trenches in a metal layer deposited on a Silicon wafer).

Notations: in this Chapter “plasmon” will refer to surface-plasmon-polariton; 'slot' to via cut in thin metal; “trench” to via cut in thick metal; 'aspect ratio' is the ratio between structure height (thickness) to its width. The coordinates system we use is: z - the propagation direction, x- the lateral (horizontal) and y – the transverse (vertical) directions.

2. The thin but infinite plasmonic trench: gap waveguide

A 1D plasmonic gap-waveguide (GW)[1,2] is essentially a thin but yet infinite (in the other 2 dimensions) trench carved in bulk metal. The GW was discussed in previous chapters and it is seemingly the generic configuration required in order to comprehend plasmonic guiding in 2D dielectric voids in-between metal layers – such as trenches, slots and channels (see a detailed review of plasmonic guiding in void structures[3]). However, as will be further discussed, the 1D and 2D void waveguides are fundamentally different and this section, devoted to the GW modes is brought about in order to emphasize these crucial differences. Yet, GW modes are frequently used in effective index (EI) approximation of various segments of the trench waveguides[4].

The dispersion relations of the GW are essentially the geometrical dispersion of a dielectric slab with the embedded chromatic dispersion of the metal layers and were derived to yield symmetric guided modes:

Plasmon: $$\mathrm{tgh}(k_P d) = -(k_M/\varepsilon_M)/(k_D/\varepsilon_D) \quad \text{(1-a)}$$

Oscillatory: $$\mathrm{tg}(k_{OS} d) = (k_M/\varepsilon_M)/(k_{OS}/\varepsilon_M) \quad \text{(1-b)}$$

and anti-symmetric guided modes:

Plasmon: $$\mathrm{ctgh}(k_P d) = -(k_M/\varepsilon_M)/(k_P/\varepsilon_D) \quad \text{(1-c)}$$

Oscillatory: $$\mathrm{ctg}(k_{OS} d) = -(k_M/\varepsilon_M)/(k_{OS}/\varepsilon_D) \quad \text{(1-d)}$$

where the transverse wavenumber in the dielectric and metal (k_P, k_M for plamonic and k_{OS}, k_M for oscillatory modes), the free space wavenumber k_0, the modal propagation constant β and the dielectric and metal permittivities (ε_D,ε_M), are related by:

$$k_D^2=\beta^2-k_0^2\cdot\varepsilon_D,\quad k_{OS}^2=k_0^2\cdot\varepsilon_D-\beta^2,\quad k_M^2=\beta^2-k_0^2\cdot\varepsilon_M \tag{2}$$

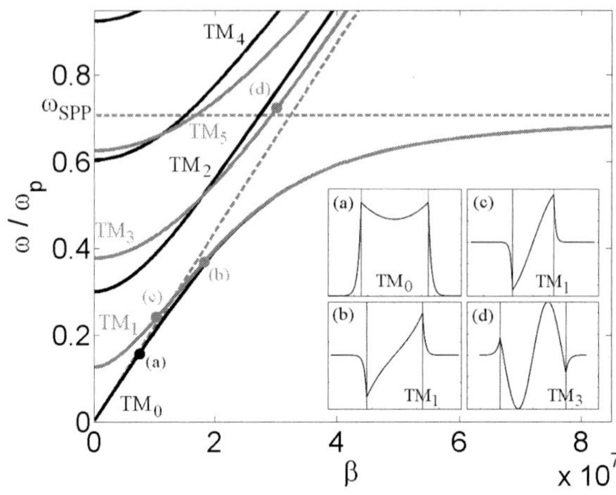

Fig. 2. Dispersion curves of TM propagating modes of a GW structure in gold for gap width = 500 nm. Even (odd) modes are black (gray) lines and the slopped dashed line is the light line. The inset is the H field distribution across the gap of various modes at selected points. The TM_0 mode is a plasmonic slow wave, while TM_1 crosses the light line for lower frequencies (from slow to fast wave –point (b) to (c)). The other modes are always fast waves.

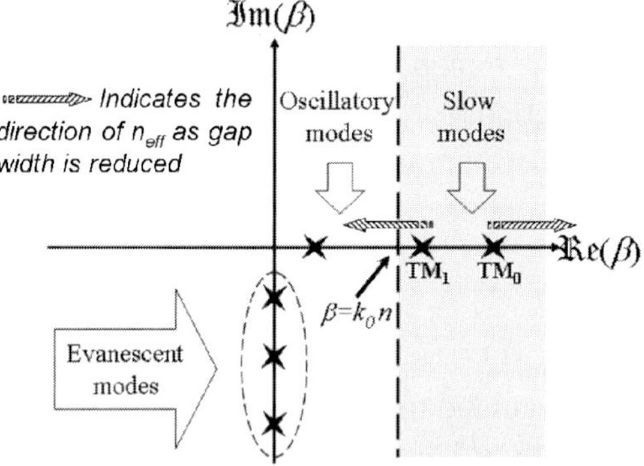

Fig. 3. The discrete set of modes supported by the GW.

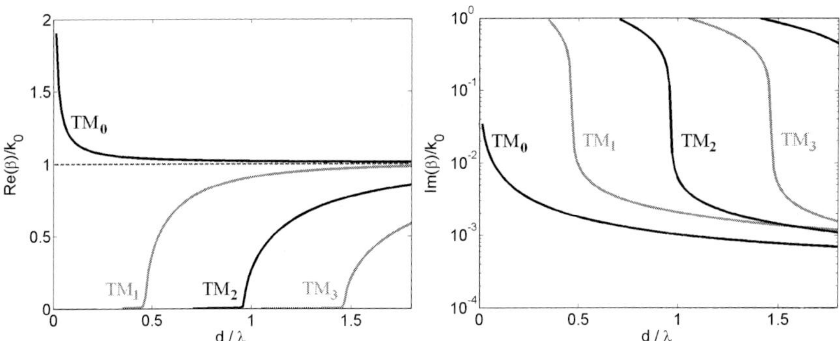

Fig. 4. Geometric dispersion of the plasmonic GW: Effective index of the various modes as a function of the gap width d (λ=1.55 μm and the metal is gold). Even (odd) modes are denoted by black (gray) lines.

The analysis of the GW yields two decoupled TE/TM mode families (The TM family is detailed in Figs. 2-4). Two distinctions from the forthcoming 2D trench waveguides are emphasized here: a. the 2D modes are essentially hybrid rather than TE/TM; b. (and the more fundamental) 1D gap has a modal structure of a closed waveguide – namely only a discrete spectrum of bound and evanescent modes is supported (Fig. 3) - while the open waveguide of the 2D trench, slot, channel, supports also continuous radiation modes spectrum and complex leaky modes which are dominant in establishing the waveguide characteristics.

The plasmonic (slow) modes of the GW are only the 2 first TM modes, while all other modes (both higher TM and all TE modes) are either oscillatory or evanescent. The two first TM modes can be interpreted as the in- and anti-phase coupling of the two single surface plasmon modes. However, for narrow gaps (below 1 wavelength) the coupling coefficient is so large that this simple coupling picture is distorted - evident by examining the trajectory of mode TM_1 (Figs. 2-4). As the thickness of the Gap is decreased below ~1 wavelength, the TM_1 plasmonic (slow) wave crosses the light line and becomes an oscillatory (fast) wave, and eventually at a thickness of ~ half wavelength this mode crosses the cutoff point to become an evanescent one. The TM_1 exhibits an additional extraordinary feature. For very thin GW, the cutoff frequency approaches the plasma frequency ω_p such that the modal dispersion curve exhibits a negative slop, namely the TM_1 becomes a backward wave[2].

Here an additional difference from the 2D structures is emphasized: the relatively simple modal structure of the gap waveguide results from mode coupling of only two infinite interfaces, while the more complex modes of a 2D trench is stemming from coupling of many substructures – including corners (wedges) as further discussed in sections 3,4. This also differentiates the plasmonic waveguides from all-dielectric waveguides where such edge and wedge guiding is absent.

The TM_0 GW mode is a pure plasmonic (slow) wave propagating only at frequencies below ω_{SPP} (except for very lossy scenarios) and it does not experience a cutoff. The TM_0 wave slows down (phase velocity decreases) which is accompanied by reduction of the group velocity as the frequency is enhanced or the gap thickness is reduced. The power flow of the mode is peculiar – having a forward propagation in the dielectric and backward propagation in the metal – which naively may question the integrity of the mode or its usefulness for describing source-mode coupling. This extreme discontinuity of the poynting vector exists also in 2D trench structures and is important in determining the group velocity of the trench modes.

3. From metal stripes to slots and trenches

In this section we dwell with via-trenches (slots) in metal, where the cladding of the structure is highly symmetrical (index of refraction of substrate and superstrate are equal). For the actual preparation of symmetrical stripes (rectangular metal), slots (rectangular dielectric structure filling a via-gap carved in thin gold layer) and trenches (same as slots but in a much thicker Au layer), we performed a lift-off patterning of a gold layer, symmetrically embedded in a polymer based dielectric cladding. Thick polymer layers (>10 μm benzocyclobutene (BCB)) were used below and above the metal layer in order to assure the symmetry of the waveguide. Tested waveguides were micrometers wide and 1 to 5 mm long. All waveguides were measured by end-fire input coupling of light (λ = 1.55 μm) by a polarization maintaining lensed fiber. The output modal intensity pattern was imaged and analyzed

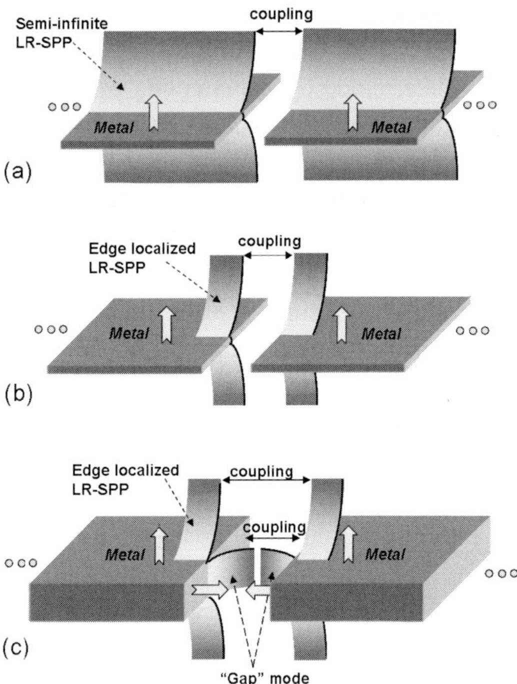

Fig. 5. Schematic illustration of the embedded guiding mechanisms for the slot (trench) modes. Block arrows indicating the polarization directions. (a) Wide slot in thin metal can be considered as a coupler between semi-infinite long range slab modes (LR-SPPs). (b) More realistically – these LR-SPPs are collapsing to edge modes that are coupled via the slot. (c) Wide trench in thick metal enables also the coupling of horizontally polarized "gap" modes, namely the modes of the vertical walls.

3.1. *From stripes to slots*

In this subsection we analyze micrometers wide slots in a thin – 20 nm thick, metal layer, which are characterized by ultralow aspect ratio (thickness/width=1:200 at the most). In principle, when preparing a slot waveguide (SW), in addition to the semi –infinite horizontal metal layers at both sides of the slot, additional metal surfaces are exposed including vertical metal walls and 4 corners (wedges), each can in principle guide plasmonic waves[5,6]. However, due to the very small aspect ratio and mainly due to the support of long range plasmon mode (LR-SPP)[7] by the horizontal metallic layers (the thin metal layer enables coupling of the top and bottom plasmons to generate the well known LR-SPP), the effects of the internal walls of the slot is reduced. Thus, the modal structure is mainly determined by the coupling of two semi-infinite LR-SPPs of the horizontal metal layers mediated by the dielectric slot (Fig. 5a). This naïve description is not enough – since it cannot readily explain the observed confinement of the mode to the slot area.

To comprehend this confinement, we examine first a related localization of the field in thin metallic stripe waveguides. The long range modes of these stripes, studied extensively in[8,9], are exhibiting a remarkable feature when going from narrow to wider dimensions[10] (Fig. 6). In addition to conventional multimode effects we observe an enhanced localization of the field at the stripe edges. This is clearly seen in Fig. 6 for the 14 and 45 micrometers wide stripes. Localization of field due to slight inhomogeneity of an extended lateral structures is known[11,12] – but here the definite step discontinuity (the edge) serves as an attractor for the localized plasmonic fields. The edge (or coupled wedges) mode was discussed also in Refs. 8,13. This is in contrast to the random localization in extended dielectric structures where edges are rather "repulsive".

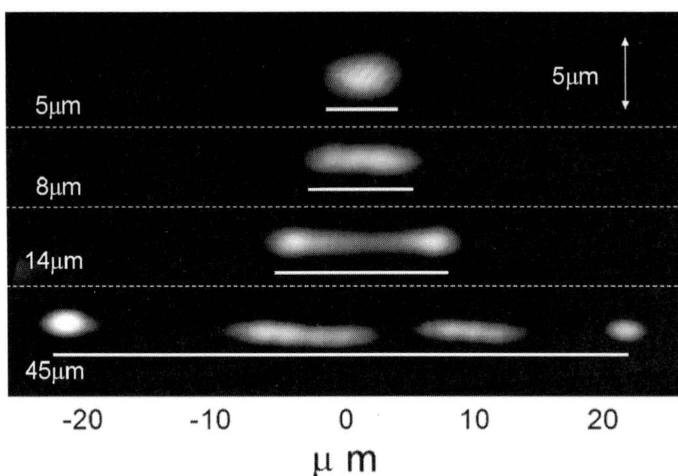

Fig. 6. Transverse intensity distributions measured from plasmonic stripe waveguides, from a narrow (5 μm) to very wide (45 μm) structure. Localization at the stripe edges is evident as going to wider stripes (each stripe size is schematically indicated by the white line) (gold layer 20 nm thick, λ=1.55 μm).

When two such (infinitely) wide stripes are separated by a few micrometers wide dielectric barrier (the slot), one expects to get coupling between the two localized fields to substantiate the slot mode (Fig. 5b).

The experimental results are conforming nicely with the above considerations (Fig. 7a). The mode is TM polarized (vertically) similar to a slab LR-SPP – while no mode launching was observed for horizontal polarization. The field exhibits localization in the slot in the form of two coupled field distributions (in-phase) near the metal edges. The mode exhibits long propagation distance - similar to the case of LR-SPPs of stripes under similar conditions.

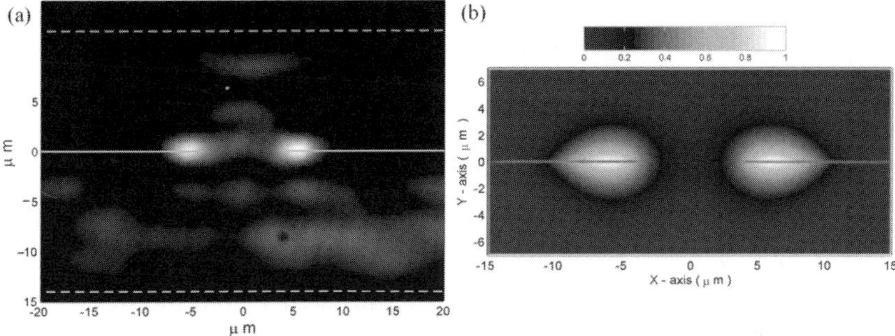

Fig. 7. Modal pattern of the plasmonic slot mode of an 8 µm wide slot in a 20 nm thin gold slab embedded in a 26 µm BCB cladding (λ=1.55 µm). The location of the metal is overlaid graphically: (a) experimental ***intensity*** pattern (vertical polarization). Dashed lines are located at the end of the polymer cladding layers; (b) calculated H_x (similar to E_y) ***field*** by the FVH-BPM.

Calculation of slot modes can be performed by a full vectorial *H*-field beam propagation method (FVH-BPM) – where the *H*-field is the preferred selection due to its continuity on the interfaces – in contrast to the discontinuity of the *E*-field. The calculated mode (Fig. 7b.) exhibits a field pattern very similar to the experimental results, and the calculated effective index: $n_{eff} = 1.5025 - j\cdot 0.00015$ (decay distance of ~0.8 mm), is typical to long range plasmonic (slow) modes.

To evaluate the slot mode confinement, we employed the figure of merit (FOM) suggested by Berini[14]. It is based on the ratio of transverse confinement and propagation distance:

$$M_2 = \frac{\Re\left(n_{eff}\right) - n_d}{-\Im\left(n_{eff}\right)} \tag{3}$$

where n_d is the dielectric index of refraction. The FOM of the slot mode was compared to LRPMs of stripes with similar thickness and cladding, resulting in the same magnitude as of a single mode LR-SPP. The FOM results for 5 μm wide stripe is M=13 while for a 8 μm wide slot M=16.

This micrometers wide plasmonic slot is a remarkable waveguide – it has a good modal confinement, long propagation length and supports a single mode also for wide slots. The H_x field (and the related E_y) is completely symmetric (across the slot and across the metal layer (Fig. 7(a)), which facilitates the excitation of this mode. Although the power is not propagating "in" the slot – it serves as an ideal long range plasmon waveguide.

3.2. *Modes of deep symmetrical trench modes*

Deep trenches are similar to the slots of the previous subsection but they are via-engraved in a much thicker gold layer (e.g. 400 nm). The dramatic change in the modal characteristics stems from the fact that as the metal becomes thicker (than the skin depth), the TM polarized single surface plasmons on both sides of the metal detach. As a result, the inner (vertical) metal interfaces become important as plasmon guiding interfaces and we expect a plasmonic mode confined within the trench, polarized perpendicular to the trench "walls" (major "TE" polarization – Fig. 5(c))[15]. The minor field component (vertical) is expected to be spatially located on the edges.

Experiments were performed by exciting a 2 mm long trench waveguide (in 400 nm thick gold layer embedded in 26 μm of BCB) by horizontally polarized light. The resulting modal patterns are shown in Figs. 8(a) and 9(a) for 6 μm and 10 μm wide trenches respectively. In both cases, this excitation resulted in a well defined intensity lobe inside the trench and smaller optical power at the vicinity of the edges of the metal layer (metal is graphically overlaid on the measured image). Pseudo-color images (Color Plate 1 at the end of the chapter) is a better reproduction of these details.

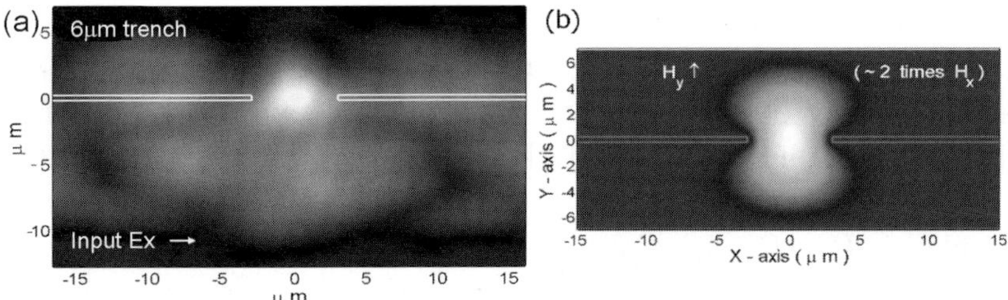

Fig. 8. Basic mode of a 6 μm wide trench excited by horizontal polarization at λ = 1.55 μm: (a) experimental ***intensity*** pattern of the optical power at the output facet; (b) H_y (similat to E_x) **major** component of the guided ***field*** calculated by FVH-BPM: n_{eff} = 1.496 + j • 0.000164. See H_x component and additional complete information in Color Plate 1.

The calculated modes – by FVH-BPM (Figs. 8(b), 9(b) and Color Plate 1), exhibit a modal pattern matching the observed trench mode, yet providing additional information related to the effective indices: n_{eff} = 1.4960 – j· 0.00016 and n_{eff} = 1.4977 – j· 0.000051 for the 6 μm and 10 *μm* wide trenches respectively, manifesting that these modes are slightly leaky-modes. The propagation length is surprisingly long: 0.75 mm and 2.4 mm for the 6 μm and 10 μm wide trenches, the latter is even longer than that of typical LR-SPPs of the complimentary (but much thinner) metal stripes. Most of the modal power is located at the mid trench central lobe and is horizontally polarized while a small fraction of the mode power, related to vertical polarization, is located along the two surfaces of the gold layer, in the vicinity of the trench, with a maximum amplitude approximately half – third of the peak field value of the central lobe.

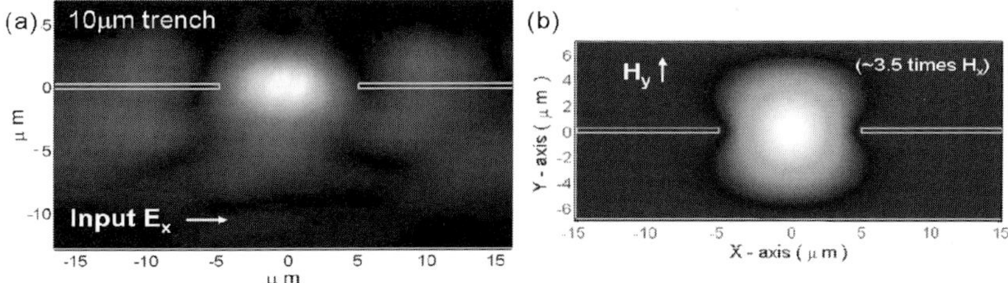

Fig. 9. Basic mode of a 10 μm wide trench excited by horizontal polarization at λ =1.55 μm: (a) experimental ***intensity*** pattern of the optical power at the output facet; (b) H_y (similar to E_x) **major** component of the guided ***field*** calculated by FVH-BPM: n_{eff} = 1.49768 + j • 0.0000513. See H_x component and additional information in Color Plate 1.

For vertically polarized input field, the same trenches exhibited different modal patterns, (Fig. 10, Color Plate 1 and the colored polarization resolved image in Color Plates 2(a)-(b)). The mode distribution in the trench has a higher lateral order (2) exhibiting two zero crossings near the metal edges of the main lob (clearly seen for the calculated H_y (E_x) in Color Plate 1 and also in the respective experimental result). The field on the gold layer surfaces is mainly vertically polarized and carries most of the power, while the central lobe is predominantly horizontally polarized. The calculated modal effective index e.g. for the 6 μm wide trench: $n_{eff} = 1.5079 - j \cdot 0.00147$ indicates a guided plasmonic (slow) mode ($n_{eff} > 1.5$) with a shorter propagation distance, similar to that of a single surface SPP (~80 μm).

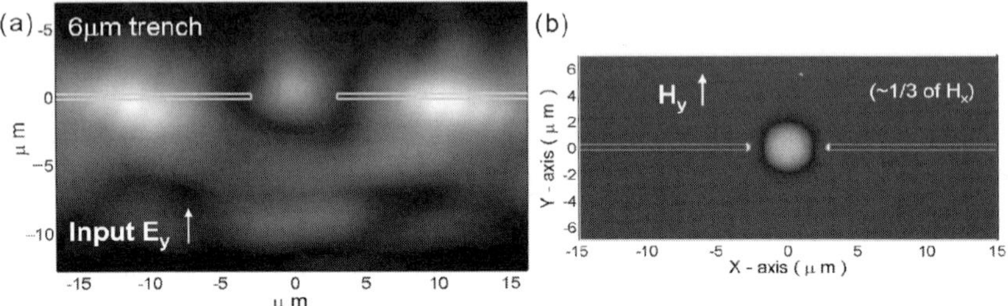

Fig. 10. Higher order mode of a 6 μm wide trench, excited by vertical polarization at λ=1.55 μm: (a) experimental ***intensity*** pattern of optical power at the output facet; (b) H_y (similar to E_x) **minor** component of the guided ***field*** calculated by FVH-BPM: $n_{eff} = 1.5079 + j \bullet 0.001468$. See H_x component as and additional information in Color Plate 1.

Trench modes with progressively higher radial order were calculated (Color Plate 3) for a 10 μm wide trench. All the modes are almost degenerated with modal effective indices: $n_{eff} \approx 1.508 - j \cdot 0.00148$ (slow modes), and very similar H_x (E_y) major component, shown in Color Plate 3, carrying most of the optical power. The various higher order modes are differentiated by their particular radial order characterizing the mid trench intensity lobe. It is interesting to get such radial symmetry, for a predominantly planar structure. These modes were not observed experimentally due to the poor coupling efficiency with the input field distribution.

The modal characteristic discussed above is consistent for trenches in the width range of 6 μm to 12 μm. However, narrower trenches (4 μm and below) supported predominantly the vertically polarized modes, where the in-trench lobe is completely eliminated as can be seen in Color Plate 2(b), for a 4 μm wide trench. It should be noted that although there is an apparent similarity to the slot mode of Sec. 3.2 the current mode is completely asymmetric (across the trench and across the metal). This is the result of the indirect coupling of the top and bottom plasmons which is mediated by the field within the trench.

3.3. *Conclusions*

Symmetric trench structures can be used to guide plasmons for long distances – similar and even better than thin metal stripes. For micrometers wide slots in very thin metal structures (thickness < skin depth), the propagating mode is predominantly the coupled plasmonic long range "edge" modes – from both sides of the slot – thus the mode is double lobed and most of the power is vertically polarized and propagating along the flat surface but laterally confined near the edges of the slot – resulting in a high quality long propagating single mode waveguide.

When the metal is thicker than the skin depth (yet the aspect ratio is still small < 10) the internal metal walls are contributing additional types of modes which are confined predominantly within the trench and have field polarization of plasmons of the inner vertical walls. Although this mode has an effective index very close to the free space value (even smaller) it is nicely confined and propagates for millimeters (even longer than long range plasmons of much thinner metal stripes). These modes are vanishing when reducing the trench width (for our parameters to less than 4 micrometers), leaving only modes similar to the previous slot case (main power on the flat surfaces near the trench edge). The latter modes are however different from the thin slot modes – because the metal thickness is large enough to disable direct (through the metal) coupling of the top and bottom plasmons. Thus these trench modes exhibit much shorter propagation length similar to that of single surface plasmon.

All of these modes cannot be evaluated by approximation such as the effective index method (EIM), since the propagating modes are in general hybrid, and the EIM cannot tackle properly also the edge localization effects.

4. Hybridization of modes in structured channels

In the previous section we dwelled with the significance of the substructures of the trench on its modal characteristics. Here we exemplify this significance by analyzing a complex structured channel (a trench which is terminated within the metal) – specifically a W shaped cross-section waveguide. This irregular shape was designed for some specific applications which are discussed elsewhere[16]. Although this structure is not a via-trench as all the other waveguides of this chapter – it is a good example for illustrating this subject. Such a W channel was prepared by employing anisotropic wet etching along <100> crystal planes of Silicon followed by a deposition of 400 nm thick gold layer.

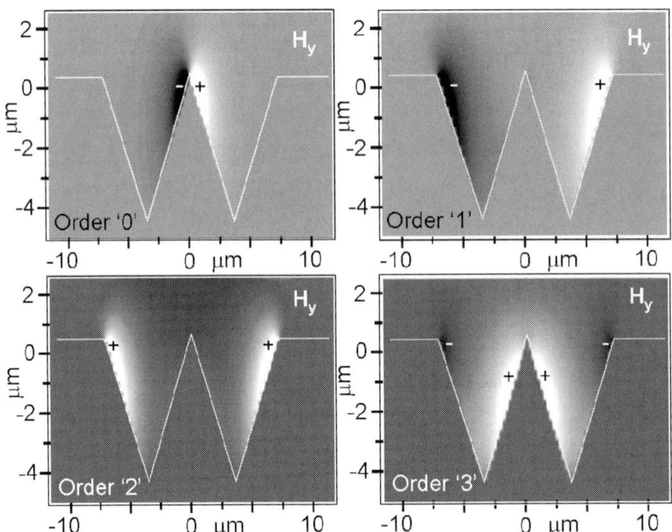

Fig. 11. Calculated H_y modal fields of the W-shaped plasmon channel, exhibiting 4 guided plasmon modes (white is +1and black is -1 for the normalized field amplitude).

Calculations of the mode structure were performed by a full vectorial H-field finite elements (FVH – FEM) solver. Whenever metallic wedges and corners with small radii are important – this mode solver surpasses the regular finite difference BPM (used in Section 3) due to much better adaptive calculations grid. Four guided plasmon modes ($n_{eff} > n_{spp}$ of a flat surface) were obtained for the air filed W groove (Fig. 11).

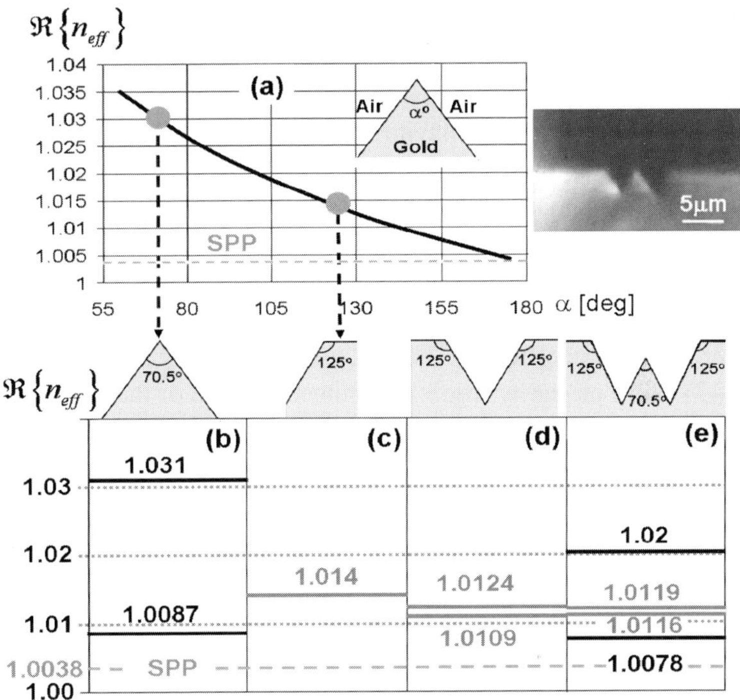

Fig. 12. The real effective index of plasmon modes for the different sub-structures comprising the W structure. Coupling is evident from the supermodes of the complete structure: (a) n_{REAL} vs. angle for a triangular gold wedge; (b) isolated 70.5° triangular gold wedge (central part of W); (c) isolated 125° triangular gold wedge (side part of the W); (d) single V-groove; (e) propagating modes of the W-wedge. The inset is an SEM photo of the W channel facet.

Using Fig. 12 we can attribute the resulting modes to various hybridizations of the modes of the W constituents, the central wedge and the edges of the V-groove on each side. For example a V groove (Fig.

12(d)) is generated by the coupling of 2 side wedges of Fig. 12(c) – showing the split of effective indices due to symmetrical and anti-symmetrical coupling. At these dimensions (the structure is few wavelengths wide) – the coupling is small (the split is small) and in addition the V bottom wedge does not contribute significantly (although pulling slightly the two modes towards lower n_{eff} values).

The zero order mode of the W waveguide (Fig. 11, n_{eff}=1.020-j·0.0015, propagation length of ~80 μm) is localized at the central wedge of the W and is similar to the anti-symmetric mode of this isolated wedge of the same angle (Re{n_{eff}}=1.031). The lowered n_{eff} is due to some coupling with the outer wedge modes having Re{n_{eff}}=1.014. The 1st and 2nd order modes are the symmetric- and anti- symmetric combinations of the modes of the two outer wedges that comprises the W, as reflected by their spatial distribution and n_{eff} values. Note that the split between the effective indices of these modes is smaller than in the V groove since they are further spatially apart in the W configuration (resulting in smaller coupling coefficient). The 3rd order mode (Fig. 12, n_{eff}= 1.0078-j·0.0017) stems primarily from the symmetric mode of the central wedge (Fig. 12b, $Re(n_{eff})$= 1.0087) with some anti-symmetric coupling to the side wedge structures – evident from a slightly larger spatial spread of the modal field and the deviation of the n_{eff} value.

Comparing these modal fields with the experimental results – we get a very good match with the hybrid field of the 3rd order mode as shown in Fig. 13. The intensity at the two polarization directions is shown: major component E_x (vertical and similar to H_y) which is symmetric about the central wedge and the minor component E_y (horizontal) anti-symmetric with an intensity notch at the wedge center). It is clear by symmetry that this mode has the best coupling to the symmetric input excitation localized at the W center.

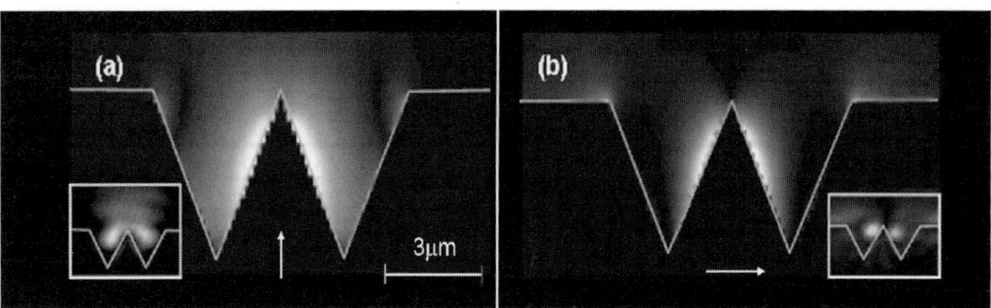

Fig. 13. Calculated intensity distribution of the 3rd order mode of the air-Au W channel waveguide: (a) vertical polarization; (b) horizontal polarization. Insets: respective experimental results (see details in Fig. 14).

The detailed measurement results of the W-shaped plasmon waveguide are depicted in Fig. 14 showing the output light distribution for a linearly polarized input aligned at an angle relative to the sample plane as to excite both polarization modes. The W modes are hybrid with horizontal and vertical components of similar magnitudes, showing clearly the anti-symmetrical lateral characteristics of the vertical component (the intensity notch at mid slot).

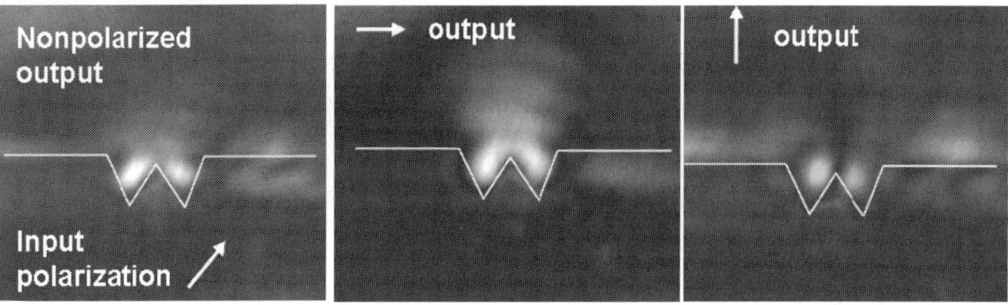

Fig. 14. Measured light distribution from a W-shaped plasmon waveguide. Output polarization in the sub figures is: random, horizontal, and vertical as indicated.

As a summary – the W channel exemplifies nicely the drastic differences between plasmonics and dielectric waveguides. Here the supported modes are clearly determined by coupling of the wedge constituents of the structure. Furthermore we can use this understanding to design plasmonic waveguides with better light confinement – e.g. by adding a small angle wedge in the mid of a flat trench waveguide. This type of modes cannot be foreseen by approximation methods such as the EIM. Only under very specific designs the wedge coupling becomes the minor effect – e.g. for a subwavelength V groove with very high aspect ratio. This "tall" structure may support a field deep within the groove – far way the side wedges such that the mode is rather determined by the coupling of the slanted metal interfaces. These channel modes were analyzed and measured[4,17,18] and in this case they do match the EIM results. This is due to the fact that the modes are generated by the interface coupling (not by wedges) and they have almost a single E component (horizontal), while wedge related modes are hybrid.

5. Nano – wide asymmetrical plasmonic trenches

One of the visions of the plasmonic circuitry field is to realize plasmonic based interconnects at the nano-scale. Thus the realization of nano-wide plasmonic waveguides on a Si wafer is a major quest. In this section we shortly discuss "edge" plasmon modes in highly asymmetrical nano-trenches. Plasmons confined in nano-trenches (slots) were discussed before (also in this book) showing the possibility of tight confinement in high aspect ratio structures with slight[4] or moderate[19] asymmetry.

We focus on nano-waveguides engraved in gold layer (50-200 nm thick) on a Silicon substrate and air as a superstrate. Under such asymmetrical conditions a metal stripe cannot be used as a plasmonic waveguide in the nano regime. Examining (experimentally) a laterally tapered stripe made in 50 nm thick gold layer over Silicon reveals a modal cutoff at a width of ~4.5 μm (Fig. 15).

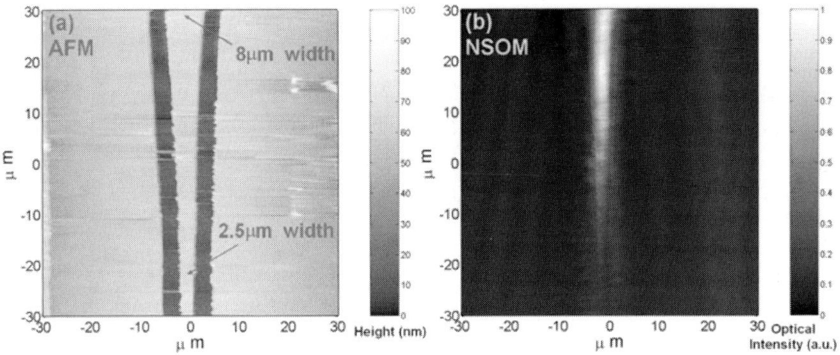

Fig. 15. Plasmonic mode supported by a tapered, 50 nm height gold stripe, at λ=1.55 μm. (a) AFM image; (b) NSOM optical image exhibiting a cutoff at width of ~4.5 μm.

But, also trenches in such an asymmetrical setting require unrealistic aspect ratios (>>1) for tightly confining the light within the trench. If one relaxes the tight confinement requirement, trench modes similar to the coupled edge modes of Section 3 are revived. The double lobed field propagating near the edges of sub wavelength slots can be used as the "long propagating" plasmons of the nanocircuitry arena. Furthermore, two coupled slot structures define, in-between, a stripe waveguide which may be reduced in dimensions into the nanometric regime. Thus a complete plasmonic coplanar circuit suite is defined (slot-line, and coplanar) which is similar to the respective microwave family.

When numerically probing for such plasmonic modes in a trench made in a thin gold layer, the FVH-FEM results verify the existence of "edge" modes for a sub-wavelength wide trench, having optical power peaks at (near) the metal edges. The respective field pattern is depicted in Color Plate 4(a) for a 250 nm wide trench, having an effective index: n_{eff} =1.0011 – j·0.004. The mode is symmetric (across the trench) quasi-TM (major vertical E component) and the two main lobes are accompanied by a secondary lateral and less confined bell shaped envelope. Horizontally polarized component does exist inside the trench, yet it is the minor one.

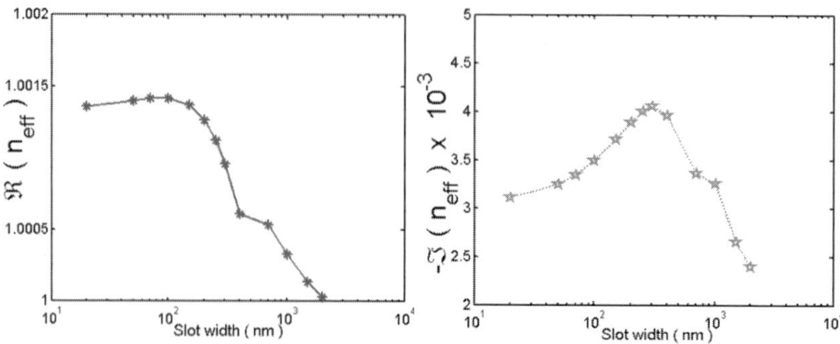

Fig. 16. Calculated effective-index of nano-trench modes as a function of the trench width. Trench are engraved in 100 nm thick gold with Silicon as substrate and air as superstrate at λ=1.55 μm

In order to track the modal existence range, n_{eff} was calculated vs. the slot width. From the results (Fig. 16) it is evident that such a coupled "edge" mode is supported in the trench for a width range in the order of 10-100 nm. Reducing the width, results in a mode which is asymptotically approaching the single surface plasmon mode of a uniform metal layer. On the other hand as the width is expanded beyond the sub-wavelength regime, the slot mode reaches a cutoff due to radiation.

Some observations on the mode characteristics are emphasized here. The use of "trench" (which by our definition applies to thick metal >> skin depth) is justified here also for much thinner metal layers. This stems from the large difference between the sub- and super- strate indices of refraction, which results in the localization of the modes on a single interface of the metal (as can be seen also in Color Plate 4). The mode – localized on the metal-air interface, has an effective index slightly larger than that of air but significantly lower than that of Silicon (3.5), thus it is formally a leaky mode. However – since it is pinned to the metal air surface, the mode is virtually isolated from the high index substrate and does not suffer radiation loss – until the trench width becomes larger than ~1 wavelength (Fig. 16), allowing the mode to tunnel efficiently to the substrate. The cutoff from above confirms that this mode exists only in the nanometers regime (for asymmetrical structure). Similar leaky modes of asymmetrical stripes were discussed in Ref. 20. For stripes the leaky mode of the air metal interface is not isolated from the substrate due to the finite lateral extent of the mode. Thus, as the stripe becomes narrower - the leakage of the modal field to the high index substrate is enhanced – until a cutoff (from below) is reached as shown experimentally above (Fig. 15).

Although the trench mode is confined near the trench edges - the lateral confinement is not tight ($Re\{n_{eff}\}-n_{air}=0.0015$) and one cannot expect to implement sharp curves or abrupt structures. These waveguides can be used primarily as the "long-haul transmission channel" of the plasmonic nano-circuit, exhibiting up to 100μm propagation length compared to tightly confined modes having propagation lengths of few micrometers[4,13,19].

As mentioned and explained above - a stand-alone gold stripe cannot support propagating modes in the asymmetrical subwavelength regime. However, nano-stripes can be formed by two adjacent nano trenches. The FVH-FEM calculations confirm, e.g. for two 250 nm wide trenches, separated by a 300 nm wide metal stripe, the existence of such stripe mode (Color Plate 4(b)). The modal effective index is similar to that of an isolated nano trench mode, and the mode peak power is clearly located on the stripe.

To verify these predictions, we examined waveguides prepared by the deposition of gold layer on Silicon and subsequently nano-patterned by focused ion beam (FIB) milling of the gold. Fabricated waveguides are highly asymmetrical – having air as a superstrate. Light (λ=1.55 μm) was injected from a polarization maintaining lensed fiber and modal field distribution was measured by near field optical microscopy (NSOM) – allowing the observation of nano-resolved field patterns[21].

The measurement results of such a nano-stripe (or 2 coupled trenches), are depicted in Fig. 17. The modal power profile was extracted from the NSOM picture, and fits the theoretical prediction (Color Plate 4(b)), having maximum optical power at the sub-wavelength stripe. The expected two lobes of the intensity pattern at the stripe are poorly resolved due to resolution limit of the NSOM tip we used (~150 nm).

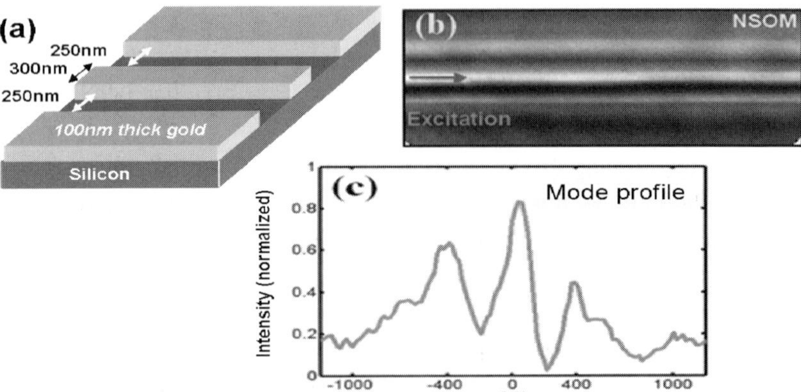

Fig. 17. Subwavelength stripe plasmon waveguide excited at λ=1.55 μm. (a) Schematics; (b) NSOM intensity image; (c) mode intensity profile.

To examine further the plasmon co-planar modal family we constructed a direct coupler between subwavelength stripe (coplanar) and trench (slot-line) waveguides. The structure (Fig. 18(a)) is comprised of a 300 nm wide stripe in between two 250 nm wide trenches. The stripe and one trench are abruptly cut while the other trench is continuing. The NSOM recording (Fig. 18(b)) manifests the features of these modes: the incoming modal field intensity is maximal on the stripe and has secondary peaks on the outer trench edges. The secondary bell shaped envelope is few micrometers wide. At the coupling plane, the field is abruptly shifted and relocalized about the center of the trench waveguide, with high coupling efficiency.

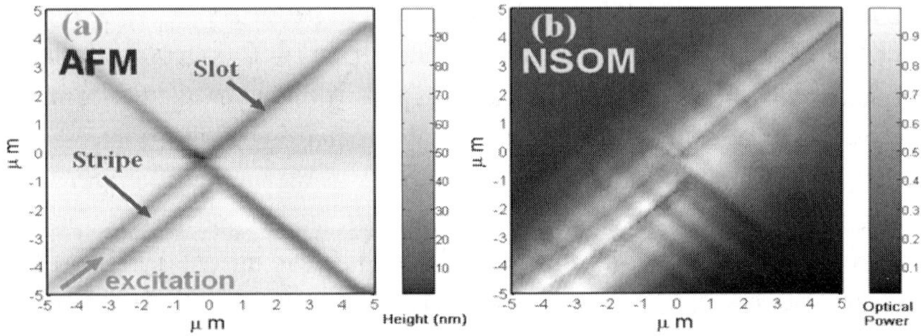

Fig. 18. Direct coupling of a nano-stripe to nano-trench. (a) SEM photo of the coupler, and (b) NSOM image acquired by launching light at λ = 1.55 μm through the stripe.

6. Conclusions

Trench waveguides are shown to be superior compared to their complimentary stripe structures both at the micrometer and nanometer regime. Micrometers wide trenches can be used to propagate confined plasmons for very long distances as a single mode even for very wide structure (larger than 10 micrometers). In the nanometers regime – highly asymmetrical trenches guide plasmons for tens of micrometers, where stripes are far beyond their modal cutoff (even their leaky modes). In most cases the guiding mechanism is related to the trench edge coupling, while specific structures are exhibiting a variety of auxiliary coupling effects (direct top-bottom plasmon coupling, trench assisted (indirect) top-bottom plasmon coupling, vertical wall coupling etc.).

The analysis presented above and its experimental validation is clearly indicating that plasmonic trench waveguides are very different from regular dielectric waveguides. The modes in most cases are determined by sub structures such as wedges, edges and their coupling, which has no parallels in dielectrics. Furthermore the simple approximations that are intensively used in the design of dielectric photonic circuits – such as the effective index method, are not valid for these modes and applicable only in some extreme cases where wedges and similar discontinuities are forced to be insignificant. But – on the positive side we are equipped here with a diverse and expanded tool box for designing guiding structures for plasmons. For example one can add into a waveguide some wedges (or reduce existing wedge angle) to increase locally the confinement – for tight bending or enhanced interaction with small volume medium; or remove wedges (enhance wedge angle) for supporting long plasmon propagation. We believe that we are still far away from the exhaustion of these possibilities in realization of unique plasmonic circuitry.

Acknowledgments

This chapter was written based on the devoted work of my graduate students: Yinon Stav (Satuby), David Arbel, Nikolai Berkowitz, Eyal Figenbaum and Pavel Ginzburg.

I would like to acknowledge also the Israeli Ministry of Science for a partial support of the work.

References

1. E. N. Economou, "Surface Plasmons in thin films", *Phys. Rev.*, vol. 182, pp. 539-554, 1969.
2. B. Prade, J.Y. Vinet, A. Mysyrowicz, "Guided optical waves in planar heterostructures with negative dielectric constant", *Phys. Rev. B*, vol. 44, pp. 13556–13572, 1991.
3. E. Feigenbaum and M. Orenstein, "Modeling of Complementary (Void) Plasmon Waveguiding", *J. of Lightwave Technology,* Vol. 25, pp. 2547-2562, 2007.
4. S. I. Bozhevolnyi, "Effective-index modeling of channel plasmon polaritons", *Opt. Express*, vol. 14, pp. 9467–9476, 2006.
5. D. F. P. Pile, T. Ogawa, D. K. Gramotnev, T. Okomoto, M. Haraguchi, M. Fukui, S. Matsuo, "Theoretical and experimental investigation of strongly localized plasmons on triangular metal wedges for subwavelength waveguiding", *Appl. Phys. Lett.*, vol. 87, 061106, 2005.
6. E. Feigenbaum, M. Orenstein "Nano Plasmon Polariton Modes of Wedge Cross Section Metal waveguide", *Opt. Express*, Vol. 12, pp. 8779-8784, 2006.
7. F. Yang, J.R. Sambles, G.W. Bradberry, "Long-range surface modes supported by thin films", *Phys. Rev. B*, vol. 44, pp. 5855–5872, 1991.
8. P. Berini, "Plasmon-polariton waves guided by thin lossy metal films of finite width: Bound modes of symmetric structures", *Phys. Rev. B*, vol. 61, pp. 10484-10503, 2000.
9. T. Nikolajsen, K. Leosson, I. Salakhutdinov, S. I. Bozhevolnyi, "Polymer-based surface-plasmon-polariton stripe waveguides at telecommunication wavelengths", Appl. *Phys. Lett.*, vol. 82, pp. 668-670, 2003.
10. Y. Satuby, M. Orenstein, "Surface Plasmon Polariton Waveguiding - From Multi-Mode Stripe to a Slot Geometry", *Appl. Phys. Lett.*, Vol. 90, pp. 251104-251107, 2007.
11. D. Natan and M. Orenstein, "Localization immunity and coherence of extended two-dimensional semiconductor vertical-cavity locked laser arrays", *JOSA B*, Vol. 14, pp. 1501–1504, 1997.
12. U. Sivan and A. Sa'ar, "Light Wave Localization in Dielectric Wave Guides", *Europhys. Lett.*, vol. 5, pp. 139-144, 1988.
13. D. F. P. Pile, T. Ogawa, D. K. Gramotnev, Y. Matsuzaki, K. C. Vernon, K. Yamaguchi, T. Okamoto, M. Haraguchi, M. Fukui, "Two-dimensionally localized modes of a nanoscale gap plasmon waveguide", *Appl. Phys. Lett.*, Vol. 87, 261114, 2005.
14. P. Berini, "Figures of merit for surface plasmon waveguides," *Opt. Express,* Vol. 14, pp. 13030-13042, 2006.
15. Y. Satuby, M. Orenstein, "Surface-Plasmon-Polariton modes in deep metallic trenches- measurement and analysis", *Opt. Express*, vol. 15, pp. 4247–4252, 2007.

16. D. Arbel and M. Orenstein, "Plasmonic modes in W-shaped metal-coated silicon grooves" *Opt. Express* (in publication).
17. I. V. Novikov, A. A. Maradudin, "Channel polaritons", *Phys. Rev. B*, vol. 66, 035403, 2002.
18. S.I. Bozhevolnyi, V.S. Volkov, E. Devaux, W. Ebbesen, "Channel Plasmon-Polariton Guiding by Subwavelength Metal Grooves", *Phys. Rev. Lett.*, vol. 95, 046802, 2005.
19. G. Veronis, S. Fan, "Guided subwavelength plasmonic mode supported by a slot in a thin metal film", *Opt. Lett.*, vol. 30, pp. 3359–3361, 2005.
20. R. Zia, M. D. Selker, M. L. Brongersma, "Leaky and bound modes of surface plasmon waveguides", *Phys. Rev. B*, vol. 71, 165431, 2005.
21. Y. Satuby, N. Berkowitch and M. Orenstein, "Coupling of Nano-Stripe and Nano-Slot Plasmonic Waveguides", *CLEO*, paper QThG2, 2007.

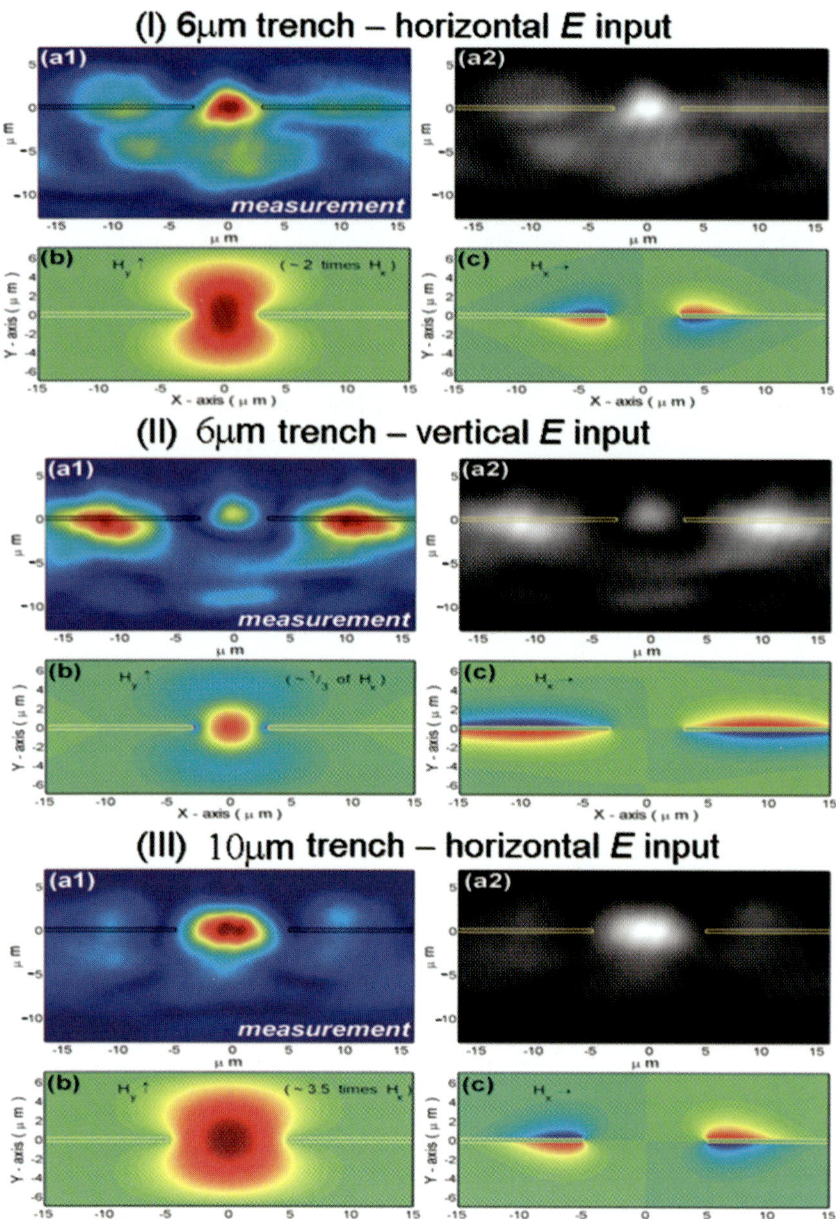

Color Plate 1. Trench mode patterns for 6 μm wide (I) horizontal and (II) vertical input E field and 10 μm wide (III) horizontal input *E* field. For each: experimental ***intensity*** (a1) pseudo-color; (a2) gray scale; calculated *H* ***field*** (b) vertical; (c) horizontal.

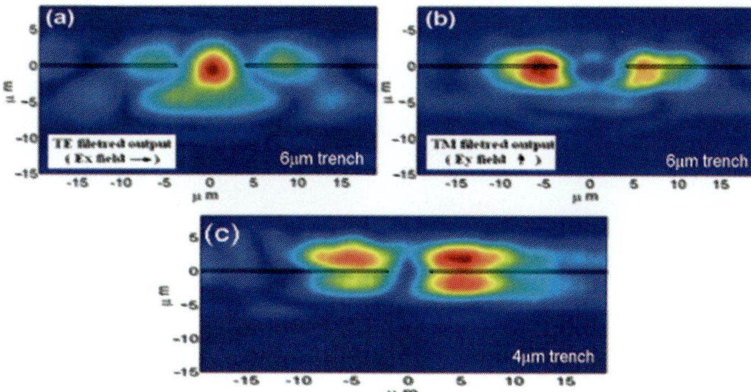

Color Plate 2. Measured mode intensity pattern for vertical polarization input: 6 μm trench (a) horizontal output polarization; (b) vertical output polarization; (c) 4 μm trench.

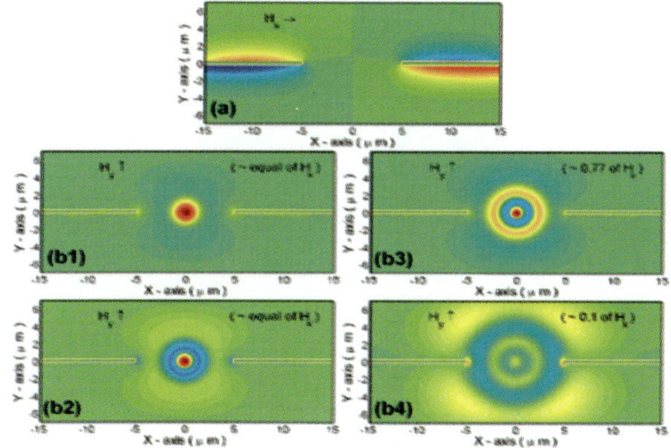

Color Plate 3. Calculated field patterns of the higher order modes of a 6 μm wide trench.

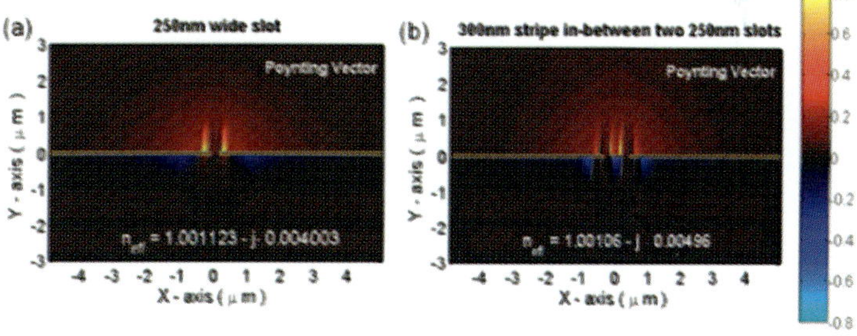

Color Plate 4. Calculated power density of plasmonic nano- trench and coplanar stripe modes.

Chapter 9

FUNDAMENTALS OF CHANNEL AND WEDGE PLASMON POLARITONS

Esteban Moreno and F. J. Garcia-Vidal

Departamento de Fisica Teorica de la Materia Condensada, Universidad Autonoma de Madrid, E-28049 Madrid, Spain

E-mail: esteban.moreno@uam.es

Sergio G. Rodrigo and L. Martin-Moreno

Departamento de Fisica de la Materia Condensada, Universidad de Zaragoza-CSIC, E-50009 Zaragoza, Spain

This chapter analyzes the electromagnetic modes supported by metallic grooves and wedges. Various approaches employed for the computation of the electromagnetic field in such structures are presented. The wave guiding properties are determined by means of rigorous electrodynamics techniques. Dispersion, modal size, shape, polarization, and propagation length of these modes is discussed. It is concluded that both channel and wedge modes are well confined and feature reasonable losses, wedge structures being favorable in terms of modal size.

1. Introduction

Channel plasmon polaritons (CPPs) are electromagnetic modes supported by grooves carved in metallic surfaces whereas wedge plasmon polaritons (WPPs) are the corresponding modes sustained by metallic wedges. One can say, without being too precise at this point, that the electromagnetic field of CPP and WPP modes is guided along the bottom of the groove or the edge of the wedge, respectively. For reasons discussed in the next paragraph, these modes have received quite a lot of attention in the last years. This chapter is devoted to a description of the guiding properties of CPPs and WPPs.

There exists interest on the behavior of electromagnetic fields next to corners and edges since a long time. To our knowledge, the first report dealing with these geometries within the context of surface plasmons was published by Maradudin and coworkers.[1] The mentioned paper considers an idealized geometry in the electrostatic approximation. The next landmark, in 2002, was a complete treatment including retardation and for realistic geometries.[2] After this year the number of works, both theoretical and experimental, reporting on CPPs and WPPs increases rapidly. We can mention the following reasons for this late interest. First, from a general point of view, in the last ten years or so the field of plasmonics has blossomed and various SPP-based waveguiding schemes are being considered. The achievement of tightly confined modal fields and long propagation lengths count among the main design goals. We shall see that CPPs and WPPs feature good confinement and a reasonable propagation length, and are therefore promising candidates. Second, the planar paradigm is preferred from a technological perspective, and the modes studied in this chapter fit well with planar metallic structures. Let us mention that in spite of the difficulties to fabricate narrow angle CPPs and WPPs, these have been already demonstrated and CPP-based functional devices have been reported.[3] Finally, from a more fundamental point of view, let us remark that edges and corners appear whenever a flat surface is folded. In this sense CPPs and WPPs constitute building blocks that show up in other kinds of plasmonic guides such as stripes, trenches, gaps, and so on. Thus, understanding the properties of CPPs and WPPs is very useful for the design of most of the remaining plasmonic waveguides considered in this book (an important exception being circular cross section waveguides).

As mentioned above, the goal of this chapter is providing an understanding of the properties of CPP and WPP modes. We shall be dealing with straight waveguides, leaving for other chapters the analysis of devices based on the electromagnetic modes studied here. We shall describe the behavior of modal field, dispersion, modal size, propagation length, and the dependence of these magnitudes on various parameters. In order to build up some intuition, we shall briefly mention in Section 2 some approximate approaches, but the main emphasis of the chapter lies on properties calculated by means of rigorous computational electrodynamics techniques. In Section 3 the modes of simple grooves and wedges are considered, and the obtained results will be employed in Section 4 to understand more realistic structures. We close the presentation with a summary of the main results.

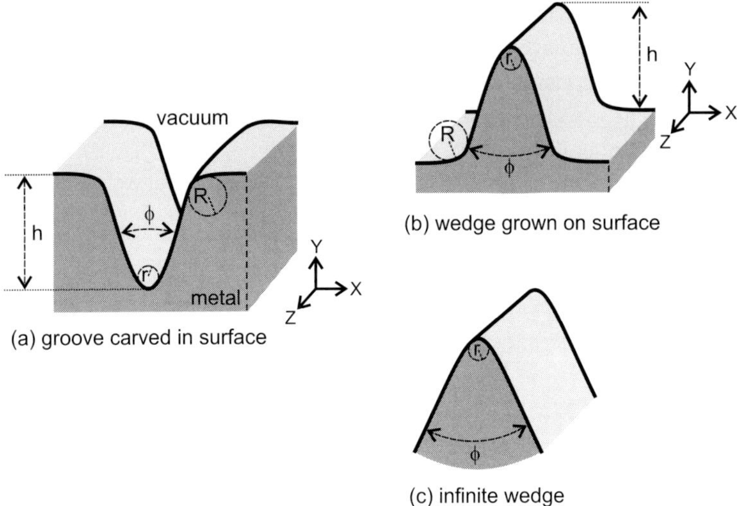

Fig. 1. Schematics of structures supporting channel and wedge plasmon polaritons.

2. Approximate approaches

The problem we are interested in is the following. Figure 1 displays examples of the structures being analyzed, and the definition of coordinates and geometric parameters. We consider structures of two different kinds: structures truncated at a certain height $y = h$ [Figs. 1(a) and (b)], and idealized non-truncated structures where $h \to \infty$ and the inclined sides of the groove or wedge are infinitely extended [*e.g.*, Fig. 1(c)]. The modes corresponding to non-truncated structures will be named CPP(∞) and WPP(∞). In most cases the lateral sides of the groove or wedge are approximately flat and form an angle denoted by ϕ. For reasons explained in Section 2.1, corners and edges shall be rounded and the corresponding radii of curvature are denoted as r or R. The chosen radii are always larger than 5 nm. With this election, the minimum feature size of the structures is sufficiently large so as to use bulk dielectric functions in the metal. Notice that, for small feature sizes (radii smaller than about 2 nm), one has to take into account more complex effects in the dielectric function of the metal. Namely, a size-dependent damping due to electron scattering at the metal surface, or/and a non-local behavior of the permittivity of the metal $\epsilon_r(\omega, \mathbf{k})$, ω and $\mathbf{k}$ being the frequency and wave vector, respectively. In addition, it should be mentioned that smaller radii are extremely difficult to fabricate.

We shall be concerned with modes propagating along the Z axis and confined in the transverse XY plane. The determination of the modes supported by the considered structures requires the resolution of Helmholtz equation for the electromagnetic fields in vacuum and metal, supplemented by the boundary conditions at the interfaces.

2.1. *The electrostatic limit*

The simplest approximation consists in disregarding retardation. In this limit Helmholtz equation is replaced by Laplace equation. There are some wedge (or groove) geometries for which the equation can be solved analytically, the easiest one being a sharp ($r = 0$) wedge, with flat sides and infinite height ($h \to \infty$), see Fig. 1(c). Laplace equation for this configuration can be separated in polar coordinates and was considered in Ref. 1. The corresponding solution has the attractive of being very simple but it has two important problems. First, the fields are divergent at the tip of the structure, and second, the computed modes do not have an standard dispersion relation $\omega = \omega(k_z)$, where ω and k_z are the frequency and modal wave vector, respectively. It was soon realized[4] that both defects are due to the infinitely sharp tip. In the following years analytical solutions for wedges with rounded edges (corresponding to parabolic[5] and hyperbolic[6] cross sections) were computed, and it was shown that the mentioned difficulties no longer occur. However, it can be seen[7] that the dispersion relation obtained within this framework is not a good approximation when it is close to the light line. This regime turns out to be the most interesting one, and for this reason the remaining approaches take into account the finite propagation speed of the electromagnetic fields.

2.2. *Geometric optics approximation and effective index method*

The geometric optics approximation (GOA) and the effective index method (EIM) have been used to understand several aspects of the behavior of electromagnetic fields in grooves and wedges. Their main merit is that they allow one to obtain important properties of the electromagnetic field and, moreover, they are not heavily numerical techniques. These techniques are described in more detail in other chapters of this book and we shall thus only sketch the core ideas behind them.[8,9]

For both approaches it is useful to consider the wedge (respectively groove) as a vertical stack of metallic films (respectively vacuum gaps in

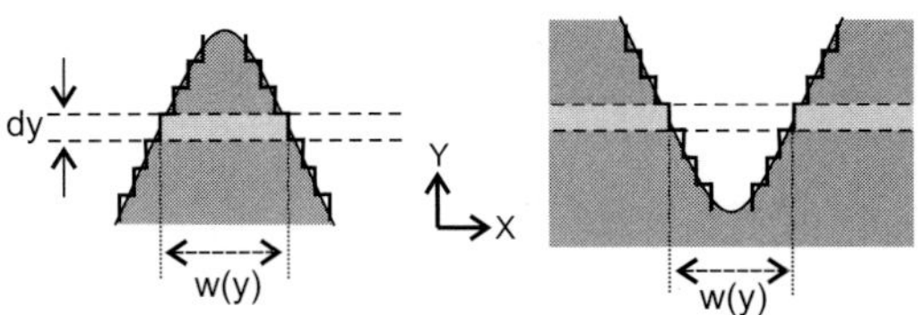

Fig. 2. Stack of variable width metallic films (left) or gaps (right).

a metal) with infinitesimal vertical thicknesses, and horizontal widths depending on the vertical coordinate: $w(y)$, see Fig. 2.

The effect of the varying width of the wedge (or groove) is taken into account by considering the modes supported by the infinitesimal slices. The dispersion relations of the first modes supported by a metallic film or gap with constant width w is plotted in Fig. 3 for several widths (for the sake of simplicity the dielectric constant of the metal is represented by a Drude model with a typical plasma frequency $f_{\rm p} = 1.91 \times 10^{15}{\rm s}^{-1}$ and no absorption).

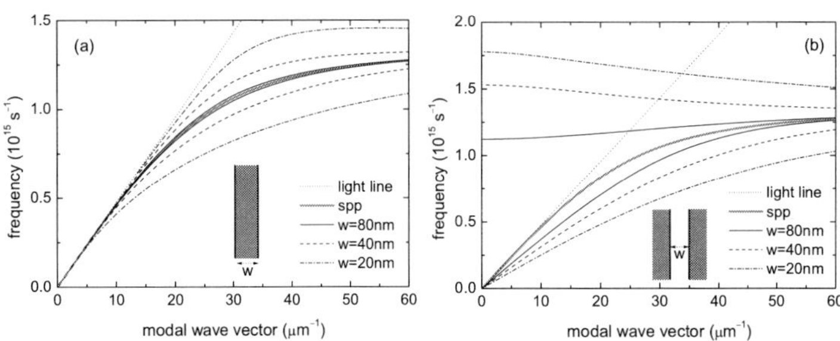

Fig. 3. Dispersion relation for plasmonic modes supported by (a) film, and (b) gap, both of width w.

As it is well known, for increasing width the SPP modes at both interfaces decouple and the modal dispersion relation tends to that of a single SPP on a flat interface. In the case of the film, those modes below the SPP curve have even parity with respect to symmetry plane[a] (they are named

[a] At this point it may be useful to dissipate a possible terminology confusion that may arise. The electromagnetic field of the mentioned modes has even parity, but the functional dependence of the various components (parallel or perpendicular to the surface) of **E** and **H** can be symmetric or antisymmetric with respect to the symmetry plane. Different conventions are used in chapter 1 and here: modes with even parity are termed *antisymmetric* in chapter 1, attending to the functional form of the electric field component perpendicular to the interface (beware also that the labels for the Cartesian axes is not the same in both chapters).

Short Range Plasmon Polaritons). On the other hand, the gap modes below the SPP curve have odd parity. For both configurations, the modes below the SPP curve become more and more separated from the dispersion relation of a SPP as the width w decreases. In other words, at a given frequency the modal effective index ($n_{\text{eff}} = k_z/k_0$, k_0 being the wave vector in vacuum) of these modes depends on the width, $n_{\text{eff}}(w)$, increasing as the width decreases. This is the necessary behavior to achieve field confinement at the bottom of the grooves (or the edge of the wedges), where the width is smaller. The reason is that the modes corresponding to a small width $w_<$ are outside the dispersion curve corresponding to a larger width $w_>$. Thus, the former modes are evanescent along the vertical direction in the gap (or film) of width $w_>$ and are therefore vertically confined.

The geometric optics approximation and the effective index method are based on the above explained behavior of the effective index of the modes supported by a gap or a film. In both approaches the groove (or wedge) is replaced by a dielectric medium the refractive index N of which is non-homogeneous and is defined as follows: (i) N is constant in the horizontal (XZ) planes, and (ii) $N(y) = n_{\text{eff}}(w(y))$. Notice that the modal effective index $n_{\text{eff}} \to \infty$ as $w \to 0$, so that $N(y)$ diverges at a height y corresponding to the structure tip, Figs. 4(a) and (b).

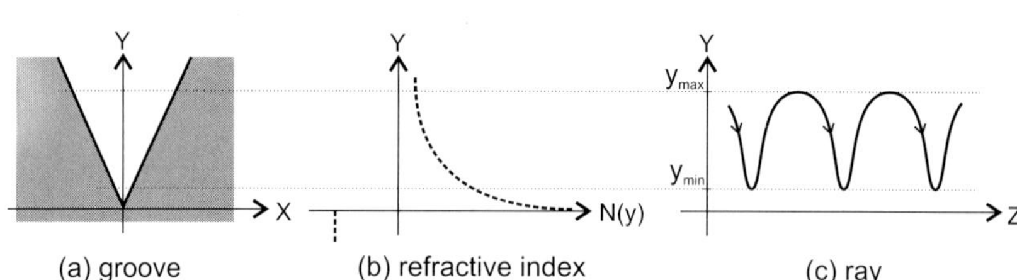

Fig. 4. Geometric optics approximation: (a) groove geometry, (b) refractive index $N(y)$ of equivalent medium, (c) ray path corresponding to guided mode.

The GOA consists in the optics of rays (representing the propagation direction of gap plasmons) in the mentioned non-homogeneous dielectric medium. It predicts the important properties that the rays bend toward the tip, slowing down while the corresponding field amplitude increases, leading to a nanofocusing effect. Within this picture a guided mode corresponds to a ray progressing along the Z coordinate while bouncing back and forth between two heights $y_{\min}, y_{\max}$. The lower height $y_{\min}$ corresponds to the point where the adiabatic approximation of GOA breaks down and the ray is reflected, Fig. 4(c). Thus, guided modes are consistent with the GOA but their properties cannot be determined within this framework.[8,10]

The EIM solves numerically the one dimensional Helmholtz equation in the mentioned non-homogeneous dielectric medium. The basic formulation assumes that the electromagnetic field has the same vector components as the gap (or film) plasmonic mode, *i.e.*, H_y, E_x, E_z, having transverse magnetic (TM) character. In this fashion a dispersion relation, propagation length, and vertical modal profile can be found.[9,11] This is a very useful approach but there are a number of issues to be improved. First, the computed dispersion relation and propagation losses are approximate. For CPPs, in most cases, the results are surprisingly good, but for WPPs the approximation is worse. This is most likely due to the fact that the assumed polarization is (approximately) the correct one for CPPs whereas it is not correct (near the tip) for WPPs (see Section 3.2). Second, the two-dimensional modal shape and polarization cannot be computed. This impedes the quantitative computation of the coupling of the mode with external incident fields. Third, for grooves truncated at a finite height $y = h$, two additional side edges appear that sustain WPP modes. This effect, that substantially modifies modal shape, size, polarization, and propagation length, cannot be accounted for within this approximation. Finally, the divergence of $N(y)$ at the tip of the structures leads to some questions (not yet definitively answered within the EIM) about the value of the field at this location.[9]

For the explained reasons, the above described approximate techniques are not completely adequate to determine the properties of CPP and WPP modes. In the rest of the chapter we analyze these properties with the help of rigorous computational electrodynamics techniques.

3. Rigorous techniques. Basic structures

There are a few rigorous methods that have been applied for the computation of CPPs and WPPs. The first one is described in an early paper dealing with retarded WPP modes.[7] The technique is accurate, but it is designed for a very specific geometry (parabolic cross section cylinder). A method valid for generic geometries (based on Green's second integral identity) was presented by Maradudin and coworkers.[2] Since then various general purpose numerical techniques have been used to study CPPs and WPPs, such as the Multiple Multipole (MMP) method,[12–14] the Finite Differences in Time Domain (FDTD) technique,[13–18] and the (FE) method.[19–22] In addition, a modal expansion method was reported in Ref. 23. The results presented from now on have been computed with the MMP method or/and the FDTD technique. The structures are made of gold in vacuum. The dielectric constant of gold is represented by a Drude-Lorentz model[24] (including absorption), which nicely fits the well known experimental data of Johnson and Christy[25] in the spectral range considered here, $\lambda \in [0.6\,\mu\text{m}, 1.6\,\mu\text{m}]$. The structures considered in this section are not truncated ($h \to \infty$).

3.1. *Dispersion*

Figures 5 and 6 render the dispersion relation corresponding to CPP(∞) and WPP(∞) modes. As corresponds to bound modes they are outside the light line and outside the dispersion curve of a SPP on a flat surface, *i.e.*, outside the shaded regions in these figures (otherwise the mode would be radiated either to free space or along the metallic interface). As mentioned in Section 2.1 the modes are very close to the SPP curve at telecom wavelengths rendering the electrostatic approximation inadequate to determine accurate values of the dispersion relation. For high frequencies the dispersion relation is far from the light line which means, as we shall see in Section 3.2, that the modes are well confined at the tip of the structures. On the other hand the modal size increases for small frequencies. Let us mention that this behavior imposes stringent conditions on the mesh size, and the size of the simulation window. For high frequencies a very fine mesh was needed (for FDTD the results were converged for a mesh size of 5 nm), whereas large simulation windows were needed for small frequencies (for MMP the size has to be much larger than the modal size, see Section 3.2). In the small frequency regime the size of the numerical problem rapidly increases, and for this reason it is very difficult to demonstrate numerically that the modes have no cut-off (and are asymptotic to the light line). However, we never found a cut-off and the exact electrostatic solutions suggest that no one exists.

Figure 5 displays the first two CPP(∞) modes but it is clear that the structure supports an infinite number of them, higher order modes having larger modal size. The modes corresponding to the curves have odd parity, as hinted in the discussion in Section 2.2. On the other hand Fig. 6 displays a single WPP(∞) mode of even parity. Actually, higher WPP(∞) modes exist, but they appear for very high frequencies, outside the scale of the diagram. These higher order modes are very much confined at the tip (in the region where the structure is rounded) but, unfortunately, they are extremely lossy and thus useless for guiding purposes.

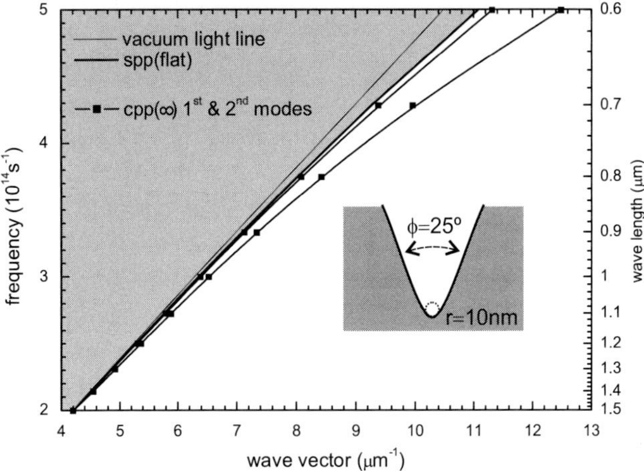

Fig. 5. Dispersion relation of CPP(∞) modes. The groove angle is $\phi = 25°$. The radius of curvature of the tip is $r = 10\,\text{nm}$.

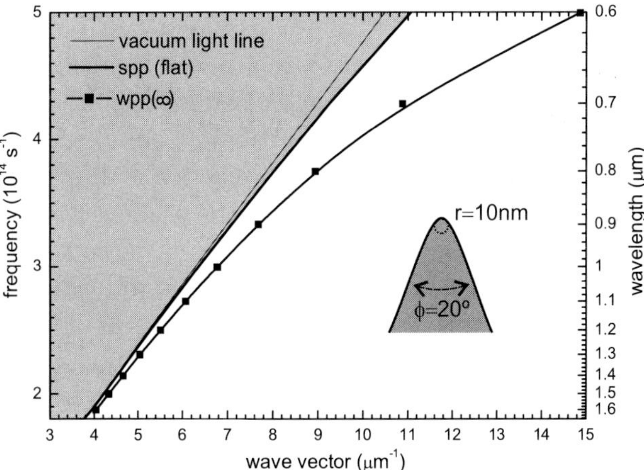

Fig. 6. Dispersion relation of a WPP(∞) mode. The wedge angle is $\phi = 20°$. The radius of curvature of the tip is $r = 10\,\text{nm}$.

3.2. *Modal shape, size, and loss*

The electric and magnetic fields corresponding to the modes in Figs. 5 and 6 are rendered in Color Plate 1 (see color plates at the end of the chapter). The wavelength is $\lambda = 0.6\,\mu$m and the parameters are described in the caption. Qualitatively, for longer wavelengths, the modal shapes are similar to those shown, but with a correspondingly larger scale. Notice that the modes have hybrid character, *i.e.*, all components of the field are non vanishing. At this wavelength, transverse and longitudinal magnetic field components have the same order of magnitude, whereas the transverse electric field component is about one order of magnitude larger than the longitudinal one. The displayed CPP(∞) modes have odd parity and the WPP(∞) mode has even parity. This means that polarization charges of different sign accumulate at both sides of a groove, whereas this does not happen for the wedge. The transverse fields of the fundamental mode very much resemble the field of a flat SPP folded at the tip. For the CPP(∞) modes the polarization is approximately horizontal matching the EIM assumption (as was mentioned in Section 2.2), but for the WPP(∞) mode the electric field points radially from the tip, so that at the very tip is vertical. The shown polarization of the fields let us expect that CPPs(∞) would have a better coupling with a plane wave propagating along the Z axis. The maximum electric field of WPP(∞) modes occurs at the very tip. In contrast, the maximum of the CPP(∞) modes does not occur at the bottom of the groove, due to the finite radius of curvature. For instance, at $\lambda = 0.6\,\mu$m the maximum electric field of the fundamental CPP(∞) mode occurs at a height of about 70 nm above the groove bottom, and at a wavelength of $\lambda = 1.4\,\mu$m at a height of about 1 μm.

The modal size as a function of the wavelength is displayed in Figs. 7 and 8. Here the modal size is defined as the transverse separation between the locations where the electric field amplitude of the mode has fallen to one tenth of its maximum value. The factor 1/10 in this definition is somehow arbitrary but it is sufficient for our mode characterization purposes. CPP(∞) modes are very well confined between the metallic walls, but they can extend quite a lot in the vertical direction, in fact the modal size of the considered CPP(∞) mode is dominated by its vertical extension. Note that at $\lambda = 1.4\,\mu$m the CPP(∞) modal size is about 6 μm. On the other hand, for the chosen parameters, the WPP(∞) mode has approximately equal horizontal and vertical sizes. It is very well confined, its size being subwavelength in the whole considered regime. For instance, at $\lambda = 1.6\,\mu$m the modal size is about 0.5 μm.

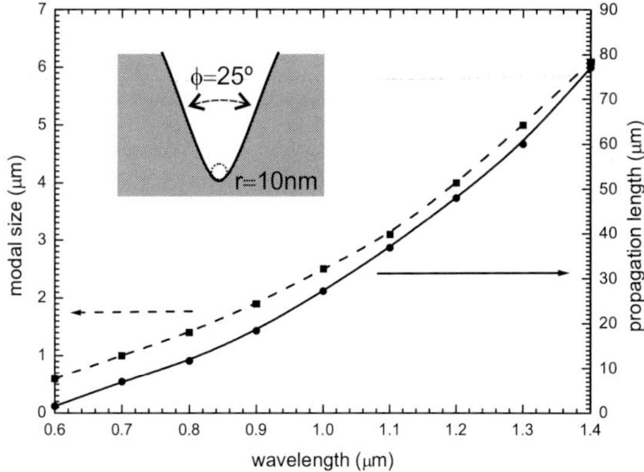

Fig. 7. Modal size and propagation length for CPP(∞) as a function of wavelength. The angle of the groove is $\phi = 25°$ and the radius of curvature of the tip is $r = 10\,\text{nm}$.

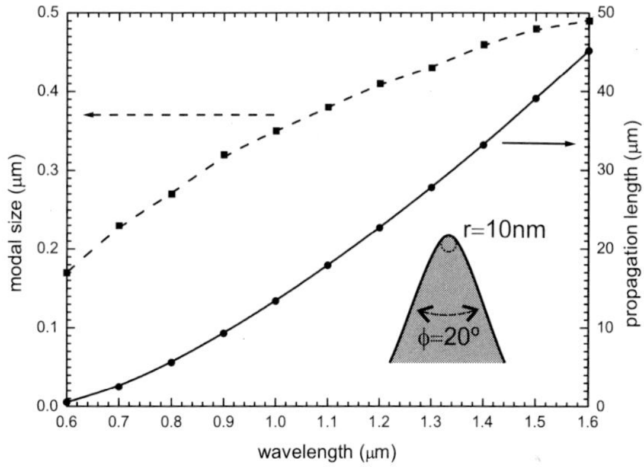

Fig. 8. Modal size and propagation length for WPP(∞) as a function of wavelength. The angle of the wedge is $\phi = 20°$ and the radius of curvature of the tip is $r = 10\,\text{nm}$.

The propagation length ($l = [2\text{Im}(k_z)]^{-1}$, k_z being the modal wave vector) as a function of wavelength is also presented in Figs. 7 and 8. As compared to the propagation length of standard long range plasmon polaritons, the propagation lengths of these modes are not very large (about $80\,\mu\text{m}$ for CPP(∞) at $1.4\,\mu\text{m}$ and about $45\,\mu\text{m}$ for WPP(∞) at $1.6\,\mu\text{m}$). Nevertheless they amount to many wavelengths so that these modes can be used for various purposes.

In the previous results the comparison of the wedge and groove modes shows that the WPP(∞) is better confined but it is lossier than the CPP(∞). At this point no conclusions should be drawn from this comparison because the structure angles, ϕ, are different in both simulations. In the next section, we will analyze the dependence of the modal properties on the geometric parameters showing that confinement and losses are very sensitive to the structure angle. Afterwards, CPPs and WPPs with the same geometry shall be compared in Section 4.

3.3. *Geometrical dependence*

In this subsection we present the dependence of the modal properties with the geometric parameters of the structures, namely angle ϕ and radius of the tip r. Figure 9 displays the ϕ-dependence for a WPP(∞) mode. As expected, it is observed that the properties of a SPP on a flat surface are recovered in the limit $\phi \to 180°$. The propagation length increases for increasing wedge angle but, at the same time, the modal size rapidly grows. This is a general property observed in all cases: there is a trade-off between low losses and tight modal confinement.

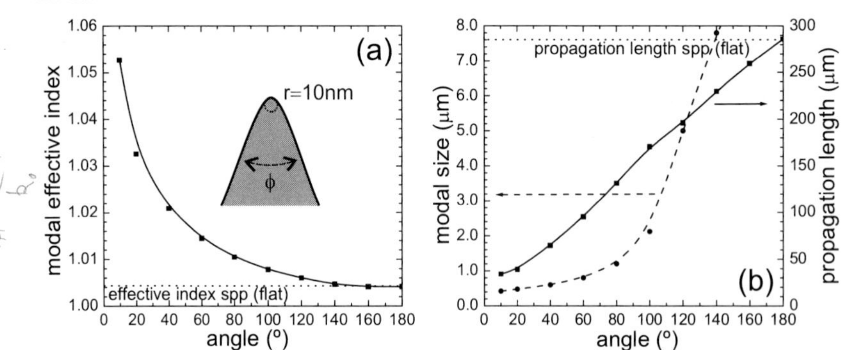

Fig. 9. Dependence of the modal parameters of WPP(∞) as a function of the wedge angle ϕ. (a) Effective index, (b) Modal size and propagation length. The radius of curvature of the tip is $r = 10\,\mathrm{nm}$. The wavelength is $\lambda = 1.5\,\mu\mathrm{m}$.

Compared to the sensitivity to the angle, the dependence on the radius of the tip is more moderate, as shown in Fig. 10. The behavior is again as expected: larger radii (for which the structure is flatter) lead to larger modal size, longer propagation length, and lower effective index.

Similar behaviors occur for CPP(∞) modes. As an example Fig. 11

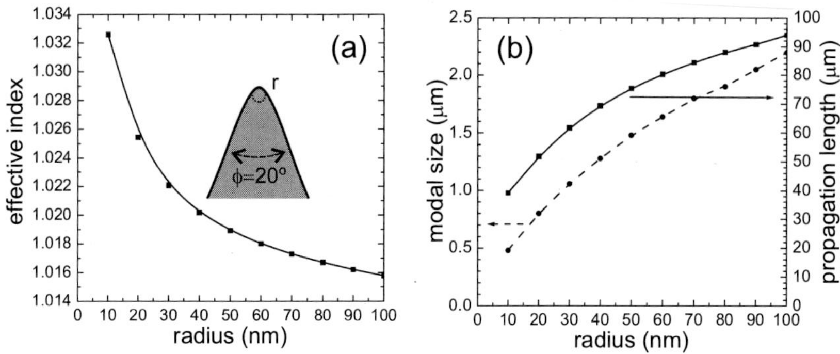

Fig. 10. Dependence of the modal parameters of WPP(∞) as a function of the tip radius r. (a) Effective index, (b) Modal size and propagation length. The angle of the wedge is $\phi = 20°$. The wavelength is $\lambda = 1.5\,\mu$m.

renders the effective index as a function of the groove angle. In this case the sensitivity to the angle is even stronger than for WPPs(∞). The modal size grows extremely fast and the computations for large angles are very demanding.

The dependence with the radius of curvature of the bottom of the groove is again more moderate. For radii between $r = 10 - 100\,$nm the modal size varies between $6.1\,\mu$m and $6.8\,\mu$m and the propagation length between $77\,\mu$m and $104\,\mu$m.

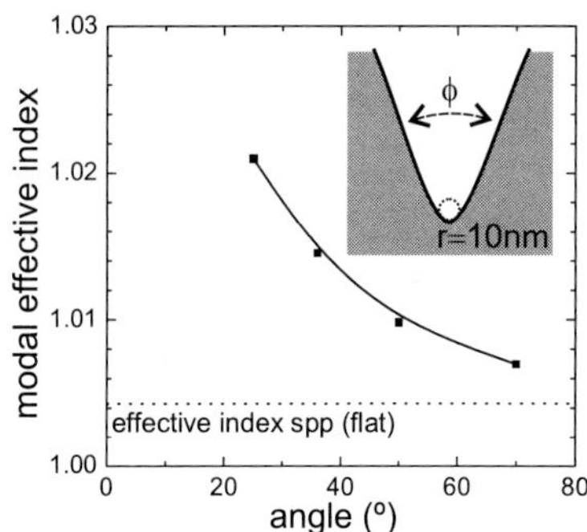

Fig. 11. Dependence of the effective index of CPP(∞) as a function of the groove angle ϕ. The radius of curvature of the tip is $r = 10\,$nm. The wavelength is $\lambda = 1.4\,\mu$m.

4. Rigorous techniques. Truncated structures

Up to now we have considered grooves and wedges as elementary building blocks, as mentioned in the Introduction (for instance, the role of wedge modes in other kinds of plasmonic waveguides can be observed in Refs. 21 and 22). For this reason infinitely high structures were computed. For sufficiently high structures it is justified to consider them infinite but, sometimes, practical limitations or design considerations require the computation of truncated structures. In the experiments both regimes have been considered.[11,17] In this section *truncated* grooves and wedges shall be analyzed. We shall focus on the telecom regime.

Figure 12 shows the dependence of the modal characteristics of a WPP as a function of its height. A new important behavior is observed, namely, the mode exhibits a cut-off height. For the chosen geometry the cut-off height is about $h_c \simeq 0.2\,\mu\text{m}$. Notice how, as the wedge height approaches the cut-off h_c and the modal effective index tends to the effective index of a SPP, modal size and propagation length increase.

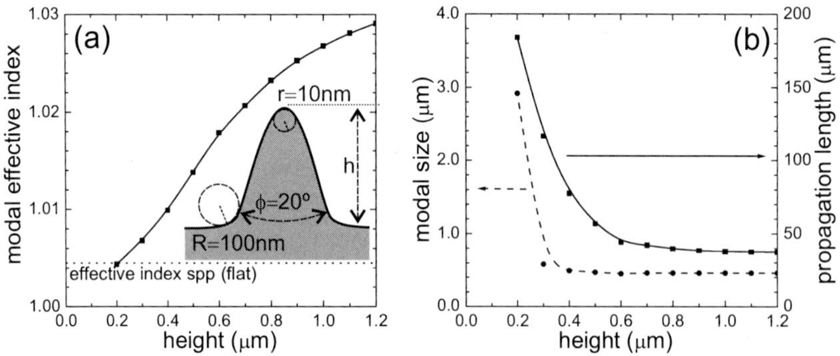

Fig. 12. Dependence of the modal parameters of WPPs as a function of wedge height h. (a) Effective index, (b) Modal size and propagation length. The angle of the wedge is $\phi = 20°$. The radius of the tip is $r = 10\,\text{nm}$. The wavelength is $\lambda = 1.5\,\mu\text{m}$.

The analogous dependence of CPPs is displayed in Fig. 13. Again, a cut-off height exists ($h_c \simeq 1.05\,\mu\text{m}$ for this configuration). The existence of a cutoff for CPPs, which is very important to achieve single-mode waveguiding, was first reported in Ref. 26. The non-monotonic behavior of the curves is related to the fact that the fields are now a hybridization of groove and wedge modes, as will be demonstrated below.

Let us now compare the guiding properties of CPPs and WPPs. We have

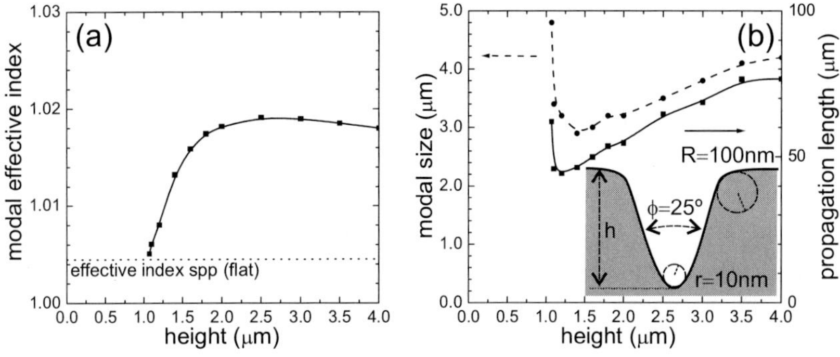

Fig. 13. Dependence of the modal parameters of CPPs as a function of groove height h. (a) Effective index, (b) Modal size and propagation length. The angle of the groove is $\phi = 25°$. The radius of the tip is $r = 10\,\text{nm}$. The wavelength is $\lambda = 1.4\,\mu\text{m}$.

seen that the geometric parameters play an important role, in particular the angle ϕ. For this reason, in the following, we shall consider grooves and wedges with the same angle ($\phi = 20°$), radius of curvature of the tip ($r = 10\,\text{nm}$), and at a wavelength of $\lambda = 1.5\,\mu\text{m}$. Color Plate 2 displays the transverse electric fields corresponding to a WPP and a CPP for the same geometry (exchanging the metal and vacuum regions). The modal size of the WPP is $0.46\,\mu\text{m}$, significantly smaller than that of the CPP ($2.5\,\mu\text{m}$). This is mainly due to the fact that, at the considered wavelength, the CPP mode is a mixture of the mode in the groove and the modes guided by the edges at both sides of the groove.[13] These edges correspond to wedges with a large angle ($\phi' = 100°$) and radius of curvature ($R = 100\,\text{nm}$), and for which the corresponding WPP modal sizes are larger, as was shown in Section 3.3. Despite the different modal sizes, the computed propagation lengths are quite similar for both modes: $37\,\mu\text{m}$ for the WPP, and $34\,\mu\text{m}$ for the CPP. Thus, at telecom wavelengths, the guiding properties of WPPs are superior to the ones offered by CPPs.

There are a number of remarks that should be made about the above discussion. First, one could expect that, in a truncated wedge, channel modes should appear at both sides of the basis of the wedge. Actually this does not happen because the angle ($\phi' = 100°$) is very large and the corresponding groove modes are beyond their cut-off. Second, as mentioned above, the size of the CPP mode in Color Plate 2(b) is dominated by the lateral wedge modes. We have chosen a value of $R = 100\,\text{nm}$ because this is consistent with the geometry in the experiments. Nevertheless, it is possible to decrease the mode size in a certain measure by reducing the value of R. For instance, for $R = 10\,\text{nm}$ the CPP modal size is about $1.6\,\mu\text{m}$, the propagation length remaining almost unaltered. Finally, the truncated grooves support modes with even parity, being essentially two coupled wedge modes. However their cut-off wavelength is much lower than that of the mode represented in Color Plate 2(b).

5. Conclusions

In summary, we have reviewed the guiding properties of the plasmonic modes supported by straight metallic grooves and wedges. We have briefly introduced two approximate methods useful to determine some of the properties of CPPs and WPPs, and then rigorous numerical results concerning both geometries have been presented. The dependence on various geometric parameters of the modal dispersion, size, polarization, and losses have been analyzed. The knowledge of these properties can be of great help for the analysis of plasmonic devices featuring edges and corners. Both CPP and WPP modes are quite well confined in the transverse plane and their losses are reasonably low. These properties make them very interesting candidates for plasmonic interconnects.

References

1. L. Dobrzynski and A. A. Maradudin, *Phys. Rev. B* **6**, 3810 (1972).
2. I. V. Novikov and A. A. Maradudin, *Phys. Rev. B* **66**, 035403 (2002).
3. S. I. Bozhevolnyi, V. S. Volkov, E. Devaux, J. Y. Laluet, and T. W. Ebbesen, *Nature (London)* **440**, 508 (2006).
4. L. C. Davis, *Phys. Rev. B* **14**, 5523 (1976).
5. A. Eguiluz and A. A. Maradudin, *Phys. Rev. B* **14**, 5526 (1976).
6. A. D. Boardman, R. Garcia-Molina, A. Gras-Marti, and E. Louis, *Phys. Rev. B* **32**, 6045 (1985).
7. A. D. Boardman, G. C. Aers, and R. Teshima, *Phys. Rev. B* **24**, 5703 (1981).
8. D. K. Gramotnev, *J. Appl. Phys.* **98**, 104302 (2005).
9. S. I. Bozhevolnyi, *Opt. Express* **14**, 9467 (2006).
10. D. K. Gramotnev and K. C. Vernon, *Appl. Phys. B* **86**, 7 (2007).
11. S. I. Bozhevolnyi, V. S. Volkov, E. Devaux, and T. W. Ebbesen, *Phys. Rev. Lett* **95**, 046802 (2005).
12. C. Hafner, *Post-Modern Electromagnetics* (Wiley, Chichester, 1999).
13. E. Moreno, F. J. Garcia-Vidal, S. G. Rodrigo, L. M. Moreno, and S. I. Bozhevolnyi, *Opt. Lett.* **31**, 3447 (2006).
14. E. Moreno, S. G. Rodrigo, S. I. Bozhevolnyi, L. M. Moreno, and F. J. Garcia-Vidal, *Phys. Rev. Lett* **100**, 023901 (2008).

15. A. Taflove and S. Hagness, *Computational Electrodynamics: The Finite-Difference Time-Domain Method* (Artech House, Boston, 2000).
16. D. F. P. Pile and D. K. Gramotnev, *Opt. Lett.* **29**, 1069 (2004).
17. D. F. P. Pile, T. Ogawa, D. K. Gramotnev, T. Okamoto, M. Haraguchi, M. Fukui, and S. Matsuo, *Appl. Phys. Lett.* **87**, 061106 (2005).
18. D. F. P. Pile, *Appl. Phys B* **81**, 607 (2005).
19. T. Apel, *Anisotropic Finite Elements: Local Estimates and Applications* (Teubner Verlag, Stuttgart, 1999).
20. M. Yan and M. Qiu, *J. Opt. Soc. Am. B* **24**, 2333 (2007).
21. J. Jung, T. Sondergaard, and S. I. Bozhevolnyi, *Phys. Rev. B* **76**, 035434 (2007).
22. I.-M. Lee, J. Jung, J. Park, H. Kim, and B. Lee, *Opt. Express* **15**, 16596 (2007).
23. E. Feigenbaum and M. Orenstein, *Opt. Express* **14**, 8779 (2006).
24. A. Vial, A.-S. Grimault, D. Macias, D. Barchiesi, and M. Lamy de la Chapelle, *Phys. Rev. B* **71**, 085416 (2005).
25. P. B. Johnson and R. W. Christy, *Phys. Rev. B* **6**, 4370 (1972).
26. D. K. Gramotnev and D. F. P. Pile, *Appl. Phys. Lett.* **85**, 6323 (2004).

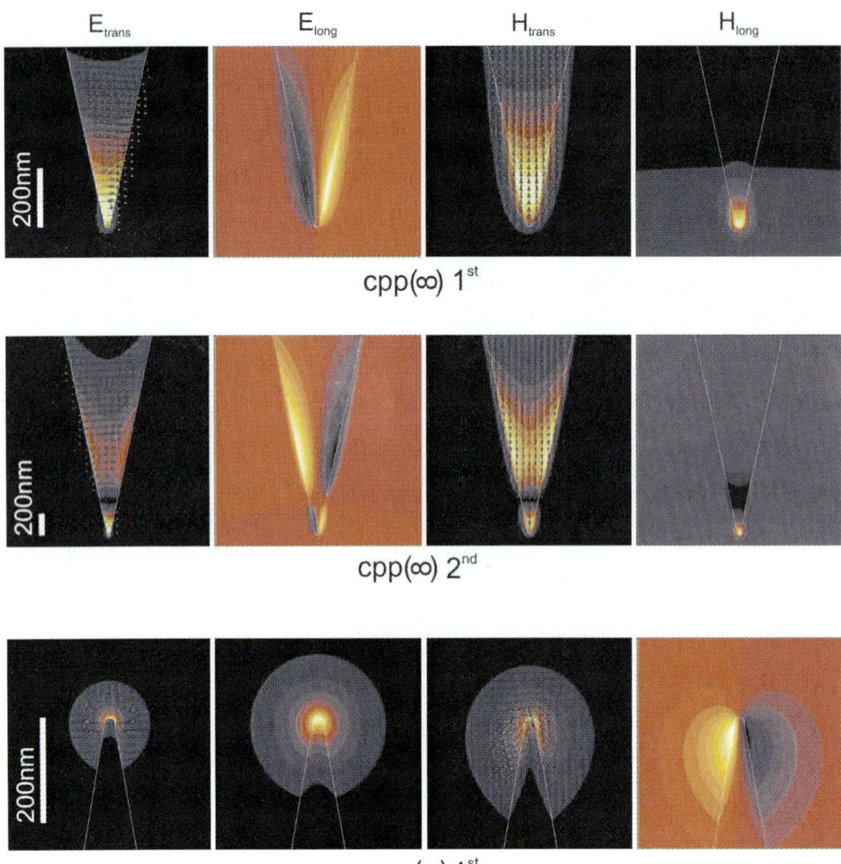

Color Plate 1. Lower modes corresponding to infinite groove [first and second rows, CPP(∞)] and wedge [third row, WPP(∞)]. The angle of the groove is $\phi = 25^\circ$ and that of the wedge is $\phi = 20^\circ$. The radius of curvature of the tip is $r = 10\,\text{nm}$ in both cases. The wavelength is $\lambda = 0.6\,\mu\text{m}$. Notice that the color scale and lateral size of the various panels is different to achieve better visualization. The various columns represent the transverse (XY) or longitudinal (Z) components of the electric and magnetic fields.

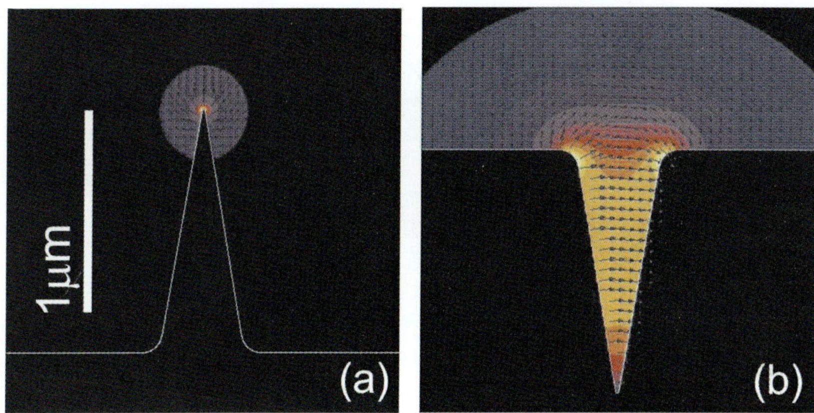

Color Plate 2. Transverse electric field of (a) WPP mode, and (b) CPP mode. The geometry of both structures is identical exchanging the metallic and vacuum regions. The angle of the structures is $\phi = 20°$. The radii of the edges are $r = 10\,\text{nm}$ (wedge tip or groove bottom) and $R = 100\,\text{nm}$ (lateral edges or grooves). The height is $h = 1.2\,\mu\text{m}$. The wavelength is $\lambda = 1.5\,\mu\text{m}$.

Chapter 10

CHANNEL PLASMON-POLARITONS IN TRIANGULAR GROOVES

Dmitri K. Gramotnev
Applied Optics Program, School of Physical and Chemical Sciences, Queensland University of Technology, GPO Box 2434, Brisbane, QLD 4001, Australia
E-mail: d.gramotnev@qut.edu.au

This chapter reviews propagation and feasible applications of strongly localized plasmons in metallic triangular grooves, including channel plasmon-polaritons (CPPs) for efficient sub-wavelength guiding and plasmon nano-focusing in tapered metal gaps. Dispersion, propagation parameters, localization, dissipation, and existence conditions of CPP eigenmodes are analyzed and discussed. Theoretical and numerical analyses of adiabatic and non-adiabatic plasmon nano-focusing in triangular grooves (tapered gaps) are reviewed, including an important link between nano-focusing and guided CPP eigenmodes.

1. Introduction

Design, investigation and optimization of efficient guiding structures with sub-wavelength localization constitutes one of the major directions of research in modern photonics, nano-optics, novel sensor technology, near-field microscopy and spectroscopy. The development of efficient optical devices, circuits and measurement techniques that would surpass their electronic counterparts requires a substantial increase in the degree of integration and miniaturization of such devices and techniques. However, conventional dielectric guiding structures (such as optical fibers, dielectric slabs, *etc*.) do not allow localization of optical waves (signals) in regions significantly smaller than the wavelength determined by refractive indices in the guiding structure. This is the so-called diffraction limit of light[1,2]. Impossibility of using conventional dielectric guiding structures for localization of guided optical signals beyond the diffraction limit is one of the major limitations on the way of miniaturization and integration of nano-optical devices, circuits and measurement techniques.

One of the main ideas for the design of effective sub-wavelength waveguides and interconnectors for integrated nano-optics is based upon the use of evanescent fields in media with negative real part of the dielectric permittivity, e.g., metals. Metallic nano-structures can support and guide a special type of electromagnetic waves coupled to collective oscillations of the free electron gas in a metal — surface plasmons[3-7]. The major benefit of surface plasmons is that they can be used to beat the diffraction limit of light and localize electromagnetic signals in nano-scale sub-wavelength regions.

Different types of guiding metallic nano-structures and surface plasmon modes have been proposed for the design of sub-wavelength waveguides, interconnectors and surface plasmon devices. In particular, these structures include chains of metal nano-particles[1,2], metal strips surrounded by dielectric or placed onto a dielectric substrate[8-13], metal nano-rods[14,15], sharp metal wedges[16-19], *etc*. However, it could be that the two best options for the design of effective plasmonic waveguides are the groove waveguides[20-32] and gap plasmon waveguides in the form of a nano-gap in a thin metal film or membrane[33-36]. Both these types of sub-wavelength plasmonic waveguides demonstrate several important features that could be superior compared to other metallic guiding nano-structures: (1) strong sub-wavelength localization of plasmons (the smaller the angle of the groove, or the smaller the width of the gap, the stronger the localization of the guided modes)[21,22,31-34], (2) relatively weak dissipation and reasonable propagation distances (~ 10 – 20 wavelengths)[21-33], (3) lower sensitivity to structural imperfections (because the plasmon field is largely surrounded by the metallic medium, and mode leakage is impeded by the fact that light does not propagate in the metal)[23,24], (4) compatibility with planar technology and feasibility of such nano-optical devices as interferometers, splitters, filters, *etc.*[23,26,27,29,30], (5) broad-band transmission[21,22,31,33,34], (6) possibility of single-mode operation[22,28], and (7) possibility of nearly 100% transmission through sharp bends[24,34,35].

Another important possibility offered by strongly localized plasmons in triangular metallic grooves (tapered gaps) is related to the efficient nano-focusing of electromagnetic energy into regions with dimensions as small as a few nanometers[37-41]. Thus a tapered gap can be regarded as an analogue of a cylindrical lens with the capability of concentration (focusing) and efficient delivery of light energy to the nano-scale. This offers unique opportunities for the development of near-field optical microscopy and spectroscopy with sub-wavelength resolution, high-resolution lithography, coupling of light into and out of photonic nano-circuits, new nano-sensors and measurement techniques.

2. Triangular grooves for sub-wavelength guiding

Channel plasmon-polaritons (CPPs) guided by metallic grooves were first analyzed by Novikov and Maradudin[20]. The method of analysis was largely analytical based on parameterization of the surface profile of the groove. As a result, dispersion and field distribution in CPPs in grooves with smooth (such as Gaussian and other) shapes were determined[20].

Subsequent numerical analyses[21,22,28,31] have confirmed that similar plasmonic modes also exist in triangular grooves on a metal surface. It was also suggested that these structures could be used for the development of new efficient sub-wavelength plasmonic waveguides with superior properties (see Sec. 1). The structure analyzed in Refs. 21, 22 and 31 are presented in Figs. 1a,b.

The numerical analysis of plasmon propagation in a groove was conducted using two different versions of the custom-made finite-difference time-domain (FDTD) algorithm[21,22,42], geometrical optics approach[38] and a commercial finite-element frequency-domain software package COMSOL[31]. To avoid numerical instabilities associated with negative permittivity ε_m, the local Drude free-electron model was incorporated into the FDTD algorithms. All three numerical approaches were used to obtain propagation distances of the CPP modes, as well as CPP wave numbers and field distributions inside the groove[21,22,31]. However, the finite element analysis (COMSOL software package) seems to be preferable for the determination of CPP wave numbers and field distributions, due to its higher spatial resolution and accuracy.

CPP modes were launched in the groove using the end-fire excitation, i.e. by a bulk wave with the wavelength λ_{vac} in vacuum, incident onto the end of the groove at $x = 0$ (Fig. 1a). If the depth of the groove $h \lesssim \lambda_{vac}$, then an opaque metal screen of 200 nm thickness was used at $x = 0$ (Fig. 1a). This is to suppress scattered bulk waves in and above the groove, which interfere with CPP modes and impede the analysis.

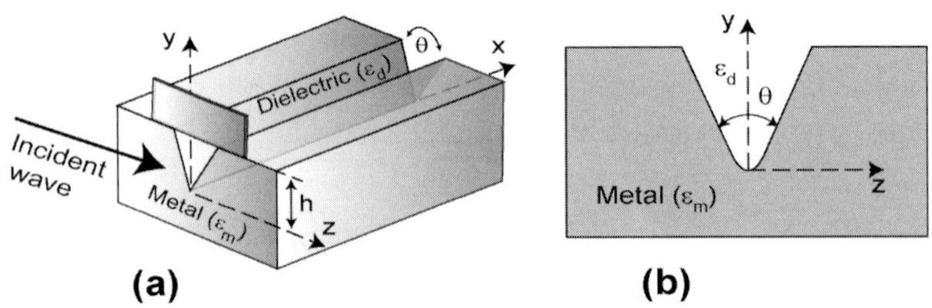

Fig. 1. (a) Triangular groove on a metal surface with the angle θ and depth h. The dielectric permittivity inside the groove ε_d is real (no dissipation in the dielectric), while the metal permittivity $\varepsilon_m = e_1 + ie_2$ is complex, where e_1 and e_2 are the real and imaginary parts of ε_m ($e_1 < 0$, $e_2 > 0$). (b) The same groove but with the rounded tip of radius r.

For example, consider a vacuum ($\varepsilon_d = 1$) triangular groove with $\theta = 30^{\circ}$ on a silver substrate. The end-fire excitation is achieved using a bulk wave incident onto the end of the groove (at $x = 0$) at the angle of 45° with respect to the x-axis (Fig. 1a). The magnetic field in the incident wave is in the (x,y) plane, $\lambda_{vac} = 632.8$ nm (He-Ne laser), and the metal permittivity[44] $\varepsilon_m = -16.22 + 0.52i$. The resultant typical distribution[21,22] of the electric field in the (x,y) plane in an infinitely deep (h is much larger than the CPP localization region) groove is shown in Fig. 2a. No metal screen was used in this case, and noticeable bulk scattered waves can be seen in Fig. 2a near $x = 0$. The major feature of Fig. 2a is the existence of a periodically changing electromagnetic field localized near the apex of the groove. This is the numerical evidence of CPP eigenmodes in the considered structure[21,22].

The periodic modulation of the field in the wave, clearly seen in Fig. 2a, is due to interference of two different CPP modes[21,22]. To analyze the mode structure of the launched signal in the groove (Fig. 2a), the field at the tip of the groove at $x > 0$ can be expanded into the Fourier integral[21,22]. The resultant Fourier amplitude as a function of wave vector is presented in Fig. 3a. The dominating wave number $q_1 \approx 1.195\times10^7$ m^{-1} corresponds to the fundamental CPP mode in the groove with the wavelength $\lambda_1 \approx 0.528$ μm.

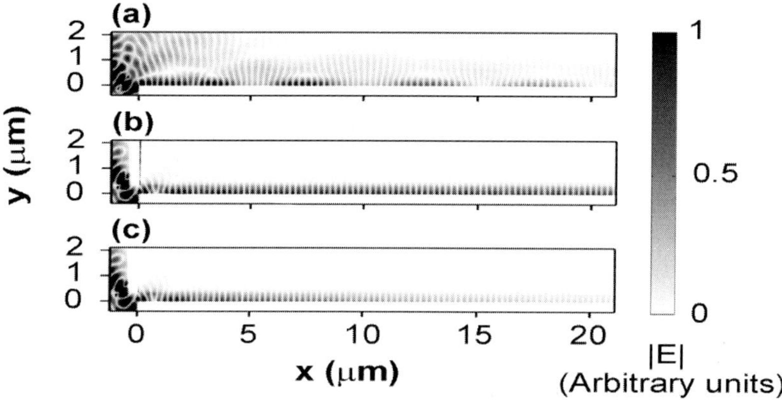

Fig. 2. (a) The distribution of the magnitude of the instantaneous electric field in a deep vacuum groove (ε_d = 1) in silver ($\varepsilon_m = -16.22 + 0.52i$), resulting from the end-fire excitation of CPP modes at $x = 0$. The magnetic field in the incident wave is in the (x,y) plane, the groove angle $\theta = 30^{\circ}$, and λ_{vac} = 632.8 nm. (b) Same as in Fig. 2a, but with h = 316.4 nm and $\varepsilon_m = -16.22$ (no dissipation in the metal). (c) Same as in Figs. 2a,b, but with h = 316.4 nm and $\varepsilon_m = -16.22 + 0.52i$. (Adapted from Ref. 22).

The other two maxima at $q_2 \approx 1.076 \times 10^7$ m^{-1} and $q_3 \approx 1.0\times10^7$ m^{-1} on the left of the main maximum in Fig. 3a correspond to the second CPP mode and surface plasmons on the sides of the groove, respectively[21]. Interference between the first (fundamental) and second CPP modes[21] must give a pattern of beats with the period $\Lambda = 2\pi/(q_1 - q_2) \approx 6.43$ μm, which is in excellent agreement with the period of modulation of the field in Fig. 2a. The number of higher modes increases with decreasing θ and/or increasing ε_d (see below)[21,31].

An alternative way of determining wave numbers of different modes is to use the eigenmode solver in the COMSOL package[31]. The results obtained by all the approaches were found to agree with each other.

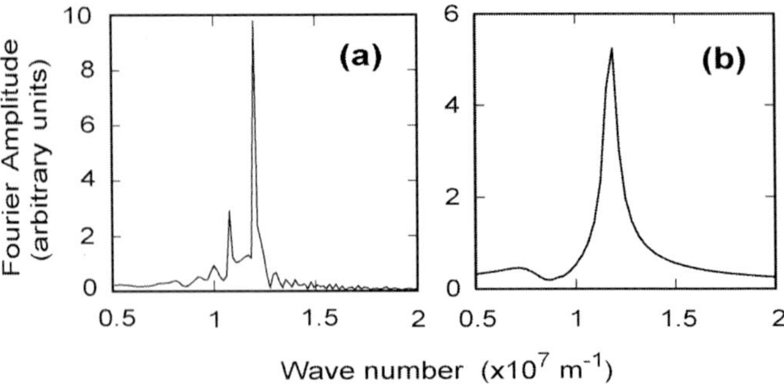

Fig. 3. The dependencies of the Fourier amplitudes for the field distributions at the tip of the groove within the interval 2 μm < x < 40 μm on wave number. (a) Infinite groove (Fig. 2a). (b) Groove of finite depth h = 316.4 nm (Fig. 2b,c). (Adapted from Ref. 22).

As was demonstrated in Ref. 22, the second CPP mode can easily be eliminated by reducing depth of the groove, thus resulting in a single-mode waveguide. It can be estimated that the penetration depth of the second CPP mode up the groove (along the y-axis) is h_{p2} ~ 800 nm (Fig. 2a), while the penetration depth for the fundamental CPP mode in the same direction is h_{p1} ~ 300 nm (Fig. 2a). If we reduce the depth of the groove to h ~ h_{p1}, then the second mode will not be supported by the structure, while the fundamental mode should still be effectively guided by the groove[22]. This is confirmed by Fig. 2b showing the field distribution in the same groove as for Fig. 2a, but with h = 316.4 nm, additional 200 nm screen, and zero dissipation in the metal (e_2 = 0). It can be seen that only the fundamental mode propagates along the groove with no visible beat pattern, which means no second CPP mode – Fig. 2b. This conclusion is also confirmed by Fig. 3b showing that the Fourier amplitude of the electric field at the tip of the groove displays only one maximum at $q_1 \approx 1.20\times10^7$ m^{-1}, which corresponds to the wave number of the fundamental CPP mode in the groove – single-mode guiding[22].

Figs. 2a and 2c also demonstrate the effect of dissipation in silver (e_2 = 0.52) on CPP modes. From these figures, the propagation distances of the fundamental CPP mode at the considered wavelength and groove parameters are 10 μm – 15 μm, which should be sufficient for plasmonic waveguides and interconnectors in integrated nano-optics with nanometer distances between optical devices and elements[21,22].

The typical dependencies of the wave numbers of the fundamental CPP modes in triangular grooves in gold and silver on groove angle θ were determined in Refs. 21 and 31, and are shown in Figs. 4a,b for λ_{vac} = 632.8 nm. Here, we assume that the tip of the triangular groove is rounded with radius of curvature $r \approx 2.24$ nm[31]. This is done to enable better convergence of the calculated field near the tip of the groove and make the analyzed structures more realistic. On the other hand, as we will see below, such small tip roundness has negligible effect on the obtained wave numbers and actual field distributions in the groove, if the groove angle is not too small (i.e., the size of the CPP localization region is mush larger than the radius of curvature of the tip).

As demonstrated in Ref. 31, several important conclusions can be drawn from Figs. 4a,b. Firstly, the wave numbers of the fundamental CPP modes for both the gold and silver grooves rapidly increase with decreasing groove angle θ for all values of the dielectric permittivity in the groove. Secondly, increasing ε_d results in a substantial increase of CPP wave number. Thirdly, increasing groove angle eventually reduces

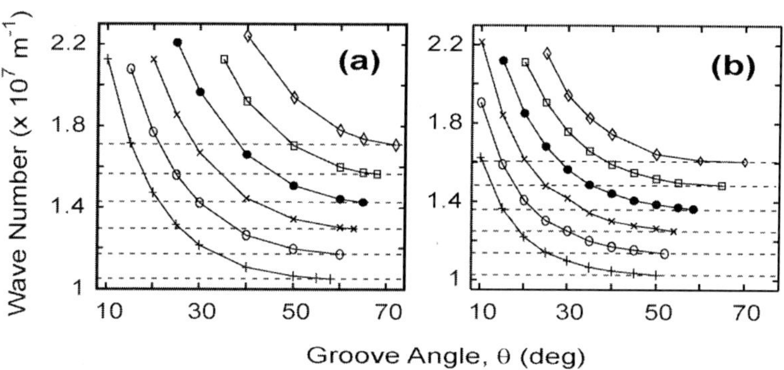

Fig. 4. Wave numbers of the fundamental CPP modes versus groove angle for a triangular groove with $h = +\infty$ and rounded tip of radius $r \approx 2.24$ nm in gold (a) and silver (b) at λ_{vac} = 632.8 nm and different values of the dielectric permittivity in the groove: ε_d =1 (+), ε_d =1.21 (○),ε_d =1.44 (×),ε_d =1.69 (●),ε_d =1.96 (□), and ε_d =2.25 (◊). The metal permittivities are $\varepsilon_m = -9.50 + 1.13i$ for gold, and $\varepsilon_m = -16.22 + 0.52i$ for silver[44]. Dotted lines correspond to wave numbers of surface plasmons on the sides of the groove far from the tip, where coupling between surface plasmons on the opposite sides of the groove can be neglected. (Adapted from Ref. 31).

the wave number of the fundamental CPP mode to that of the surface plasmons at the sides of the groove far away from the tip.

A CPP mode can only exist if its wave vector is larger than those for bulk waves in the dielectric filling the groove and surface plasmons on the isolated metal-dielectric interface (i.e., on the sides of the groove far from the tip). If this condition is not satisfied and, for example, the wave vector of a CPP mode is smaller than that of the surface plasmons on the sides of the groove, then the CPP mode should leak into the surface plasmons, and thus ceases to exit as an eigenmode. This is because in this case Snell's law is satisfied for the CPP mode and surface plasmons on the sides of the groove, traveling at an angle with respect to the x-axis, such that the tangential (along the x-axis) component of the wave vector of the surface plasmon is equal to the wave vector of the CPP mode. This means that such surface plasmons will be generated by the propagating CPP mode, and such a CPP cannot be regarded as a structural eigenmode. Similar situation occurs when the wave vector of the CPP mode is smaller than that of the bulk waves in the dielectric filling the groove.

Therefore, the groove angles at which the curves in Figs. 4a,b intersect the respective horizontal dashed lines can be regarded as critical angles θ_{c1}. CPP guided eigenmodes can only exist at groove angles that are smaller than these upper critical angles: $\theta < \theta_{c1}$ (see Refs. 21 and 31). Thus, Figs. 4a,b also determines the dependencies of the upper critical groove angles on dielectric permittivity ε_d in the groove. If $\theta \geq \theta_{c1}$, the respective CPP mode leaks into surface plasmons on the side of the groove, and thus it stops being a structural eigenmode of the groove. No analytical equation for the upper critical angle has been obtained so far, but its values can be determined numerically by finding groove angles at which the wave numbers of the respective CPP modes are equal to those of surface plasmons at the sides of the groove (Figs. 4a,b).

The upper critical angle θ_{c1} for the fundamental CPP mode in a triangular groove increases with increasing dielectric permittivity in the groove[31] (Figs. 4a,b). Similar situation occurs for other (higher) CPP modes with the only difference that the values of the upper critical angles for higher CPP modes decrease with increasing order of the mode[31]. It has also been demonstrated that the upper critical angle θ_{c1} depends only weakly on radius of curvature r of the tip of the groove[31]. For example, if

$\varepsilon_d = 2.25$, changing radius of the tip from 0.5 nm to 5 nm results in less than ~ 1% variation of the CPP wave number near the critical angle[31]. Nevertheless, it is worth noting that the actual values of θ_{c1} determined by the curves in Figs. 4a,b may contain some computational errors associated with numerical difficulties near the cut-off.

The typical dispersion curves for CPPs in triangular infinitely deep ($h = +\infty$) vacuum ($\varepsilon_d = 1$) grooves in silver[21], obtained by means of the FDTD algorithm, are presented in Fig. 5a. It can be seen that for all θ, decreasing λ (increasing ω) results in increasing mode localization due to an increased difference between the CPP wave vector and wave vector of surface plasmons (especially for smaller angles θ – Fig. 5a). As highlighted in Ref. 21, the dispersion of CPP modes in triangular metal grooves is solely due to the frequency dependence of the metal permittivity ($-23.81 \leq e_1 \leq -10.02$ for Fig. 5a). If it were not for dispersion of $\varepsilon_m(\omega)$, CPPs would have been non-dispersive, because changing wavelength would have only resulted in a proportional scaling of the field distribution in the triangular groove[21].

It is well-known that dispersive optical media, such as metals, are associated with noticeable dissipation of electromagnetic waves[45]. The effect of plasmon dissipation in metallic grooves has already been considered above in Figs. 2a,c. More detailed numerical results showing the effect of dissipation on CPP modes and their propagation distances in metallic grooves[31] are presented in Fig. 5b. Significant reduction of the propagation distance with increasing dielectric permittivity in the groove and/or decreasing groove angle θ can be seen from Fig. 5b (see also Refs. 21 and 31). Both these effects are explained by the general tendency of increasing localization of the CPP mode with increasing ε_d and decreasing θ (which is caused by increasing wave number of the CPP mode – Fig. 4). Increased localization of the CPP mode results in increased portion of its energy propagating in the dissipative metal (because of diminished plasmon penetration depth into the surrounding dielectric), which naturally leads to increasing dissipation. Note, however, that for the structural parameters used for Fig. 5b, dissipative effects in the fundamental CPP mode are still sufficiently weak to ensure propagation within at least several plasmon wavelengths[31]. For example, the smallest propagation distance shown in Fig. 5b corresponds to ~ 12 plasmon wavelengths, while the largest propagation distance corresponds to ~ 85 plasmon wavelengths[31]. These propagation distances are regarded to be sufficient for sub-wavelength plasmonic interconnectors between optical devices on the nano-scale.

The concept of CPP localization already appeared in the discussions of Figs. 2, 4 and 5 (see also Refs. 21 and 22). This includes the general tendency of increased localization (decreased region of localization) with increasing CPP wave vector, which occurs when θ is decreased, ε_d is increased, $|e_1|$ is decreased, and/or ω is increased (Figs. 4 and 5a). However, the consistent determination and analysis of CPP localization requires the detailed knowledge of the field distribution in the groove.

The typical field distribution inside the groove of finite depth h in the fundamental CPP mode is presented in Fig. 6a[22]. Usually, the field is maximal close to the tip inside the groove, but not exactly at the tip[20,22,28,31,32]. If the groove depth h is not too large (but is rather of the order of the penetration depth of the CPP mode up the groove), the evanescent field may extend outside the groove, producing a weak dome-like feature (Fig. 6a). This dome field rapidly decreases in intensity with increasing depth of the groove[22].

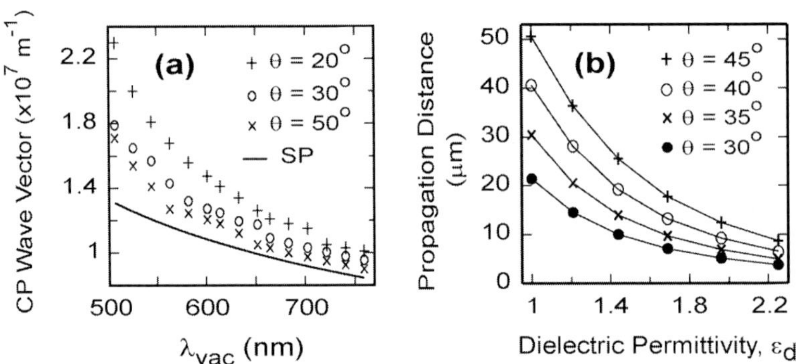

Fig. 5. (a) The typical dispersion curves for the fundamental CPP mode in the silver-vacuum triangular groove with $h = +\infty$ and three different angles θ. The solid curve represents dispersion of surface plasmons on an isolated smooth silver-vacuum interface. (b) The dependencies of propagation distance (at which the electric field amplitude decreases e times) for the fundamental CPP mode on dielectric permittivity ε_d inside the infinitely deep silver groove at λ_{vac} = 632.8 nm for the four different groove angles. (Adapted from Refs. 21 and 31).

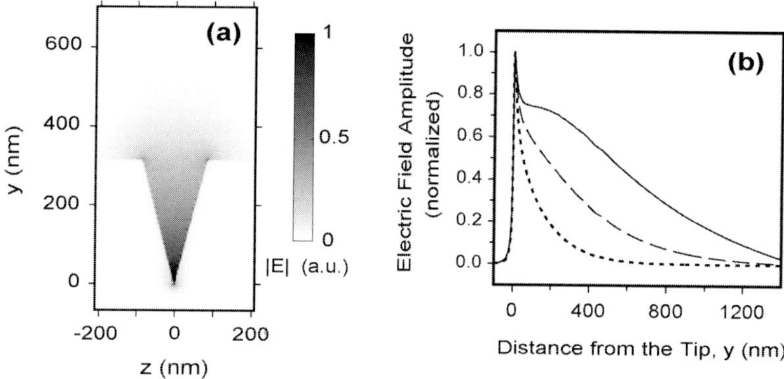

Fig. 6. (a) The typical field distribution in a cross-section of a vacuum ($\varepsilon_d = 1$) groove in silver with $h = 316.4$ nm that is close to the penetration depth of the fundamental CPP mode up the groove ($h \approx h_{p1}$)[22]. The other parameters are the same as for Figs. 2 and 3. (b) The dependencies of the magnitude of the electric field amplitude in the fundamental CPP modes in the middle of the groove on distance from the rounded tip (Fig. 1b) for the silver groove with $\theta = 45^{\circ}$, $\lambda_{vac} = 632.8$ nm, radius of the tip $r \approx 2.24$ nm, $\varepsilon_m = -16.22 + 0.52i$, and three different values of the dielectric permittivity in the groove: $\varepsilon_d = 1$ (solid curve), $\varepsilon_d = 1.44$ (dashed curve), and $\varepsilon_d = 2.25$ (dotted curve)[32]. All the three curves are normalized to the maximum of the field in the groove. The value $y = 0$ corresponds to the rounded tip of the groove (Fig. 1b). (Adapted from Refs. 22 and 31).

Based upon the obtained field distributions, it has been estimated[21] that for the fundamental mode in a silver-vacuum triangular groove with $\theta = 30^{\circ}$ and $\lambda_{vac} = 632.8$ nm, the typical penetration depth of the field (at the $1/e$ level) along the sides of the groove is ~ 170 nm from the tip, while across the groove (i.e., in along the z-axis) the localization of the wave is ~ 120 nm (with the typical propagation distance of ~ 15 μm[21,22]). This demonstrates the feasibility of plasmonic interconnectors using CPP modes for integrated nano-optics applications.

However, such a simplified determination of localization of CPP modes is not always accurate. Even if $h = +\infty$ (infinite groove), the decay of the field up the groove in the fundamental CPP mode is often far from being a simply exponential[20,28,31,32] (Fig. 6b). The complex pattern of decaying field in the fundamental CPP mode with increasing distance from the tip (Fig. 6b), makes questionable the usual definition of the localization region using the distance at which the amplitude of the field drops e times. Moreover, because of the triangular shape of the groove, the region where most of the CPP energy is localized may not coincide with the region where the field (or the square of the field) is localized at the $1/e$ level. Indeed, regions with weaker field further away from the tip may still contribute significantly to the overall energy in the mode, because of increased volume of these regions, due to the triangular shape of the groove. On the contrary, the volume with the strongest field (near the tip) is small, which reduces its contribution to the overall energy in the mode, despite the large field amplitude. Therefore, the direct use of the dependencies in Fig. 6b for the determination of mode localization may be ambiguous and misleading. At the same time, physically reasonable determination of localization regions is highly important for the optimal design of plasmonic waveguides.

In accordance with Ref. 31, for accurate determination of CPP localization, we use an approach in which the region of mode localization along the y-axis is determined by the distance L from the tip, such that the region $y < L$ contains $(1 - 1/e)$, i.e., approximately 63% of the overall energy in the mode. The resultant dependencies of localization distance L on dielectric permittivity ε_d in the silver groove with the tip radius $r \approx 2.24$ nm are presented in Fig. 7a for the vacuum wavelength $\lambda_{vac} = 632.8$ nm and three different groove angles.

As expected, increasing dielectric permittivity in the groove results in decreasing L, which means increasing localization of the CPP mode[31]. For example, if $\theta = 40^{\circ}$ and $\varepsilon_d = 1$, we have $L \approx 350$ nm, while for $\varepsilon_d = 2.25$, mode localization is ~ 4 times stronger: $L \approx 80$ nm (Fig. 7a). As indicated above, decreasing θ also results in increasing CPP localization (Fig. 7a). However, this is accompanied by increased dissipation (see Fig. 5b and the related discussions). Thus the information presented in Figs. 5 – 7 represents a basis for the determination of a reasonable trade-off between the achievable sub-wavelength localization and required propagation distance in a plasmonic interconnector.

The effect of rounded tip of a triangular groove on CPP wave numbers[31] is demonstrated by Fig. 7b. This figure demonstrates that the impact of the rounded tip increases with increasing dielectric permittivity in the groove, and this is related to increasing localization of the CPP modes. The sharp maximum at ~ 5 nm from the tip on the solid curve in

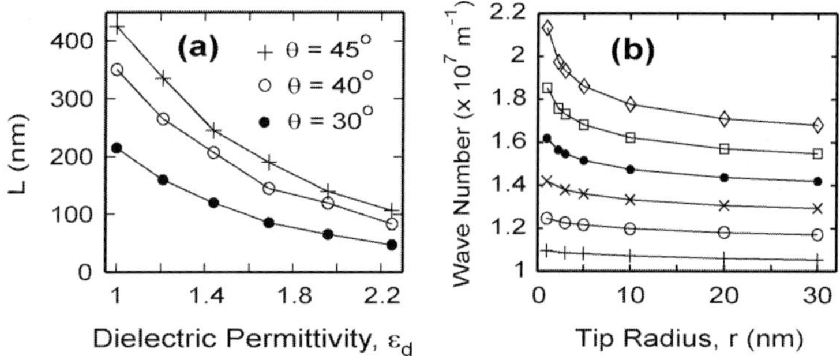

Fig. 7. (a) The dependencies of size L of the localization region (along the y-axis) for the fundamental CPP mode in the triangular groove in silver with the tip radius $r \approx 2.24$ nm at $\lambda_{vac} = 632.8$ nm on dielectric permittivity ε_d in the groove for three different groove angles. (b) The dependencies of the wave numbers of the fundamental CPP mode on radius of the tip of the 30° triangular groove in silver ($\varepsilon_m = -16.22 + 0.52i$) at the vacuum wavelength $\lambda_{vac} = 632.8$ nm and different values of the dielectric permittivity in the groove: $\varepsilon_d = 1$ (+), $\varepsilon_d = 1.21$ (○),$\varepsilon_d = 1.44$ (×),$\varepsilon_d = 1.69$ (●),$\varepsilon_d = 1.96$ (□), and $\varepsilon_d = 2.25$ (◊). (Adapted from Ref. 31).

Fig. 6b disappears when the radius of the tip is increased above ~ 5 nm. However, at the considered groove angle $\theta = 30°$, this does not have a very significant effect on CPP wave numbers, if the dielectric permittivity in the groove is small (Fig. 7b). Though the sharp maximum of the field at $y = y_{max}$ is relatively high, the width of the groove (and thus the volume with the strong field) is small, because of the proximity to the tip. Therefore, the contribution of the sharp field maximum (solid curve in Fig. 6b) to the overall energy and localization of the CPP mode is insignificant, because the portion of energy corresponding to this maximum is small compared to the overall energy in the mode. If the permittivity in the groove is increased, then the relative portion of the CPP energy corresponding to this field maximum also increases (compare solid and dotted curves in Fig. 6b). Therefore, the effect of rounded tip is more notable in this case (see the higher curves in Fig. 7b).

In addition to the discussed theoretical and numerical investigations of CPPs in triangular grooves[20-24,28,31,32], substantial efforts have also been focused onto the experimental observation, investigation and application of CPP modes for the design of efficient sub-wavelength plasmonic waveguides and nano-optical devices[25-27,29,30]. In particular, the first experimental observation of CPPs in triangular grooves fabricated by means of focused ion beam lithography in gold was conducted by Bozhevolnyi, *et al*[25] at telecommunication wavelengths (1425–1620 nm). This was the first experimental confirmation of the earlier theoretical predictions[21,22] of strong sub-wavelength localization of CPP modes in triangular grooves and their significant propagation distances (over tens of microns). Subsequent investigation was focused onto the successful design of a range of sub-wavelength photonic devices such as large angle S-bends and Y-splitters with low radiation losses[26,30], Mach-Zehnder interferometers and waveguide-ring resonators[27,30], ring resonator add-drop multiplexers and compact Bragg grating filters[29,30].

As a result of these theoretical and experimental efforts, it is possible to suggest that CPP modes in triangular metallic grooves are one of the best currently available options for the design of efficient plasmonic integrated circuit components, interconnectors and nano-optical devices.

3. Plasmonic groove waveguides with sharp bends

Design of integrated nano-optical components and circuits with high degree of integration requires the development of sub-wavelength interconnectors and couplers that will be capable of efficient guiding and transmitting optical signals within such a circuit through large-angle bends with intricate configurations. In addition, development of nano-optical devices such as resonators, interferometers, Y-splitters, *etc.* will also require detailed understanding and optimization of plasmon propagation through large-angle sharp bends and couplers. Recent experimental efforts in this direction have confirmed the efficiency of groove plasmonic waveguides in transmitting optical signals through large-angle S-bends, interferometers, and ring resonators[26,27,29,30].

Nevertheless, these promising results still leave open the question about optimization of bends in plasmonic waveguides. For example, until recently, nearly 100% transmission through a sharp 90° bend could only be achieved in photonic crystal waveguides[47]. This was because a guided mode in a photonic crystal waveguide must have a frequency from the band gap[47], which was the reason for zero radiation losses from the bend – such radiation simply cannot propagate in the photonic crystal surrounding the bend. In addition, for nearly 100% transmission to occur, the bend should be specially designed, so that the waves reflected back into the first arm of the bend cancel each other as a result of destructive interference. The combination of these two factors ensures nearly 100% transmission of light through the bend in a photonic crystal waveguide. Sadly, these advantages of photonic crystal waveguides are difficult to use for the design of sub-wavelength integrated optical components. Photonic crystal waveguides cannot provide sub-wavelength localization, because spatial periods of such a crystal must be of the order of the wavelength of the guided mode[47].

Therefore, it is very interesting that some of the plasmonic waveguides which do provide sub-wavelength localization may exhibit similar features to those of the photonic crystal waveguides. For example, as demonstrated in Sec. 2, the plasmonic waveguide in the form of a triangular groove with relatively small taper angle $\theta \sim 20^{\circ} - 40^{\circ}$ can provide strong sub-wavelength localization of the CPP modes. On the other hand, if a sharp bend is introduced in such a groove, then radiation bend losses will be impeded (similar to a photonic crystal), because the light cannot propagate in the metallic medium largely surrounding the bend. The only way how radiation loss can occur in such a sharp bend is up the groove, which is in the direction that is essentially perpendicular to the plane of the bend. This by itself is a factor reducing radiation bend losses. In addition, bulk waves cannot propagate in a metallic gap of width that is much smaller than the wavelength, and this is the case near the tip of a triangular groove where the CPP field is strong (especially if the groove angle is relatively small and localization is strong). On the other hand, further away from the tip, where the gap width is sufficiently large for bulk waves to propagate, the CPP field is already evanescently small, and this makes leakage into bulk waves inefficient. The smaller the angle θ, the less efficient should be CPP leakage into bulk waves (or surface plasmons) at groove non-uniformities (such as bends or Fabry-Perot mirrors – see Sec. 4 below). Thus we are led to a suggestion that, at the same level of localization, CPP modes should be more tolerant to structural imperfection than other strongly localized modes, for example, such as wedge plasmons[16-18,32,46]. The discussed mechanisms are expected to severely impede radiation losses in from a sharp bend in a plasmonic groove waveguide. Thus it is possible to expect that an appropriate design of the bend, resulting in cancellation of back-reflected CPPs should lead to nearly 100% transmission through the bend.

The first numerical confirmation of this expectation, as well as the proof that plasmonic waveguides may result in nearly 100% transmission through a sharp 90° bend (similar to photonic crystals), combined with strong sub-wavelength localization (contrary to photonic crystals), was presented in Ref. 24. In this section, we will review the consideration conducted in Ref. 24 to demonstrate practical ways for optimizing sharp bends in sub-wavelength plasmonic groove waveguides to achieve nearly 100% transmission of strongly localized optical signals. Conditions for minimum reflection and radiation losses at the bend will be investigated numerically using the FDTD algorithm[21,22,24] (see Sec. 2). Dissipation in the considered waveguides will be sufficiently low to ensure significant propagation distances of the localized plasmon in each of the bend arms.

The analyzed structures are presented in Figs. 8a-c. Two triangular grooves on a silver surface form a sharp 90° bend (Fig. 8a-c). Figs. 8a,b show the structure with a defect in the form of a wedge metal pillar in the corner of the bend. Fig. 8c shows the bend without the pillar defect. The combination of CPP fundamental and 2nd modes (see Fig. 2a) is launched in the first arm of the bend by means of the end-fire excitation, i.e. by considering a beam incident onto the first arm of the bend from the left (Fig. 8d). The depth of the grooves was 2.1 μm, which is significantly larger than the penetration depth of CPP up the groove $L \sim 200$ nm[21,22,31] (therefore, no screen shown in Fig. 1a was used).

Fig. 8d shows the typical distribution of the magnitude of the electric field in the groove with the pillar defect (Figs. 8a,b) at the distance of 60 nm above the tips of the grooves. Insignificant decrease of the field in the middle of the second arm of the bend (Fig. 8d) is mainly caused by beats resulting from interference of two different CPP modes launchedby the end-fire excitation (Fig. 2a)[21,24]. Dissipation has an insignificant contribution to this decrease because of relatively large propagation distances for the CPP modes of ~ 10 μm[21,22]. Reflections from the end of the second groove (edge of the computational window) were insignificant due to use of absorbing boundary conditions[24,43].

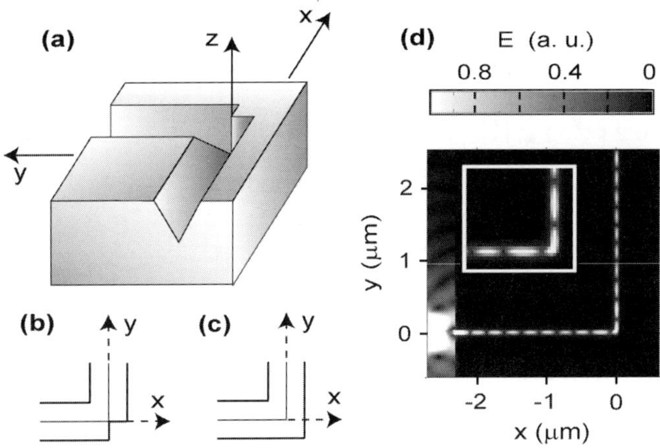

Fig. 8. A groove sub-wavelength waveguide with a 90° bend on a metal substrate/film: (a) three-dimensional view, (b,c) top views; (a,b) the bend with a wedge pillar defect, (c) the bend without the pillar defect. (d) The top view of the field distribution (magnitude of the electric field) in the vacuum ($\varepsilon_d = 1$) groove at the height of 60 nm above the tip with the sharp 90° bend and the pillar defect (Fig. 8a). The groove angles in both the arms of the bend are the same: $\theta = 30^\circ$, $\lambda_{vac} = 632.8$ nm, $\varepsilon_m = -16.22 + 0.52i$ (silver)[44]. The insert zooms into the field distribution near the bend. (Adapted from Ref. 24).

In order to obtain quantitative estimates for the transmissivity T of CPPs through the sharp bend, the following two structures were considered[24] — one with the sharp bend (Fig. 8a,b), and the other with the straight groove of the same length as the combined length of the two arms of the bend. We averaged the amplitude of the field in the second arm of the bend (Fig. 8d) over 5 wavelengths of the fundamental mode (approximately one period of the beat pattern[21]), and in the straight groove at the same distance (along the groove) from the point of the end-fire excitation. The ratio of the resultant average CPP amplitude in the second arm of the bend to that in the straight groove was taken as the amplitude transmissivity T through the bend[24].

As was found in Ref. 24, in the considered bend with the pillar defect (Figs. 8a,b), $T \approx 0.97$ with the numerical error ~ 5%. Thus, practically 100% transmission through a sharp bend can be achieved in a plasmonic sub-wavelength groove waveguide with the pillar defect in the bend

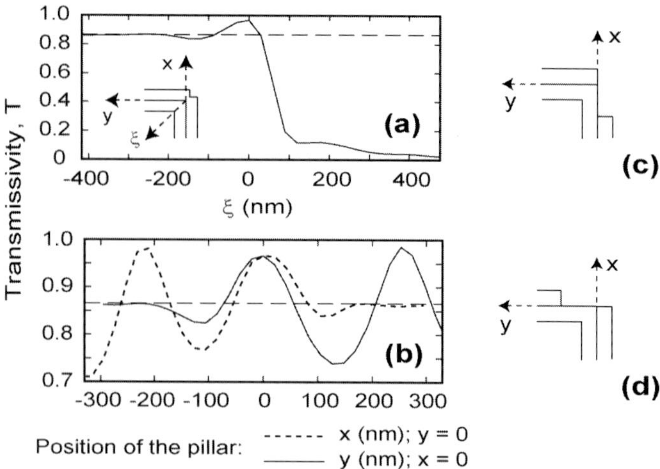

Fig. 9. (a) The dependencies of the amplitude transmissivity T through the bend as a function of position of the corner of the pillar defect on the ξ-axis (the origin of ξ corresponds to $x = 0$, $y = 0$). The insertion in (a) shows the pillar defect shifted along the negative ξ-direction, i.e., the pillar is reduced in size. (b) The dotted curve shows the dependence of T on extension of the pillar into the first arm of the bend (Fig. 9c), i.e., on x-coordinate of the pillar corner at the fixed $y = 0$. The solid curve gives the dependence of T on extension of the pillar into the second arm of the bend (Fig. 9d), i.e., on y-coordinate of the pillar corner at the fixed $x = 0$. The dashed lines in both the plots give T in the absence of the pillar defect (Fig. 8c). (Adapted from Ref. 24).

(Fig. 8a,b). Remarkably, if we remove the pillar defect (Fig. 8c), then the transmissivity is significantly reduced: $T \approx 0.86$. Simultaneously, noticeable reflection from the bend back into the first arm is also observed in the form of significant perturbations of the field in this arm. Therefore, the pillar defect is essential for achieving ~ 100% transmission of CPP through a sharp 90° bend in a groove plasmonic waveguide[24]. The optimal position of the pillar is such that its corner is exactly in the middle of both the grooves, i.e., at $x = 0$, $y = 0$ (Fig. 8b)[24].

If we change the size of the pillar defect by shifting its corner from the considered position ($x = 0$ and $y = 0$) in the middle of the grooves, then T changes significantly[24] (Figs. 9a,b). For example, shifting the corner of the pillar defect along the ξ-axis, so that the size of the defect is increased, results in a rapid reduction of T to about zero (Fig. 9a). In this case, a substantial increase in reflectivity occurs from the bend back into the first arm. Shifting the corner of the pillar in the opposite direction (i.e., along the negative ξ-direction), eventually results in the removal of the pillar defect from the bend (Fig. 8c), which corresponds to $T \approx 0.86$ (Fig. 9a)[24]. The same situation occurs if the pillar corner is shifted in the negative y-direction or positive x-direction. This also results in decreasing the pillar defect, its eventual removal from the bend, and the same value for the transmissivity $T \approx 0.86$ (Fig. 9b).

If the pillar is extended into the first or the second arms of the bend (Figs. 9c,d), then the transmissivity experiences oscillations as a function of pillar position with the period of ~ 200 nm (Fig. 9b)[24]. In other words, nearly 100% transmissivity ($T \approx 1$) is also obtained when the pillar is extended into either of the arms of the bend by $\Delta l \approx 200$ nm. In this case, a section of the groove with the angle $\theta_1 = 15^\circ$ and $\Delta l \approx 200$ nm length is formed. The wavelength of the fundamental CPP mode in this groove $\lambda_{15} \approx 400$ nm[21], i.e., $\Delta l \approx \lambda_{15}/2$. This suggests that nearly 100% transmissivity through the bend is related to destructive interference of CPPs reflected back into the first arm from the different sides of the pillar defect[24]. Indeed, if such a destructive interference (and $T \approx 1$) occurs when the corner of the pillar defect is in the middle of the bend (at $x = 0$ and $y = 0$), then extension of the pillar into one of the arms by $\Delta l \approx \lambda_{15}/2$ should result in additional phase shift of 2π between the reflected CPPs, and we must again have a maximum of $T \approx 1$ (Fig. 9b)[24]. The minimums of T at $\Delta l \approx \lambda_{15}/4 \approx 100$ nm (Fig. 9b) correspond to constructive interference of the reflected CPPs[24]. Here, only reflection and interference of the fundamental mode was considered, because the second CPP mode has significantly smaller amplitude and has little effect on values of T.

Fig. 10 shows the distribution of the magnitude of the electric field along the grooves forming the bend with the pillar defect with its corner in the middle of both the grooves[24] (i.e., at $x = 0$ and $y = 0$). As expected in the beginning of this Section, no significant bend losses in the form of bulk electromagnetic waves can be seen near the bend in Fig. 10 (which is one of the reasons for nearly 100% transmission through the bend). At the same time, an interesting feature that can be seen from Fig. 10 is the noticeable leakage from the bend in the form of a wave propagating along the triangular tip of the pillar upwards from the bottom of the groove[24]. This is another type of strongly localized plasmonic mode that is called wedge plasmon[18,19,32,46] propagating along the sharp 90° wedge of the pillar defect and having typical penetration depth into vacuum of ~ 60 nm and ~ 200 nm along the sides of the metal pillar[24].

The leaking wedge plasmon (Fig. 10) could provide an interesting opportunity for tapping into a CPP mode in a groove with the bend. Such tapping may aid in the development of three-dimensional photonic circuits and sub-wavelength couplers. As was shown in Ref. 24, the ratio of the amplitudes of the wedge plasmon and CPP is ~ 0.2. Due its relatively small amplitude and strong localization near the wedge, the energy losses associated with the wedge plasmon appear to be negligible, and this effect does not seem to have any noticeable impact on the overall transmissivity through the bend (Figs. 9a,b). It is interesting that the considered tapping effect appears to be the strongest when the corner of the pillar defect is in the middle of the grooves[24] (Figs. 9a,b).

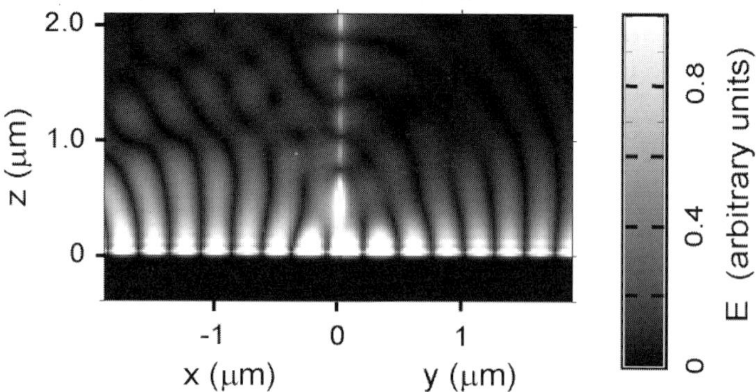

Fig. 10. The distribution of the magnitude of the electric field in the middle of the grooves along the direction of CPP propagation for the groove with the corner of the pillar defect in the middle of the bend (at $x = 0$, $y = 0$) – Fig. 9a,b. The left half of the field distribution is in the (x,z) plane (incident CPP) up to the corner of the pillar defect, and the right half of the distribution is in the (y,z) plane (transmitted CPP) from the corner of the pillar defect. The corner of the pillar defect is in the middle of the picture with the strongly localized wedge plasmon resulting in energy tapping from the bend. Weak decrease of the field in the second arm of the bend is mainly due to interference of two different CPP modes generated in the groove by means of the end-fire excitation[21]. The weak periodic field pattern at noticeable distances from the tip, especially in the first arm of the bend is the residual effect of the end-fire excitation. (Adapted from Ref. 24).

Thus we have demonstrated that nearly 100% transmission through sharp 90° bends is not unique to photonic crystal waveguides, but is also achievable in sub-wavelength plasmonic groove waveguides[24]. As has been discussed, the major physical reasons for achieving nearly 100% transmission through a sharp bend are similar for photonic crystal and plasmonic waveguides. In particular, this can be achieved if the radiation losses are suppressed by the media / structure surrounding the bend (in the case of plasmonic waveguides this is the metal), while the bend should be specifically designed (optimized) to ensure destructive interference of the reflected waves in the first arm of the bend.

It is interesting to note that plasmonic waveguides with strong sub-wavelength localization of the mode field may have an additional reason for high transmission through a sharp or gradual bend. The stronger the localization of the guided mode, the less efficient should be its coupling into bulk waves in the surrounding media. This is because increasing mode localization results in decreasing its speed and effective impedance of the guiding structure. Radiation losses may only occur when a guided mode is coupled into a bulk wave, i.e., undergoes a transmission from a medium with a small impedance (groove) into a medium with a much larger impedance (vacuum or dielectric filling the groove). Such a process is typically highly inefficient. This is similar to low efficiency of transmission of sound from the air into a solid — due to a much larger acoustic impedance of the solid, the acoustic energy will be almost totally reflected from the solid even at normal incidence. In the same way, if a strongly localized CPP mode (with much smaller speed and effective structural impedance) is incident onto a bend, it will not be able to efficiently transfer energy into bulk scattered waves because of much larger optical impedance in the surrounding dielectric (or vacuum). This consideration is applicable not only for CPP modes, but also for any other strongly localized plasmonic modes in sub-wavelength waveguides. Therefore, increased transmission through bends should be a general feature of all plasmonic waveguides with strong localization (if the waves reflected from the bend are cancelled by destructive interference). At the same time, the additional advantage of CPP modes for achieving high transmission through sharp bends is related to the discussed fact that in triangular grooves radiation losses are additionally impeded by the metal effectively surrounding the bend.

The obtained numerical results and conclusions will be important for the design and optimization of integrated plasmonic circuits and nano-optical components, though full experimental confirmation of these results has not been achieved so far.

4. Nano-scale Fabry-Perot resonator in groove waveguides

Substantial and successful recent efforts in the experimental observation of CPP modes in triangular grooves and development of the related applications, such as Mach-Zehnder interferometers and waveguide-ring resonators[27,30], ring resonator add-drop multiplexers, compact Bragg grating filters[29,30], *etc.*, have demonstrated high suitability of these waveguides for integrated nano-optics, optical information and signal processing. However, the first theoretical evidence that nano-optical devices such as nano-scale resonators and filters can be successfully designed using strongly localized CPP modes in triangular groove was presented in Ref. 23. In this reference, a nano-scale Fabry-Perot interferometer / resonator was described and analyzed using the 3D FDTD algorithm[21]. Consideration of such a resonator is especially interesting, because this could be one of the first steps towards the development of nano-lasers.

A CPP Fabry-Perot resonator / cavity is shown in Fig. 11a. Two semitransparent metallic membranes (mirrors) of the same thickness l_m are inserted perpendicular to the triangular groove in the silver substrate, so that the distance between these membranes is W (Fig. 11a).

CPP modes were launched by the end-fire excitation in the silver-vacuum groove with the cavity (Fig. 11a). The incident wavelength used for the end-fire excitation was λ_{vac} = 632.8 nm (He-Ne laser), which corresponds to the silver permittivity[44] $\varepsilon_m \approx -16.22 + 0.52i$. The mirrors of the cavity were also assumed to be made of silver. The groove depth $h \approx 1.6$ μm was determined by the size of the computational window, which was significantly larger than the penetration depth of the fundamental CPP up the groove (~ 200 nm – see Sec. 2 and Refs. 21, 22). Therefore, the results corresponded to an infinitely deep groove[23].

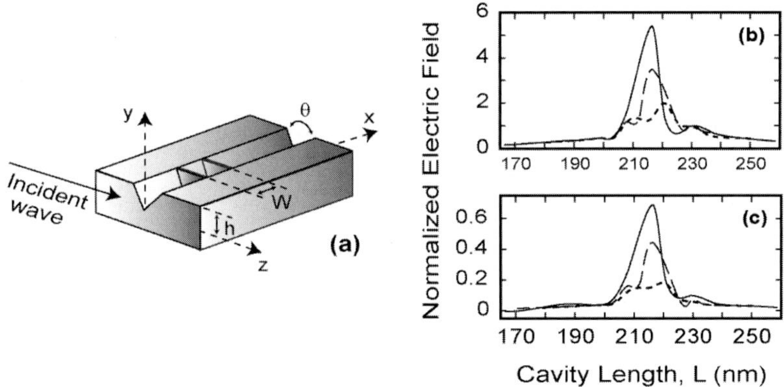

Fig. 11. (a) The sub-wavelength groove waveguide on a metal surface with two metal membranes (mirrors) forming a Fabry-Perot cavity of length W. CPP modes are launched by the end-fire excitation using a bulk wave incident at the angle of 45° to the x-axis. The magnetic field in the incident wave is in the (x,y) plane. (b,c) The dependencies of the amplitudes of the electric field: (b) in the middle of the cavity, and (c) in the wave transmitted through the cavity on cavity length L for the silver-vacuum structure at the vacuum wavelength λ_{vac} = 632.8 nm in the absence of dissipation (the dielectric permittivity of the metal is real: $\varepsilon_m = -16.22$) at three different moments of time $t_1 = 1.22 \times 10^{-13}$ s (dotted curves), $t_2 = 2.44 \times 10^{-13}$ s (dashed curves), and $t_3 = 4.88 \times 10^{-13}$ s (solid curves). The thickness of the cavity mirrors l_m = 42.2 nm, $\theta = 30°$. The electric field amplitudes are normalized to the amplitude of the incident wave at the point of consideration in the absence of the cavity. (Adapted from Ref. 23).

The size of the FDTD grid cells outside the cavity was $\lambda_{vac}/30$ (i.e., ≈ 21.1 nm) along the x-axis (Fig. 11a). However, the grid cells along the x-axis inside the cavity and the membranes (mirrors) were $\lambda_{vac}/300$ and $\lambda_{vac}/150$, respectively. Along the y-axis and z-axis, the grid spacing was $\lambda_{vac}/60$ everywhere in the computational window. This ensured sufficient accuracy and convergence of the obtained numerical results[23].

Figs. 11b,c present the numerical dependencies of the normalized amplitudes of the electric field in the middle of the cavity (Fig. 11b), and in the wave transmitted through the cavity (Fig. 11c) at zero dissipation in the metal ($e_2 = 0$) and three different moments of time after switching the incident wave on at $t = 0$. It is clear that the amplitudes of the field in the cavity and behind it resonantly increase in time at the cavity length[23] $W = W_r \approx 216.5$ nm. The solid curves approximately correspond to the steady-state case, whereas the dashed and dotted curves correspond to strongly non-steady-state regimes. The resonant behavior of the field inside and behind the cavity demonstrates that the considered simple arrangement with just two membranes (mirrors) across the groove can indeed work as a Fabry-Perot resonator/interferometer. Further increase of the thickness of the mirrors of the interferometer results in a rapid increase of the height and sharpness of the observed resonance[23].

In particular, this suggests that there is no significant leakage of the energy from the cavity in the form of surface and/or bulk waves. This is the further confirmation of the conclusion discussed at the beginning and in the end of Sec. 3 about low efficiency of CPP coupling into bulk and surface waves in the presence non-uniformities in the groove (such as a metal membrane). The smaller the groove angle θ, the less the efficiency of CPP leakage into bulk waves (or surface plasmons) at groove non-uniformities (e.g., such as Fabry-Perot mirrors — Fig. 11a).

The effect of dissipation of CPPs in the metal and Fabry-Perot mirrors[23] is demonstrated by Figs. 12a,b showing the dependencies of the steady-state amplitudes of the electric field in the cavity (Fig. 12a) and behind it (Fig. 12b) on cavity length W for the silver permittivity $\varepsilon_m = -16.22 + 0.52i$ (other parameters are the same as for Figs. 11b,c).

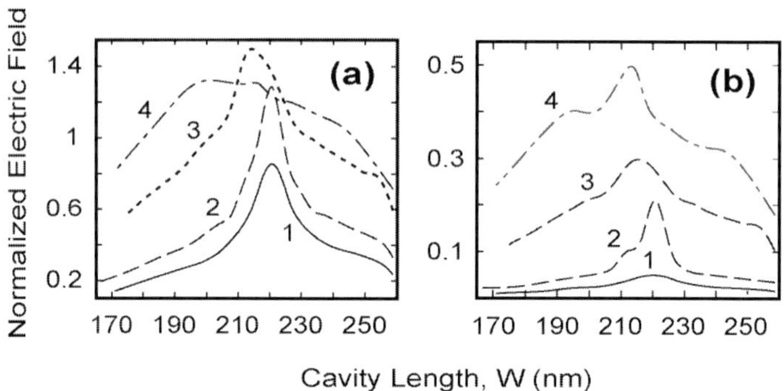

Fig. 12. The steady-state dependencies of the amplitudes of the electric field: (a) in the middle of the cavity and (b) in the wave transmitted through the cavity on cavity length W in the presence of dissipation ($\varepsilon_m = -16.22 + 0.52i$) at different thicknesses of the cavity mirrors: (1) $l_m = 50.6$ nm, (2) $l_m = 42.2$ nm, (3) $l_m = 25.3$ nm, (4) $l_m = 16.9$ nm. $\theta = 30^{\circ}$, $\lambda_{vac} = 632.8$ nm. The electric field amplitudes are normalized to the amplitude of the incident wave at the point of consideration in the absence of the cavity (membranes). (Adapted from Ref. 23).

Understandably, the resonance in Figs. 12a,b is significantly lower compared to that in Fig. 11b without dissipation. If the mirror thickness is increased, then the leakage from the cavity through the mirrors is reduced, but the effect of dissipation in the metal inside the cavity is increased (due to increased relaxation time). As a result, the resonance becomes lower and broader (curves 1 in Figs. 12a,b). On the other hand, if the thickness of the mirrors is decreased, the effect of dissipation inside the cavity is reduced, but the leakage through the mirrors is increased. As a result, the resonance also becomes weaker and broader (curves 3, 4 in Fig. 12a,b). Therefore, there exists an optimal membrane thickness ($l_{mo} \approx 42.2$ nm) at which the resonance is the strongest and sharpest[23] (curves 2 in Fig. 12a,b). At this optimal membrane thickness, the full width half maximum of the resonance curve is ~ 10 nm, which corresponds to the quality factor of the cavity[23] $Q \sim 100$.

If the incident wave is switched on at the moment of time $t = 0$, then Fig. 13a presents the time evolution of the field amplitudes in front of, inside, and behind the cavity for the two different cavity lengths: $W = W_r \approx 220.7$ nm (the minimum resonant cavity length) and $W = 170.1$ nm for the same mirror thicknesses $l_m = 42.2$ nm. Note that these specific values of the cavity widths (and mirror thicknesses) were dictated by the specific choice of the size of the grid cells in the FDTD simulation — only an integer number of grid cells fit within the cavity and a mirror[23]. Curve 1 in Fig. 13a represents the field incident onto the cavity as a result of the end-fire excitation at $x = 0$, and it is the same for both the considered cavities. The oscillations of curve 1 at small t are due to repeated reflections of the front of the switched (at $t = 0$) pulse from the first mirror and the point of the end-fire excitation.

Fig. 13b demonstrates a significant resonant increase of the electric field amplitude inside the cavity with the length $W = W_r \approx 220.7$ nm. It can be seen that the cavity contains half the CPP wavelength with the two nodes of the standing pattern being close to the mirrors of the cavity (compare also curves 2 and 4 in Fig. 13a). This is what should be expected for a Fabry-Perot cavity with the minimal resonant length. Also expectedly, the comparison of curves 3 and 5 in Fig. 13a reveals drastic differences in the transmitted CPP amplitudes in the presence (curve 3) and in the absence (curve 5) of the resonance.

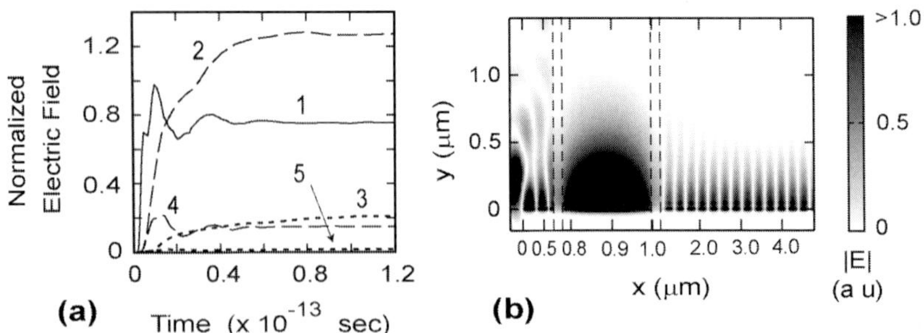

Fig. 13. (a) The time dependencies of the normalized amplitudes of the electric field in front of the cavity (solid curve), in the middle of the cavity (dashed curves), and behind the cavity (dotted curves) at two different cavity lengths: $W = W_r \approx 220.7$ nm (curves 1 – 3), and (b) $W = 170.1$ nm (curves 1, 4, 5). $l_m = 42.2$ nm, and the other structural parameters are the same as for Figs. 12a,b. (b) The distribution of the electric field in the (x,y) plane inside the groove, cavity and the mirrors at the resonant cavity length $W = W_r \approx 220.7$ nm and $l_m = 42.2$ nm. The scale along the x-axis inside the mirrors and the cavity is 5 and 10 times larger than outside the cavity. These regions are shown by the 4 dashed lines. The end-fire excitation occurs at $x = 0$. (Adapted from Ref. 23).

As can be seen from the presented field distribution in Fig. 13b, there is hardly any leakage from the cavity in the form of bulk and/or surface waves up the walls of the cavity (see also Sec. 3). Thus, vertical membranes inserted into the groove do not result in an uncontrollable scattering of the guided CPP mode. As explained in Sec. 3, the physical interpretation for this effect is largely related to the strong sub-wavelength localization of the fundamental CPP mode in the considered structure (within the region of ~ 200 nm from the tip[21,22,31]). Therefore, the field and the energy in the mode are significant only in the region where the width of the groove is smaller than ~ 100 nm, which is much less than the length of the bulk wave (632.8 nm). As a result, generation of bulk waves in such a narrow gap is strongly impeded. Therefore, leakage from the cavity in the form of scattered bulk and surface waves is negligible (Fig. 13b). The smaller the groove angle, the smaller the expected leakage losses from the cavity.

As shown in Sec. 2, the end-fire excitation in the groove launches two different CPP modes causing strong beats in the field structure along the groove[21,22]. Therefore, both the modes are generated in Fig. 13b in front of the cavity. In addition, the field in front of the cavity also contains bulk scattered waves resulting from the end-fire excitation (Fig. 13b). Nevertheless, only one fundamental mode is effectively transmitted through the cavity of the resonant length $W = W_r \approx 220.7$ nm — Fig. 13b (this is also confirmed by Fourier analysis of the output). Thus the cavity works as an efficient mode-selective and frequency-selective filter for CPP signals[23]. In particular, Fig. 13b demonstrates that we can also achieve single-mode propagation in a groove waveguide by means of a Fabry-Perot cavity, instead of making the depth of the groove such that the higher modes are not guided by the structure[22].

In conclusion, the obtained results demonstrate that efficient nano-scale optical devices (such as Fabry-Perot resonators and filters) are feasible in groove plasmonic waveguides. Moreover, these devices could be fabricated simply by using compact elements, such as nano-sized mirrors or membranes, similar to how it is done in the conventional optics. In combination with the results of Sec. 3, this is also a demonstration that manipulation and processing of CPP signals does not necessarily require use of such extended elements as gratings which are essential for manipulation of guided signals in dielectric fibers and waveguides. This may be regarded as a notable advantage for the design of highly integrated nano-optical circuits and components.

5. Nano-focusing in triangular metallic grooves

Another major physical phenomenon associated with plasmon propagation in triangular grooves, which is also linked to strongly localized CPP modes (see below), is focusing (concentrating) light energy, into spatial regions as small as a few nanometers. This effect called nano-focusing is far beyond the capabilities of the conventional optics which is restricted by the diffraction limit of light, and is not capable of focusing light into a region that is significantly smaller than the wavelength[1,2]. Nano-focusing thus offers unique opportunities for the efficient delivery of electromagnetic energy to nano-scale regions (particles) and even separate molecules, development of near-field optical microscopy and optical imaging with sub-wavelength resolution[48-56], high-resolution lithography[57], coupling of light into and out of photonic nano-circuits, new sensors and detection techniques based on surface-enhanced Raman spectroscopy[49,58], *etc.*

Different metal structures have been suggested for nano-focusing of plasmons. These include sharp metal tips[49,59-62], dielectric conical tips covered in metal film[48,50,51], pyramidal tips covered in metal film with a nano-aperture[53], nano-particle lenses[58], sharp triangular grooves and nano-wedges[37-41,63-65], *etc.* Only special types of plasmon can experience nano-focusing — those whose wave numbers increase with decreasing size of the metallic structure. For example, in a metallic gap there can exist two gap plasmon modes with the symmetric and anti-symmetric charge distribution across the gap[38]. The mode with the symmetric charge distribution (symmetric mode) has a cut-off gap width below which this mode does not exist. On the contrary, the anti-symmetric gap mode can exist at arbitrarily small gap widths, and its wave vector increases to infinity when the gap width tends to zero. This is caused by the attractive interaction of the opposite charges across the gap. As a result, if we consider a gap with slowly changing width (tapered gap), then the anti-symmetric plasmon propagating in the direction of the taper experiences gradual decrease of its wavelength, speed, and localization region. Thus we have plasmon nano-focusing in tapered grooves[37-41].

One of the major features of nano-focusing is the possibility of concentration and strong local field enhancement in regions with dimensions of a few nanometers[37-41,60-65]. This opens unique opportunities for observation of non-linear plasmonic effects, spectroscopic analysis with sub-wavelength resolution, development of new sensors and measurement techniques. Physical limits of nano-focusing are established by the atomic structure of matter, spatial dispersion and Landau damping[45] which is typically notable at structural dimensions of about a few nanometers[60]. Thus the theory of nano-focusing based on the approximation of continuous electrodynamics[37-41,60-65] is applicable only down to the dimensions of the focusing structure (gap width, rod diameter, or wedge/film thickness) of ~ 2 – 4 nm.

Two major regimes of plasmon nano-focusing have been identified and investigated. These are adiabatic[38,41,60,61,63,64] and non-adiabatic[39-41,62] nano-focusing. Adiabatic nano-focusing occurs when the taper angle of the focusing metallic structure (e.g., gap) is sufficiently small, so that the plasmon propagation parameters do not change significantly within one plasmon wavelength[38,41,60,63,64]. This approximation is also known in the literature as the geometrical optics or WKB (in the terminology of quantum mechanics) approximation. During the adiabatic nano-focusing the plasmon does not experience any significant reflections from the taper. Negligibility of such reflections can be regarded as the applicability condition for the adiabatic approximation. Non-adiabatic nano-focusing occurs in triangular grooves with sufficiently large taper angle, such that plasmon reflections off the taper cannot be neglected.

The analyzed structure is presented in Fig. 14a. A deep triangular groove with the taper angle θ is formed in a metal with the complex permittivity $\varepsilon_m = e_1 + ie_2$ ($e_1 < 0$, $e_2 > 0$). The coordinate axes are as indicated in the figure. The dielectric permittivity ε_d of the material inside the groove is real and positive, and $|e_1| > \varepsilon_d$ (the existence condition for surface and gap plasmons[4]).

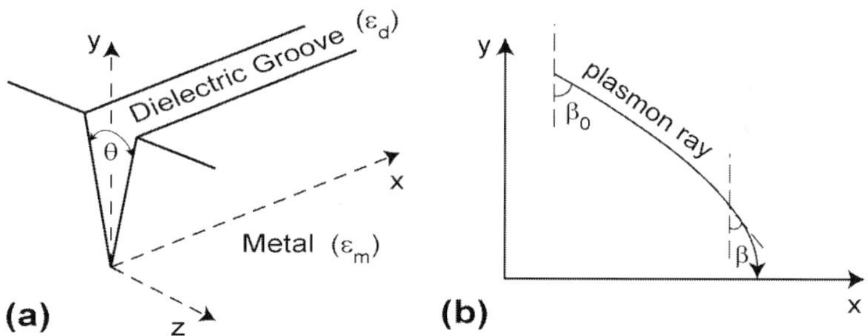

Fig. 14. (a) The nano-focusing dielectric groove structure with the taper angle θ. ε_d is the permittivity of the dielectric filling the groove, and ε_m is the metal permittivity. (b) A ray representing the direction of propagation of an anti-symmetric gap plasmon in the groove in the adiabatic approximation. β_0 is the angle of propagation of the gap plasmon at large distances from the tip of the groove, where the coupling between the two surface plasmons representing the gap plasmon is negligible. (Adapted from Ref. 38).

As the anti-symmetric gap plasmon propagates towards the tip of the groove, it will experience changing width of the groove (gap). This means that the effective dielectric permittivity of the structure $\varepsilon_{eff} = (q/k_0)^2$ also changes. Here, q is the wave number of the anti-symmetric gap plasmon (depending on gap width h), and $k_0 = \omega/c$, where ω is the wave frequency, and c is the speed of light in vacuum. Propagation of the gap plasmon can be considered in the adiabatic approximation, if variations of the wave vector of the plasmon are small within distances of the order of one wavelength in the structure[38,45,60]:

$$|d(Q_{1y}^{-1})/dy| << 1, \tag{1}$$

where Q_{1y} is the y-component of the real part of the wave vector $\mathbf{q} = \mathbf{Q}_1 + i\mathbf{Q}_2$ of the gap plasmon in the groove (Fig. 14b), and dissipation is assumed to be small, so that $Q_2 << Q_1$. If condition (1) is satisfied for all values of y, then the anti-symmetric plasmon can be approximated by a plane wave at any point along the direction of its propagation (along a plasmon ray – Fig. 14b), and the groove can be regarded as a gap with slowly varying width. The distribution of the field, dispersion and dissipation of the gap plasmon will then be approximately the same as for the plasmon in a uniform gap of the width that is equal to the local width of the tapered gap at the considered distance from the tip[38].

The wave number of the anti-symmetric gap plasmon is given by the dispersion relation[38]:

$$\tanh\left(\frac{\alpha_1 h}{2}\right) = -\frac{\alpha_2 \varepsilon_d}{\alpha_1 \varepsilon_m}, \tag{2}$$

where

$$\alpha_1 = \sqrt{q^2 - k_0^2 \varepsilon_d}, \quad \alpha_2 = \sqrt{q^2 - k_0^2 \varepsilon_m}, \tag{3}$$

and h is the local width of the triangular groove:

$$h = 2y \tan(\theta/2). \tag{4}$$

In the analysis of plasmon nano-focusing, we typically consider that dissipation in the metal is weak ($e_2 << |e_1|$ and $Q_2 << Q_1$), so that plasmons can propagate significant distances in the groove and experience noticeable local field enhancement[38-41]. If this is the case, then in the asymptotic regime near the tip of the groove, i.e. where $h \to 0$, Eq. (2) could be simplified as[38]:

$$Q_1 \approx -2\varepsilon_d/(e_1 h). \tag{5}$$

In particular, Eq. (5) suggests that decreasing h to zero results in increasing wave number of the gap plasmon to infinity, which is equivalent to decreasing plasmon wavelength to zero. As a result, both the phase and group velocities of the anti-symmetric plasmon[38] $v_{ph} = \omega/Q_1 \to 0$ and $v_g = \partial\omega/\partial Q_1 \approx he_1^2(2\varepsilon_d de_1/d\omega)^{-1} \to 0$, if $h \to 0$ (or $y \to 0$). This means that the anti-symmetric gap plasmons asymptotically stop at the tip of the groove[38], similarly to strongly localized plasmons focused by a sharp metal cone[60].

The typical numerical dependencies of the wave number Q_1 of the anti-symmetric plasmon on gap width h, obtained from Eq. (2), are presented in Fig. 15a for different permittivities of the metal and dielectric filling the triangular groove. In particular, it can be seen that decreasing h and/or increasing ratio $\varepsilon_d/|e_1|$ result in a monotonic increase of the wave number of the antri-symmetric gap plasmon.

As the plasmon propagates towards the tip of the groove, the x-component Q_{1x} of the real part of its wave vector $\mathbf{Q}_1$ must remain the same at all points on the plasmon ray (Fig. 14b), because of Snell's law

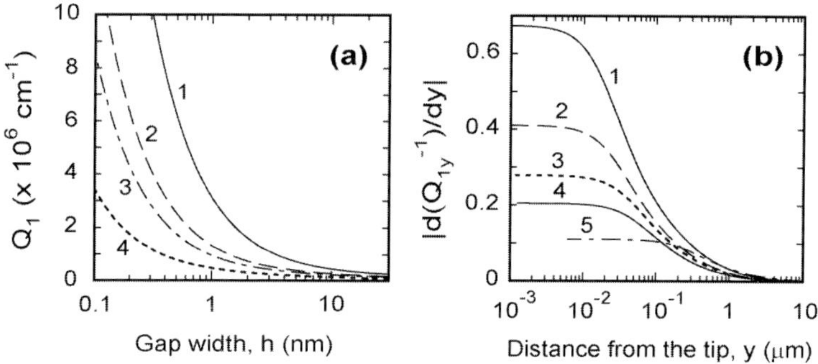

Fig. 15. a) The dependencies of the wave number Q_1 of the anti-symmetric gap plasmon on gap width h for the following parameters: (1) $e_1 = -6.5$, $\varepsilon_d = 1$, $\lambda_{vac} = 459.2$ nm, (2) $e_1 = -16$, $\varepsilon_d = 1$, $\lambda_{vac} = 632.8$ nm, (3) $e_1 = -58.8$, $\varepsilon_d = 2.5$, $\lambda_{vac} = 1.127$ μm, (4) $e_1 = -58.8$, $\varepsilon_d = 1$, $\lambda_{vac} = 1.127$ μm. b) The dependencies of the left-hand side of inequality (1) on y-coordinate for the groove with the angle $\theta = 2^\circ$ for the following parameters: (1) $e_1 = -96.6$, $\varepsilon_d = 2.5$, $\lambda_{vac} = 1.631$ μm, (2) $e_1 = -58.8$, $\varepsilon_d = 2.5$, $\lambda_{vac} = 1.127$ μm, (3) $e_1 = -16$, $\varepsilon_d = 1$, $\lambda_{vac} = 632.8$ nm, (4) $e_1 = -58.8$, $\varepsilon_d = 5$, $\lambda_{vac} = 1.127$ μm, (5) $e_1 = -16$, $\varepsilon_d = 2.5$, $\lambda_{vac} = 632.8$ nm. Curve 3 in (b) also approximately corresponds to the other structure with $e_1 = -58.8$, $\varepsilon_d = 2.5$, $\lambda_{vac} = 1.127$ μm, and $\theta = 1^\circ$. The metal permittivities correspond to silver at the indicated wavelengths[44]. (Adapted from Ref. 38).

and uniformity of the groove along the x-axis. Therefore, because Q_1 increases to infinity near the tip of the gap (Fig. 15a), the y-component Q_{1y} of $\mathbf{Q}_1$ must also increase to infinity. Therefore, irrespectively of the initial incidence angle β_0 (as long as $\beta_0 < \pi/2$), the wave vector of the anti-symmetric plasmon near the tip is always perpendicular to the x-axis, which makes the plasmon ray turn perpendicular to the tip of the gap (Fig. 14b). It is interesting to note that in the adiabatic approximation and approximation of continuous electrodynamics the optical path for the plasmon to the tip of the groove is infinite. Therefore, it will take an infinite interval of time for the plasmon to reach the tip, and it will eventually dissipate in the metal with no noticeable reflections. Certainly, this is only an approximation, and in reality the considered theory is limited by the atomic structure of matter and spatial dispersion[45,60].

Fig. 15b presents the y-dependencies of the left-hand side of condition (1) for the considered silver-dielectric structure with the groove angle $\theta = 2^{\circ}$. It is useful to note that though condition (1) assumes that $|d(Q_{1y}^{-1})/dy|$ must be much smaller than 1, as we will see below, the adiabatic approximation is actually applicable under a less restrictive condition that can be written as[41,64]

$$|d(Q_{1y}^{-1})/dy| \lesssim 1 \qquad (6)$$

This is the condition that determines taper angles of the groove at which plasmon reflections from the taper are negligible (see below for the comparison of the approximate and rigorous theories of nano-focusing[39,41]). Therefore, it can be seen that for all the structures corresponding to curves 1 – 5 in Fig. 15b the adiabatic approximation is applicable at all distances from the tip of the groove. It is also interesting that the adiabatic approximation is always applicable at large distances from the tip. This is because at large values of y the two surface plasmons forming the anti-symmetric gap plasmon become uncoupled, and variations of the gap width due to the taper hardly affects their dispersion (as long as the width of the gap is larger than the penetration depth of the surface plasmons into the dielectric).

In the asymptotic regime near the tip of the groove (i.e., where $y \to 0$), condition (6) can be significantly simplified. Using Eqs. (4) and

(5), taking into account that near the tip $Q_{1y} \approx Q_1$ (Fig. 14b), and considering small groove angles $\theta << 1$, condition (6) can be reduced as[38]:

$$\left| \frac{d(Q_{1y}^{-1})}{dy} \right| \approx -\frac{\theta e_1}{2\varepsilon_d} \lesssim 1, \tag{7}$$

or

$$\theta \lesssim \theta_{c2} = -2\varepsilon_d/e_1, \tag{8}$$

where θ_{c2} is the critical angle of the groove[38], such that the adiabatic approximation is applicable for angles that are smaller than θ_{c2} (the reason for using index 2 will be explained below).

For example, for the silver-vacuum triangular groove at λ_{vac} = 632.8 nm (He-Ne laser), we have[38]: $\theta_{c2} \approx 7^\circ$. As follows from Eq. (8), the critical angle θ_{c2} increases with increasing ε_d and/or decreasing e_1. Thus decreasing frequency should result in decreasing θ_{c2} (because this leads to increasing magnitude of the real part of the metal permittivity).

A plasmon ray, i.e. a curve parallel to the direction of the wave vector of the plasmon at every point, is determined by the angle β between this ray and the y-axis (Fig. 14b). At a arbitrary value of y, this angle can be determined from Snell's law which suggests[38] that the x-component of the wave vector of the plasmon must be independent of y:

$$Q_{01}\sin\beta_0 = Q_1\sin\beta, \tag{9}$$

where Q_{01} is the real part of the wave number $q_0 = Q_{01} + iQ_{02}$ of the gap plasmon at $y \to +\infty$. Because at $y \to +\infty$ the two plasmons on the groove sides are uncoupled, q_0 is the wave number of the plasmon on the isolated interface[4] between the media with the permittivities ε_d and ε_m:

$$q_0 = \frac{\omega}{c}\sqrt{\frac{\varepsilon_d \varepsilon_m}{\varepsilon_d + \varepsilon_m}}. \tag{10}$$

If dissipation is weak, i.e. $Q_2 << Q_1$, then a plasmon ray can be determined by the following algorithm[38]. The real part Q_1 of the plasmon wave number as a function of gap width h is approximately determined by Eq. (2) with ε_m, α_1, and α_2 being replaced by e_1 and α_{01}, and α_{02}, respectively, where

$$\alpha_{10} = \sqrt{Q_1^2 - k_0^2 \varepsilon_d}\,, \quad \alpha_{20} = \sqrt{Q_1^2 - k_0^2 e_1}\,. \tag{11}$$

Choose an incidence angle $0 \le \beta_0 < \pi/2$, at which the gap plasmon propagates with respect to the y-axis at sufficiently large distances from the tip of the groove, where the two surface plasmons forming the incident anti-symmetric gap plasmon are approximately uncoupled (Fig. 14b). Then the angle β between the plasmon ray and the y-axis at any point on the ray is given by Eq. (9). Therefore, if we know the position of some point with the coordinates (x,y) on the ray, then the position of the next point with the coordinated $(x + dx,\ y + dy)$ on the same ray is determined by the angle β at the point (x,y) and the value of dy. Thus, the plasmon ray can be determined if we know the position of some initial point on this ray. Suppose that this point is chosen on the y-axis. Then the typical plasmon rays for different incidence angles β_0 and different groove angles are presented in Fig. 16.

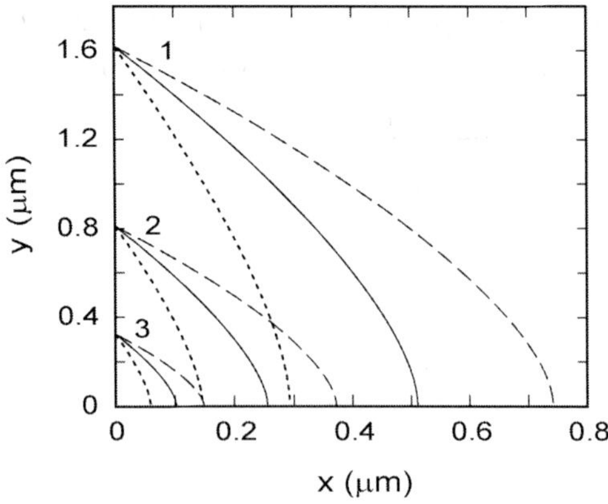

Fig. 16. Typical rays for anti-symmetric plasmons in the silver-vacuum triangular grooves with the angles: 1) $\theta = 1^\circ$, 2) $\theta = 2^\circ$, and 3) $\theta = 5^\circ$. Three different angles of incidence are considered: Dashed curves: $\beta_0 = 75^\circ$, Solid curves: $\beta_0 = 45^\circ$, and Dotted curves: $\beta_0 = 25^\circ$. The initial y-coordinates for all the rays correspond to the same width of the groove (gap): $h \approx 28.3$ nm. The permittivities are for the silver-vacuum structure at $\lambda_{vac} = 632.8$ nm: $\varepsilon_d = 1$, $e_1 = -16$. (Adapted from Ref. 38).

In accordance with the above explanation about the y-component of the plasmon wave vector turning to infinity at the tip of the groove, all the rays in Fig. 16 are orthogonal to the x-axis at $y = 0$.

As shown in Ref. 38, if a gap plasmon propagates towards the tip along a ray, then in the absence of dissipation the y-component of the energy flux in this plasmon must be constant along the ray:

$$S_0 \cos\beta_0 = S \cos\beta \,, \tag{12}$$

where S_0 is the initial energy flux in the anti-symmetric gap plasmon at a large distance from the tip of the groove (i.e. the flux in the two uncoupled surface plasmons), and S and β are the energy flux and angle of propagation of the plasmon at an arbitrary point on the plasmon ray[38].

In the presence of dissipation, Eq. (12) is not valid, because energy conservation should also involve dissipation of the plasmon along the ray. Therefore, a special numerical procedure has been developed for the determination of plasmon amplitude as a function of distance from the tip of the groove[38]. Consider two points on the plasmon ray (x,y) and $(x + dx, y + dy)$, that are separated by the distance dx_p (x_p is the axis parallel to the ray at the point (x,y)). Suppose that the angle between the ray and the y-axis at the point (x,y) is β. Then the variation in the energy flux in the plasmon between the two points consists of the two parts:

$$dS = dS_0 + dS_d, \tag{13}$$

where dS_0 is the variation of the energy flux in accordance with Eq. (12), caused by changing angle β (here, the index "0" indicates that the flux variation is calculated at zero dissipation), and dS_d is the variation of the energy flux due to dissipation of the plasmon as it propagates the distance dx_p along the ray.

Following Refs. 39 and 41, we determine dS_0 and dS_d and substitute them into Eq. (13) to determine the overall variation in the energy flux in the plasmon as a result of its propagation from a point (x,y) to the point $(x + dx, y + dy)$. This gives us the new energy flux at the point $(x + dx, y + dy)$, and thus the amplitude of the plasmon at this point. This immediately results in a numerical algorithm for calculating plasmon amplitude as a function of y along the plasmon ray, if this amplitude is known at some reference point. The value of dx is calculated from a chosen value of dy and the angle β at the point (x,y), which is determined by Eq. (9). For more detail see Refs. 38 and 41.

The resultant numerical dependencies of the amplitudes of the electric and magnetic fields in the plasmon on distance from the tip of the groove are presented in Figs. 17a,b for tapered gaps in silver. One of the most important features of this figure is that, as the anti-symmetric plasmon propagates towards the tip of the gap, both the amplitudes of the magnetic and electric fields go through a minimum and then start to increase with decreasing distance to the tip (though the rates of this increase depend upon groove angle and the initial direction of plasmon propagation). Further behavior of the electric and magnetic fields appear to be different — the electric field continues to asymptotically increase to infinity (in the approximation of continuous electrodynamics) near the tip (Fig. 17a), while the magnetic field goes through a maximum and then asymptotically tends to zero (Fig. 17b). In addition, the typical local field enhancements for the electric field are much larger than those for the magnetic fields[41] (compare Figs. 17a and 17b). These differences in the behavior of the magnetic and electric fields in the asymptotic regime

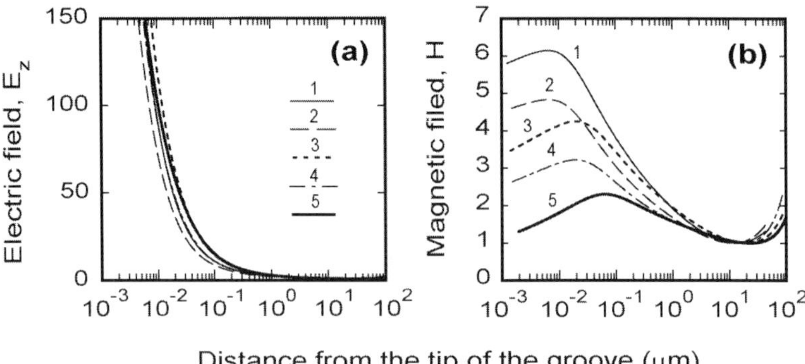

Fig. 17. The y-dependencies of the amplitudes of the electric (a) and magnetic (b) fields in the middle of the gap for the anti-symmetric gap plasmons (λ_{vac} = 632.8 nm) incident onto the tip of the vacuum triangular gap in silver. All the amplitudes are normalized to the amplitudes at the local minimums (at ~ 10 – 30 μm). Metal permittivity: $\varepsilon_m = -19.3 + 0.66i$, $\varepsilon_d = 1$, and (1) $\beta_0 = 0$, $\theta = 4^\circ$, (2) $\beta_0 = 45^\circ$, $\theta = 4^\circ$, (3) $\beta_0 = 0$, $\theta = 2^\circ$, (4) $\beta_0 = 45^\circ$, $\theta = 2^\circ$, (5) $\beta_0 = 0$, $\theta = 1^\circ$. (Curves 1 and 4 are nearly identical in (a)). (Adapted from Refs. 38 and 41).

near the groove demonstrate that strongly localized plasmons are predominantly electrostatic waves with the major part of their energy stored in the electric field.

However, the described behavior of the electric field in the plasmon experiencing nano-focusing in a triangular groove (Fig. 17a) is typical for small dissipation rate in the metal. In order to investigate the dependence of the electric field enhancement on dissipation in the metal, we consider the fixed real part of metal permittivity ($e_1 = -19.3$) and change the imaginary part. The resultant dependencies of the electric field amplitude on distance to the tip of the groove for normal plasmon incidence ($\beta_0 = 0$) are shown in Fig. 18a.

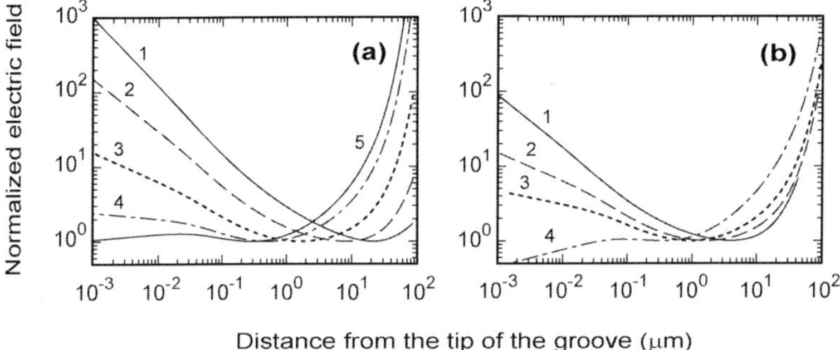

Fig. 18. The y-dependencies of the magnitude of the electric field amplitudes $|E| = |E_z|$ of the anti-symmetric gap plasmon in the middle of the silver-vacuum triangular groove with $\beta_0 = 0$, $\varepsilon_d = 1$, $\lambda_{vac} = 632.8$ nm, and $e_1 = -19.3$. (a) Fixed groove angle $\theta = 2^\circ$, but different imaginary parts of the metal permittivity: (1) $e_2 = 0.66$, (2) $e_2 = 2$, (3) $e_2 = 4$, (4) $e_2 = 6$, (5) $e_2 = 7$. (b) Fixed imaginary part of the metal permittivity $e_2 = 4$, but different groove angles: (1) $\theta = 4^\circ$, (2) $\theta = 2^\circ$, (3) $\theta = 1.5^\circ$, and (4) $\theta = 1^\circ$. Curves 3 in (a) and 2 in (b) are identical. (Adapted from Ref. 41).

As can be seen from Fig. 18a, the infinite (in the approximation of continuous electrodynamics) increase of the electric field amplitude occurs only if e_2 is smaller than a critical value e_{2c} (in Fig. 18a, $e_{2c} \approx 6.5$). If $e_2 > e_{2c}$, then the electric field amplitude near the tip monotonically decreases to zero (curve 5 in Fig. 18a). At the critical dissipation (and only at this critical dissipation) in the metal, the amplitude of the electric field near the tip asymptotically tends to a constant. Similarly, Fig. 18b demonstrates that at a fixed level of dissipation in the metal, there exists a critical taper angle θ_c, such that when $\theta > \theta_c$, the amplitude of the electric field tends to infinity at the tip (curves 1 – 3 in Fig. 18b), while if $\theta < \theta_c$, the amplitude of the electric field at the tip asymptotically tends to zero[41] (curve 4 in Fig. 18b).

It is also interesting that near the tip, the asymptotic behavior of the logarithm of the electric field amplitude is linear in logarithm of distance from the tip (Figs. 18a,b), which means that the asymptotic behavior of the electric field amplitude is the power law in distance from the tip[41]:

$$E \approx Cy^{-\mu}, \quad \text{if } y \to 0, \tag{14}$$

where C and μ are some constants depending on the structural parameters ($-\mu$ is the slope and $\ln(C)$ is the intercept of the linear asymptotic dependencies in Figs. 18a,b.). If $e_2 < e_{2c}$ (or $\theta > \theta_c$), then $\mu > 0$, and the amplitude of the electric field tends to infinity at the tip. If $e_2 > e_{2c}$ (or $\theta < \theta_c$), then $\mu < 0$, and the amplitude of the electric field tends to zero at the tip. If $e_2 = e_{2c}$ (or $\theta = \theta_c$), then $\mu = 0$ and the amplitude of the electric field at the tip is non-zero and finite (Fig. 18a)[41].

Using the linear asymptotic behavior of the electric field amplitude in the log-log scale near the tip of the groove (Figs. 18a,b), a very simple analytical relation linking e_{2c} and θ_c was derived in Ref. 41:

$$\theta_c = \frac{2e_{2c}\varepsilon_d}{e_1^2}. \tag{15}$$

For each value of e_2, this equation determines the critical angle θ_c, and vice versa. The reason for using both θ and e_2 with the indices "c" is because if Eq. (15) is satisfied, then both these quantities are equal to their critical values. Note also that Eq. (15) is correct for all possible initial angles of incidence of the plasmon onto the tip ($0 \le \beta_0 < \pi/2$). For the considered example, Eq. (6) gives: $e_{2c} \approx 6.5$ for Fig. 18a, and $\theta_c \approx 1.23^\circ$ for Fig. 18b, which is in excellent agreement with the presented numerical results.

If the groove angle is increased, then conditions (1), (6) and (8) will eventually be breached, and we will have non-adiabatic nano-focusing for which plasmon reflections from the taper can no longer be neglected.

The analysis of such non-adiabatic nano-focusing in a triangular groove has been conducted in Refs. 39 – 41. The comparison of the results obtained from the adiabatic theory of nano-focusing and those from the rigorous finite element analysis based on the FDTD algorithm and finite element analysis (COMSOL software package)[39,41] is shown in Fig. 19.

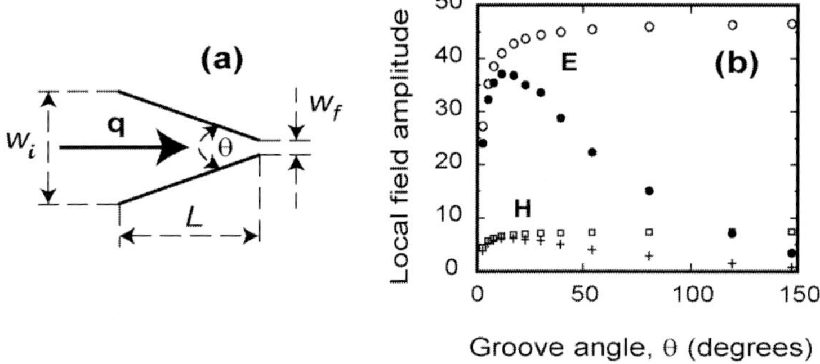

Fig. 19. (a) A tapered gap with the entry width w_i, exit width w_f, taper angle θ and the overall length L. (b) The dependencies of the magnetic and electric field amplitudes in the middle of the gap at the exit of the taper on angle θ, calculated using the rigorous finite-element analysis (crosses and filled circles) and adiabatic approximation (squares and empty circles). The angle of incidence $\beta_0 = 0$, $\lambda_{vac} = 632.8$ nm, $\varepsilon_m = -19.3 + 0.66i$, $w_i = 1.2$ μm, $w_f = 2$ nm. All the dependencies are normalized to the corresponding amplitude (of the electric and magnetic fields) at the entry of the taper. (Adapted from Refs. 39 and 41).

In particular, it can be seen that both the electric and magnetic fields experience maximal enhancement at approximately the same optimal taper angle $\theta_{opt} \approx 14^{\circ}$ (see crosses and filled circles in Fig. 19b). At the same time, the optimal enhancement of the electric field appears to be much stronger than that of the magnetic field, which is in agreement with the previous analysis conducted in the adiabatic approximation (see above). It can also be seen that the adiabatic approximation gives a good agreement with the rigorous numerical dependencies for the taper angles that are smaller than the optimal angle θ_{opt}. This is the demonstration that this approximation is sufficiently accurate in a relatively broad range of taper angles, even if formal adiabatic condition (1) is not very well satisfied. This was the reason for using less restrictive conditions (6) or (8) instead of (1) for more accurate determination of the applicability limits for the adiabatic approximation[41,64]. In the considered structure, condition (6) (or (8)) gives $\theta \lesssim 7^{\circ}$, which is in good agreement with the results shown in Fig. 19b.

The presented in this section consideration of plasmon nano-focusing in a triangular groove offers an interesting and useful insight[38] into the previously considered strongly localized CPP eigenmodes in the same structure (see Sec. 1). The groove can be regarded as a kind of a waveguide for anti-symmetric gap plasmons[38] with gradually changing effective permittivity ε_{eff}. This effective permittivity can be defined as[38] $\varepsilon_{eff}(y) = [Q_1(y)c/\omega]^{1/2}$. It increases with decreasing distance to the tip of the groove, due to increasing Q_1 (see Eq. (5) and Fig. 15a). A CPP mode could then be represented by an anti-symmetric gap plasmon successively reflecting from the tip of the groove and the turning point (simple caustic) — Fig. 20. The wave number for the CPP mode is equal to the x-component of the wave vector of the anti-symmetric gap plasmon $q_{CPP} = Q_{1x}$. The turning point y_t is then determined by the condition $Q_1(y_t) = q_{CPP}$. At this point, the gap plasmon representing the CPP mode propagates parallel to the tip of the groove (Fig. 20).

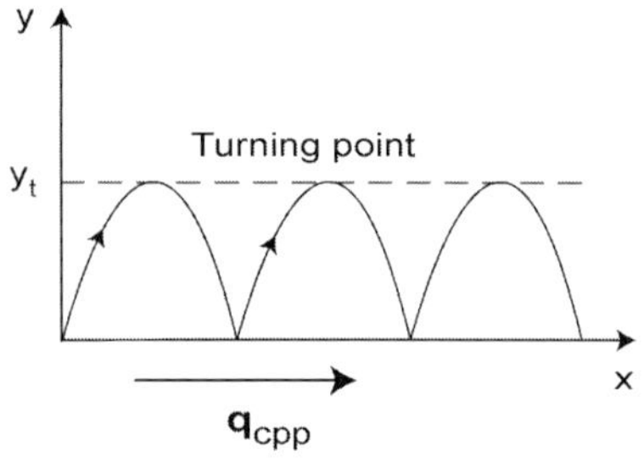

Fig. 20. Geometrical optics representation of a CPP mode in a metallic triangular groove with the tip at $y = 0$. The anti-symmetric gap plasmon with the wave vector $\mathbf{Q}_1(y)$, representing the CPP mode, is successively reflected from the tip of the groove and the turning point (caustic); the CPP wave number $q_{CPP} = Q_{1x}$. (Adapted from Ref. 38).

The difference between this situation and the previously considered nano-focusing of anti-symmetric gap plasmons is that in the case of a CPP mode there is no incident gap plasmon. The gap plasmon representing the CPP mode should be confined to a region close to the tip of the groove and cannot leave this region, because CPP modes are non-radiative structural eigenmodes[20-25,28,31,32]. This is very similar to a non-radiative mode guided by a dielectric slab.

Based on this representation of a guided CPP mode, an attempt has been made to use the adiabatic (geometrical optics) approximation to describe these modes[38], similar to how it was done previously for wedge acoustic modes[66,67]. However, as demonstrated in Ref. 38, contrary to the case with wedge acoustic modes[66,67], this approach is not applicable for CPP modes, because it leads to a diverging Bohr-Sommerfeld quantization condition[38]. Physically, this is related to the fact that if the adiabatic approximation is valid (i.e. conditions (6) and (8) are satisfied), then an anti-symmetric plasmon propagating towards the tip of the groove cannot be reflected back as required by the representation in Fig. 20. This means that the corresponding CPP mode must have infinite localization near the tip of the groove ($q_{CPP} = +\infty$), i.e. it does not exist.

It follows from here that the conditions for adiabatic nano-focusing (6) and (8) are antagonistic to the existence of a localized plasmonic eigenmode in the groove. On the other hand, if the adiabatic conditions (6) and (8) are not satisfied near the tip of a triangular groove, then the integral in the Bohr-Sommerfeld quantization condition and its divergence at $y = 0$ do not have a physical meaning[38]. In this case, the anti-symmetric gap plasmon representing the CPP mode does not reach the tip of the groove, but is rather is reflected back from the taper (due to non-adiabaticity), as required by the representation in Fig. 20. Therefore, localized CPP modes can only exist in a groove if the adiabatic conditions (6) and (8) are not satisfied. Thus, the adiabatic approximation cannot be used for the determination of dispersion of CPP modes (contrary to wedge acoustic modes[66,67]), and other methods have to be used instead (see Sec. 1 and Refs. 20 – 25, 28, 31, and 32).

This consideration is also important for understanding why the maximum of the field in a CPP mode occurs at some small distance from the tip (see Sec. 2 and Fig. 6b). This is because such a mode is formed by the anti-symmetric gap plasmon experiencing successive reflections from the tip and turning point. Reflections from the tip occur due to non-adiabaticity of the structure, so that the anti-symmetric plasmon does not reach the tip of the groove, and is reflected from the taper. This results in an interference pattern between incident and reflected gap plasmons, giving the field maximum at some distance from the tip.

We can now understand the reasons for using index 2 for the critical taper angle θ_{c2} in Eq. (8). This is not only the critical angle that determines applicability of the adiabatic theory of nano-focusing. This is also the second critical angle determining the existence condition of the localized CPP modes in a triangular metallic groove[38,41]. However, while the adiabatic theory of nano-focusing is applicable only if $\theta \lesssim \theta_{c2}$, localized CPP modes can exist only if this condition is breached[38,41], i.e., within the range $\theta_{c2} < \theta < \theta_{c1}$, where θ_{c1} was determined in Sec. 1.

It is very interesting that this consideration and the existence of the lower critical angle of the tapered structure, determining the existence of strongly localized plasmonic eigenmodes is not only the feature of CPP modes in triangular metallic grooves. It is rather a general feature of a number of other structures, such as sharp metal wedges[63], or thin tapered films on a dielectric substrate[64], *etc*. In all these cases, existence of strongly localized plasmonic eigenmodes is antagonistic to adiabatic nano-focusing. This provides an important link between the adiabatic nano-focusing and localized CPP modes in triangular grooves.

References

1. J. R. Krenn, Nature Mater. **2**, 210 (2003).
2. S. A. Maier, et al. Nature Mater. **2**, 229 (2003).
3. 1. V.M. Agranovich and D.L. Mills, Eds., *Surface Polaritons* (North-Holland, Amsterdam, 1982).
4. 2. H. Raether, *Surface Plasmons* (Springer-Verlag, Berlin, 1988).
5. 3. A.D. Boardman, Ed., *Electromagnetic Surface Modes* (John Wiley & Sons, NewYork, 1982).
6. 4. S. A. Maier, *Plasmonics: Fundamentals and Applications* (Springer, Berlin, 2007).
7. W.L. Barnes, A. Dereux, and T.W. Ebbesen, *Nature*, **424**, 824-830 (2003)
8. R. Charbonneau, P. Berini, E. Berolo, E. Lisicka-Shrzek, *Opt. Lett.* **25**, 844 (2000).
9. P. Berini, *Phys. Rev.* **B61**, 10484 (2000); *Phys Rev.* **B63**, 125417 (2001).
10. T. Nikolajsen, K. Leosson, I. Salakhutdinov, S. Bozhevolnyi, *Appl. Phys. Lett.* **82**, 668 (2003).
11. B. Lamprecht, J. R. Krenn, G. Schider, H. Ditlbacher, M. Salerno, N. Felidj, A. Leitner, F. R. Aussenegg, *Appl. Phys. Lett.* **79**, 51 (2001).

12. G. Schider, J. R. Krenn, A. Hohenau, H. Ditlbacher, A. Leitner, F. R. Aussenegg, W. L. Schaich, I. Puscasu, B. Monacelli, G. Boreman, *Phys. Rev.* **B68**, 155427 (2003).
13. J. R. Krenn, B. Lamprecht, H. Ditlbacher, G. Schider, M. Salerno, A. Leitner, F. R. Aussenegg, *Europhys. Lett.*, **60**, 663 (2002).
14. J. Takahara, S. Yamagishi, H. Taki, A. Morimoto, T. Kobayashi, Optics Lett. **22**, 475 (1997).
15. T. Onuki, Y. Watanabe, K. Nishio, T. Tsuchiya, T. Tani, T. Tokizaki, *J. Microscopy* **210**, 284 (2003).
16. A. D. Boardman, G. C. Aers, R. Teshima, *Phys. Rev. B.* **24**, 5703 (1981).
17. T. Yatsui, M. Kourogi, M. Ohtsu, *Appl. Phys. Lett.* **79**, 4583 (2001).
18. D. F. P. Pile, T. Ogawa, D. K. Gramotnev, T. Okomoto, M. Haraguchi, M. Fukui, S. Matsuo, *Appl. Phys. Lett.* **87**, 061106 (2005).
19. D. F. P. Pile, D. K. Gramotnev, M. Haraguchi, T. Okamoto, M. Fukui, *J Appl. Phys.* **100**, 013101 (2006).
20. I. V. Novikov, A. A. Maradudin, *Phys. Rev. B*, **66**, 035403 (2002).
21. D.F.P. Pile and D.K. Gramotnev, *Optics Letters*, **29**, 1069-1071 (2004).
22. D. K. Gramotnev, D. F. P. Pile, *Appl. Phys. Lett.*, **85,** 6323-6325 (2004).
23. D.F.P. Pile, D.K. Gramotnev, *Appl. Phys. Lett.* 2005, **86**, 161101.
24. D. F. P. Pile, D. K. Gramotnev, *Opt. Lett.*, **30**, 1186-1188 (2005).
25. S. I. Bozhevolyni, V. S. Volkov, E. Devaux, T. W. Ebbesen, *Phys. Rev. Lett.*, **95**, 046802 (2005).
26. V.S. Volkov, S.I. Bozhevolnyi, E. Devaux, T.W. Ebbesen, *Appl. Phys. Lett.*, **89**, 143108 (2006); *Optics Express*, **14**, 4494-4503 (2006).
27. S. I. Bozhevolyni, V. S. Volkov, E. Devaux, J.-Y. Laluet, T. W. Ebbesen, *Nature*, **440**, 508-511 (2006).
28. E. Moreno, F. J. Garcia-Vidal, S. G. Rodrigo, L. Martin-Moreno, S. I. Bozhevolnyi, *Opt. Lett.*, **31**, 3447–3449 (2006).
29. V.S. Volkov, S.I. Bozhevolnyi, E. Devaux, J.-Y. Laluet, T.W. Ebbesen, *Nano Lett.*, **7**, 880 (2007).
30. S. I. Bozhevolyni, V. S. Volkov, E. Devaux, J.-Y. Laluet, T. W. Ebbesen, *Appl. Phys. A*, **89**, 225–231 (2007).
31. K. C. Vernon, D. K. Gramotnev, D. F. P. Pile, *J. Appl. Phys.*, 2008 (accepted).
32. M. Yan, M. Qiu, *J. Opt. Soc. Am. B*, **24**, 2333-2342 (2007).
33. D. F. P. Pile, T. Ogawa, D. K. Gramotnev, Y. Matsuzaki, K. C. Vernon, K. Yamaguchi, T. Okamoto, M. Haraguchi, M. Fukui, *Appl. Phys. Lett.*, **87**, 261114 (2005).
34. G. Veronis, S. Fan, *Opt. Lett.*, **30**, 3359-3361 (2005).
35. L. Liu, Z. Han, S. He, *Optics Express*,**13**, 6645-6650 (2005).
36. G. Veronis, S. Fan, *Appl. Phys. Lett.*, **87**, 131102 (2005).
37. Kh.V. Nerkararyan, *Phys. Lett. A* **237,** 103 (1997).
38. D.K. Gramotnev, *J. Appl. Phys.* **98**, 104302 (2005).
39. D.F.P. Pile, D.K. Gramotnev, *Appl. Phys, Lett.*, **89**, 041111 (2006).
40. P. Ginzburg, D. Arbel, M. Orenstein, *Opt. Lett.*, **31**, 3288-3290 (2006).

41. D.K. Gramotnev, D.F.P. Pile, M.W. Vogel, X. Zhang, *Phys. Rev. B*, **75**, 035431 (2007).
42. D. F. P. Pile, *Appl. Phys. B*, **81**, 607 (2005).
43. G. Mur, *IEEE Trans. Electromagnetic Compatibility* **40**, 100 (1998).
44. E. D. Palik, *Handbook of Optical Constants of Solids* (Academic, New York, 1985).
45. L. D. Landau, E. M. Lifshitz, *Electrodynamics of Continuous Media*. (Pergamon Pr, 1984).
46. E. Moreno, S. G. Rodrigo, S. I. Bozhevolnyi, L. Martin-Moreno, F. J. Garcia-Vidal, *Phys. Rev. Lett.*, **100**, 023901 (2008).
47. J. D. Joannopoulos, R. D. Meade, J. N. Winn, *Photonic Crystals: Molding the Flow of Light*, Princeton University Press, Princeton (1995).
48. L. Novotny, D.W. Pohl, B. Hecht, *Ultramicroscopy*, **61**, 1-9 (1995).
49. N. Anderson, A. Bouhelier, L. Novotny, *J. Opt. A*, **8**, S227 (2006).
50. A. Bouhelier, J. Renger, M.R. Beversluis, L. Novotny, *J. Microsc.*, **210**, 220-224 (2003).
51. K.V. Nerkararyan, T. Abrahamyan, E. Janunts, R. Khachatryan, S. Harutyunyan, *Phys. Lett. A*, **350**, 147-149 (2006).
52. D. Mehtani, N. Lee, R.D. Hartschuh, A. Kisliuk, M.D. Foster, A.P. Sokolov, F. Cajko, I. Tsukerman, *J. Opt. A*, **8**, S183-S190 (2006).
53. S. Kawata, M. Ohtsu, M. Irie, *Nano-Optics* (Springer, Berlin New York, 2002).
54. A. V. Zayats, I. I. Smolyaninov, *J. Opt. A*, **5**, S16-S50 (2003).
55. A. V. Zayats, I. I. Smolyaninov, A. A. Maradudin, *Physics Reports*, **408**, 131-314 (2005).
56. S. Kawata, *Near-Field Optics and Surface Plasmon-Polaritons* (Top. Appl. Phys. **81**; Springer, Berlin, 2001).
57. W. Srituravanich, N. Fang, C. Sun, Qi. Luo, X. Zhang, *Nano Lett.* **4**, 1085 (2004).
58. K. Li, M. I. Stockman, D. J. Bergman, *Phys. Rev. Lett.* **91**, 227402 (2003).
59. A. J. Babadjanyan, N. L. Margaryan, K. V. Nerkararyan, *J. Appl. Phys.*, **87**, 3785-3788 (2000).
60. M. I. Stockman, *Phys. Rev. Lett.*, **93** (13), 137404 (2004).
61. M. W. Vogel, D. K. Gramotnev, *Phys. Lett. A*, **363**, 507–511 (2007).
62. N. A. Issa, R. Guckenberger, *Plasmonics*, **2**, 31-37 (2007).
63. D. K. Gramotnev, K. C. Vernon, *Appl. Phys. B*, **86**, 7-17 (2007).
64. K. C. Vernon, D. K. Gramotnev, D. F. P. Pile, *J. Appl. Phys.*, **101**, 104312 (2007).
65. M. Durach, A. Rusina, M. I. Stockman, K. Nelson, *Nano Lett.*, **7**, 3145-3149 (2007).
66. V. V. Krylov, Sov. Phys. Tech. Phys. **35(2)**, 137 (1990).
67. V. V. Krylov, Sov. Phys. Acoust. **36(2)**, 176 (1989).

Chapter 11

NANOPHOTONIC COMPONENTS UTILIZING CHANNEL PLASMON POLARITONS

Valentyn S. Volkov and Sergey I. Bozhevolnyi

Department of Sensors, Signals and Electrotechnics, University of Southern Denmark, Niels Bohrs Allé 1E, 5230 Odense M, Denmark
E-mail:vsv@sense.sdu.dk

Eloïse Devaux and Thomas W. Ebbesen

Laboratoire des Nanostructures, ISIS, Université Louis Pasteur, 8 allée Monge, BP 70028, 67083 Strasbourg, France

Photonic components are superior to electronic ones in terms of operational bandwidth but the diffraction limit of light poses a significant challenge to the miniaturization and high-density integration of optical circuits. The main approach to circumvent this problem is to exploit the hybrid nature of surface plasmon polaritons (SPPs), which are light waves coupled to free electron oscillations in a metal that can be laterally confined below the diffraction limit using subwavelength metal structures. Channel SPP modes-channel plasmon polaritons (CPPs) are electromagnetic waves that are bound to and propagate along the bottom of V-shaped grooves milled in a metal film. They are expected to exhibit useful subwavelength confinement, relatively low propagation loss, single-mode operation and efficient transmission around sharp bends. Here we report the design, fabrication and characterization of CPP-based subwavelength waveguide components operating at telecom wavelengths: straight subwavelength grooves, Y-splitters, Mach–Zehnder (MZ) interferometers, waveguide–ring (WR) resonators, WR-resonator based add-drop multiplexers and compact Bragg grating filters. We demonstrate that CPP guides can indeed be used for efficient guiding as well as large-angle bending and splitting of radiation, thereby enabling the realization of ultra-compact plasmonic components and paving the way for a new class of integrated optical circuits.

1. Introduction: getting into the groove

Ever increasing pace of miniaturisation and integration in optics requires further progress in fabrication and integration of several photonic components on a common planar substrate with the purpose of realizing various functionalities such as guiding, bending, splitting filtering, multiplexing and demultiplexing of optical signals. To support these functions integrated optics involves a number of technologies with different technological platforms and material systems including semiconductors,[1] silica,[2] silica-on-silicon materials,[3] polymers[4] and organo-mineral materials[5] among others. The size and density of optical device employing these technologies is nonetheless limited by the diffraction limit of light that does not allow localization of electromagnetic waves in regions noticeably smaller than half the wavelength in the structure.[6] Another limitation is the typical guiding geometry (e.g. planar waveguides are limited in their geometry because of radiation leakage at sharp bends). These circumstances result in a much lower level of integration and miniaturization of optical devises, compared to that achieved in modern microelectronics.

A new approach to circumvent this problem, which suggests employing surface plasmon polaritons (SPPs) that are light waves coupled to oscillations of free electrons in a metal, meets the requirements of low-cost, simplicity of planar fabrication, combined with good performance. The SPP field decay exponentially into both media and reach maximum at the interface, a circumstance that makes SPPs extremely sensitive to interface properties.[7] The possibility of using SPPs for miniature photonic circuits has attracted a great deal of attention to the field of SPPs during the last years.[8-14] The main issue in this context is to strongly confine the SPP field in the cross section perpendicular to the SPP propagation direction (smaller cross sections ensure smaller bend losses and higher densities of components), while keeping relatively low propagation losses. It has been shown using numerical simulations that nanometer-sized metal rods can support extremely confined SPP modes propagating though only over hundreds of nanometers.[15] Similar properties were expected[16, 17] and indeed found[18] for the electromagnetic excitations supported by chains of metal nanospheres. SPP propagation along metal stripes[19-21] and channels in periodically corrugated regions (as well as, in strongly scattering nonabsorbing random media)[22-24] has also exhibited propagation losses increasing drastically with the decrease in the stripe/channel width. Quite recently, SPP gap waveguides based on the SPP propagation between metal surfaces with varying separation[25] or material properties[26] have been suggested. It should be noted that, in general, the SPP confinement is achieved primarily by decreasing the SPP spatial extent into dielectric, thereby increasing the portion of SPP power being absorbed by metal, so that the choice of optimum guiding configuration is subject to trade-off with many intricate issues yet to be elucidated.

Channel plasmon polaritons (CPPs) are SPP modes that are bound to and propagate along the bottom of grooves cut into a metal.[27,28] They were predicted to exhibit subwavelength transverse confinement, relatively low propagation loss[29] and single mode operation[30,31] as well as efficient transmission around sharp bends.[32] To better understand the concept of channel plasmons, let us first consider gap plasmon modes guided between two closely spaced metal surfaces so that surface plasmons associated with the individual metal surfaces become coupled. The dispersion equation is well known[25,33] and its solution can be easily obtained, even when the metal dielectric function is complex. The main feature in this context is that, with the decrease of the slit width, the effective refractive index of a symmetric SPP combination increases (while the propagation length decreases), starting from the value corresponding to the individual SPP.[25,34,35] Gap plasmons can be easily squeezed to subwavelength dimensions. Smaller gaps result in shorter-wavelength gap plasmons, although at the expense of greater propagation losses.[25] Because the mode field is nearly constant and close to its maximum for subwavelength gaps, however, the propagation loss determined by the portion of field penetrating the metal is as small as it can be for a given confinement.

We now consider a straight groove cut into metal with a V-shaped cross section. Light tends to concentrate in regions where its wavelength is shortest, ensuring plasmon field confinement at the bottom of the groove. Extensive numerical simulations predicted that channel plasmons can be employed for efficient transmission along straight grooves and around sharp bends.[29-32] It is important to note that the CPP guiding in V-grooves is counterintuitive: while a certain groove depth (for a given groove angle) is required to support a CPP mode, the CPP guiding approaches cut-off and the CPP mode field stretches farther out of the groove when the groove angle increases. Our previous study on CPP guiding[34,35] allowed us to identify the range of groove parameters for low-loss and well-confined (below the light wavelength) single-mode CPP guiding at telecom wavelengths as well as the corresponding fabrication conditions.

Our experiments showed that the CPPs at telecom wavelengths propagate over tens of micrometers along grooves in gold and exhibit strong subwavelength confinement[34] along with low bend losses in large-angle *S*-bends and *Y*-splitters,[36,37] thereby enabling the realisation of ultra-compact plasmonic components such as Mach–Zehnder interferometers and waveguide-ring (WR) resonators.[38] Quite recently, we have reported the fabrication and investigation of CPP-based wavelength selective components: a WR-resonator based add-drop multiplexer and a compact Bragg grating filter (BGF).[39]

In this chapter, the results of our investigations of CPP guiding by V-grooves cut into gold are reviewed, demonstrating efficient large-angle bending and splitting of radiation as well as WR-resonators and BGFs. The paper is organized as follows. In Section 2 the samples investigated and the experimental arrangement are described. Section 3 is devoted to the experimental results featuring the SNOM images of straight and bent V-groove channels and their interpretation. The chapter is terminated with our conclusions offered in Section 4.

2. Experimental arrangement

We have first applied the effective-index (EI) approximation (which is one of standard approaches used for waveguide analysis in photonics[40]) to analyze the CPP guiding at 1.55 μm in gold V-grooves. It was found that the CPP effective index increases while the propagation length decreases with the decrease of the groove angle θ.[34,35] Note that the CPP mode index N_{eff} determines the mode confinement, since the mode penetration depth in air (above the sample surface) is given by $d_{pen}=(\lambda/2\pi)\cdot(N_{eff}^{\ 2}-1)^{-0.5}$, so that a larger effective index corresponds to a smaller penetration depth and thereby better field confinement in depth (in width, the CPP mode is confined within the groove walls). Here, better confinement is achieved at the expense of larger damping (shorter propagation length). We have also found that the CPP propagation length starts to rapidly increase when the groove depth decreasing approaches a certain (cut-off) value, below which no CPP mode could be found. Since, at the same time, the CPP effective index approaches that in air, the increase of propagation length signifies the tendency of the CPP mode field being extended progressively outside the groove. Finally, the corresponding results obtained for the groove angle of 25° at telecom wavelength 1.55μm signified that one can expect to achieve the single-mode V-groove guiding (for the groove depth d in the range of ~ 0.9-1.9 μm).[34,35,41]

For our experiments on CPP guiding along straight channels, two ~ 460 μm-long grooves have been fabricated (using a focused ion-beam (FIB) milling technique) in a 1.9 μm-thick gold layer deposited on a substrate of fused silica covered with an 80 nm-thick indium-tin-oxide layer [Figs. 1(a) and 1(b)]. The groove profiles turned out being similar to triangular profiles, especially for groove N2. It is also seen that the groove walls are somewhat rough, a feature that can influence the CPP propagation by causing the CPP scattering out of the groove. The widths and depths were measured as w ~ 0.5 and 0.6 μm and d ~ 0.9 and 1 μm for grooves N1 and N2, respectively. We have also evaluated the angles at the groove bottoms as ~ 16° and ~ 25°. Taking into account limitations of the EI approximation and inaccuracies in the groove characterization, we concluded that both grooves support only the fundamental CPP mode and that groove N2 can be close to cut-off.

The experimental setup employed for the CPP observation and characterization is essentially the same as that used in our experiments with photonic crystal waveguides.[42,43] It consists of a collection scanning

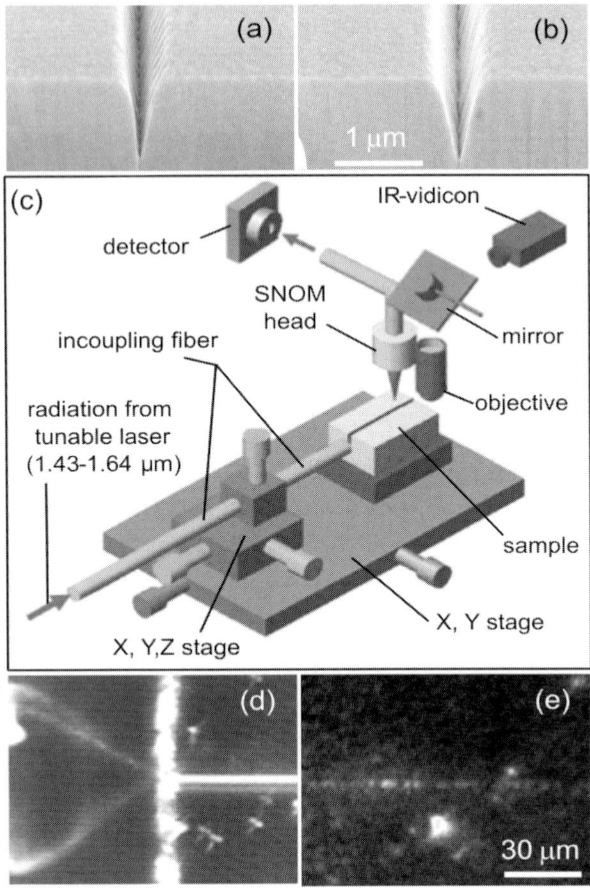

Fig. 1. Scanning electron microscope (SEM) images of grooves (a) N1 and (b) N2. (c) Schematic layout of the experimental setup. Optical microscope images of (d) the coupling arrangement (the tapered fiber is seen to the left and the illuminated groove to the right) and (e) the light propagation (λ = 1.55 μm) along groove N2 (from left to right) at a distance of ~ 200 μm from the in-coupling groove edge.

near-field optical microscope (SNOM) with an uncoated fiber tip used as a probe and an arrangement for launching tunable (1425–1640 nm) TE/TM-polarized (the electric field is parallel/perpendicular to the sample surface plane) radiation into a groove by positioning a tapered-lensed polarization maintaining single-mode fiber [Figs. 1(c) and 1(d)]. The adjustment of the in-coupling fiber was accomplished when monitoring the light propagation along the surface with help of a far-field microscopic arrangement [Fig. 1(c)]. The idea was to judge upon the CPP excitation and propagation by observing its scattering out of the surface plane by groove imperfections.

3. Experimental results

We have found that the track of radiation propagating along a groove was clearly visible only with groove N2 and only for TE polarization of incident light [Fig. 1(e)]. This track was distinguishable for distances of up to ~ 300 μm from the in-coupling groove edge and in the whole range of laser tunability. The far-field observations have confirmed the expected polarization properties of the guided radiation[34,44] and demonstrated its (relatively) low dissipation. Following these experiments (that include also adjusting the incoupling fiber position to maximize the coupling efficiency) we moved the whole fiber-sample arrangement under the SNOM head and mapped the intensity distribution near the surface of groove N2 with the SNOM fiber tip. The tip was scanned along the sample at a constant distance of a few nanometers maintained by shear force feedback, and the radiation collected by the fiber was detected with a femtowatt *InGaAs* photo receiver.

3.1. *CPP guiding in straight V-grooves*

Topographical and near-field optical images of efficient CPP guiding by groove N2 were recorded at the distance of ~ 200 μm from the in-coupling groove edge (to decrease the influence of stray light, i.e., the light that was not coupled into the CPP mode) and in the whole range of laser tunability (Fig. 2). Appearance of the optical images is similar to those obtained with photonic crystal waveguides[43] featuring efficient mode confinement (in the cross section) at the groove and intensity variations along the propagation direction (Fig. 3). The latter can be most probably accounted for by the interference between the CPP mode and

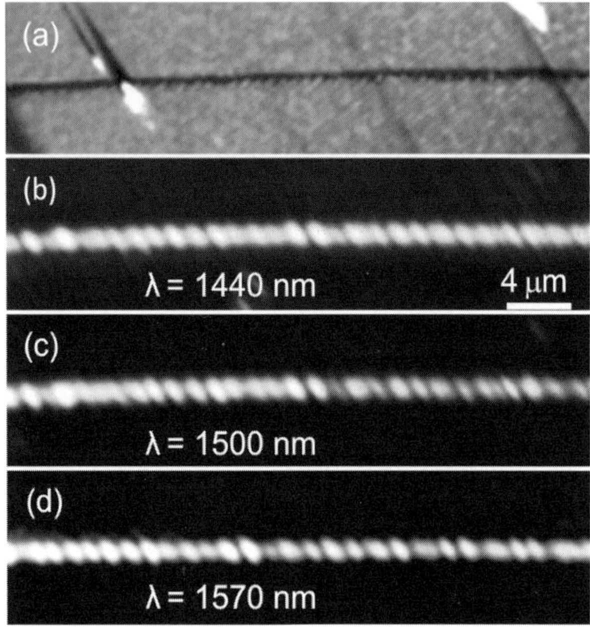

Fig. 2. (a) Topographical and near-field optical images (36 × 9 μm^2) taken with groove N2 at $\lambda \cong$ (b) 1.44, (c) 1.5, and (d) 1.57 μm. The CPP propagation is from left to right.

scattered (including stray light) field components.[42] However, the preliminary analysis of the spatial spectra for different wavelengths has shown that the situation is quite complicated, because these spectra exhibit several features that are independent on the light wavelength and

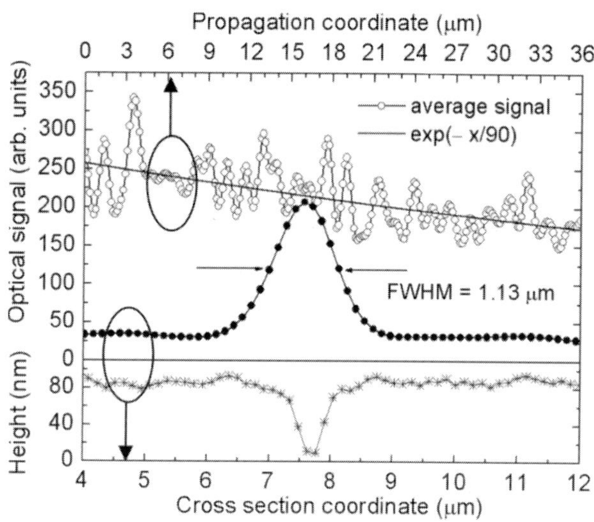

Fig. 3. Cross sections of the topographical (stars) and near-field optical (filled and open circles) images of Fig. 2 [Figs. 2(a) and 2(c), respectively] averaged along 10 lines along the propagation direction or perpendicular to it. The exponential dependence fitted by the least-square method to the signal dependence along the propagation direction is also shown.

might be related to the roughness of groove walls [Fig. 1(b)]. Leaving the detailed analysis of this phenomenon for future studies, we would like to mention that, despite the aforementioned signal variations, we could characterize the average full-width-at-half-maximum (FWHM) of the CPP mode and its propagation length (Fig. 3). It turned out that, in the wavelength range of 1425–1640 nm, the FWHM is practically constant (1.15 ± 0.1 μm) whereas the CPP propagation length (determined from the 36 μm-long SNOM images) was varying between ~ 90 and 250 μm, depending on both the coupling arrangement and wavelength used. Since the gold dispersion alone is too small to account for these variations, we attribute those to poor reproducibility of the fiber coupling that was adjusted using far-field observations [Fig. 1(e)] and that strongly influences the contribution to the detected signal from the scattered radiation.

It should be noted that we have also recorded optical images in constant plane mode at different heights above the sample surface. These images recorded (with high resolution) at λ ≅ 1440 nm for different tip-surface distances are shown in Fig. 4. It was seen that the images retained the appearance not only when changing from shear-force feedback mode to constant height mode (indicating the absence of topographical artifacts) but also when further increasing the tip-surface distance up to 400 nm.

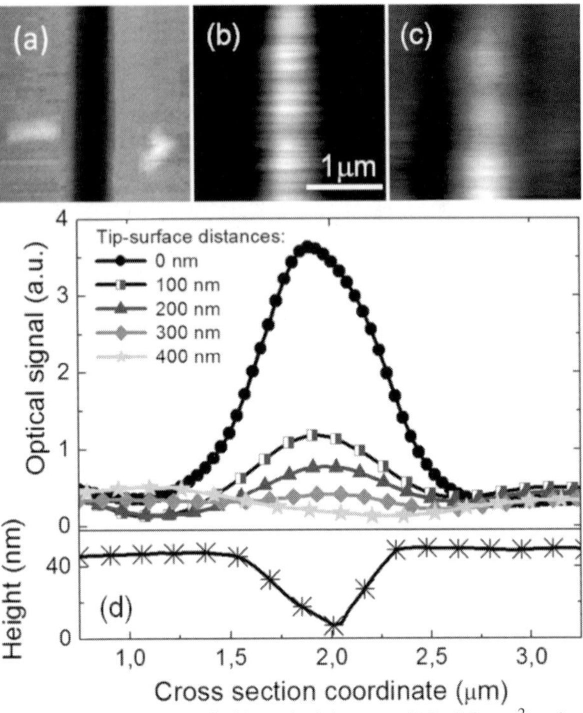

Fig. 4. (a) Topographical and near-field optical images (2.3×2.3 μm^2) taken with groove N2 at $\lambda \cong 1.44$ μm at different tip-surface distances: (b) with shear-force feedback (a few nanometers) and (c) 100 nm. (d) Dependence of the average (over a few lines) optical signal on the tip-surface distance by the use of cross sections of the corresponding near-field optical images recorded at different tip-surface distances.

The average optical signal was found to decrease rapidly with increasing distance to only ~ 100 nm above the sample surface. The images became blurred [Fig. 4(c)] exhibiting low-contrast signal variations, confirming that the main contribution to the signal detected at the groove originated indeed from the CPP (evanescent) field. One can also notice that the signal does not go to zero outside the groove indicating the presence of a homogeneous background at the level of ~ 15 % compared to the maximum signal. This background signal is also present in the optical images recorded on constant plane mode at different heights up to 400 nm above the sample surface [Fig. 4(d)]. However, it is not clear to what extent it influences the signal measured in the middle of a groove with the SNOM fiber tip being actually below the sample surface by ~ 40–100 nm, depending on the tip shape.

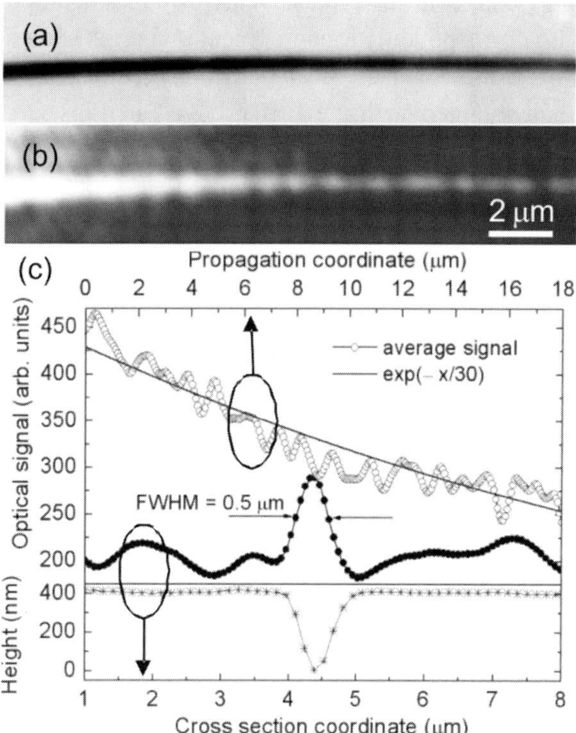

Fig. 5. (a) Topographical and (b) near-field optical images (36 × 9 μm^2) taken with groove N1 at $\lambda \cong 1.5$ μm. The CPP propagation is from left to right. (c) Cross sections of the topographical (stars) and near-field optical (filled and open circles) images shown in 5(a) and 5(b) averaged along 10 lines along the propagation direction or perpendicular to it. The exponential dependence fitted by the least-square method to the signal dependence along the propagation direction is also shown.

The same experiment has also been performed on groove N1 ($w \sim 0.5\mu m$, $d \sim 0.9\mu m$ and $\theta \sim 16°$). The far-field observations and SNOM experiments conducted with this waveguide structure confirm that the excitation of a CPP mode is less efficient to propagate light as the angle of the V-groove decreases. During the far-field adjustment it was found that, contrary to the previously investigated groove N2, the track of radiation propagating along groove N1 was not visible (the circumstance which made the procedure of the light coupling into CPP mode even more complicated and might also have influenced the quality of the recoded SNOM images). The typical SNOM image recorded at $\lambda \cong 1500$ nm (along with corresponding topographical image) is shown in Fig. 5. It is seen that the contrast in the near-field optical image is considerable lower than that obtained with the first groove [cf. Figs. 2(c) and 5(b)]. One can also see that the intensity of the CPP mode in the groove is strongly damped with the distance (along the CPP propagation direction) to the V-groove channel termination. Quantitative description of the field intensity distribution in groove N1 has been obtained by taking an average of several cross sections of the optical image along (and perpendicular to) the CPP propagation direction. The resulting average (along few lines) cross sections are presented in Fig. 5c. It was found that the FWHM was very close to $\sim 0.5 \pm 0.1\ \mu m$ (that is considerably smaller than that obtained for groove N2) whereas the CPP propagation length was found to be close to $\sim 30\ \mu m$.

The experimental observations reported above are consistent with the preceding discussion of the main properties of CPP modes. The fact that we could not observe the efficient CPP guiding with groove N1 is most probably related to its shape [Fig. 1(a)]. This groove has a rather small angle ($\sim 16°$) at its bottom that results in an increase of the CPP confinement and a decrease of the CPP propagation length as compared to groove N2. The first feature would make the imaging efficiency smaller[34] and the second feature would drastically decrease the CPP field to be imaged. Considering groove N2 ($d \sim 1\ \mu m$ and $\theta \sim 25°$), it is expected that the fundamental CPP mode can be quite close to cut-off.[34] This means that its propagation length can easily be close to 100 μm or even longer, with the CPP field extending out of the groove. The latter would facilitate its observation both with a far-field microscope (weakly confined mode is easier to scatter by surface features) and the SNOM (the detection efficiency of a fiber probe increases for lower spatial frequencies).[45,46]

3.2. *Bend loss in CPP guides: bend curvature or defect limited?*

In this subsection, we review our results obtained on the bend loss characterisation (at telecom wavelengths) of V-groove waveguides for CPPs comprising smoothly curved S-bends (that is, curved bends connecting two parallel waveguides offset with respect to each other), Y-splitters composed of two mirrored S-bends, and Mach–Zehnder (MZ) interferometers composed of two consecutive Y-splitters.[38] Our previous

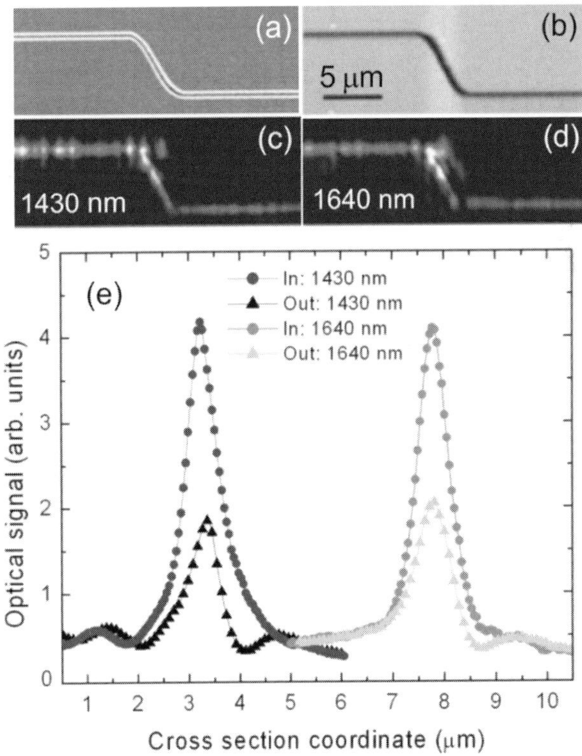

Fig. 6. (a) SEM image along with (b) topographical and near-field optical images (24 × 9 μm^2) taken at $\lambda \cong$ (c) 1.43, and (d) 1.64 μm. The CPP propagation is from left to right. (e) Cross sections obtained with the optical images shown in (c) and (d) by using ~ 3-μm-long averaging windows along straight groove regions before and after the S-bend.

study on CPP guiding[34,41] allowed us to identify the range of groove parameters for low-loss and well-confined (below the light wavelength) single-mode CPP guiding at telecom wavelengths as well as the corresponding fabrication conditions.

The sample containing V-grooves with the angles close to ~ 26° and depths of ~ 1.1 μm has been fabricated in a 1.8-μm-thick layer of gold deposited on a substrate of fused silica. The design of S-bends was based on sine curves, allowing for continuous curvature throughout the bend.[47] Anticipating (relatively) low loss for S-bends with small curvature radii (because of the subwavelength CPP confinement) [34] we have chosen the S-bend design that connects two 5-μm-offset waveguides over a distance of 5 μm (the smallest curvature radius is in this case ~ 2.25 μm). In general, such a choice is subject to a trade-off between the propagation loss and the bend loss, as the former decreases and the latter increases for shorter bends.

Topographical and near-field optical images of CPP guiding by the groove containing the S-bend were recorded at the distance of ~ 160 μm from the in-coupling groove edge and in the whole range of laser tunability (Fig. 6). One can see that the optical signal related to the CPP propagation along the groove decreases noticeably after the S-bend. We can further identify the following channels for the associated CPP bend loss: (i) the modal reflection that gives rise to the interference fringes in the intensity distribution before the bend, (ii) the light scattering out of the bend in the forward direction that was clearly seen for all wavelengths [see e.g., Figs. 6(c) and 6(d)], and (iii) corresponding propagation loss owing to CPP guiding along the bent groove. Note that the light scattering in the bend region results in scattered field components propagating (in air) away from the sample surface. These propagating components are detected with the SNOM fiber tip much more efficiently than the evanescent field components, e.g., associated with the CPP groove mode.[48] This circumstance should be borne in mind in the course of interpretation of SNOM images. Lateral cross sections of the SNOM images obtained in the course of this investigation showed the average FWHM of the CPP mode to be between 0.8 and 1 μm, being almost independent on the wavelength for a given groove, i.e. values that are even slightly smaller than those obtained previously for straight

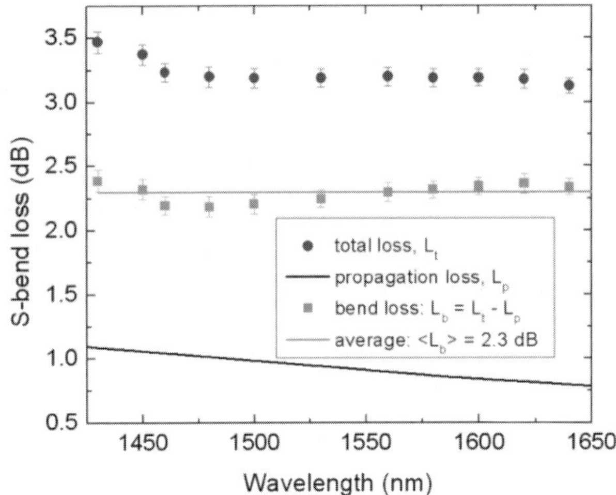

Fig. 7. Total insertion loss determined from cross sections of the near-field optical images (see Fig. 6) together with the propagation loss calculated for ~ 12-μm-long groove region (the groove area between two cross sections made at the input and at the output of double bends) and corresponding bend loss as functions of the light wavelength.

grooves.[34] The fact that the fundamental CPP mode width is wavelength independent can be explained by the circumstance that this field is squeezed by and reaches its maximum at the groove walls[28-30] that set the mode lateral extent (apart from those modes that are very close to cut-off extending significantly out of the groove – these modes should also exhibit very large propagation losses).

The average cross sections of near-field optical images (obtained at different wavelengths) made before and after the S-bend provide a direct way of the total insertion loss evaluation. Fig. 6(e) shows the cross sections obtained with the optical images shown in Fig. 6 by averaging over ~ 3-μm-long straight groove regions. The bend transmission T_S was found slightly increasing with the wavelength i.e., from $T_s \cong 0.45$ at the wavelength of 1430 nm to $T_s \cong 0.49$ at 1640 nm. It is clear that at least some of the power loss, which is estimated being at the level of ~ 3.2 dB in the wavelength range 1430-1640 nm, should be attributed to the propagation loss due to the internal damping. One should also bear in mind the loss contribution due to scattering out of the groove plane by roughness and imperfections of groove walls. Evaluating the propagation

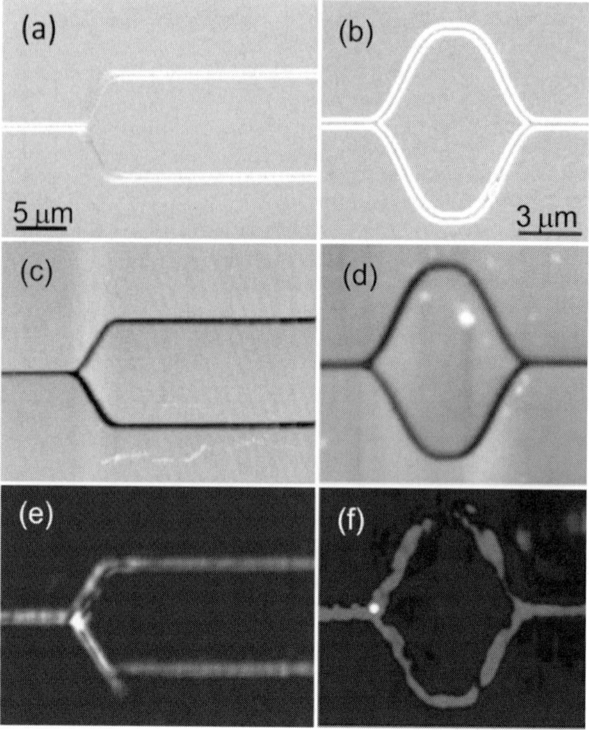

Fig. 8. (a, b) SEM images, along with (c, d), topographical and (e, f), near-field optical images of the Y-splitter (image size: 30 × 23 μm^2) and the MZ interferometer (image size: 13 × 11 μm^2) taken at $\lambda \cong 1.6$ μm.

loss[34] for a ~ 12-μm-long groove region (the area between two cross sections)[36] allowed us to determine the bend loss that includes also the scattering loss contribution. Figure 7 shows that the average bend loss of the S-bend is close to ~ 2.3 dB per double bend in the wavelength range 1430-1640 nm. However, it can be seen that the losses were increasing slightly at short wavelengths (1430-1460 nm). We would like to point out that the loss level attained already in these experiments indicates that the CPP waveguide circuits having relatively large (tens of degrees) bend angles can be realized with relatively small (< 3 dB) bend losses, allowing for high integration level.

The next set of structures to be investigated included Y-splitters and MZ interferometers. Our SNOM investigations showed that all fabricated structures performed well over the whole range of laser tunability, exhibiting single-mode and subwavelength guiding with overall losses of a few dB, demonstrating thereby robustness to small variations in structural parameters. Typical SNOM images of a Y-splitter and a MZ interferometer, along with the corresponding scanning electron microscope (SEM) images, are displayed in Fig. 8. The near-field optical images [Figs. 8(e) and 8(f)] demonstrate very efficient and well-confined CPP guiding throughout the structures consisting of splitters and bends characterized by an average bend angle of 45°. The low loss level observed is a direct (though not straightforward) consequence of the subwavelength CPP field confinement, as discussed earlier.[34] Lateral cross sections of the SNOM images obtained in the course of this investigation showed the average FWHM of the CPP mode to be between ~ 0.8 and 1 μm. An interesting feature observed in most optical images is the signal enhancement in a splitting region [white dot in Figs. 8(e) and 8(f)]. We believe that the two main contributions to this effect are the out-of-plane scattering resulting in propagating field components (whose detection efficiency is larger than that of evanescent waves, such as CPPs)[48] and the above-mentioned counterintuitive feature of CPP guiding. In the splitting region, the groove width is nearly doubled, causing the CPP guiding to approach cut-off, the effective centre of CPP mode field to move upwards (out of the groove) and, thereby, the detected optical signal to increase.

Quantitative evaluation of the insertion loss was made by considering average cross sections before and after the component in question, that is, Y-splitters or MZ interferometers (Fig. 9). These cross sections were obtained with the optical images shown in Fig. 8 by averaging within, 3-μm-long windows placed in straight groove regions. One obtains for the transmission T of the Y-splitter and MZ interferometer the following ranges: $0.82 < T_Y < 1$ and $0.4 < T_{MZ} < 0.49$, respectively, using the output-to-input ratios of maxima of the corresponding signal distributions (Fig. 9). Similar values were found for the Y-splitter and MZ interferometer transmissions at other wavelengths exhibiting (unsystematic) variations within 20% that can probably be attributed to different background levels (determined by positioning of the incoupling fibre with respect to the input groove).

The fact that the investigated Y-splitter showed next to zero insertion loss is remarkable, considering the fact that the smallest curvature radius used was only ~ 1.5λ.This demonstrates the potential of using similar groove structures in sophisticated plasmonic circuits, as well as the practical feasibility of their fabrication with sufficiently high quality. Our investigations have also revealed that these structures exhibit rather high tolerance to topological imperfections on the nanometre scale, that is, there seems to be little perturbation due to accidental presence of nanoparticles [for example, one particle sitting in the lower arm of MZ interferometer is clearly seen in Fig. 8(b)]. Note that, on the other hand, we have stumbled over somewhat contradictory measurements of the bend loss in the investigated structures. Indeed, Y-splitters (composed of two mirrored S-bends) were found being virtually lossless,[38] whereas the corresponding S-bends exhibited the wavelength independent (in the range of 1430–1640 nm) loss of 2.3 dB.[36] On the other hand, one would expect for the split loss to be about twice as large as the bend loss, both being wavelength dependent. It is also clear that the quality of groove structures (e.g., the level of wall roughness) is of paramount importance for loss characterization. With this in mind, we have fabricated (and characterized) high-quality V-grooves (d ~ 1.1 μm and θ ~ 30°)

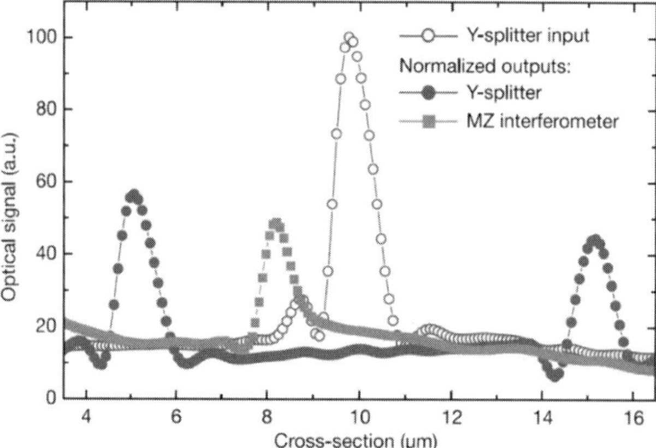

Fig. 9. Normalized cross-sections obtained with the optical images shown in Figs. 8(e) and 8(f) by using,3-μm-long averaging windows along straight groove regions. The optical signal detected with the MZ interferometer was normalized to be in the input channel the same as that detected in the input channel of the Y-splitter. a.u., arbitrary units.

comprising smoothly curved (with *subwavelength* curvature radii) S-bends and Y-splitters. Special care has been taken to fabricate V-grooves waveguides with smooth walls and prevent contamination of the completed structures with dust particles. In order to better isolate the bend loss contribution to the total loss, we have increased twice the offset in S-bends compared to the previous experiments,[36,38] choosing the S-bend design that connects two 10 μm offset waveguides (over the distance of 5 μm) with the smallest curvature radius being in this case ≅ 0.83 μm.

Typical SNOM images of the S-bend and Y-splitter recorded at $\lambda \cong$ 1600 nm along with the corresponding SEM and topographical images are displayed in Fig. 10. The near-field optical images demonstrate efficient and well-confined CPP guiding throughout the S-bend [Fig. 10(e)] and Y-splitter [Fig. 10(f)]. It is also seen that the optical signal modulations clearly noticeable before and after the bend area are not that pronounced in the middle bend sections. At present, we cannot offer a complete explanation of this feature, but it could be related to a finite depth of tip penetration into the groove that is influenced by the groove orientation with respect to the scan direction. Lateral cross sections of these images obtained by averaging over a few lines along straight groove regions in the course of this investigation showed the FWHM of the CPP mode to be close to ~ 900 nm. The S-bend transmission T_s was found rapidly increasing with the wavelength from $T_s \cong 0.51$ at the wavelength of 1425 nm to $T_s \cong 0.64$ at 1460 nm, leveling at $T_s \cong 0.82$ in the range of 1600–1640 nm. Similar features, albeit with smaller transmission, were found for CPP guiding through the Y-splitter composed of the corresponding S-bends, which was also markedly improving with the increase of the light wavelength. The optical image cross sections made before and after the Y-splitter area (for the wavelength of 1600 nm) demonstrate the splitting efficiency being close to ~ 50/50 together with well-confined (to the groove area) mode intensity distributions [Fig. 10(f)]. Finally, the results of the aforementioned procedure carried out for the investigated S-bend and Y-splitter (for both arms) are summed up in Fig. 11 together with propagating loss calculated for a 12-μm-long groove region (the groove area between two cross sections made at the input and at the output of double bends).[37]

Our results indicate that the power loss in the S-bend determined (from the near-field optical images) for long wavelengths (1600–1640 nm) is very close to the evaluated CPP propagation loss, implying that

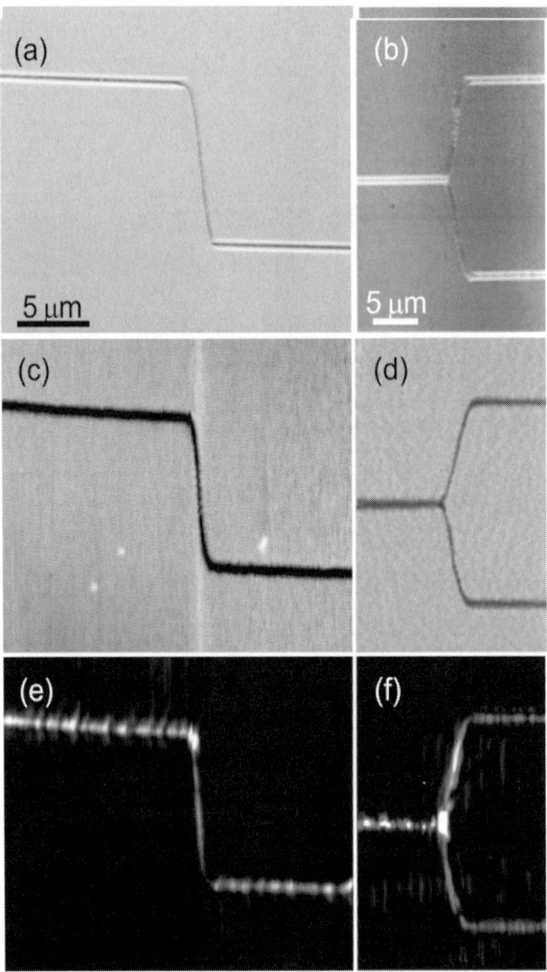

Fig. 10. (a, b) SEM images, along with (c, d), topographical and (e, f), near-field optical images of the S-bend (image size: 26 × 21 μm^2) and the Y-splitter (image size: 20 × 30 μm^2) taken at $\lambda \cong 1.6\ \mu m$.

the (pure) bend loss is very close to 0 dB in this wavelength range. The increase in the bend loss for shorter wavelengths can be attributed, at least partially, to the light scattering by groove imperfections in the bend area (it has been observed that fabrication required smaller steps in the focused ion-beam position to ensure the same quality as compared to straight grooves). In principle, the bend loss contains the fundamental contribution due to the phase mismatch in the curved region[49] that might also increase for shorter wavelengths, if the mode confinement is only weakly dependent on the wavelength (as was observed in these and previous experiments).[34,36-39] Note that, in general, the CPP confinement is expected to improve for shorter wavelength.[28,34] The power loss in the Y-splitter was found to be larger than that in the S-bend as one would have expected, showing the same trend to decrease for longer wavelengths (Fig. 11). The residual loss of ~ 0.5 dB (with the propagation loss being subtracted) in the range of 1600–1640 nm could have been attributed to the (pure) splitting loss in the Y-splitter region if the groove structure were perfect. Unfortunately, the upper arm of the Y-splitter [Fig. 11(b)] contains visible imperfections causing additional scattering seen in the enhanced brightness of this arm on the near-field optical image [Fig. 11(f)]. These imperfections might also be responsible for a smaller CPP intensity observed in the deformed arm[37] and, thereby, for the residual loss. Note that the investigated previously Y-splitter (albeit with a larger curvature radius) did not exhibit noticeable loss.[38] Overall the results obtained indicate that, apart from the CPP propagation loss, the main loss incurred in the investigated groove structures might be due to structural imperfections (at least in the range of 1600–1640 nm) and not due to the phase mismatch caused by small radii of groove curvature. However, further investigations, including accurate modeling of CPP propagation along curved grooves, are needed to verify this conjecture.

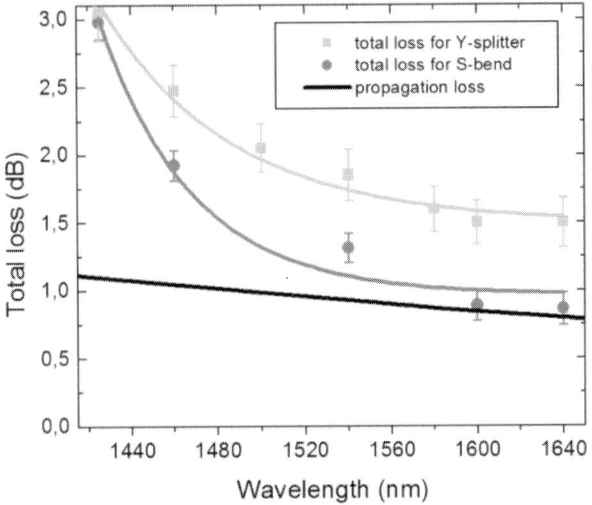

Fig. 11. Total insertion loss determined from cross sections of the near-field optical images (similar to ones shown in Fig. 10) obtained for the investigated S-bend and Y-splitter (for both arms) together with the propagation loss calculated for ~ 12-μm-long groove region (the groove area between two cross sections made at the input and at the output of double bends). The solid lines are shown merely to guide the eyes.

3.3. *CPP - based photonic components*

The results obtained with the above set of structures encouraged us to further exploit the potential of CPPs and to realize more functional plasmonic components. We have fabricated (using FIB milling in a 1.8-μm-thick gold layer) various WR resonator-based photonic components consisting of V-grooves with the angles close to ~ 25° and depths of ~ 1.3 μm. The set of structures to be investigated included straight reference guides, WR resonators (with different ring radiuses) and a WR resonator-based add-drop multiplexer (WRR-ADM) arrangement that is expected to demonstrate relatively high drop efficiency.[50] In the design of WR resonators, we used 5- and 10-μm-radius rings and the *S*-bends that connect two 5-μm-offset waveguides over a distance of 5 μm [Figs. 12(a) and 12(b)]. The intention with using *S*-bends after the coupling regions and laterally displacing the output channels was to decrease the influence of radiation scattered in the coupling regions on the signal detected in the output channels. The main system parameters are transmission and coupling coefficients, t and k, and transmission around the ring, $\alpha \exp(i\theta)$ [see inset in Fig. 12(a)]. The transmission through a WR resonator under the condition that a single unidirectional mode of the ring resonator is excited is given by[51]

$$T = \frac{\alpha^2\eta^2 + |t|^2 - 2\alpha\eta|t|\cos(\theta+\phi)}{1+\alpha^2|t|^2 - 2\alpha|t|\cos(\theta+\phi)},$$
$$t = |t|\exp(i\phi),\ \ |t|^2 + |k|^2 = \eta,\ \ \theta = \frac{2\pi}{\lambda} N_{eff}\, 2\pi R \tag{1}$$

where η is the overall coupling efficiency, which is influenced by the

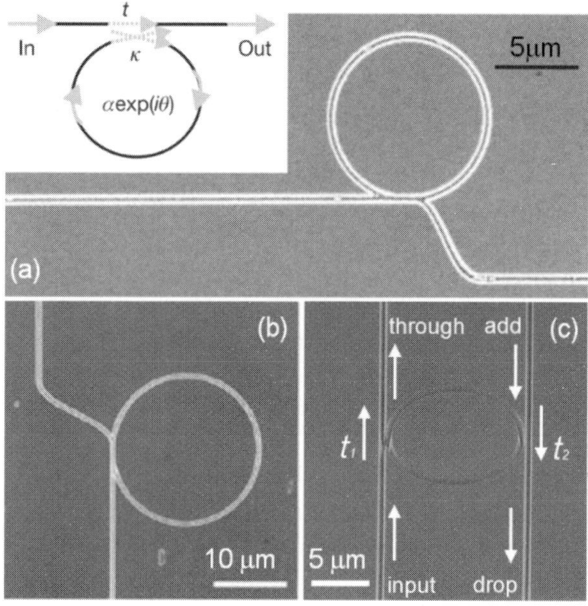

Fig. 12. (a) and (b) SEM images of the WR resonators (with 5- and 10-μm-radius rings used) schematically shown in the inset in (a) (t and k are the transmission and coupling coefficients, respectively and $\alpha\exp(i\theta)$ denotes the amplitude transmission around the ring). (c) SEM image of the fabricated WRR-ADM composed of two triangular grooves (milled in gold) combined with 5-μm-radius WR resonator.

radiation (power) loss at the coupling region, R is the ring radius, N_{eff} is the CPP effective index, and λ is the light wavelength in air. It is seen that the transmission is periodic with respect to the phase θ accumulated by the CPP mode per circulation, implying the possibility of wavelength filtering. At resonance ($\theta + \phi = 2\pi m$, m is integer), the transmitted power vanishes at the condition of critical coupling, $\alpha\eta = |t|$, allowing for dramatic influence of the ring transmission α on the WR resonator transmission and other interesting effects.[51]

In the design of WRR-ADM, we used a 5-μm-radius ring and two identical straight V-groove waveguides [Fig. 12(c)]. The groove profiles turned out to be similar to triangular ones. It is also seen that the groove walls are somewhat rough, a feature that can influence the CPP propagation by causing the CPP scattering out of the groove. The main system parameters are transmission coefficients, t_1 and t_2, through the coupling regions [see Fig. 12(c)] and the transmission around the ring, $\alpha\exp(i\theta)$. The power transmissions in the through (P_T) and drop (P_D) channels of WRR-ADM under the condition that a single unidirectional mode of the ring resonator is excited are given by[52]

$$P_T = \frac{(\alpha t_2)^2 + t_1^2 - 2\alpha t_1 t_2 \cos(\theta)}{1 + (\alpha t_1 t_2) - 2\alpha t_1 t_2 \cos(\theta)},$$
$$P_D = \frac{(1 - t_1^2)(1 - t_2^2)\alpha}{1 + (\alpha t_1 t_2) - 2\alpha t_1 t_2 \cos(\theta)}, \tag{2}$$
$$\theta = \frac{2\pi}{\lambda} N_{eff} 2\pi R$$

where we neglected losses in the system except for during the propagation in the ring. It is seen that the power transmissions are out of phase and periodic with respect to the phase θ accumulated by the CPP mode per circulation, implying the possibility of wavelength adding and dropping. Finally, all fabricated plasmonic structures were characterized with a collection scanning near-field optical microscope (SNOM). Our main interest was in verifying the possibility of realization of waveguide-ring resonator-based plasmonic components, achieving thereby the wavelength dependent transmission of CPPs.

As a first example, we have chosen a WR resonator, which is interesting owing to its wavelength response and non-trivial dependence on the coupling between straight and ring waveguides.[51] We have already noticed during the far-field adjustment that, contrary to the previously investigated structures (for example *S*-bends and *Y*-splitters),[37,38] the transmitted power depended on the wavelength of light used. We then recorded SNOM images for different wavelengths in the laser tunability range for two nominally identical WR resonators. The typical topographical and corresponding near-field optical images obtained at different wavelengths are shown in Color Plate 1, exhibiting a clear difference in the transmitted CPP power. The corresponding input and output average cross sections were used in order to evaluate the normalized output, that is, transmission, for the investigated WR resonators as a function of the light wavelength. The data obtained for both structures were fitted with the dependence calculated to equation (1), demonstrating quite good agreement [Color Plate 1(d)]. When fitting, we used the CPP parameters as previously calculated[34] (N_{eff} = 1.015, α = 0.77, obtained from the calculated CPP propagation length and ring circumference, neglecting the bend loss), assumed the transmission coefficient to be real, that is, $\phi = 0$, and allowed for small variations in the ring radius R. The fitted values were found to be $\eta \approx$ 0.78, $t \approx$ 0.33 and $R \approx$ 5.13 μm, which are consistent with the previous experiments (for example, the overall coupling efficiency η is relatively high, implying low radiation loss, similar to that seen previously with *Y*-splitters and *MZ* interferometers).[38] Note that the achieved maximum-to-minimum transmission ratio is quite large (on average, >3) allowing for wavelength filtering with a bandwidth of ~ 40 nm.

We have also characterized a WR resonator, which is similar to that shown above but with a ring radius (R = 10 μm) twice as large [see Fig. 12(b)]. Our main interest in conducting the experiments with a WR resonator having a twice larger ring radius R than in the previous experiment was in verifying the scaling in the wavelength dependent response of the WR resonator, which is expected to exhibit the same (e.g. maximum or minimum) transmission when tuning the wavelength over the interval $\delta\lambda \approx \lambda^2/(2\pi R N_{eff})$, with N_{eff} being the CPP effective refractive index.[38] Typical SNOM images of the WR resonator recorded at two close wavelengths corresponding to the maximum and minimum transmission are shown in Color Plate 2 along with the corresponding topographical image. We have observed that the CPP intensity in the output channel of the investigated WR resonator changed between its maximum [Color Plate 2(b)] and minimum [Color Plate 2(c)] levels when the wavelength was tuned by ~ 20 nm. The expected value of wavelength variation required for such a change is $\delta\lambda \approx 18$ nm, which is close enough to that obtained in our experiments. The accurate characterisation of this WR resonator required acquiring the SNOM images in a broad wavelength range, quantifying the WR resonator transmission at each wavelength and comparing thus obtained values with the theoretical dependence [see Color Plate 2(d)] in a manner similar to that previously used.

Finally, we have investigated a WR-resonator based add-drop multiplexer (WRR-ADM)) arrangement, for which we used a 5-μm-radius ring and the same V-groove waveguides [see Fig. 12(c)]. The typical near-field optical images obtained with the transmitted (through) light power at maximum and at minimum are shown in Color Plate 3 along with the experimental WRR-ADM transmission spectra obtained for the through and drop channels and the fitted curves corresponding to the modeled WRR-ADM response. The output-to-input ratio was obtained by using cross sections of near-field optical images averaged along several lines. The error bars are estimated from the maximum variation in the ratio when placing the averaging windows at different positions along the input and output channels. The data obtained for both output channels were fitted with the dependences calculated according to equation (2), demonstrating quite good agreement [Color Plate 2(d)]. When fitting, we used the following CPP parameters: $N_{eff} = 1.015$, $\alpha = 0.71$ (which are close to the values previously used[38] and calculated by using the effective-index approximation, neglecting the bend loss)[35] and allowed for small variations in the ring radius R. The fitted values were found to be $t_1 \approx 0.31$, $t_2 \approx 0.78$, and $R \approx 5.12$ μm, which are consistent with the previous experiments[38] and with the investigated WRR-ADM configuration. In particular, we noticed that the ring-channel separation was smaller (by ~ 100 nm) at the input-through channel compared to that at the add-drop channel,[39] implying that the transmission should be larger

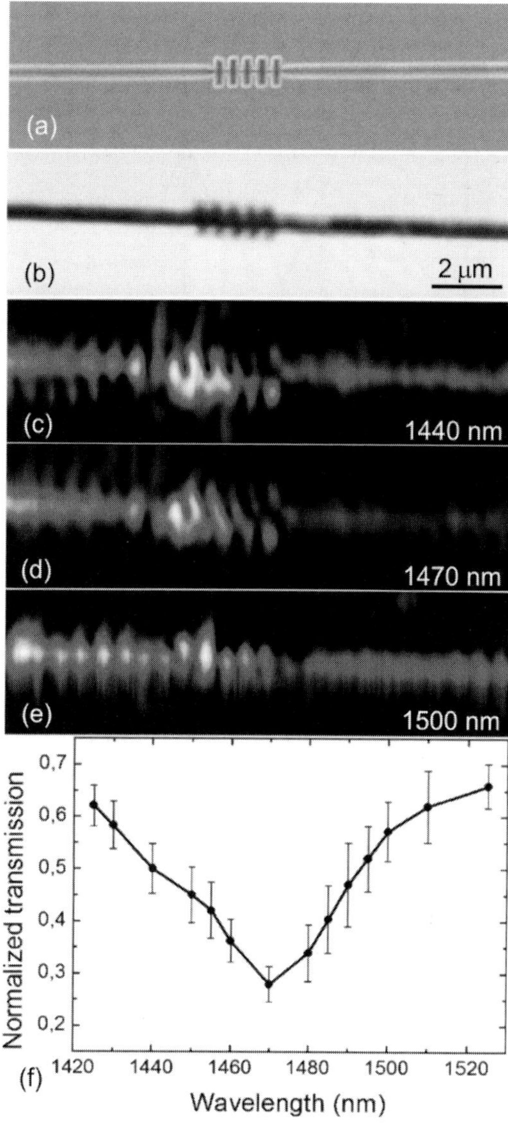

Fig. 13. (a) SEM and (b) topographical and near-field optical images ($15 \times 4\ \mu m^2$) obtained with the investigated BGF at $\lambda \cong$ (c) 1.44, (d) 1.47, and (e) 1.5 μm. (f) BGF transmission spectra representing the output-to-input ratio obtained by using average (along 10 lines) cross-sections of near-field optical images taken perpendicular to the CPP propagation direction before and after the BGF. The error bars are estimated from the maximum variation in the ratio when placing the averaging windows at different positions along the groove regions (before and fter the grating).

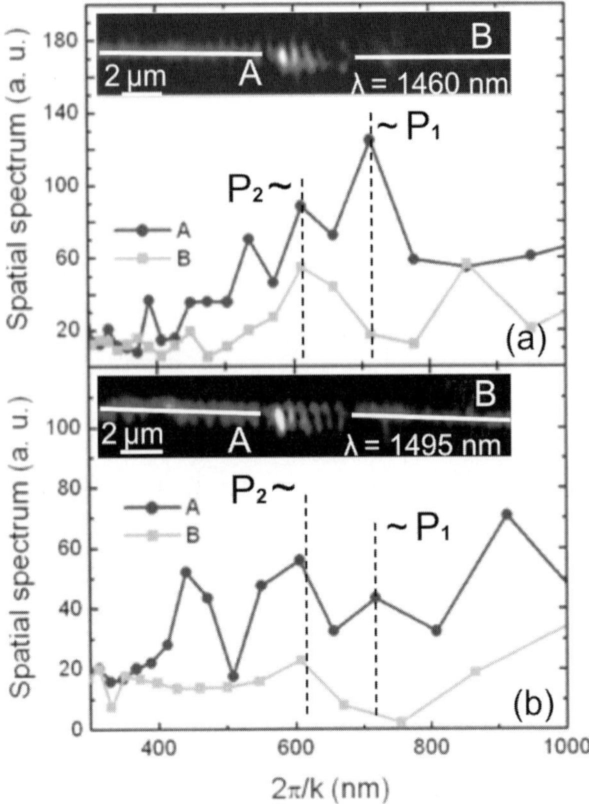

Fig. 14. Spatial-frequency spectra obtained with near-field optical images (shown in the insets) straight regions of the groove before and after (along the lines marked with A and B) the BGF at at $\lambda \cong$ (a) 1.46, and (b) 1.495 μm. Positions of the main peaks dominating the spectrum are indicated with dashed lines. The size of the inset optical image is 20 × 4 μm^2.

in the latter, i.e. t_2 should be larger than t_1, as indeed was deduced during the fitting procedure. Such a sensitivity of the WRR-ADM performance with respect to the system parameters implies stringent requirements on the fabrication, a feature which is, in fact, typical also for (conventional) compact photonic components based on high-index-contrast (photonic wire) waveguides.[53] Note that the suppression of the dropped wavelength in the trough channel is quite large (on average, ~ 5), allowing for efficient wavelength filtering with the bandwidth of ~ 40 nm.

Our next step was to try out one of the simplest wavelength-sensitive waveguide component, a Bragg grating filter (BGF). A conventional waveguide BGF incorporates typically a periodic stack of layers made of materials with different optical refractive indexes.[3] The CPP propagating along a metal groove features the subwavelength confinement and effective refractive index larger than that in air,[28-31,35] implying the impedance mismatch with the propagating in free space modes. Our idea was to periodically mill deep and wide wells across a V-groove, forcing the CPP mode to "jump" over free space junctions [Fig. 13(a)] and to experience periodic reflections (as well as out-of-plane scattering) that can eventually add up to the efficient reflection and filtering at the Bragg wavelength. It was also clear that a compromise should be sought between the filter efficiency and scattering loss, both being proportional to the number of periods. A 750-nm-period BGF configuration consisting of five 500-nm-long and 1-μm-wide wells was fabricated using the usual design of V-grooves (with the angles close to ~ 25° and depths of ~ 1.2 μm).

The SNOM images recorded for different light wavelengths demonstrate clearly the filtering effect, showing strong attenuation in transmission of the CPP at the wavelength of ~ 1470 nm (Fig. 13), which is close to the double BGF period as expected. Note that the near-field optical images are rather complicated reflecting interference between the propagating CPP (incident on the BGF), the CPP reflected by the BGF structure, and free-space propagating field components scattered at the fiber-CPP coupling and by the BGF well-groove junctions. The scattering processes occur at all wavelengths, resulting in the signal increase inside the BGF region and the BGF insertion loss, whereas the Bragg reflection is most pronounced at the filter wavelength of 1470 nm, leading to stronger signal oscillations before and inside the BGF as well as to the attenuation in transmission.[38] We have further evaluated the transmission spectra (the corresponding output-to-input ratio as a function of the light wavelength) for the investigated BGF by making the optical image average cross sections perpendicular to propagation direction [Color Plate 2(f)]. It is seen that the transmission dip is surprisingly well pronounced and narrow for such a short (3.75-μm-long) BGF. When applying a periodic-layered media approach[54] for modeling the BGF, we could not get a reasonable agreement between the measured and simulated transmission spectra no matter what refractive indexes were used. At present, we cannot thereby offer an explanation of the measured spectrum, but it could be related to more complicated physical phenomena (e.g., cavity effects) involved in the process of CPP interaction with the periodic structure of wells. However, we would rather desist from definite conclusions before conducting further detailed investigations, including accurate modeling of the CPP propagation along metal grooves containing periodic wells, which are needed to verify this conjecture.

To verify the occurrence of the Bragg reflection of the incident CPP at the transmission minimum, we have studied the spatial-frequency spectra in the near-field optical images cut along the CPP propagation direction. The typical spatial-frequency spectra obtained before and after the BGF area for two different wavelengths (Fig. 14) exhibit significant differences in heights of the peak approximately corresponding to half of the CPP wavelength (peak P_1 at ~ 712 nm). A very pronounced peak P_1 dominating the spectrum for the incoming CPP at $\lambda \cong 1460$ nm (that is close to the Bragg resonance) might be explained by the efficient CPP reflection from the periodic structure [Fig. 14(a)]. This peak is virtually absent in the spectrum of the optical signal distribution recorded after the BGF for both wavelengths and is barely seen for the incoming spatial spectrum obtained at $\lambda \cong 1495$ nm, implying weak CPP reflection [Fig. 14(b)]. On the other hand, our analysis of the spatial spectra for different wavelengths has shown that the situation is quite complicated, because these spectra exhibit also several other features that are independent of the light wavelength [e.g., spectral peaks P_2 and P_3 on Fig. 14(a)] and are probably related to the roughness of groove walls.[34]

4. Conclusions

In summary, we have realized, and characterized the CPP propagation along subwavelength-sized metal grooves. Using SNOM imaging, we have demonstrated low-loss (propagation length~100 μm) and well confined (mode width ~ 1 μm) CPP guiding at telecom wavelengths (1425–1640 nm) along triangular (~ 0.6 μm-wide and ~ 1 μm-deep) grooves in gold. We have also reviewed the results of our investigations of CPP guiding by V-grooves cut into gold, demonstrating efficient large-angle bending and splitting of radiation as well as waveguide-ring resonators and Bragg grating filters. It should be stressed that, in contrast to the CPP-based structures, the degree of light confinement in dielectric waveguides, including those based on the photonic band-gap effect, is fundamentally limited by the light wavelength in the dielectric used. By using the CPP based photonic components, we have demonstrated a WR resonator with an insertion loss of < 3 dB and a footprint of only 200 μm^2 ($\sim 100\lambda^2$) that can be used for wavelength filtering with a bandwidth of$\sim$ 40 nm as well as a compact (only 3.75-μm-long) BGF suppressing the transmitted CPP power by a factor of $\sim$ 3 at the Bragg wavelength. The performance of the considered structures could be further improved by perfecting the quality of fabricated grooves (especially walls and junctions) and optimizing geometrical parameters with respect to the trade-off between the mode confinement and the propagation loss. We can envisage further development of this concept for miniature bio-sensors and for ultracompact plasmonic interconnects that can be naturally integrated with electrical circuits in away similar to that recently demonstrated with long-range SPPs.[55]

Acknowledgments

The authors gratefully acknowledge the help of Jean-Yves Laluet (Université Louis Pasteur, France) in fabrication of the samples used in this work. This research was carried out in the framework of the European Network of Excellence, PLASMO-NANO-DEVICES (FP6-2002-IST-1-507879), and the FTP project (Contract No. 274-07-0258) supported by the Danish Research Agency.

References

1. T. Morioka, H. Takahara, S. Kawanishi, K. Uchiyama and M. Saruwatari, *Electron. Lett.*, **32**, 840 (1996).
2. M. Oguma, T. Kitoh, Y. Inoue, T. Mizuno, T. Shibata, M. Kohtoku and Y. Hibino, *IEEE J. of Lightwave Technol.*, **22**, No. 3, 895 (2004).

3. D. L. Lee, *Electromagnetic principles of integrated optics* (John Wiley & Sons, Inc., New York, 1986).
4. L. Eldada and L. W. Shacklette, *IEEE J. Sel. Top. Quantum Electron.*, **6**, 54 (2000).
5. P. Coudray, P. Etienne and Y. Moreau, *Material Science in Semiconductor Processing*, **3**, 331 (2000).
6. B. E. A. Saleh and M. C. Teich, *Fundamentals of photonics,* (John Wiley & Sons, Inc., New York, 1991).
7. H. Raether, *Surface plasmons* (Springer-Verlag, Berlin, 1988).
8. S. I. Bozhevolnyi, I. I. Smolyaninov and A. V. Zayats, *Phys. Rev. B*, **51**, 17916 (1995).
9. S. I. Bozhevolnyi and F. A. Pudonin, *Phys. Rev. Lett.*, **78**, 2823 (1997).
10. W. L. Barnes, A. Dereux and T. W. Ebbesen, *Nature*, **424**, 824 (2003).
11. S. A. Maier and H. A. Atwater, *J. Appl. Phys.* **98**, 011101 (2005).
12. S. I. Bozhevolnyi and V. M. Shalaev, *Photonics Spectra*, **40**, No. 1, 58, (2006).
13. F. Lopez-Tejeira, S. G. Rodrigo, L. Martin-Moreno, F. J. Garcia-Vidal, E. Devaux, T. W. Ebbesen, J. R. Krenn, I. P. Radko, S. I. Bozhevolnyi, M. U. Gonzalez, J. C. Weeber and A. Dereux, *Nature Phys.*, **3**, 324, (2007).
14. R. Zia, J. A. Schuller, A. Chandran and M. L. Brongersma, *Materials Today*, **9**, 20 (2006)
15. J. Takahara, S. Yamagishi, H. Taki, A. Morimoto and T. Kobayashi, *Opt. Lett.*, **22**, 475 (1997).
16. M. Quinten, A. Leitner, J.R. Krenn and F. R. Aussenegg, *Opt. Lett.*, **23**, 1331 (1998).
17. M. L. Brongersma, J. W. Hartman and H. A. Atwater, *Phys. Rev. B*, **62**, 16356 (2000).
18. S. A. Maier, M. L. Brongersma, P. G. Kirk, S. Meltzer, A. A. G. Reguicha and H. A. Atwater, *Adv. Mater.*, **13**, 1501 (2001).
19. J.-C. Weeber, J. R. Krenn, A. Dereux, B. Lamprecht, Y. Lacroute and J. P. Goudonnet, *Phys. Rev. B*, **64**, 045411 (2001).
20. J. R. Krenn and J. C. Weeber, *Phil. Trans. R. Soc. A*, **362**, 739 (2004).
21. R. Zia, M. D. Selker and M. L. Brongersma, *Phys. Rev. B*, **71**, 165431 (2005).
22. S. I. Bozhevolnyi, J. Erland, K. Leosson, P. M. W. Skovgaard and J. M. Hvam, *Phys. Rev. Lett.*, **86**, 3008 (2001).
23. S. I. Bozhevolnyi, V. S. Volkov, K. Leosson and A. Boltasseva, *Appl. Phys. Lett.*, 7**9**, 1076 (2001).
24. S. I. Bozhevolnyi, V. S. Volkov and K. Leosson, *Phys. Rev. Lett.*, **89**, 186801 (2002).
25. K. Tanaka and M. Tanaka, *Appl. Phys. Lett.* **82**, 1158 (2003).
26. B. Wang and G. P. Wang, *Appl. Phys. Lett.*, **85**, 3599 (2004).
27. J. Q. Lu and A. A. Maradudin, *Phys. Rev. B*, **42**, 11159 (1990).
28. I.V. Novikov and A. A. Maradudin, *Phys. Rev. B*, **66**, 035403 (2002).
29. D. F. P. Pile and D. K. Gramotnev, *Opt. Lett.*, **29**, 1069 (2004).

30. D. K. Gramotnev and D. F. P. Pile, *Appl. Phys. Lett.*, **85**, 6323 (2004).
31. E. Moreno, F. J. Garcia-Vidal, S. G. Rodrigo, L. Martin-Moreno and S. I. Bozhevolnyi, *Opt. Lett.*, **23**, 3447, (2006).
32. D. F. P. Pile and D. K. Gramotnev, *Opt. Lett.*, **30**, 1186 (2005).
33. I. P. Kaminow, W. L. Mammel and H. P. Weber, *Appl. Opt.*, **13**, 396 (1974).
34. S. I. Bozhevolnyi, V. S. Volkov, E. Devaux and T. W. Ebbesen, *Phys. Rev. Lett.*, **95**, 046802 (2005).
35. S. I. Bozhevolnyi, *Opt. Express*, **14**, 9467 (2006).
36. V. S. Volkov, S. I. Bozhevolnyi, E. Devaux and T. W. Ebbesen, *Opt. Express*, **14**, 4494 (2006).
37. V. S. Volkov, S. I. Bozhevolnyi, E. Devaux and T. W. Ebbesen, *Appl. Phys. Lett.*, **89**, 142108 (2006).
38. S. I. Bozhevolnyi, V. S. Volkov, E. Devaux, J.-Y. Laluet and T. W. Ebbesen, *Nature*, **440**, 508 (2006).
39. V. S. Volkov, S. I. Bozhevolnyi, E. Devaux J.-Y. Laluet and T. W. Ebbesen, *Nano Lett.*, **7**, 880 (2007).
40. A. B. Buckman, *Guided-Wave Photonics* (Saunders College Publishing, New York, 1992).
41. S. I. Bozhevolnyi, V. S. Volkov, E. Devaux, J.-Y. Laluet and T. W. Ebbesen, *Appl. Phys. A*, **89**, 225 (2007).
42. S. I. Bozhevolnyi and V. S. Volkov, *Phil. Trans. R. Soc. London, Ser. A*, **362**, 757 (2004).
43. V. S. Volkov, S. I. Bozhevolnyi, P. I. Borel, L. H. Frandsen and M. Kristensen, *Phys. Rev. B*, **72**, 035118 (2005).
44. Since the main component of the SPP electric field is oriented perpendicular to the metal surface,[7] the CPP electric field tends to be oriented perpendicular to the groove walls and thereby, for small groove angles, parallel to the sample surface.
45. I. P. Radko, S. I. Bozhevolnyi and N. Gregersen, *Appl. Opt.*, **45**, 4054 (2006).
46. N. Gregersen, B. Tromborg and S. I. Bozhevolnyi, *Appl. Opt.*, **45**, 8739 (2006).
47. A. Kumar and S. Aditya, *Microwave Opt. Technol. Lett.* **19**, 289 (1998).
48. S. I. Bozhevolnyi, B. Vohnsen and E. A. Bozhevolnaya, *Opt. Commun.*, **172**, 171 (1999).
49. D. Marcuse, *IEEE J. Quantum Electron.*, **29**, 12 (1993).
50. S. Xiao, L. Liu and M. Qiu, *Opt. Express*, **14**, 2932 (2006).
51. A. Yariv, *Electron. Lett.*, **36**, 321 (2000).
52. J. M. Choi, R. K. Lee and A. Yariv, *Opt. Lett.*, **27**, 1598 (2002).
53. F. Xia, L. Sekaric and Y. Vlasov, *Nature Photon.*, **1**, 65 (2007)
54. A. Yariv and P. Yeh, *Optical waves in crystals* (John Wiley & Sons, Inc., New York, 1984).
55. T. Nikolaijsen, K. leosson and S.I. Bozhevolnyi, *Appl. Phys. Lett.*, **85**, 5833 (2004).

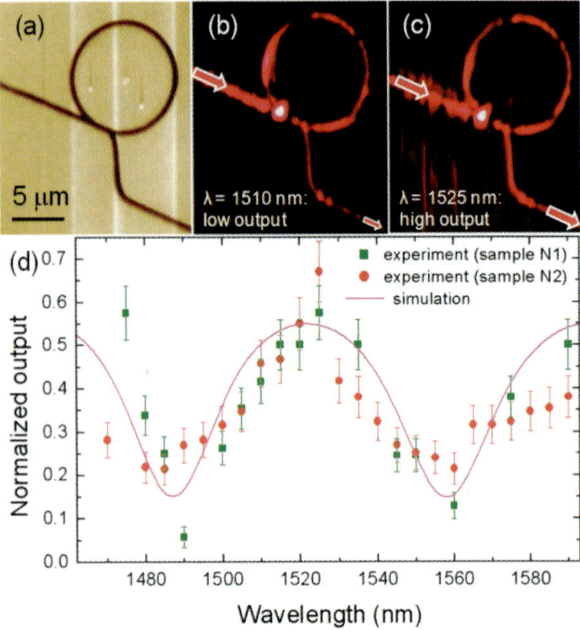

Color Plate 1. (a) Topographical and near-field optical images of the WR resonator (with a 5-μm-radius ring used) recorded at $\lambda \cong$ (b) 1.51, and (c) 1.525 μm. (d) The output-to-input ratio obtained by using average cross-sections of optical images [similar to those shown in (b) and (c)] as a function of the light wavelength for two nominally identical WR resonators [one is shown in Fig. 12(a)], along with the fitted curve corresponding to the response of the WR resonator (see text). The error bars are estimated from the maximum variation in the ratio when placing the averaging windows at different positions along the input and output channels.

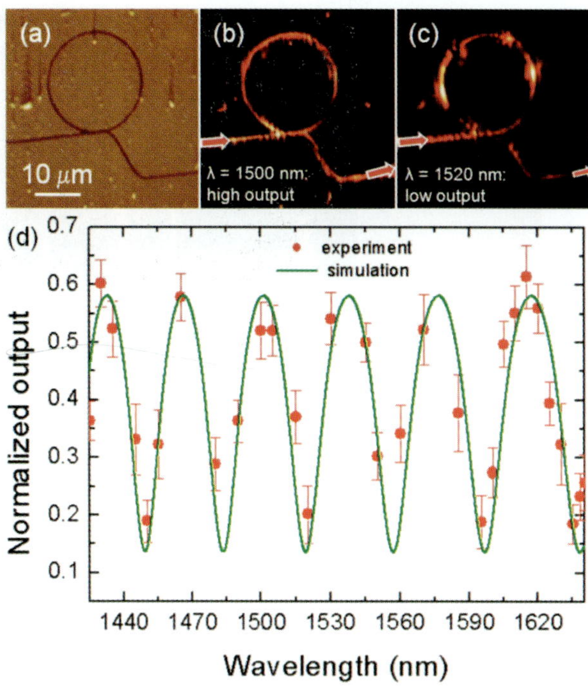

Color Plate 2. (a) Topographical and near-field optical images of the WR resonator (with a 5-μm-radius ring used) recorded at $\lambda \cong$ (b) 1.5, and (c) 1.52 μm. (d) The output-to-input ratio obtained by using average cross-sections of optical images [similar to those shown in (b) and (c)] as a function of the light wavelength along with the fitted curve corresponding to the response of the WR resonator. The error bars are estimated from the maximum variation in the ratio when placing the averaging windows at different positions along the input and output channels.

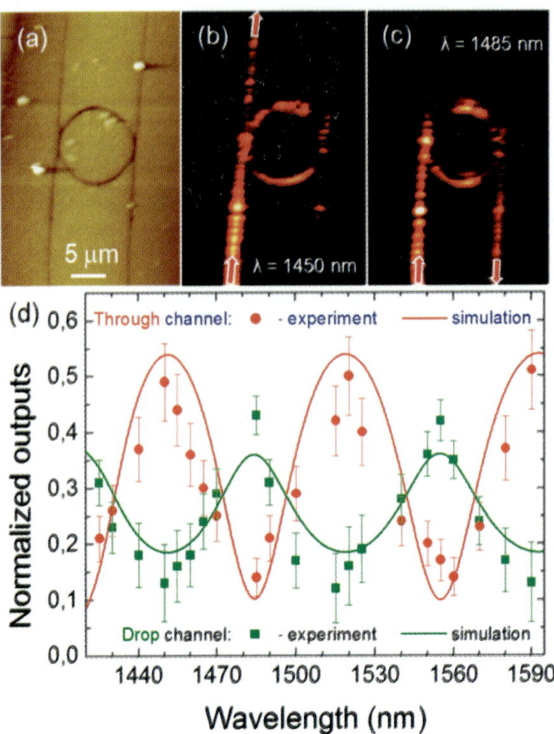

Color Plate 3. (a) Topographical and near-field optical images of the WRR-ADM (with a 5-μm-radius ring used) recorded at $\lambda \cong$ (b) 1.45, and (c) 1.485 μm. (d) WRR-ADM transmission spectra obtained for the through and drop channels along with the fitted curves corresponding to the modelled WRR-ADM response (see text). The error bars are estimated from the maximum variation in the ratio when placing the averaging windows at different positions along the input and output channels.

Chapter 12

ADIABATIC CONCENTRATION AND COHERENT CONTROL IN NANOPLASMONIC WAVEGUIDES

Mark I. Stockman

Department of Physics and Astronomy, Georgia State University Atlanta, Georgia 30303, USA

We describe ideas and results of controlling surface plasmon polaritons (SPPs) in nanoplasmonic waveguides. The adiabatic concentration of the optical energy in tapered nanoplasmonic waveguides is presented as a universal approach to transfer of energy and coherence from macro- or microscale to the nanoscale. Coherent control allows for directed and dynamically-defined concentration of the optical energy on nanometer-femtosecond spatio-temporal scale.

1. Introduction

Nanooptics and nanoplasmonics are experiencing presently a period of explosive growth and attracting a great interest. Nanoplasmonics deals with electronic processes at the surfaces of metal nanostructures, which are due to electronic excitations called surface plasmons (SPs) that can localize optical energy on the nanoscale.[1–3] The nanoplasmonic processes can potentially be the fastest in optics: their shortest evolution times are defined by the inverse spectral width of the region of the plasmonic resonances and are on the order of 100 as.[4] The relaxation times of the SP excitations are also ultrashort, in the $10 - 100$ fs range.[5–9] Such nanolocalization and ultrafast kinetics make plasmonic nanostructures promising for various applications, especially for the ultrafast computations, data control and storage on the nanoscale.

These and potentially many other applications require precise control over the optical excitations of the nanostructures in time and space on the femtosecond-nanometer scale. Such a control cannot be imposed by far-field focusing of the optical radiation because the diffraction limits its dimension to greater than half wavelength. In other words, the optical radiation does not have spatial degrees of freedom on the nanoscale. There is a different class of approaches to control a system on nanoscale based on plasmonic nanoparticles or waveguides brought to the *near-field* region of the system. Among these we mention: the tips of scanning near-field optical microscopes,[1] adiabatic plasmonic waveguides,[10] nanowires,[11,12] plasmonic superlenses[13] or hyperlenses.[14] In all these cases, massive amount of metal is brought to the vicinity of the plasmonic nanosystem, which will produce strong perturbations of its spectrum and SP eigenmodes, cause additional optical losses, and adversely affect the ultrafast dynamics and energy nanolocalization in the system. This nanowaveguide approach also may not work because of the excitation delocalization due to the strong interaction (capacitive coupling) at the nanoscale distances for optical frequencies.

We have proposed[15] a principally different approach to ultrafast optical control on the nanoscale based on the general idea of coherent control. The coherent control of the quantum state of atom and molecules is based on the directed interference of the different quantum pathways of the optical excitation,[16–25] which is carried out by properly defining the phases of the corresponding excitation waves. This coherent control can also be imposed by an appropriate phase modulation of the excitation ultrashort (femtosecond) pulse.[22,26–28] Shaping the polarization of a femtosecond pulse has proven to be a useful tool in controlling quantum systems.[29]

Our initial idea[15] has been subsequently developed theoretically[29–32] and confirmed experimentally.[33–35] In this coherent control approach, one sends from the far-field zone a shaped pulse (generally, modulated by phase, amplitude, and polarization) that excites a wide-band packet of SP excitations in the entire nanosystem. The phases, amplitudes, and polarizations of these modes are forced by this shaped excitation pulse in such a manner that at the required moment of time and at the targeted nanosite, these modes' oscillations add in phase while at the other sites and different moments of time they interfere destructively, which brings about the desired spatio-temporal localization. Theoretically, the number of the effective degrees of freedom that a shaped femtosecond pulse may apply to a nanoplasmonic system is on the order of its quality factor Q (i.e., the number of coherent plasmonic oscillations that system undergoes before dephasing). In the optical region for noble metals $Q \sim 100$, providing a rich, ~ 100-dimensional space of controlling parameters. The coherent control approach is non-invasive: in principle, it does not perturb or change the nanosystem's material structure in any way.

The coherent control approach can be employed to control not only nanosystems, which support localized SP modes, but also extended nanostructured systems that carry surface plasmon polaritons. In this case, the phase and polarization modulation of the excitation radiation can determine the propagation pathway in a nanoplasmonic waveguide splitters.[32,36]

Even more promising is the recent idea of the full coherent control on the nanoscale.[37] The idea of this approach is a nanoplasmonic counterpart of the Active Phased Array radar (APAR), also called Active Electronically Steered Array (AESA) radar. It also incorporates ideas of the adiabatic concentration of the optical energy on nanoscale (see Sec. 2) and of time-reversal (back-tracing).

The full coherent control on the nanoscale is achieved by placing an array (possibly, a linear array) of nanoparticles on a SPP nanoplasmonic waveguide. Each such particle may be, in particular, a nanosphere, nanoshell, nano-indentation of the surface, or simply a fragment of the surface roughness. These nanoparticles are positioned far away from each other so they can independently be excited, each by its own shaped (phase-, amplitude-, and polarization-modulated) ultrashort excitation pulse. As a result, there will be a packet of SPPs propagating from each of the nanoparticles in a non-directed way. However, as a result of the interference of these SPP waves, a propagating wavefront will be established. By a proper phase modulation, one can force all these SPP waves at every frequency to interfere coherently at a certain point of the nanoplasmonic waveguides at a defined time, creating un ultrashort and nanolocalized spike of the optical fields, i.e., fully controlled spatio-temporal localization.[37] The required phase modulation of the each of the controlling pulses can be found using the principle of back-propagation (or, time-reversal). Such a system can be called NAPAR (Nanoplasmonic Active Phased Array "Radar"). We will describe this approach in Sec. 3.

2. Adiabatic concentration of optical energy on nanoscale

2.1. *Introduction to adiabatic concentration in nanoplasmonics*

A central problem of nanooptics is the commonly known impossibility to concentrate (focus) optical radiation on the nanoscale because the wavelength of light is on the microscale, many orders of magnitude too large. This problem appear to be fundamental because the electric and magnetic component of the electromagnetic field exchange energy on distances of half wave length, which is on order of a micron for optical frequencies, i.e., three orders of magnitude larger than the desired nanoscale. However, this limitation is removed in nanoplasmonics, because SPs are electrostatic excitations where the electric field and electron kinetic energy are exchanged in the process of optical oscillations.

Nevertheless, any attempts to couple a laser source to the nanoscale through, e.g., tapered fibers (see, e.g., Refs. 38–41) or by focusing on metal tips[42,43] lead to a tremendously low energy efficiency: only a very small part of the excitation energy is transferred to the nanoscale. This is due to the fact that the absorptions, scattering, and other effective cross sections of such nanosystems are still much less than the minimum size of the focal spot, which is on the microscale.

In this Section, we show that such impossibility is not a law of nature: we introduce a way to focus and concentrate the optical radiation on the nanoscale, in principle, without major losses by *exciting the propagating surface plasmon polaritons (SPPs) and then adiabatically, but as rapidly as possible, transforming them into localized surface plasmons (SPs).* The latter are purely electric oscillations (without significant magnetic component) that can and do nano-localize.[3]

This possibility of the adiabatic transformation of the propagating SPPs, which are electromagnetic waves, into standing, non-propagating SPs has been introduced in Refs. 10,44. These publications considered various graded systems for the SPP adiabatic concentration of optical energy: a SPP waveguide consisting of a semi-infinite metal covered by a semi-infinite dielectric with graded dielectric permittivity, a dielectric nanolayer on massive metal, and a tapered plasmonic waveguide. These systems are discussed in detail below in this Section along with the adiabatic concentration of SPPs in metal nanolayers surrounded by dielectrics and dielectric nanolayers surrounded by metal. We note that efficient adiabatic nanofocusing of SPPs has been shown theoretically also in SPP waveguides that made of nanogrooves on metal.[45] Such metal grooves have been shown experimentally to be effcient SPP waveguides.[46]

Specifically, the adiabatic transformation from the far to near zone can be done in the following way. Light waves can be transformed with high efficiency into SPPs using, e.g., a grating coupler or the well-known Kretschmann geometry. These SPPs propagate along a graded nanoplasmonic waveguide along which parameters slowly (adiabatically) change in such a way that the phase velocity of the SPPs tends to zero in the vicinity of some point at a finite distance. It is required that the excited SPP mode exists without a cut-off in the entire range of these parameters including this stopping point. Then the adiabatic theorem predicts that the wave will propagate without back-reflection and scattering into three dimensions (3D) to the stopping point. Thus this wave will adiabatically slow down and asymptotically stop at this point. This will result into the adiabatic conversion of the propagating SPPs into the stopped, quasielectrostatic SPs. At this stopping point the wavelength of the SPPs asymptotically tends to zero. This is in sharp contrast to massive (quasi)particles whose wavelength becomes infinite at the stopping point, which violates the adiabatic approximation. Note that the adiabatic approximation is called the Wentzel-Kramers-Brillouin or quasiclassical approximation in quantum mechanics.[47]

As a result of the above-described process, it is possible to adiabatically convert with high efficiency the propagating SPPs into the standing waves, which are SPs localized on the nanoscale. Being adiabatic to avoid back-reflection and scattering, this conversion nevertheless should be as rapid as possible to prevent the absorption losses in the metal. The efficiency of transformation along with maximum value of the achieved local optical fields are determined by the length of the adiabatic waveguide and the possible minimum value of the adiabatic parameter, and are limited only by the dielectric losses in the graded polaritonic waveguide.

We will consider below some examples of such systems and processes. A a metal half-space covered with semiconductor with dielectric permittivity increasing in the direction of the SPP propagation is considered in Sec. 2.2. Thin metal wedge covered with dielectric where SPPs propagate toward the sharp edge is subject of Sec. 2.6. The most energy-efficient of the graded systems considered below is a thin metal cone embedded in a dielectric, which is a subject of Sec. 2.7.

2.2. *Adiabatic stopping of SPPs at interface of metal with graded semiconductor*

Consider a semi-infinite metal occupying a half-space of $z < 0$, whose dielectric function $\varepsilon_m(\omega)$ is uniform in space, where ω is the optical excitation frequency. The upper half-space is occupied by a dielectric or undoped semiconductor whose permittivity $\varepsilon_1(\mathbf{r})$ smoothly depends on the coordinates $\mathbf{r} = (x, y)$ in the plane (xy) of the interface. Such spatial dependence is achieved, e.g., in the well-developed semiconductor heterostructures. Using this smoothness, we will employ the eikonal approximation,[48] also called Wentzel-Kramers-Brillouin (WKB) or quasi-classical approximation[49]).

We consider SPPs as transversally-magnetic (TM) waves. For such waves, the magnetic field H_x at a plane interface between uniform metal and dielectric is given by

$$H_x(y,z,t) = H_x^{(0)} \exp\left(iky - \kappa|z| - i\omega t\right) \ , \tag{1}$$

$$E_{y,z}(y,z,t) = E_{y,z}^{(0)} \exp\left(iky - \kappa|z| - i\omega t\right) , \tag{2}$$

where $H_x^{(0)}$ and $E_{y,z}^{(0)}$ are constant amplitudes of the corresponding field components; k is the wave vector that is directed along the y axes, and κ is the corresponding evanescent exponent. We are using the a definition of the coordinate system where the interface is in the xy plane, and the normal to it is in the z direction. We introduce a two-dimensional (2D) coordinate vector $\boldsymbol{\rho}$ in the interface plane with the x and y components only. We assume that the permittivity of the dielectric is a slowly varying function $\varepsilon_d(\boldsymbol{\rho})$. The WKB solution for the in-plane magnetic field is

$$\mathbf{H}(\mathbf{r},z,t) = \mathbf{H}_0(\mathbf{r}) \exp\left\{[-\kappa(\mathbf{r})|z| + ik_0\phi(\boldsymbol{\rho})] - i\omega t\right\} \ , \tag{3}$$

and κ is the local (at a given $\boldsymbol{\rho}$) value equal to skin-depth exponent κ_m in the metal and evanescent exponent κ_d in the surrounding dielectric (semi-conductor). The eikonal $\phi(\boldsymbol{\rho})$ is related to the effective index $n(\boldsymbol{\rho})$ by the standard eikonal equation[48]

$$\left[\frac{\partial\phi(\boldsymbol{\rho})}{\partial\boldsymbol{\rho}}\right]^2 = n^2(\boldsymbol{\rho}) \ . \tag{4}$$

Here, $n(\boldsymbol{\rho})$ is the effective 2D refraction index of SPPs that is locally (at a given $\boldsymbol{\rho}$) given by an expression

$$n(\boldsymbol{\rho}) = \left[\frac{\varepsilon_m(\omega)\varepsilon_d(\boldsymbol{\rho})}{\varepsilon_m(\omega)+\varepsilon_d(\boldsymbol{\rho})}\right]^{\frac{1}{2}} \ . \tag{5}$$

The exponents are given by

$$\kappa_m = k_0\left[-\frac{\varepsilon_m^2(\omega)}{\varepsilon_m(\omega)+\varepsilon_d(\boldsymbol{\rho})}\right]^{\frac{1}{2}} \ , \tag{6}$$

$$\kappa_d = k_0\left[-\frac{\varepsilon_d^2(\boldsymbol{\rho})}{\varepsilon_m(\omega)+\varepsilon_d(\boldsymbol{\rho})}\right]^{\frac{1}{2}} \ , \tag{7}$$

The surface index $n(\boldsymbol{\rho})$ has a singularity (branch-cut edge) at an SPP stopping point in the complex plane of ω defined by an equation

$$\varepsilon_m(\omega) + \varepsilon_d(\boldsymbol{\rho}) = 0 \ . \tag{8}$$

The real stopping points are given by a related equation

$$\mathrm{Re}\varepsilon_m(\omega) + \varepsilon_d(\boldsymbol{\rho}) = 0 \ , \tag{9}$$

beyond which the SPPs do not propagate. This equation in a general case at a given ω defines a line ("caustic") in the plane of ρ where the SPPs are stopped and accumulated.

The adiabatic parameter is defined as the relative change of the SPP wavelength at a distance of this wavelength,

$$\delta = \left| \frac{\partial}{\partial \boldsymbol{\rho}} \frac{1}{k_0 n(\boldsymbol{\rho})} \right| = \left| \frac{1}{k_0 n^2(\boldsymbol{\rho})} \frac{\partial n(\boldsymbol{\rho})}{\partial \boldsymbol{\rho}} \right| \ . \tag{10}$$

For the eikonal approximation to be valid, this adiabatic parameter should be small, which for the conventional optics would be violated at the stopping point. However, for the SPPs at this critical point, δ is finite,

$$\delta = \left| \frac{\partial \varepsilon_d(\boldsymbol{\rho})}{\partial \boldsymbol{\rho}} \Big/ \left(k_0 \varepsilon_m \sqrt{\mathrm{Im}\, \varepsilon_m} \right) \right| \ . \tag{11}$$

Due to this finite value, δ can always be made arbitrarily small by choosing small enough grading $|\partial \varepsilon_d(\boldsymbol{\rho})/\partial \rho|$. In such a case, in contrast to the conventional quantum mechanics of massive particles, the eikonal (WKB) approximation is valid in the *entire* space, including the stopping point and beyond. Mathematically, this is due to the presence of losses that shift the exact stopping point from the real axis into the complex plane. Physically, this means that close to the stopping point, the SPPs accumulate but then they are absorbed, never actually reaching the true singular point where the WKB approximation would have been violated. Close to the stopping point, the group velocity of the SPPs is proportional to the distance to this point; therefore the traveling time to reach this point diverges logarithmically. This is a general properties of the adiabatic nanooptics of the SPPs at interfaces that we will encounter over and over again for various graded plasmonic waveguides described in the this Section below.

The SPP propagation losses decrease with decrease of $\mathrm{Im}\, \varepsilon_m$, which can be achieved by reducing excitation frequency ω. On the other hand, in that region, $|\mathrm{Re}\, \varepsilon_m|$ becomes large. This implies that at the stopping point (9) the required permittivity of the dielectric is large, $\varepsilon_d = -\mathrm{Re}\, \varepsilon_m = 5 - 15$ for $\omega = 2 - 3$ eV. Such values are typical for semiconductors, e.g., $\mathrm{Al}_x\mathrm{Ga}_{1-x}\mathrm{As}$.[50]

Below, in numerical illustrations, we set $\omega = 2.5$ eV and reduced vacuum wavelength $\lambda = 1/k_0 = 80$ nm. As a metal, we consider silver as

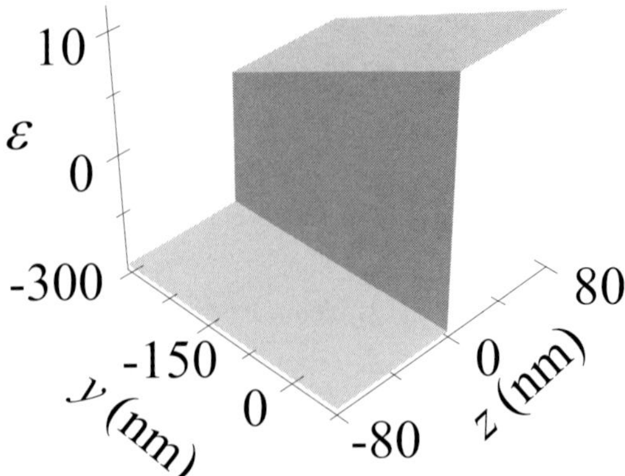

Fig. 1. Spatial dependence of dielectric constant as a function of coordinates $\boldsymbol{\rho} = (x, y)$ in the xy interface plane.

having the lowest losses in the optical spectral region.[51] As an example, we consider SPPs propagating along the y direction. Assume that the dielectric (semiconductor) component possesses permittivity with a constant gradient of $\varepsilon_{dy} = \partial\varepsilon_d/\partial y$ in this direction, as illustrated in Fig. 1.

Then the eikonal is a function of only y that be found analytically as

$$\phi(y) = -\frac{1}{\varepsilon_{dy}}\Big\{ \left[\varepsilon_d\varepsilon_m(\varepsilon_d + \varepsilon_m)\right] + \frac{1}{2}\varepsilon_m^{3/2}\ln\left[2\varepsilon_d + \varepsilon_m + 2\varepsilon_d^{1/2}(\varepsilon_d + \varepsilon_m)^{1/2}\right]\Big\}\,, \tag{12}$$

where $\varepsilon_d = \varepsilon_d(y)$.

The pre-exponential $H_0(\mathbf{r})$ can be found from the conservation of the energy flux utilizing the Pointing vector, which yields

$$H_0 \propto \left(\sqrt{-\varepsilon_m/\varepsilon_d^3} - \sqrt{-\varepsilon_d/\varepsilon_m^3}\right)^{-1/2}\,. \tag{13}$$

Close to the stopping point, this pre-exponential becomes large, describing enhancement of the local field by a factor $\sim (|\varepsilon_m|/\varepsilon_1)^{3/4}Q$, where $Q = -\mathrm{Re}\varepsilon_m/\mathrm{Im}\varepsilon_m$ is the quality factor of the SP resonance. This enhancement describes the accumulation of SPP's at the stopping point. An instantaneous solution determined by Eqs. (3)-(13) is shown in Fig. 2, where we can clearly see the stopping of SPP's and the corresponding field enhancement at $y = 0$. Due to the adiabaticity condition, the wave in Fig. 2(a) is not a standing one: it runs from the left to the right, toward

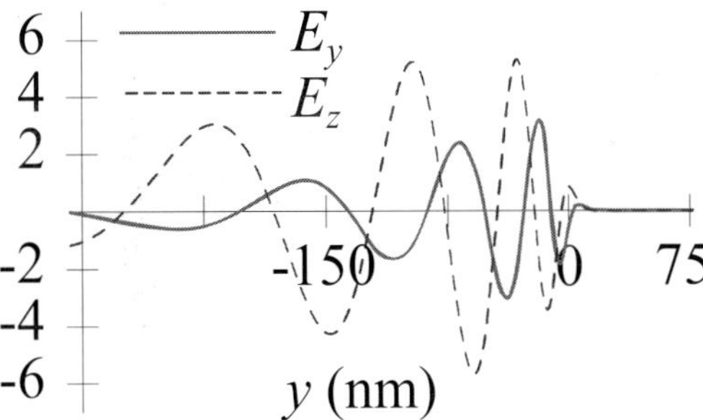

Fig. 2. In the vicinity of the stopping point, instantaneous value of optical electric field along the propagation direction y. The solid line is for tangential electric field E_y and the dashed line is for field E_z normal to the interface (the fields are in arbitrary but comparable units).

the stopping point, where it is not reflected back, but adiabatically stopped, enhanced, and, further in space, absorbed.

The adiabatic parameter should not be too small, otherwise the stopping would be slow and energy losses in the metal high [cf.: for the case of Fig. 2, $\delta \approx 0.1$]. Therefore, this field of phenomena can be called *rapid adiabatic nanooptics.* For the model under consideration, we assume that the dielectric (semiconductor) permittivity is linearly graded at length L of the propagation interval; at the end of this interval $\varepsilon_d = |\varepsilon_m|$. Thus $\partial\varepsilon_d/\partial y \approx |\varepsilon_m|/L$. Taking this into account, from Eq. (12) we find that this rapidity limitation is

$$\delta > \frac{\sqrt{\operatorname{Im}\varepsilon_m}}{|\varepsilon_m|} . \tag{14}$$

Substituting expression (11) for adiabatic parameter δ and adding the condition of adiabaticity $\delta \ll 1$, we can write a condition on the propagation length L of the SPPs in the adiabatic nanooptics

$$\frac{|\varepsilon_m|}{\operatorname{Im}\varepsilon_m} \gtrsim k_0 L \gg \frac{1}{\sqrt{\operatorname{Im}\varepsilon_m}} . \tag{15}$$

Here the right inequality expresses the condition of adiabaticity, and the left inequality describes the requirement of the rapidity. Note that it is possible to satisfy this double inequality only of the quality factor of the SP resonance is high, i.e., $|\varepsilon_m| \gg \operatorname{Im}\varepsilon_m$. We also estimate minimum size l_m of the energy concentration region as $l_m \sim \delta\lambda$; it is directly determined by the adiabaticity parameter.

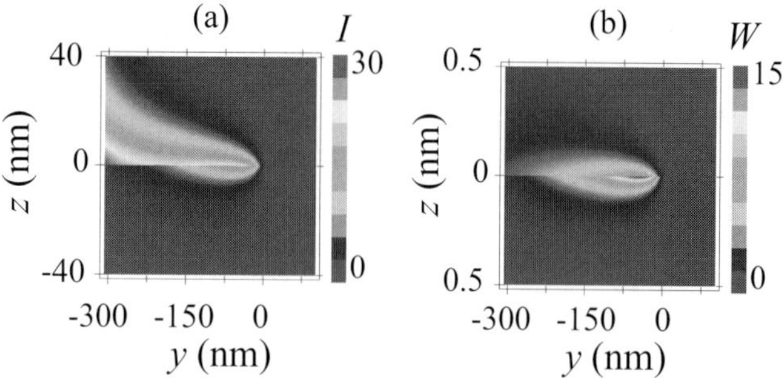

Fig. 3. For graded dielectric, spatial distributions of the local optical intensity I (a) and optical energy density W (b) (arbitrary units, linear density-coding scales). The distributions are shown in the vertical cross-section (yz) plane.

In Fig. 3(a), we show the spatial distribution of local optical intensity I defined as $I(\mathbf{r}) = |\mathbf{E}(\mathbf{r})|^2$, which determines the rates of optical responses of the materials. As the radiation propagates toward the stopping point, the local optical intensity becomes dramatically localized at the metal-semiconductor interface, bound to a layer of hight of < 10 nm at the stopping point ($y = 0$).

We display the local optical energy density W

$$W(\mathbf{r}) = \frac{1}{4\pi} \left\{ \mathrm{Re}\, \frac{\partial[\omega\varepsilon(\mathbf{r},\omega)]}{\partial\omega} |\mathbf{E}(\mathbf{r},\omega)|^2 + |\mathbf{H}(\mathbf{r},\omega)|^2 \right\} . \tag{16}$$

in Fig. 3(b). The verttical (in yz plane) distribution of the energy is completely different from that of the local field intensity, which is localized mostly in the dielectric (semiconductor) component, except for the vicinity of the stopping point — cf. panel (a). In contrast as panel (b) shows, the energy tends to localize in the metal. The energy density has sharp nanofocus at the stopping point in all three dimensions where it is localized in both the metal and semiconductor components. Thus this graded semiconductor-metal interface system provides an efficient way to delivery the optical energy to nanoscale-size region of high-permittivity semiconductor and metal, which may be useful in applications.

2.3. *General layered system*

For the sake of convenience and self-containment, we will briefly summarize some results regarding SPPs in layered systems. The problem of the SPP modes in media consisting of metal and dielectric planar layers has been considered by Economou in a pioneering paper.[52] Here we present the corresponding theory in the appropriate detail. We start below with a general layered planar system. In what follows, we will consider a metal film embedded in dielectric.

Consider a general system containing N planar layers of materials with different permittivities. We number these layers by an index $i = 1, 2, \ldots, N$, with the corresponding layer permittivities denoted as ε_i. We position the coordinate system as the following: the xy coordinate plane is parallel to the layers, and the z axis is at the normal to these planes. We will always consider the first layer ($i = 1$) to be infinite in the direction $z \to -\infty$, and the last layer ($i = N$) to be infinite in the direction of $z \to \infty$; the thicknesses of the remaining layers we denote as d_i, $i = 2, \ldots, N-1$. We assume that there exists at least one metal-dielectric interface in this system that supports SPPs.

We seek SPP as a solution of the electromagnetic wave equations that is bound to these layered system and has evanescent behavior in the infinite layers as a transverse magnetic (TM) mode where the magnetic field in an ith layer has only x component and is expressed as

$$H_x^{(i)}(y,z) = \left(A_i e^{\kappa_i z} + B_i e^{-\kappa_i z}\right) e^{iky} \ , \tag{17}$$

where A_i, B_i, κ_i, and k are constants to be determined from equations and boundary conditions. Substituting this in the wave equation, we obtain for each layer,

$$k^2 - \kappa_i^2 = k_0^2 \varepsilon_i \ , \tag{18}$$

where, to remind, $k_0 = \omega / c$.

The sign of κ_i is arbitrary, and we will define it imposing a condition $\mathrm{Re}\kappa_i > 0$. Under this sign definition, the condition of the evanescent behavior becomes

$$B_1 = 0 \ , \quad A_N = 0 \ . \tag{19}$$

Exactly as for the case of a single interface, the wave vector k is the same for all the layers as follows from the continuity of H_x at the interfaces. For electric field, the Maxwell curl equation yields

$$E_y = \frac{i}{k_0 \varepsilon_i} \frac{\partial H_x}{\partial z} \ , \tag{20}$$

From this, we obtain the parallel electric field component in an ith layer as

$$E_y^{(i)}(y,z) = \frac{\kappa_i}{\varepsilon_i k_0} \left(A_i e^{\kappa_i z} - B_i e^{-\kappa_i z}\right) e^{iky} \ , \tag{21}$$

and for the normal (transverse) component a general equation is valid:

$$E_y = \mp \frac{i\kappa_i}{k_0 \varepsilon_i} H_x \ , \tag{22}$$

The normal component is given by a general equation

$$E_z = \frac{k}{k_0 \varepsilon_i} H_x \ , \tag{23}$$

form which it follows that it copies the magnetic field.

Now let is determine the number of degrees of freedom (i.e., the difference between the number of unknown variables and the number of independent equations) for the system of equations that we obtain. The Maxwell boundary condition require the continuity of H_x and E_y at each of the $N-1$ interfaces in the system, which supplies $2(N-1)$ equations. Plus, there are N wave equations (18). Thus, overall we have $3N-2$ equations. Among the unknown variables, there are $2(N-2)$ amplitudes A_i and B_i in the finite layers ($i = 2, \ldots, N-1$), plus two amplitudes (A_1 and B_N) in the semi-infinite layers, plus N evanescent exponents (κ_i), plus the wave vector k. Overall, this sums up to $3N-1$ variables. Subtracting the number $3N-2$ of equations from this, we conclude that there is *one* degree of freedom. This has an absolutely clear meaning. The *linear* system of equations for the amplitudes A_i and B_i, which is obtianed by applying the continuity boundary conditions for fields (17) and (17), does not define the total magnitude of the fields for a SPP mode, but only relative amplitudes. Therefore one of these amplitudes is undetermined, and the rest of them are expressed in terms of that one. This situation is characteristic of any *linear* eigenmode problem.

Eliminating the evanescent exponents κ_i in terms of the wave vector k from the boundary conditions, we obtain a linear system of equations for amplitudes A_i and B_i. For this system to have a nontrivial solution, its determinant should be zero, which gives a characteristic equation for the wavevector k. To be specific, we consider below in this Section a three-layer system where the thickness of the mid layer is d. Carrying out the above-described procedure, we immediately obtain the characteristic equation in a form

$$\exp\left(2k_0 \, d \, \varepsilon_2 \, u_2\right) = \frac{(u_1 - u_2)(u_3 - u_2)}{(u_1 + u_2)(u_3 + u_2)} \ , \tag{24}$$

where we introduced for each layer a quantity u_i according to the definition

$$u_i = \frac{1}{\varepsilon_i} \sqrt{\left(\frac{k}{k_0}\right)^2 - \varepsilon_i} \equiv \frac{\kappa_i}{k_0 \varepsilon_i} \ . \tag{25}$$

Note that Eq. (24) defines a relation between k and ω, where k enters the problem explicitly through coefficients u_i, and ω enters both through parameter k_0 and the frequency dependence of dielectric permittivities ε_i.

By an identity transformation, transcendental Eq. (24) can be given another, equivalent form, which may be more convenient in some analytical or numerical computations,

$$\tanh\left(k_0\, d\, \varepsilon_2\, u_2\right) = -\frac{u_2(u_1+u_3)}{u_2^2+u_1u_3}\,. \tag{26}$$

For the sake of reference, we give the expressions for non-zero coefficients A_i and B_i that define SPP electromagnetic fields:

$$\begin{aligned} A_1 &= 2\exp\left[-\frac{1}{2}k_0d(u_2\varepsilon_2+u_3\varepsilon_3)\right]u_2(u_2-u_3)\,,\\ A_2 &= \exp\left[-\frac{1}{2}k_0d(u_1\varepsilon_1+u_3\varepsilon_3)\right](u_1+u_2)(u_2-u_3)\,,\\ B_2 &= \exp\left[-\frac{1}{2}k_0d(u_1\varepsilon_1-2u_2\varepsilon_2+u_3\varepsilon_3)\right](u_1+u_2)(u_2+u_3)\,,\\ B_3 &= 2\exp\left[-\frac{1}{2}k_0d(u_2\varepsilon_2-u_1\varepsilon_1)\right]u_2(u_1+u_2)\,, \end{aligned} \tag{27}$$

We remind that these equations provide only the relative magnitudes of the coefficients; there is also an undetermined general factor that depends on the intensity and phase of the physical SPP wave.

2.4. *Symmetric layered system*

There is especial interest in the symmetric three-layer system where the two semi-infinite layers are made of the same material (metal or dielectric), and the middle layer is of a different material (dielectric or metal, respectively). Physically, the special place of such systems is due to the fact the SPP at the two interfaces $1-2$ and $2-3$ have exactly the same frequency (spectral degeneracy). Therefore they strongly interact and mix (hybridize). The problem solution is facilitated by the appearance of an exact symmetry: parity. With respect to reflection in the plane of symmetry, which we for certainty assume to be at $z=0$, the fields are either even or odd functions of z. It can be shown that E_z and H_x have always the same parity with respect to the z reflection, and E_y has opposite. It is conventional to associate the parity of a SPP mode with that of E_z. The even SPPs are also often called symmetric, and the odd ones antisymmetric.

For a symmetric system, $\varepsilon_1 = \varepsilon_3$, and the characteristic equation (26) simplifies to acquire a form

$$\tanh\left(k_0\, d\, \varepsilon_2\, u_2\right) = -\frac{2u_1\, u_2}{u_1^2 + u_2^2}\ . \tag{28}$$

To better understand properties of the corresponding solutions and their relation to parity, we re-derive the corresponding characteristic equation from the very beginning using the parity as a good quantum number. Starting with the even (symmetric) SPPs, we impose a specific boundary condition at the symmetry plane $z = 0$,

$$E_y|_{z\,=\,0} = 0\ . \tag{29}$$

After that we only use boundary conditions of the continuity of E_y and H_x at the $1-2$ interface (the conditions at the $2-3$ interface are redundant), which following the procedure described above in Sec. 2.3 brings about a characteristic equation

$$\tanh\left(\frac{1}{2}\, k_0\, d\, \varepsilon_2\, u_2\right) = -\frac{u_1}{u_2} \tag{30}$$

describing only the even (symmetric) SPPs.

Similar derivation can immediately be effected for the odd SPPs, which satisfy the boundary condition at the symmetry plane

$$H_x|_{z\,=\,0} = 0\ . \tag{31}$$

This leads to the characteristic equation

$$\tanh\left(\frac{1}{2}\, k_0\, d\, \varepsilon_2\, u_2\right) = -\frac{u_2}{u_1}\ , \tag{32}$$

which specifically describes only the odd (antisymmetric) SPP modes.

To compare these defined-parity characteristic equations (30) and (32) with generic Eq. (28), we invoke a hyperbolic trigonometry identity

$$\tanh 2\theta = \frac{2\tanh\theta}{1 + \tanh^2\theta}\ . \tag{33}$$

From this, setting in the left hand side of Eq. (28) $\theta = \frac{1}{2}\, k_0\, d\, \varepsilon_2\, u_2$, we immediately see that the solutions of Eqs. (30) and (32) satisfy Eq. (28) automatically. Thus these two definite-parity characteristic equations together are equivalent to the generic equation (28).

Equivalent characteristic equations with the exponentials in the left-hand sides, which may be more convenient in some cases, are

$$\exp\left(k_0 d u_2 \varepsilon_2\right) = \mp\frac{u_1 - u_2}{u_1 + u_2}\ . \tag{34}$$

The corresponding non-zero coefficients of fields (17), (21), and (23) are

$$A_1 = \pm B_3 = u_2 \ , \quad A_2 = \pm B_2 = \frac{1}{2} \exp\left[\frac{1}{2} k_0 d \left(u_2 \varepsilon_2 - u_1 \varepsilon_1\right)\right] (u_1 + u_2) \ . \tag{35}$$

In these expressions, the upper sign refers to the symmetric SPPs and the lower sign to the antisymmetric SPPs.

2.5. *Adiabatic SPP concentration in quasistatic regime*

In the case of a plasmon polaritonic waveguide whose size (in the direction normal to the SPP propagation) is graded, properties of the adiabatic concentration of optical energy can be described in a quite general form applicable to various waveguide geometries. This is possible due to the fact that in the limit where the SPP reduced wavelength along the propagation direction k^{-1} becomes much smaller than vacuum reduced wavelength λ, the behavior of the SPPs becomes universal. Their wavelength k^{-1} is so small that λ is irrelevant. In such a case, k^{-1} is determined by and on order of the normal size of the waveguide which is on the nanoscale, i.e., much smaller tham λ. Because λ disapears from the solution, so does also the speed of light. Hence, the solution becomes quasistatic. All the quantities of the dimensionality of length become on order of normal waveguide size d. Such universality allows one to present a general, universal description based on this scaling.

Consider a plasmon polaritonic waveguide where a SPP can propagate in one dimension (in our convention along the y coordinate) but is restricted in the normal direction by a size, which we assume to be on the nanoscale. For instance, for a metal nanolayer embedded into dielectric or dielectric nanolayer embedded into metal this size is by order of magnitude is the thickness d of the nanolayer and on the same order as the reduced wavelength k^{-1} of the SPPs. A SPP solution for such nanolayers is quasistatic: the magnetic component of the electromagnetic field is much smaller than the electric component.

We consider the thin metal or dielectric layers and similar systems for which there is no cut off: the SPP excitation branch exists for d arbitrarily small (limited, however, by Landau damping to $d \gtrsim v_F/\omega$). Then the properties of the adiabatic concentration can be established in a quite general form.

We will consider systems where the dielectric losses of the SPPs are not too large. Then in the first approximation, the energy of the SPPs is conserved, and so should be its total flux. In what follows, we will not be interested in specific coefficients but will trace only the scaling with the layer thickness d. The conservation of energy is reflected in the conservation of the energy flux as

$$v_g \int \mathbf{E}^2 \mathrm{d}xz = \text{const} , \tag{36}$$

where we remind that v_g is the group velocity. Because the only relevant parameter of the problem with the dimensionality of length is d (the radiation vacuum wavelength is many orders of magnitude too large), then the group velocity asymptotically (for nano-thin systems) scales as

$$v_g \propto d . \tag{37}$$

We will consider two type of systems: (i) thin layers, which provide one-dimensional (1d) confinement of the fields, and (ii) cones or prisms, which provide the pwo-dimensional (2d) confinement. Correspondingly, the integral in Eq. (36) can be estimated as $|E|^2 d^\nu$, where $\nu = 1, 2$ is dimensionality of the waveguide, as discussed above. From this, taking (36) and (37) into account, we obtain the scaling of the electric field inside the polaritonic waveguide with its characteristic size d as

$$E \propto d^{-\frac{1+\nu}{2}} . \tag{38}$$

The wave propagation in the WKB approximation is described by the eikonal $\phi = \int k(y)\, \mathrm{d}y$, where $k(y)$ is the wave vector at the coordinate y along the waveguide. The universal quasistatic scaling for the nano-thin waveguides implies that k is inversely proportional to the only relevant parameter of the problem with the dimensionality of length, which is d, i.e.,

$$k(y) = \frac{k_a}{d(y)} , \tag{39}$$

where k_a is a complex constant, and $d(y)$ is the waveguide grading, i.e., the dependence of its thickness on the longitudinal coordinate y. As a result, both the phase velocity and group velocity of the SPPs scale with nanolayer thickness as

$$v_p = \frac{\omega}{\mathrm{Re}\, k_a} d , \quad v_g = \left(\frac{\partial \mathrm{Re}\, k_a}{\partial \omega} \right)^{-1} d . \tag{40}$$

This scaling along with Eq. (38) brings about the dependence of the field along the waveguide as

$$|E| \propto d^{-\frac{1+\nu}{2}} \exp\left[-\mathrm{Im}\, k_a \int \frac{\mathrm{d}y}{d(y)} \right] . \tag{41}$$

Assuming for simplicity a constant grading of the waveguide, $d' = \partial d/\partial y = \text{const}$,

$$|E| \propto d^{-\frac{1+\nu}{2} - \frac{\operatorname{Im} k_a}{d'}} \ . \tag{42}$$

This is a very general result that follows from scaling properties of the quasistatic solution. Interestingly enough, it shows that the field increases along the waveguide if the grading exceeds a certain critical value,

$$-d' > 2\frac{\operatorname{Im} k_a}{1+\nu} \ . \tag{43}$$

From Eq. (42), it follows that field intensity I and energy density W also asymptotically obey a universal scaling

$$I \propto W \propto d^{-(1+\nu)-2\frac{\operatorname{Im} k_a}{d'}} \ . \tag{44}$$

Note that $-d' > 0$, and that the the adiabatic parameter

$$\delta = \left|\frac{d(k^{-1})}{dy}\right| = \frac{1}{k_a}|d'| \ . \tag{45}$$

This parameter is finite along the entire waveguide including its apex. Thus the WKB can be applicable for the entire system, in contrast to the quantum mechanics where it is violated at the stopping (turning) points. It is important to understand how this situation can take place. The answer is in the travel time needed for the SPPs to reach the apex that plays the role of such a stopping point. This travel time T_t for SPPs at an initial coordinate point y_i to the final coordinate point y_f is

$$T_t = \int_{y_i}^{y_f} \frac{dy}{v_g} \propto \ln\frac{y_i}{y_f} \ , \tag{46}$$

where we took into account scaling (37) and assumed for simplicity a constant grading, i.e., $d' = \text{const}$. It is obvious that T_t logarithmically diverges for $y_f \to 0$. This implies that it takes SPPs infinite time to travel to the apex, which is a singular point. This is the underlying reason why the WKB solution is valid for the entire waveguide.

Because k_a is a (complex) constant whose modulus is on order of 1 [see, e.g., Eq. (48)], the adiabaticity condition $\delta \ll 1$ reduces to $|d'| \ll 1$. We aslo see that in this asymptotic regime, the adiabatic parameter is always finite. Also, the "'best" solution is such that the adiabatic parameter is as large as possible compatible with the adiabaticity, because it minimizes the propagation time and, correspondingly, the dielectric losses. But this also implies that this parameter should be constant in the asymptotic region

($k \sim d^{-1} \gg k_0$); the grading, correspondingly, should also be constant, $d' = \text{const}$.

For the graded waveguide in the shape of a wedge or a cone, $-d' = \tan\theta$, where θ is the angle at the tip of this wedge or cone. Thus condition (43) shows that the grade angle θ of the waveguide should be large enough,

$$\theta > \theta_{min} = \arctan\left(2\frac{\mathrm{Im}\, k_a}{1+\nu}\right) . \tag{47}$$

If condition (43) or (47) is violated, then the fields in the waveguide will monotonously decay along the waveguide. If the losses in the waveguide are large, this inequality or (43) may contradict to the adiabaticity condition $|d'| \ll 1$.

The value of k_a depends on the specific model. For instance, for a nano-thin metal layer in dielectric environment or nanolayer of dielectric embedded in a metal, we obtain

$$k_a = \ln\left[\frac{\varepsilon_m(\omega) - \varepsilon_d}{\varepsilon_m(\omega) + \varepsilon_d}\right] . \tag{48}$$

This expression is valid for the symmetric SPPs in the case of the dielectric nanolayer in metal and for the antisymmetric SPPs for the case of the metal nanolayer in dielectric (the modes of the opposite symmetry in each of these two cases are highly lossy and will not be considered). Note that because $\mathrm{Re}\,\varepsilon_m + \varepsilon_d < 0$, the real part of the argument of the logarithmic function is positive. Therefore, $\mathrm{Im}\, k_a$ originates only from the existence of $\mathrm{Im}\,\varepsilon_m$, i.e., from the dielectric losses in the metal, which is natural to expect.

We show in Fig. 4 dependence of minimum angle $\theta_m in$ on frequency of SPPs for both vacuum and a dielectric with $\varepsilon_d = 3$ for silver and gold as a metal. To remind, the adiabaticity condition requires $\tan\theta_{min} \ll 1$, so the value of θ_{min} should not be large. One can reasonable assume that a value acceptable from the requirements of adiabaticity is $\theta_{min} \lesssim 20$ deg. This is satisfied for silver in wide frequency range of the near-ir and entire visible spectrum, $\hbar\omega \leq 3.1$ eV; for vacuum this condition is actually more relaxed. In contrast, for gold the adiabaticity requirement is stronger, leading to $\hbar\omega \leq 2.1 eV$, thus limiting to the near-ir and red-yellow spectral region. The reason for this difference is the *ds* band absorption in gold that is in the green-blue spectral region, while for silver it is in the near uv.

Let us note that formally the results presented above are valid for all types of SPPs in symmetric planar waveguides: even or odd. However, in the region of high k, the frequencies of the even (symmetric) modes for the metal nanolayer guide and of the odd (antisymmetric) modes for the

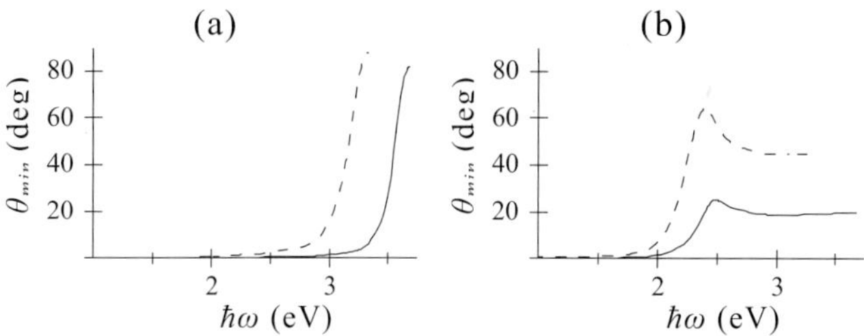

Fig. 4. Minimum grading angle θ_{min} (degree) at which the adiabatic concentration causes increase of SPP field toward the tip of the plasmon polaritonic waveguide. This angle is displayed as a function of the SPP frequency (eV). The calculations are applicable to a wedge-shaped nanolayer of metal in dielectric or dielectric nano-wedge in metal. The metal is silver (a) and gold (b). Dielectric permittivity $\varepsilon_d = 1$ (solid curves) and $\varepsilon_d = 3$ (dashed curve). Each curve is shown within the spectral range of the existence of the corresponding SPP mode, i.e., for $\omega \leq \omega_{sp}$.

dielectric nanolayer guide are high, leading to high losses and, correspondingly, to large values of θ_{min} that contradict to the adiabaticity. Thus, the results presented above in this Section are valid for and applicable to the lower frequency SPP modes. These "good" modes are the odd (antisymmetric) SPPs in the metal nanolayers and the even (symmetric) modes in the dielectric nanolayer in the metal waveguides.

Another system of interest is a metal conical (tapered cylinder) waveguide introduced in Ref. 10 — see also Sec. 2.7. In this case, as a measure of the transverse size of the waveguide d we choose the local cone radius R that depends on the longitudinal (along the propagation direction) coordinate z. The formulas given above in this Section apply with the substitution $d \rightarrow R$ and for $\nu = 2$. However, the value of coefficient k_a for a TM_0 SPP mode in a tapered cylinder changes to[10]

$$k_a = 2\left\{\frac{\varepsilon_d}{\varepsilon_m(\omega)}\frac{1}{\ln\left[\frac{\varepsilon_d}{4\varepsilon_m(\omega)}\right] + 2\gamma}\right\}^{1/2}, \tag{49}$$

where $\gamma \approx 0.577216$ is the Euler constant. Note that this analytical expression is an approximation valid in the long-wavelength portion of the optical spectrum where $-\varepsilon_m(\omega) \gg \varepsilon_d$.

Using this function in Eq. 47, we have obtained the spectral dependence of minimum angle θ_{min} for which the adiabatic increase of the SPP field intensity in the direction of the cone tip will take place. For smaller angles, the linear losses in the metal will overcome the adiabatic concentration

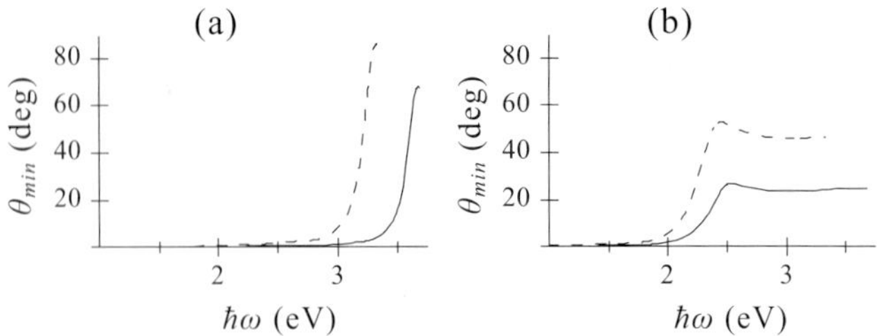

Fig. 5. The same as in Fig. 4 but for nanocone as a SPP waveguide, for TM_0 mode. Angle θ in this case is an angle between the cone axis and its surface.

and the SPP intensity will monotoneously decrease along the cone. Such dependences for silver and gold are displayed in Fig. 5.

First, we note that these dependences for the nanocone are remarkably similar, practically identical to the corresponding dependences for metal nanowedges — cf. Fig. 4. This is despite the fact that the underlying coefficients k_a, which solely define these spectral dependences, are utterly different by form — cf. Eqs. (48) and (49). Take into account, however, that angle θ_{min} in Eq. (48) is the full angle at the tip of the wedge, while in Eq. (49) it is the angle between the cone axis and its surface, i.e., the half of the full constituent angle at the tip of the cone.

Returning to Fig. 5, for the case of the cone, the same as for the wedge, the silver provides much better a system where the plasmonic concentration can be expected for smaller angles in a wider spectral range. For gold, the required minimum angle increases dramatically toward the 2.2 eV edge of the *ds* band absorption, which dramatically limits its usefulness. Note that large required angles contradict to the adiabaticity requirement $\tan\theta \ll 1$.

In conclusion of this Section, we discuss the limits of the adiabatic energy concentration in SPP waveguides. The optical losses, as we discussed above is the most important limiting factor within the macroscopic local electrodynamics that limits the efficiency of the energy delivery to the nanoscale. It also limits the minimum angle for which the field enhancement at the tip of the waveguide occurs and can make the adiabatic regime unattainable whatsoever — see the discussion of Eqs. (48) and (49). Important limiting effects that become important at the small scale are the spatial dispersion of the dielectric responses and Landau damping. These effects are described by spatially dispersive, nonlocal dielectric function. Though there has been some work done on the nonlocal effects in the adiabatic SPP concentration,[53] this area is still mostly unexplored.

On a qualitative level, the nonlocality of the dielectric responses and Landau damping become important when the localization size of the SPPs becomes on order of the correlation length associated with the Fermi velocity v_f of electrons. Because in the adiabatic following regime, this localization length is on order of local size d of the waveguide, then the condition that these effects are important is

$$d \sim k^{-1} \lesssim v_f/\omega \ . \tag{50}$$

In practical terms, in the optical region the effects of Landau damping and dielectric nonlocality become important for d of order of a few nanometers. This critical size increases when the frequency decreases toward the red and infrared spectral region.

There are also other phenomena limiting the adiabatic energy concentration on the nanoscale, which become important when the size of the waveguide d becomes very small, that can only be described in a fully microscopic theory. These are electron spill out and the related underscreening. These effects have recently been considered for the problem of the surface enhanced Raman scattering (SERS) in the framework of the local density approximation (LDA) of the density functional theory (DFT).[54–58] The application of this or other microscopic methods to the problem of the adiabatic transformation and concentration of the optical energy on the nanoscale is still ahead.

2.6. *Adiabatic SPP stopping in metal wedge*

In this Section we consider the adiabatic transformation and concentration of energy in a metal wedge embedded in dielectric media. We start with a symmetric problem: the embedding dielectric is uniform. In this case, the SPP modes are characterized by parity. Inspection of the corresponding dispersion relations for the even (symmetric) and odd (antisymmetric) SPPs (30) and (32) shows that it is the antysymmetric mode that has velocity that decreases when the layer becomes thin, which is the prerequisite for the adiabatic following by the SPPs of the layer thickness.

Correspondingly, the starting point of our consideration is the characteristic equation for the antisymmetric SPPs (32). It provides the antisymmetric SPP dispersion relation as a complex function $k(\omega, d)$ of frequency and the metal layer thickness. As an example, we show in Fig. 6 (a) the dependence of the wavevector on the layer thickness for a silver nanolayer in vacuum. Note that the ratio $n = k/k_0$ that is shown in the graphs is

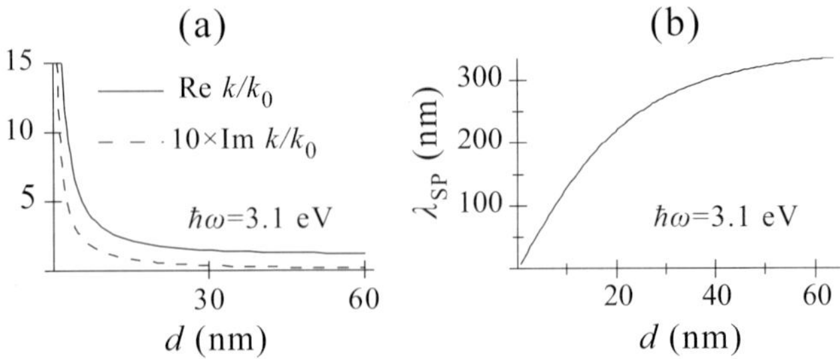

Fig. 6. (a) For a silver nanofilm in vacuum, dependence of the real part of the SPP effective refraction index, $\mathrm{Re}\, k/k_0$, on the layer thickness d (solid line). The same dependence for the imaginary part of the SPP index, $\mathrm{Im}\, k/k_0$, is displayed by the dashed line multiplied by a factor of 10. (b) Dependence of the SPP wavelength, $\lambda_{SP} = 2\pi/\mathrm{Re}\, k$. The dependences are calculated for the antisymmetric mode of frequency $\omega = 3.1$ eV.

the effective refraction index for the antisymmteric SPPs. As we see from this figure, both the real and imaginary part of this effective index increas when the layer thickness is in nanometer range and decreases. As we know from the general scaling theory of the adiabatic following presented above in Sec. 2.5, this dependence is $n \propto 1/a$ — see Eq. (39). This scaling should take place for the local wedge thickness less or on order of the skin depth in the metal, $d \lesssim l_s \sim 20$ nm. Such a behavior is actually in a very good agreement with the curves in Fig. 6.

The imaginary part of the wavevector, as we see from Fig. 6 (a) is small relative to its real part, $\mathrm{Im}\, k/\mathrm{Re} k \lesssim \frac{1}{20} - \frac{1}{30}$. Thus the antisymmetric SPPs remain good propagating waves for even for very thin nanolayers. This relatively weak decay and their high index makes them useful for ultramicroscopy. The resolution (both in-plane and out-of-plane) that can be obtain with such modes is determined by their wavelength λ_{SP} displayed in Fig. 6 (b). For instance, for the wedge tapered down to 4 nm, $\lambda_{SP} \approx 30$ nm, which is in agreement with the scaling of their reduced wave length $\lambda_{SP} \sim d/k_a \sim d$. The corresponding resolution is ~ 15 nm. It is sometimes called an "X-ray resolution at optical frequencies".

From the dispersion relation, one can calculated the phase velocity and group velocity of the SPP. They are shown in Fig. 7 (a) as function of the nanolayer thickness. The phase velocity for relatively thick nanolayers ($d \ll l_s$) is close to speed of light, while the group velocity is significantly less than that. For thin nanolayers ($d \lesssim l_s$), both these velocities tend to zero proportionally d, in accord with the general scaling (40). Thus, the adiabatic following of the nanolayer by the SPPs provides an

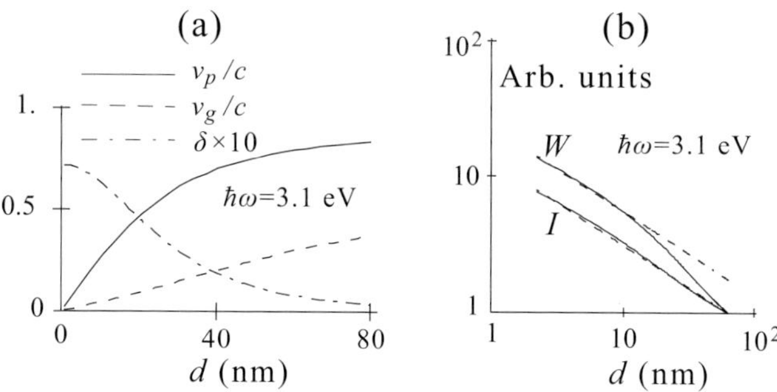

Fig. 7. (a) For silver wedge in vacuum, displayed as functions of wedge local thickness d plotted are: phase velocity v_p (solid curve), group velocity v_g (dashed curve), and adiabatic parameter δ multiplied by a factor of 10 (dash-dot curve); (b) Dependences of intensity I on the vacuum side of the silver surface and energy density W on its metal side as a function of the wedge local thickness d. The results of WKB theory are shown by solid lines, and those of the scaling approximation are denoted by the dashed lines. Note the double logarithmic scale.

efficient (with relatively low losses) wide-band approach to obtaining the "slow light".

The adiabatic parameter δ (multiplied for the sake of the readability by a factor of 10) is also shown in Fig. 7 (a). It is finite in the entire region of propagation, including the vicinity of the stopping edge, in acord with the scaling law (45). This examples uses grading of the wage $d' = -1/30$. For this value of grading, the WKB parameter is small even at the edge, $\delta \leq 0.07$. Correspondingly, the WKB approximation is valid for the entire system.

Now let us consider the fields. The WKB solutions can be obtained from the exact expressions by multiplying them by the corresponding exponential of the eikonal, $\exp(i\phi)$ and a pre-exponential $\propto J^{-1/2}$, where J is the energy flux that allows one to satisfy the energy conservation condition. We give below the corresponding expressions explicitly in the general case, without assuming that the system is symmetric. For layer 1, which is the semiinfinite ($z < 0$) dielectric underlying the metal wedge, we have

$$\begin{aligned} H_x^{(1)} &= \frac{1}{\sqrt{J}} A_1 \exp\left[\kappa_1 z + i\phi(y)\right] , \\ E_z^{(1)} &= \frac{1}{\sqrt{J}} \frac{k}{k_0\varepsilon_1} A_1 \exp\left[\kappa_1 z + i\phi(y)\right] , \\ E_y^{(1)} &= \frac{1}{\sqrt{J}} \frac{\kappa_1}{k_0\varepsilon_1} A_1 \exp\left[\kappa_1 z + i\phi(y)\right] , \end{aligned} \qquad (51)$$

where eikonal

$$\phi(y) = \int k(d(y))\mathrm{d}y \ , \tag{52}$$

and $k = k(d)$ is a function of the local thickness d, and so is $\kappa_1 = \kappa_1(d)$ as given by the general relation (18), where, by definition, $k_0 = \omega/c$.

For layer 2, i.e., the metal wedge $d(y) \geq z \geq 0$, the solutions are

$$\begin{aligned} H_x^{(2)} &= \frac{1}{\sqrt{J}}\left[A_2 \exp\left(\kappa_2 z\right) + B_2 \exp\left(-\kappa_2 z\right)\right] \exp\left[i\phi(y)\right] \ , \\ E_z^{(2)} &= \frac{1}{\sqrt{J}}\frac{k}{k_0\varepsilon_2}\left[A_2 \exp\left(\kappa_2 z\right) + B_2 \exp\left(-\kappa_2 z\right)\right] \exp\left[i\phi(y)\right] \ , \\ E_y^{(2)} &= \frac{1}{\sqrt{J}}\frac{\kappa_2}{k_0\varepsilon_2}\left[A_2 \exp\left(\kappa_2 z\right) - B_2 \exp\left(-\kappa_2 z\right)\right] \exp\left[i\phi(y)\right] \ . \end{aligned} \tag{53}$$

Finally, the fields for the semi-infinite upper dielectric layer 3 ($z > d(y)$) are

$$\begin{aligned} H_x^{(3)} &= \frac{1}{\sqrt{J}} B_3 \exp\left[-\kappa_2 z + i\phi(y)\right] \ , \\ E_z^{(3)} &= \frac{1}{\sqrt{J}}\frac{k}{k_0\varepsilon_1} B_3 \exp\left[-\kappa_3 z + i\phi(y)\right] \ , \\ E_y^{(3)} &= \frac{1}{\sqrt{J}}\frac{\kappa_3}{k_0\varepsilon_3} B_3 \exp\left[-\kappa_3 z + i\phi(y)\right] \ , \end{aligned} \tag{54}$$

The coefficients of these equations can be found in general case, but they are cumbersome and we do not display them.

The WKB recipe to find the pre-exponential $1/\sqrt{J}$ is the following. At the first stage one neglects the imaginary part of the eikonal. Fields (51)-(54) should then be used to compute the energy density (16). Finally, energy flux J, which determines the pre-exponential, is computed as

$$J(y) = v_g \int_{-\infty}^{\infty} W(y,z)\mathrm{d}z \ . \tag{55}$$

As everywhere, $v_g = \partial\omega/\partial k$ is the group velocity [see a numerical illustration in Fig. 7 (a)], which depends on the local thickness $d = d(y)$.

As a result of elementary but bulky computations, we obtain

$$J = v_g \left\{ \frac{k_0 |A_1|^2}{2\mathrm{Re}\,\kappa_1} \left(1 + \frac{|k|^2 + |\kappa_1|^2}{k_0^2 \varepsilon_1} \right) \right.$$
$$+ \frac{k_0}{2\mathrm{Re}\,\kappa_2} \left[1 + \mathrm{Re} \left(\frac{\partial \omega \varepsilon_2(\omega)}{\partial \omega} \right) \frac{|k|^2 + |\kappa_2|^2}{k_0^2 \varepsilon_2} \right]$$
$$\times \left\{ |A_2|^2 \left[\exp\left(2\mathrm{Re}\,\kappa_2 d \right) - 1 \right] + B_2|^2 \left[1 - \exp\left(-2\mathrm{Re}\,\kappa_2 d \right) \right] \right\}$$
$$+ \frac{k_0}{\mathrm{Im}\,\kappa_2} \mathrm{Re} \left(\frac{\partial \omega \varepsilon_2(\omega)}{\partial \omega} \right) \frac{|k|^2 - |\kappa_2|^2}{|k_0^2 \varepsilon_2|^2} \mathrm{Im} \left\{ AB^* \left[\exp\left(2i\mathrm{Im}\,\kappa_2 d \right) - 1 \right] \right\}$$
$$\left. + \frac{k_0 |B_3|^2}{2\mathrm{Re}\,\kappa_3} \left(1 + \frac{|k|^2 + |\kappa_3|^2}{k_0^2 \varepsilon_3} \right) \left[1 - \exp\left(-2\mathrm{Re}\,\kappa_3 d \right) \right] \right\} . \tag{56}$$

Substituting this into the expressions for fields (51)-(54) and computing eikonal (52) by numerical integration, we obtain the required WKB solution. Note that the fields (51)-(54) should be multiplied by a common normalization multiplier that defines the energy of a specific solution.

Returning back to Fig. 7 (b), we compare the local field intensity I and energy density W of an SPP mode obtained from this solution with the scaling prediction (44) for a silver nanowedge with grading $d' = -1/30$, which is well over the critical value needed for the adiabatic concentration to take place. The only adjustable parameter in such a comparison is the mode normalization coefficient mentioned above in the previous paragraph. Note that these quantities are displayed as functions of the local metal wedge thickness d. The scaling agreement is very good for W in the asymptotic region, i.e. for d significantly less than skin depth $l_s \approx 25$ nm of the metal, as expected. What is surprising, the local intensity on the outside surface of the metal scales almost perfectly in the entire region.

The distributions of the local electric fields in the plane normal to wedge surface in the direction of propagation is shown in Fig. 8 for a silver nanowedge with $d' = -1/30$. These distributions demonstrate well the general qualitative properties of the adiabatic transformation and concentration of energy on the nanoscale. For the wedge thickness large enough, $d \gg l_s$, which is the case in the left-hand side of the panels, the fields are delocalized outside the metal at distances greater than the wavelength in vacuum; the normal field, E_z, inside the metal is weak. This behavior is similar to skin effect in a semi-infinite metal. In this region, the SPP wavelength in the direction of propagation is smaller than the vacuum wavelength but still is on the microscale rather than on the nanoscale. As

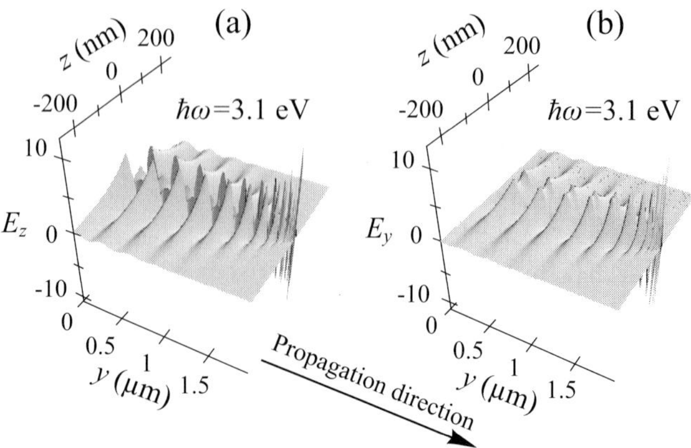

Fig. 8. Local electric fields of SPP antisymmetric mode with frequency $\omega = 3.1$ eV where panel (a) displays normal field E_z, and panel (b) shows the longitudinal field E_y. The wedge material is silver, and the embedding medium is vacuum. The wedge's slope tangent is $\tan\theta = 1/30$.

the SPPs propagate toward the edge (to the right in the figure), the local thickness of the wedge becomes comparable with the skin depth, and then the adiabatic following sets on. Both the wavelength in the propagation direction and the evanescent length at the normal to the surface become on order of the local wedge thickness. This nanolocalization is accompanied by a sharp increase in the amplitude of the fields as the SPP approach the edge in the right-hand side of the panels. Also, in this region, the skin effect becomes inefficient: the entire thickness of the metal is filled with the fields.

Now let us turn to the spatial distribution of the local electric field intensity defined here as

$$I(y,z) = |E_z(y,z)|^2 + |E_y(y,z)|^2 \,. \tag{57}$$

This distribution in the plane of propagation normal to the interfaces (the yz plane) is displayed in Fig. 9 (a). Here we can clearly see that at the widest part of the wedge, where its thickness significantly exceeds the skin depth, the fields are delocalized in the normal plane at micrometer-scale distances. The thin skin layers at the surfaces and low field area in the middle of the wedge are clearly evident. As the SPPs propagate toward the edge (to the right in the figure), their fields become more localized at teh surfaces and start to penetrate the bulk of the metal. When the adiabatic following is fully developed (for $y \gtrsim 1.5\mu$m), the entire cross section of the

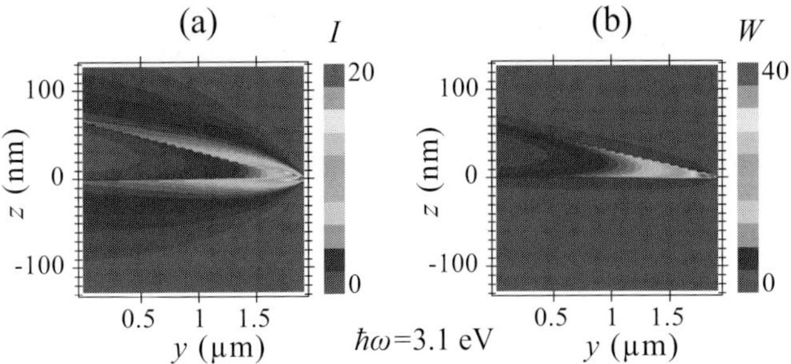

Fig. 9. For antisymmetric SPP mode of silver wedge in vacuum, distributions in the normal plane of propagation (yz) of local field intensity I (a) and energy density W (b). Frequency is $\omega = 3.1$ eV, wedge material is silver. Color coding bars are shown to the right of the plots along with the corresponding scales.

metal is uniformly filled with the fields and their localization length outside of the metal is comparable with the local thickness of the wedge. This is in the full qualitative agreement with the picture produced by the scaling theory — see Sect. 2.5. In the process of propagation toward the edge, the intensity at the outer surface of the metal (where it is maximum) is increased by a factor of ~ 10 [corresponding to the change of the coding (false) color in the plot from light blue to intense red].

Consider the adiabatic concentration of energy density $W(y, z)$ (16) illustrated in Fig. 9 (b). In contrast to the intensity that is mostly localized in the dielctric component (vacuum in this case) [cf. panel (a)], the energy density is much more concentrated in the metal due to the high magnitude of the metal dielectric permittivity. When the SPPs propagate toward the sharp wedge, the metal wedge becomes uniformly filled with the electromagnetic energy density. Toward the wedge, within the range of the plot, the energy density increases by more than a factor of 10. Thus the metal wedge is an efficient conductor and adiabatic concentrator of the SPP energy.

It is of interest to evaluate qualitatively what is the fraction of the electromagnetic energy flux propagating in the metal (to be denoted below as J_m) with respect to that propagating in the dielectric (to be denoted as J_d). It is most interesting and also easiest to do in the quasistatic regime (for this specific plot, such a regime is established for $y \gtrsim 1\ \mu$m).

In the quasistatic regime, the field coefficients can be found in a simplified form. From these, we immediately obtain expressions for the electric fields (the magnetic field in the quasistatic regime is small by parameter

k_0/k and can be neglected). For the embedding dielectric, we obtain

$$E_z(y,z) = E_y(y,z) = \pm\varepsilon_m \exp\left[-k|z| + i\phi(y)\right] , \tag{58}$$

where signs $\pm$ refer to the upper and lower half spaces. For the metal wedge, we obtain

$$\begin{aligned} E_z(y,z) &= (\varepsilon_d + \varepsilon_m)\sinh(kz)\exp\left[i\phi(y)\right] , \\ E_y(y,z) &= (\varepsilon_d + \varepsilon_m)\cosh(kz)\exp\left[i\phi(y)\right] . \end{aligned} \tag{59}$$

Reminding again, there is an arbitrary normalization coefficient that these eigenmode fields should be multiplied by to obtain observable, physical fields, which defined the total phase and intensity of the SPP.

With these fields, it is easy to find the ratio of the electromagnetic energy fluxes in the metal and dielectric

$$\frac{J_m}{J_d} = \frac{\int_0^{d/2} W(y,z)\mathrm{d}z}{\int_0^{\infty} W(y,z)\mathrm{d}z} . \tag{60}$$

Computing this ratio, one can safely neglect a small imaginary part of the metal dielectric function. We obtain a simple, closed expression

$$\frac{J_m}{J_d} = -\frac{1}{\varepsilon_m}\frac{\partial\left[\omega\varepsilon(\omega)\right]}{\partial\omega}\frac{(\varepsilon_d + \varepsilon_m)^2}{(\varepsilon_d - \varepsilon_m)^2} . \tag{61}$$

Note that this ratio does not depend on local thickness d of the wedge, as expected in the asymptotic quasielectrostatic regime. Numerical evaluation of this ratio for the present system shows that in the entire plasmonic range, between 1 and 3 eV, this ratio varies very little, just between 0.9 and 1.6. Thus, the electromagnetic (or, rather, quasielectrostatic) energy is split approximately equally between the metal and dielectric components of the plasmonic waveguide. Note that typically in the plasmonic region $|\varepsilon_m| \gg \varepsilon_d$. In such a case, Eq. (61) further simplifies

$$\frac{J_m}{J_d} = -\frac{1}{\varepsilon_m}\frac{\partial\left[\omega\varepsilon(\omega)\right]}{\partial\omega} . \tag{62}$$

This expression does not depend on the dielectric at all, and it is numerically very close to 1 for silver.

Returning to Fig. 9 (b), we note that the graphics renditions visually overestimates the fraction of the optical energy propagating in the metal, which appears to be significantly greater than 1. This is due to the fact that the dielectric fields are spread over the two (upper and lower) semi-infinite dielectric regions, and also in both of these regions they are delocalized somewhat greater than the fields in the metal. In teh actuality, as we know from Eqs. (61) and (62), the SPP optical energy in the quasielectrostatic regime propagates in approximately equally fractions in the metal and dielectric constituents of the waveguide.

Finally, it is of significant interest to consider an asymmetric SPP waveguide constituted by a metal wedge surrounded on both sides by different dielectrics. In particular, this is a case for the metal wedge on a dielectric substrate surrounded by vacuum.

Qualitatively, the underlying physics of the SPP adiabatic energy transformation and concentration is somewhat different from the symmetric case. In the present case, for a thick ($d \gg l_s$) portion of the wedge, the SPP propagating at the lower and upper interfaces for the same ω have different wave vectors and very weakly interact and hybridize with each other (i.e., it is a case of weak coupling). The situation changes for $d \lesssim l_s$ when the coupling and mixing of the SPPs at both the interfaces become strong. In this case, we select the high-k mode that is asymptotically (for a very strong coupling) antisymmetric, slow SPP branch. This one provides the best adiabatic control and concentration of energy.

The distribution of the local field intensity for such a slow SPP mode in the normal plane of propagation (the yz plane) is illustrated in Fig. 10 (a) for $\varepsilon_1 = 3$ (a typical glass), $\varepsilon_3 = 1$ (vacuum), and silver as the metal. As we see from the figure, the mode selected is initially (in the left-hand side of the figure) localized predominantly at the lower (metal/dielectric) interface and is mostly a SPP at the interface of the semi-infinite silver and the dielectric (glass). As it propagates to the right (toward the edge) it does become much more concentrated in space with the localization length on order of a few nanometers. However, this localization is not accompanied by an enhancement in its intensity. In fact, this intensity decays by a factor of ~ 10 due to the losses. This is due to the much smaller component of the adiabatically-controlled odd mode in this SPP because of the much weaker coupling between the two interfaces.

The behavior of the optical energy density in this system is illustrated in Fig. 10 (b). In contrast to panel (a), the energy is much more concentrated in the metal, similar to what takes place for the symmetric wedge waveguide discussed above in conjunction with Fig. 9. The energy density actually decays significantly less than the local field intensity due to its better localization: along the propagation pass it is attenuated only by a factor of ≈ 3. Thus even an asymmetric SPP waveguide is an efficient adiabatic concentrator of optical energy.

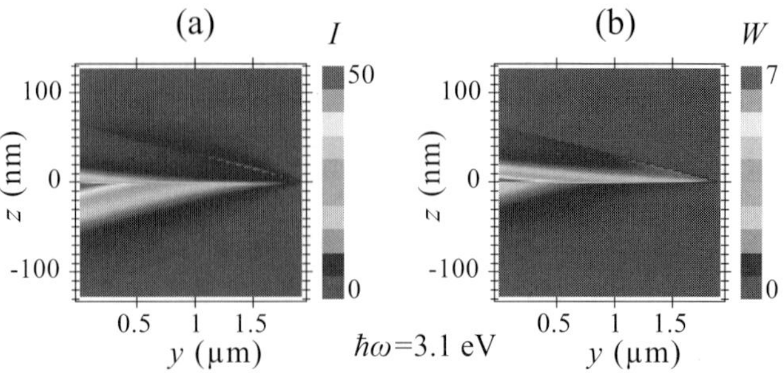

Fig. 10. Same as in Fig. 9 but for an asymmetric nanowedge SPP waveguide: $\varepsilon_1 = 3$, and $\varepsilon_3 = 1$.

2.7. *Adiabatic concentration and stopping of SPPs in tapered nanoplasmonic cylinder*

Consider a nanoplasmonic waveguide that consists of a metal nanowire whose axis coincides with the coordinate z axis and whose dielectric function $\varepsilon_m(\omega)$ is uniform in space, where ω is the optical excitation frequency. The radius $R(z)$ of this nanowire is a smooth function of z and is assumed to decrease from microscale for z large negative to a nanoscale size at $z \to 0$, as discussed above, see Fig. 11(a). This wire is surrounded by a dielectric medium with dielectric constant ε_d. Using the smoothness of dependence $R(z)$, we will employ the eikonal approximation[48] also called Wentzel-Kramers-Brillouin (WKB) or quasiclassical approximation in quantum mechanics.[49]

We consider an axially uniform SPP mode that is a TM wave whose magnetic field has the ϕ polarization, and electric field $\mathbf{E}$ has both transverse (radial) component E_r and longitudinal component E_z. In the eikonal (WKB) approximation, this field has the form

$$\mathbf{E}(\mathbf{r}, z, t) = \mathbf{E}_0(\mathbf{r}) A(z) \exp[ik_0\varphi(\mathbf{r}) - i\omega t] \tag{63}$$

where $\mathbf{r}$ is a two-dimensional (2D) vector in the xy plane, $k_0 = 1/\lambda$, and $A(z)$ is a slow-varying pre-exponential factor, to be determined later in this Section. From the Maxwell equations, using the corresponding boundary conditions at the interface, for the SPP guided mode, we find the eikonal as $\varphi = k_0 \int n(z) dz$, where $n(z)$ is the effective surface index of the plasmonic waveguide at a point z, which is determined by the equation

$$\frac{\varepsilon_m}{\kappa_m} \frac{I_1(k_0\kappa_m R)}{I_0(k_0\kappa_m R)} + \frac{\varepsilon_d}{\kappa_d} \frac{K_1(k_0\kappa_d R)}{K_0(k_0\kappa_d R)} = 0\ , \tag{64}$$

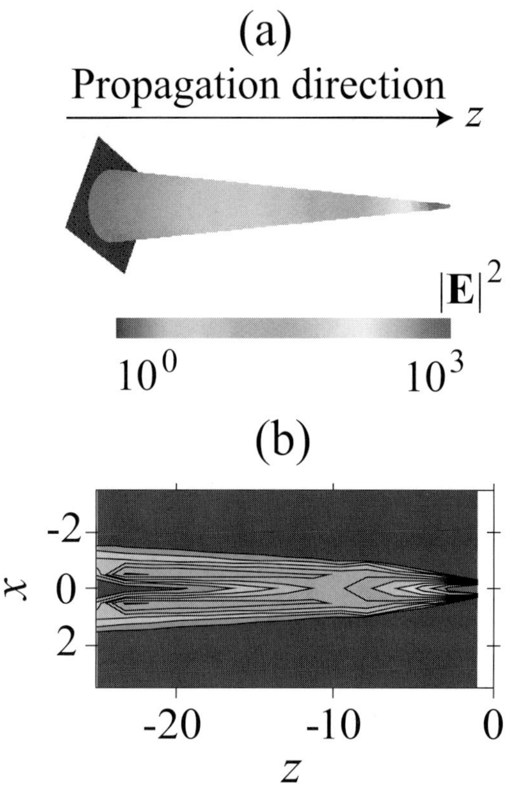

Fig. 11. (a) Geometry of the nanoplasmonic waveguide. The propagation direction of the SPP's is indicated by the arrow. Intensity $I(\mathbf{r}) = |\mathbf{E}(\mathbf{r})|^2$ of the local fields relative to the excitation field is shown by color. The scale of the intensities is indicated by the color bar in the center. (b) Local electric field intensity $I(\mathbf{r})$ is shown in the longitudinal cross section of the system. The coordinates are indicated in the units of the reduced radiation wavelength in vacuum, $\lambda = 100$ nm. The radius of the waveguide gradually decreases from 50 to 2 nm.

where I_p and K_p ($p = 0, 1$) are the modified Bessel functions; the complex decrements of the field in the metal and dielectric are: $\kappa_m = \sqrt{n^2 - \varepsilon_m}$ and $\kappa_d = \sqrt{n^2 - \varepsilon_d}$. This equation determines n as a function of the local wire radius R, which together with the grading dependence $R = R(z)$ defines the required effective index $n(z)$. Under the conventional plasmonic condition $\mathrm{Re}\varepsilon_m < -\varepsilon_d$, Eq. (64) has nearly real solutions corresponding to the propagating SPP's. For a thick wire ($k_0 R \gg 1$), the solution is, understandably, the same as for the flat surface, $n = \sqrt{\varepsilon_m \varepsilon_d / (\varepsilon_m + \varepsilon_d)}$. For a thin, nanoscale-radius wire ($k_0 R \ll 1$) with logarithmic precision, we have

$$n(R) \approx \frac{1}{k_0 R}\sqrt{-\frac{2\varepsilon_d}{\varepsilon_m}}\left[\ln\sqrt{-\frac{4\varepsilon_m}{\varepsilon_d}}-\gamma\right]^{-1}, \tag{65}$$

where $\gamma \approx 0.57721$ is the Euler constant. Note that at the tip $n \to \infty$, and SPP's do asymptotically stop, i.e., both the phase velocity $v_p = c/n$ and group velocity $v_g = c[d(n\omega)/d\omega]^{-1}$ tend to zero $\propto k_0 R$ for $k_0 R \to 0$. The point $R = 0$ (or $z = 0$) is an essential singularity. The time to reach this point $\propto \int n(R)dR \propto -\ln(k_0 R) \to \infty$ diverges logarithmically.

The eikonal parameter (also called WKB or adiabatic parameter) is defined as $\delta = \left|R' d(k_0 n)^{-1}/dR\right|$, where $R' = dR/dz$ is the wire grading. For the applicability of the eikonal (WKB) approximation, it necessary and sufficient that $\delta \ll 1$. At the nanoscale tip of the wire, which is the critical site for the adiabaticity (eikonal approximation applicability), from Eq. (65) we obtain

$$\delta \approx \left|R'\sqrt{-\frac{\varepsilon_m}{2\varepsilon_d}}\left[\ln\sqrt{-\frac{4\varepsilon_m}{\varepsilon_d}}-\gamma\right]\right|.$$

Thus, δ stays finite at the tip and can be made small enough by choosing sufficiently small grading R', so *the eikonal approximation is valid for the entire wire waveguide, including the stopping point at the tip.* This conclusion does not rely or significantly depend on $\mathrm{Im}\varepsilon_m$ (the optical losses in the system).

In our example, the tapered nanoplasmonic waveguide is a silver cone in vacuum with $R' = -0.02$; its angle of opening is 0.04 radian, as shown in Fig. 11(a). The vacuum reduced wavelength of the excitation radiation is $\lambda = 100$ nm, which corresponds to red light of $\lambda = 630$ nm. The SPPs are efficiently excited at the wide end of the waveguide nanowire by using, e.g., grating or Kretschmann[59,60] geometry and propagate to the tip as indicated. This propagation causes accumulation of the SPP energy at the tip and the corresponding increase of the local fields by more than three orders of magnitude. As shown in Fig. 11 (b), the intensity of the local optical field is sharply concentrated in 3D in a nanolayer at the surface of the metal, which is a signature of SPPs. In this figure, as everywhere in this Section, we show all lengths in units of λ, so the sizes range from the micro- to nanoscale. The hot spot of local fields is created in a nanosize region at the very tip. If it were not a plasmonic nanowire waveguide, but the conventional tapered optical fiber supporting guided photonic modes, than there would be a cut-off at some waveguide radius beyond which the propagation is not possible:[61] the wave is reflected back with only a short evanescent tail in the forward direction; no field enhancement would occur, and only an exponentially small part of the incident energy would reach the tip.

It was suggested earlier[62] that the propagation of SPPs toward the tip can produce energy concentration. It was also noted[63] that at the cut-off point, the guided photonic modes of an optical fiber could couple to the plasmonic modes of its metal coating causing further transfer of optical energy to the tip. However, no role of the adiabatic slowing-down and stopping of SPPs was previously elucidated. It is feasible that the observed[41] high efficiency of a metal tip on aperture probe is due to the proposed effect of the adiabatic accumulation.

The physical reason that the nanoplasmonic waveguide is an efficient energy concentrator can be inferred from Fig. 12(a). Both the phase and group velocity of SPPs asymptotically tend to zero toward the nanotip. Consequently, the SPPs are slowed down and adiabatically stopped at $z \to 0$, which leads to their accumulation at the tip. Correspondingly, in Fig. 12 (b) the local optical field is oscillating in space with progressively decreasing wavelength and its amplitude increasing by more than an order of magnitude. The highest enhancement is in fact limited only by the minimum tip size that can be considered on the basis of continuous electrodynamics. Importantly, being adiabatic to prevent the back reflection and 3D scattering, this process should be as rapid as possible to prevent losses in the metal.

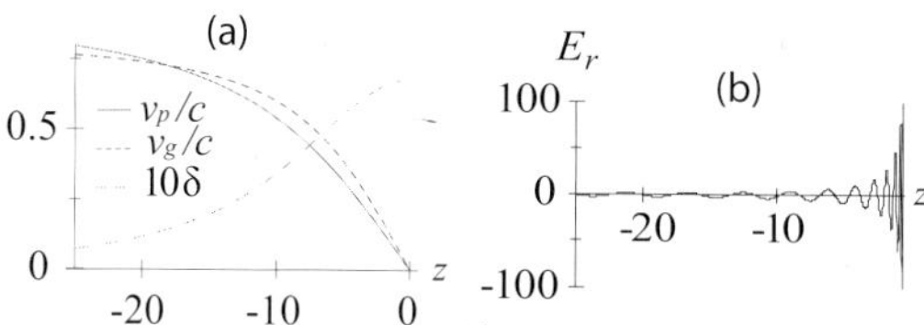

Fig. 12. (a) Phase velocity v_p, group velocity v_g, and adiabatic parameter δ (scaled by a factor of 10) are shown as functions of the plasmonic waveguide radius. (b) Radial optical electric field at the surface of the metal nanowire waveguide in the units of the excitation field against the coordinate (in the propagation direction).

In Fig. 12 (a), along with the phase velocity v_p and group velocity v_g, we show also the adiabatic (WKB) parameter δ. It is of principal importance that this adiabatic parameter does stay finite and small, $\delta \leq 0.07$, throughout the entire system, ensuring the global applicability of the eikonal approximation, including the essentially singular point at $z = 0$.

The SPP electric fields are found from the Maxwell equations in eikonal (WKB) approximation in the form:

$$E_z(r,z) = \theta(R-r)I_0(k_0\kappa_m r) + \theta(r-R)BK_0(k_0\kappa_d r) ,$$
$$E_z(r,z) = \theta(R-r)i\frac{n}{\kappa_m}I_1(k_0\kappa_m r) + \theta(r-R)i\frac{n}{\kappa_d}BK_1(k_0\kappa_d r) , \qquad (66)$$

where $B = I_0(k_0\kappa_m R)/K_0(k_0\kappa_m R)$, and $\theta(\dots)$ denotes the Heaviside θ-function. To determine the pre-exponential $A(z)$ in Eq. (63), we use the energy flux conservation in terms of the Pointing vector integrated over the normal (xy) plane, obtaining

$$A \propto \mathrm{Re}\left[\frac{n^*\varepsilon_m^*}{|\kappa_m|^2}|K_0(k_0\kappa_d R)|^2\int_0^R |I_1(k_0\kappa_m r)|^2\, rdr + \frac{n^*\varepsilon_d^*}{|\kappa_d|^2}|I_0(k_0\kappa_m R)|^2\int_R^\infty |K_1(k_0\kappa_d r)|^2\, rdr\right]^{-\frac{1}{2}} , \qquad (67)$$

where all the spatially-varying quantities, n, κ_m, and κ_d, are functions of local radius R of the wire, as originally given by Eq. (64). The required dependence $A = A(z)$ is obtained by substituting the grading relation $R = R(z)$. We indicate only the proportionality of A: the total scale of A is undetermined by the equations and is defined by the total power of the propagating SPP wave. This completes the eikonal (WKB) solution.

The intensity of the optical electric fields has already been discussed in conjunction with Fig. 11 where they are shown on the logarithmic scale. For this example and below, we set the minimum radius of the wire to be $R_{min} = 0.02\lambda = 2$ nm to avoid effects of the spatial dispersion of the dielectric response that are important at shorter distances, cf. Refs.,[64,65] and the maximum radius (at $z = -25\lambda = -2.5\ \mu$) to be $R_{max} = 0.5\lambda = 50$ nm.

In Fig. 13, we display the amplitudes of the local optical fields in the cross section of the system for the normal and longitudinal (with respect to the axis) components of the optical electric field. In Fig. 13, far from the tip of the nanoplasmonic waveguide, the optical electric field is mostly transverse, extending in vacuum to distances $\sim\lambda$ where most of the SPP field is propagating. The longitudinal field in the metal is very small, proportional to a factor of $|\varepsilon_d/\varepsilon_m| \ll 1$, as should be from the boundary conditions. Therefore, the guide itself is clearly seen in panel (a) as the acute triangular region of low fields. As SPPs move toward the tip, the SPP fields start

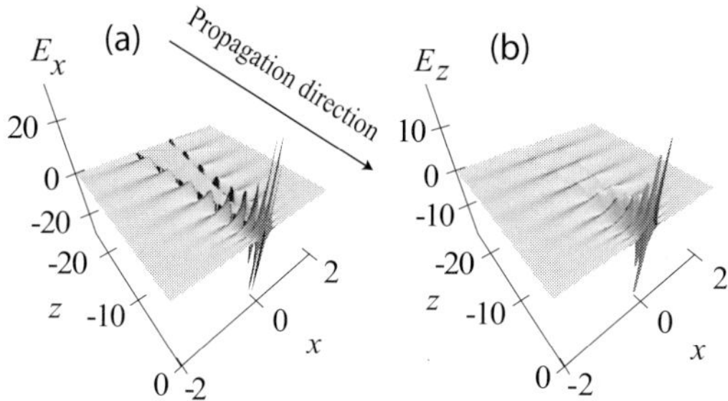

Fig. 13. Snapshot of instantaneous fields (at some arbitrary moment $t = 0$): Normal component E_x (a) and longitudinal component E_z (b) of the local optical electric field are shown in the longitudinal cross section (xz) plane of the system. The fields are in the units of the far-zone (excitation) field.

to localize at the metal surface, and simultaneously, their wavelength is progressively reducing and amplitude growing. Because the very tip is not included, the singularity point of the fields does not show in these figures. Even with this truncation, the field magnitudes grow significantly at small $|z|$. The transverse x component grows by an order of magnitude as the SPPs approach the tip of the guide, while the longitudinal z component, which is very small far from the tip, grows relatively much stronger. Close to the tip, both these components are on the same order of magnitude, as is expected for the localized excitations. This growth in magnitude is concurrent with the energy localization in 3D and the significant reduction of the wavelength, which are due to the dramatic slowing down of the SPPs. Note the SPPs in Fig. 13 are not standing but running waves; the fields shown represent an instantaneous snapshot of these waves.

In Fig. 14, we show the spatial behavior of relative intensity $I(\mathbf{r})$ and energy density[48] $W(\mathbf{r}) = \{d[\omega\varepsilon(\mathbf{r},\omega)]/d\omega\}\,|\mathbf{E}|^2$ of the local optical electric field. The intensity grows by more than three and energy density by four orders of magnitude at the tip. If these fields were used to induce SERS, it would be enhanced by seven to eight orders of magnitude. The further enhancement of SERS by many orders of magnitude can be achieved by positioning a resonant nanolens[66] at the tip of the nanoplasmonic waveguide. Note that $|\mathbf{E}|^2$ is concentrated at the outer surface of the metal nanowire, with the exponential decay far from this surface. In contrast, a metal nanoparticle excited by an external field would produce dipolar local fields where $|\mathbf{E}|^2 \propto r^{-6}$. At the same time, $I(\mathbf{r})$ is significantly larger and *localized inside the metal plasmonic waveguide where most of the SPP energy is propagating* due to the large value of $d[\omega\varepsilon(\mathbf{r},\omega)]/d\omega$ for metals.

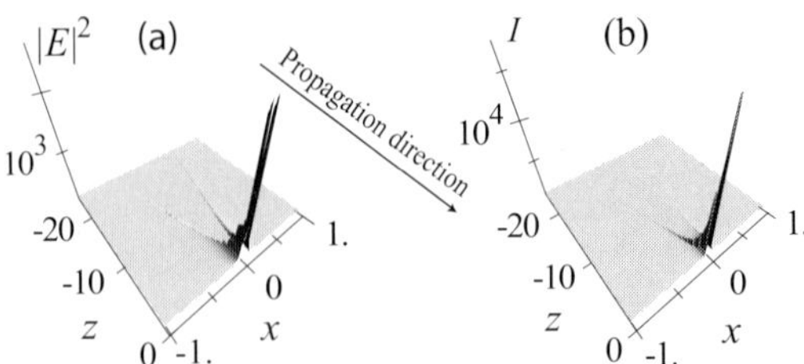

Fig. 14. Mean (time averaged) intensity $I(\mathbf{r})$ (a) and the energy density $W(\mathbf{r})$ (b) of the local optical electric field in the xy plane of the system. The magnitudes are relative to those of the excitation wave.

At the foundation of the above-described high-efficiency coupling of the far-field radiation to the near-field is the adiabatic slow-down of the running, propagating SPPs and their gradual conversion into standing, SP-like modes. The 3D energy concentration occurs at the tip of a smoothly tapered metal nanoplasmonic waveguide. This causes the local field increase by three orders of magnitude in intensity and four orders in energy density. The stopping of SPPs is asymptotic, i.e., they need logarithmically divergent time to reach the tip, which mathematically is the point of an essential singularity. Similar phenomena are likely to exist for a hollow tapered waveguides, in particular subwavelength holes. The rapid adiabatic nanofocusing promises to find various applications in nanooptics and nanotechnology where greatly enhanced local optical fields are required, in particular, for probing, spectroscopy, detection, and modification on the nanoscale in physics, chemistry, biology, electrical engineering, etc.

2.7.1. *Experimental Observations of Adiabatic SPP Concentration*

Following the publication of the original theoretical proposal,[10,44] there has been several experimental confirmations of this concept. The adiabatic concentration of SPPs manifested by enhanced local optical fields around a tip of the nanoplasmonic tapered planar wedge have been recently observed.[67,68] This energy concentration is accompanied by greatly enhanced nonlinear- optical effects. A very original method using upconverted far-field fluorescence of rear-earth ions have been used in these studies to visualize the *near-field* optical energy. The principal possibility of such a method is based on the fact the upconverted wavelength is much shorter than the fundamental wavelength in vacuum; thus far-field observation of the upconverted radiation allows one to see the near fields of the SPPs at the fundamental frequency. The observed phenomena are in a good qualitative agreement with the above theory. However, he quantitative comparison is difficult because of the planar geometry (metal wedge on a substrate) in the experiment and the cylindrical one in the theory.

An independent confirmation has been obtained in an experiment using a free-standing silver cone.[69] In this experiment, the SPPs are excited using a grating coupler cut into a side of the cylinder. A strongly enhanced emission has been observed in the far-field using a confocal optical microscope from the tip of the cone. It is likely that this emission is caused by outcoupling of light from the cone due to the some roughness of the silver surface. Very promising is combination of the adiabatic SPP concentration with the electron emission, which, similar to Ref. 70 will allow for the creation of nanoscale sources electron pulses. In combination with the ideas of the coherent control,[15,31,37] it is possible to make this nanoscale electron source also ultrafast, producing femtosecond electron pulses. Such sources can have a wide range of applications in science and engineering.

3. Nanoplasmonic active phased array (NAPAR)

3.1. *Introduction to spatio-temporal coherent control*

Two novel areas of optics have recently attracted a great deal of attention: nanooptics and ultrafast optics. One of the most rapidly developing directions in ultrafast optics is quantum control, in which coherent superpositions of quantum states are created by excitation radiation to control the quantum dynamics and outcomes.[22,25,71,72] Of special interest are coherently controlled ultrafast phenomena on the nanoscale where the phase of the excitation waveform along with its polarization provides a functional degree of freedom to control the nanoscale distribution of energy.[15,31–34,36,73] Spatiotemporal pulse shaping permits one to generate dynamically predefined waveforms modulated both in frequency and in space to focus ultrafast pulses in the required microscopic spatial and femtosecond temporal domains.[74,75]

In this Section, we theoretically describe a method of full coherent control on the nanoscale proposed in Ref. 37. In this method, a spatio-temporally modulated waveform is launched in a graded nanostructured system, specifically a wedge. Its propagation from the thick (macroscopic) to the thin (nanoscopic) edge of the wedge and the concurrent adiabatic concentration provide a possibility to focus the optical energy in *nanoscale* spatial and femtosecond temporal regions. This method unifies three approaches that individually have been developed and experimentally tested. The coupling of the external radiation to the surface plasmon polaritons (SPPs) propagating along the wedge occurs through an array of nanoobjects (nanoparticles or nanoholes) that is situated at the thick edge of the wedge. The phases of the SPPs emitted (scattered) by individual nanoobjects are determined by a spatio-temporal modulator. The nanofocusing of the SPPs occurs due to their propagation toward the nanofocus and the concurrent adiabatic concentration.

The coupling of the external radiation to SPPs and their nanofocusing have been observed – see, e.g., Refs. 76,77. The second component of our approach, the spatio-temporal coherent control of such nanofocusing has been developed.[74,75] The third component, the adiabatic concentration of SPPs also has been recently observed.[67,69] The idea of adiabatic concentration is described above in Sec. 2. It is based on adiabatic following by a propagating SPP wave of a graded plasmonic waveguide, where the phase and group velocities decrease while the propagating SPP wave is adiabatically transformed into a standing surface plasmon (SP) mode. A new quality that is present in our approach is a possibility to arbitrary move the nanofocus along the nanoedge of the wedge. Moreover, it is possible to superimpose any number of such nanofoci simultaneously and, consequently, create any distribution of the nanolocalized fields at the thin edge of the wedge.

The idea of the spatio-temporal coherent control is analogous to that of the synthetic aperture radar (SAR) and conceptually similar active phased array radar (APAR), or active electronically scanned array (AESA) radar widely used in military and civilian radar systems. This idea can be introduced using a schematic shown in Fig. 15. An APAR consists of active oscillators that act as dipole antennas (shown by bold short vertical segments). Each of these antennas generates a pulse of radiation whose phase is controlled elctronically. If the phases are equal, as in Fig. 15(a), the beam produced by the interference of the waves emitted by each of the antennas is straight with the wavefront parallel to the array base. If there is a linear phase shift, as in panel (b), the beam is steered toward the emitters with retarded phases. For a parabolic-type modulation of the phases along the base line with a maximum phase delay in the center, a focused beam is formed as in panel (c). And finally, a linear combination of the linear and focusing phase shifts leads to the simultaneous steering and focusing as shown in Fig. 15(d). Due to the superposition principle, any superposition of the beams can be created by the correspondingly superimposing

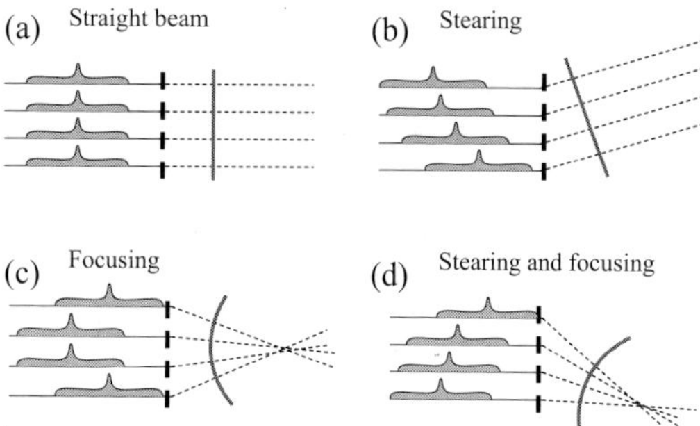

Fig. 15. Schematic of the Active Phased Array Radar (APAR). The short solid elements indicated the active dipoles (oscillators) of the antenna array. The dashed horizontal lines show the ray propagation in the far field of the array. The solid vertical red line shows the wavefront of the radar pulse. The individual pulses fed to (for a passive array) generated by the array oscillators (for an active array) are indicated by the waveforms shown above the corresponding rays. (a): Case of the straight beam (wavefront parallel to the array base) where there is no phase delay between the oscillators. (b) Beam steering: There is a phase delay between the oscillator pulses increasing linearly from bottom to the top. (c) Beam focusing: the phase delays are minimum at the center and maximum at the edges to form a spherical wavefront. (d) Beam staring and focusing: the phase delays are a combination of a linear shift and parabolic modulation.

the phase modulations, provided that the number of the active antennas is large enough.

The idea of this full coherent control is APAR transferred to the nanoplasmonics, which can be called Nanoplasmonic Active Phased Array "Radar" (NAPAR). This can be complemented with the adiabatic nanofocusing to further increase the concentration of the optical energy. To introduce it consider a schematic in Color Plate 1. As shown, a nanoplasmonic wedge contains a line of nano-size scatterers (shown as nanospheres) located at the thick edge and parallel to it, i.e. in the x direction in Color Plate 1(b). These scatters play the role if the emitting dipoles in NAPAR and can be either nanoparticles of various shapes and compositions, nanoholes, or elements of the surface roughness.

Consider first monochromatic light irradiating these nanoparticles or nanoholes that scatter and couple it into SPP wavelets, shown as red arc segments in Color Plate 1. Every such a scatterer emits SPPs in all directions; there is, of course, no favored directionality of the scattering. However, we assume that the excitation radiation and, correspondingly, the scattered wavelets of the SPP are coherent, and their phases smoothly vary in space along the thick edge, i.e., in the x direction. The SPP wavelets emit-

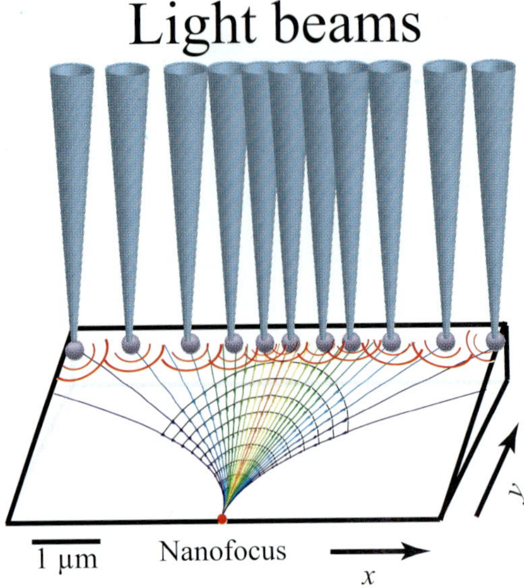

Color Plate 1. Schematic of the full coherent control on the nanoscale or nanoplasmonic active phased array radar (NAPAR). The system consists of plasmonic (silver) nanowedge with nanoparticles (silver nanospheres) positioned at its thick edge. Light beams focused on each of the nanospheres are shown by cones. The omni-directed SPP waves emitted by these excited nanospheres are shown by the red arc segments. The wavefront form by the interference is shown bu concave black curves. Trajectories of SPP rays propagating from the thick to sharp edge of the wedge are shown by color curves, with color coding the initial coordinates. The SPP frequency used in actual computations is $\hbar\omega = 2.5$ eV.

ted by different scatterers will interfere, which in accord with the Huygens-Fresnel principle leads to formation of a smooth wavefront of the SPP wave at some distance from the scatterers in the far SPP field. This distance, of course, must be much greater than the spacing of the scatterers.

Such wavefronts are shown in Color Plate 1(b) with concave black curves. The energy of the SPP is transferred along the rays, which are lines normal to the wavefronts, shown by the colored lines. By the appropriate spatial phase modulation of the excitation radiation along the line of scatterers over distances of many SPP wavelengths, these wavefronts can be formed in such a way that the rays intersect at a given point, forming a nanofocus at the thin (sharp) edge of the wedge, as shown schematically in Color Plate 1(b). Diffraction of the SPP waves will lead to a finite size of this focal spot that we will estimate later in this Section.

By changing the spatial phase profile of the excitation radiation, this focal spot can be arbitrarily moved along the thin edge. This focusing and adiabatic concentration, as the SPPs slow down approaching the sharp edge, will lead to the enhancement of the intensity of the optical fields in the focal region. This dynamically-controlled concentration of energy is a plasmonic counterpart of a large phased antenna array (also known as an aperture synthesis antenna or beamformer), widely used in radar technology and radio astronomy.[78] Now we can consider excitation by spatiotemporally shaped ultrashort pulses.[74,75] The field produced by them is a coherent superposition of waves with different frequencies whose amplitudes and phases can arbitrarily vary in space and with frequency. This modulation can be chosen so that all the frequency components converge at the same focal spot at the same time forming an ultrashort pulse of the nanolocalized optical fields.

Turning to the theory, consider a nanofilm of metal in the xy plane whose thickness d in the z direction is adiabatically changing with the coordinate-vector $\boldsymbol{\rho} = (x, y)$ in the plane of the nanofilm. Let $\varepsilon_m = \varepsilon_m(\omega)$ be the dielectric permittivity of this metal nanofilm, and ε_d be the permittivity of the embedding dielectric. As we discussed in Sec. 2.4, due to the symmetry of the system, there are odd and even (in the normal electric field) SPPs. It is the odd SPP that is a slow-propagating, controllable mode. The dispersion relation for this mode defining its effective index $n(\boldsymbol{\rho})$ is given by Eq. 32, which can be explicitly rewritten as

$$\tanh\left(\frac{1}{2}k_0 d(\boldsymbol{\rho})\sqrt{n(\boldsymbol{\rho})^2 - \varepsilon_m}\right) = -\frac{\varepsilon_d\sqrt{n(\boldsymbol{\rho})^2 - \varepsilon_m}}{\varepsilon_m\sqrt{n(\boldsymbol{\rho})^2 - \varepsilon_d}}, \tag{68}$$

where $k_0 = \omega/c$ is the radiation wave vector in vacuum.

Let $\boldsymbol{\tau}$ be a unit tangential vector to the SPP trajectory (ray). It obeys an equation of ray optics[48] $n\,(d\boldsymbol{\tau}/dl) = \partial n/\partial\boldsymbol{\rho} - \boldsymbol{\tau}\,(\boldsymbol{\tau}\partial n/\partial\boldsymbol{\rho})$, where l is the length along the ray.

Now let us consider a nanofilm shaped as a nanowedge as in Color Plate 1(b), see also theory of Sec. 2.6. In such a case, $n = n(y)$, and these trajectory equations simplify as $n\,(d\tau_y/dl) = \tau_x^2\,(dn/dy)$, $n\,(d\tau_x/dl) = -\tau_x\tau_y\,(dn/dy)$. From these, it follows that $n_x \equiv \tau_x n = \text{const}$. The SPP wave vector, related to its momentum, is $\mathbf{k}(\boldsymbol{\rho}) = k_0 n(\boldsymbol{\rho})\boldsymbol{\tau}$; this is the conservation of k_x (the transverse momentum). This allows one to obtain a closed solution for the ray. The tangent equation for the ray is $dx/dy = \tau_x/\tau_y$, where $\tau_y = \sqrt{1 - n_x^2/n^2}$. From this, we get an explicit SPP ray equation as

$$x - x_0 = \int_{y_0}^{y} \left(\frac{n(y')^2}{n_x^2} - 1 \right)^{-1/2} dy' \,, \tag{69}$$

where $\boldsymbol{\rho}_0 = (x_0, y_0)$ is the focal point where rays with any n_x converge. To find the trajectories, as $n(y)$ we use the real part of effective index (68), as WKB suggests.

When the local thickness of the wedge is subwavelength ($k_0 d \ll 1$), the form of these trajectories can be found analytically. Under these conditions, dispersion relation (68) has an asymptotic solution

$$n = \frac{n_a}{k_0 d} \,, \quad n_a = \ln \frac{\varepsilon_m - \varepsilon_d}{\varepsilon_m + \varepsilon_d} \,. \tag{70}$$

Substituting this into Eq. (69), we obtain an explicit SPP ray equation $\left(x - x_0 - \sqrt{\bar{n}_a^2/n_x^2 - y_0^2} \right)^2 + y^2 = \bar{n}_a^2/n_x^2$, where $\bar{n}_a = n_a/(k_0 \tan\theta)$, and $\tan\theta$ is the slope of the wedge. Thus, each SPP ray is a segment of a circle whose center is at a point given by $x = x_0 + \sqrt{(\bar{n}_a/n_x)^2 - y_0^2}$ and $y = 0$. This analytical result is in agreement with Color Plate 1(b). If the nanofocus is at the sharp edge, i.e., $y_0 = 0$, then these circles do not intersect but touch and are tangent to each other at the nanofocus point.

As an example we consider a silver[51] nanowedge illustrated in Color Plate 1(b) whose maximum thickness is $d_m = 30$ nm, the minimum thickness is $d_f = 4$ nm, and whose length (in the y direction) is $L = 5$ μm. Trajectories calculated from Eq. (69) for $\hbar\omega = 2.5$ eV are shown by lines (color used only to guide eye); the nanofocus is indicated by a bold red dot. The different trajectories correspond to different values of n_x in the range $0 \le n_x \le n(L)$. In contrast to focusing by a conventional lens, the SPP rays are progressively bent toward the wedge slope direction.

The eikonal is found as an integral along the ray $\Phi(\boldsymbol{\rho}) = \int_{\boldsymbol{\rho}_0}^{\boldsymbol{\rho}} \mathbf{n}(\boldsymbol{\rho}) d\boldsymbol{\rho}$. Consider rays emitted from the nanofocus [Color Plate 1(b)]. Computed from this equation, the phases of the SPPs at the thick edge of the wedge (for $y = L$) are shown in Color Plate 2(a) as functions of the coordinate x along the thick edge. The colors of the rays correspond to the visual perception of the ray frequencies. The gained phase dramatically increases toward the blue spectral region, exhibiting a strong dispersion. The extinction for most of the frequencies except for the blue edge, displayed in Color Plate 2(b), is not high.

Now consider the evolution of the field intensity along a SPP ray. For certainty, let SPPs propagate along the corresponding rays from the thick

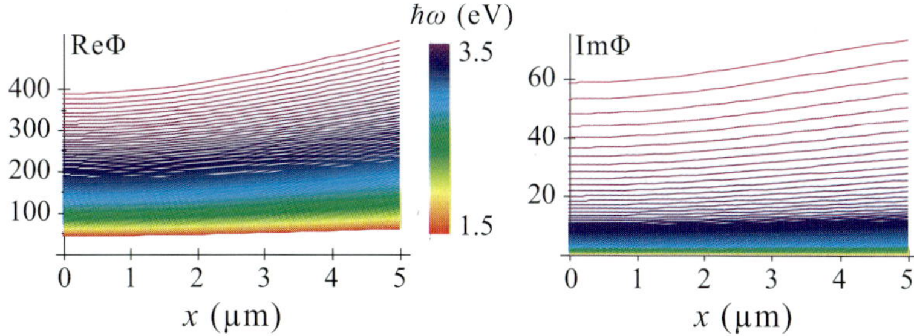

Color Plate 2. (a) Phase (real part of eikonal Φ) acquired by a SPP ray propagating between a point with coordinate x on the thick edge and the nanofocus, displayed as a function of x. The rays differ by frequencies that are color coded by the vertical bar. (b) The same as (a) but for extinction of the ray ($\operatorname{Im}\Phi$).

edge of the wedge toward the nanofocus as shown in Color Plate 1(b). In the process of such propagation, there will be concentration of the SPP energy in all three directions (3d nanofocusing). This phenomenon differs dramatically from what occurs in conventional photonic ray optics.

To describe this nanofocusing, it is convenient to introduce an orthogonal system of ray coordinates whose unit vectors are $\boldsymbol{\tau}$ (along the ray), $\boldsymbol{\eta} = (-\tau_y, \tau_x)$ (at the surface normal to the ray), and $\mathbf{e}_z$ (normal to the surface). The concentration along the ray (in the $\boldsymbol{\tau}$ direction) occurs because the group velocity $v_g = [\partial(k_0 n)/\partial\omega]^{-1}$ of SPP asymptotically tends to zero (for the antisymmetric mode) for $k_0 d \to 0$ as $v_g = v_{0g} d$ where $v_{0g} = \text{const.}$[10] This contributes a factor $A_\parallel = 1/\sqrt{v_g(d)}$ to the amplitude of an SPP wave.

The compression of a SPP wave in the $\mathbf{e}_z$ (vertical) direction is given by a factor of $A_z = \left(\int_{-\infty}^{\infty} W \mathrm{d}z\right)^{-1/2}$, where W is the energy density of the mode. Substituting a standard expression[48] for W, one obtains explicitly

$$\begin{aligned} A_z = \Bigg(& \frac{1}{8\pi} \exp\left(\operatorname{Re}\kappa_d d\right) \Bigg\{ \frac{\sinh\left(\operatorname{Re}\kappa_m d\right)}{\operatorname{Re}\kappa_m \left|\sinh\left(\kappa_m d/2\right)\right|^2} \\ & \times \left[1 + \frac{d(\omega \operatorname{Re}\varepsilon_m)}{d\omega} \frac{|n|^2 + |\kappa_m|^2}{|\varepsilon_m|^2} \right] - \frac{\sin\left(\operatorname{Im}\kappa_m d\right)}{\operatorname{Im}\kappa_m \left|\sinh\left(\kappa_m d/2\right)\right|^2} \\ & \times \left[1 + \frac{d(\omega \operatorname{Re}\varepsilon_m)}{d\omega} \frac{|n|^2 - |\kappa_m|^2}{|\varepsilon_m|^2} \right] \\ & + \frac{2}{\operatorname{Re}\kappa_d} \left[1 + \frac{|n|^2 + |\kappa_d|^2}{\varepsilon_d} \right] \Bigg\} \Bigg)^{-1/2}, \end{aligned} \tag{71}$$

where $\kappa_m = k_0\sqrt{n - \varepsilon_m}$ and $\kappa_d = k_0\sqrt{n - \varepsilon_d}$. Note that the intensity distribution in Color Plate 1(a) is $I \propto \left(A_{\|}A_z\right)^{-2}$.

To obtain the compression factor $A_\perp$ for the $\boldsymbol{\eta}$ direction), we consider conservation of energy along the beam of rays corresponding to slightly different values of n_x. Dividing this constant energy flux by the thickness of this beam in the $\boldsymbol{\eta}$ direction, we arrive at

$$A_\perp = \left\{\left(1 - \frac{n_x^2}{n^2}\right)^{1/2} \int_{y_0}^{y} \frac{1}{n(y')}\left[1 - \frac{n_x^2}{n(y')^2}\right]^{-3/2} \mathrm{d}y'\right\}^{-1/2} . \tag{72}$$

The ray amplitude thus contains the total factor which describes the 3d adiabatic compression: $A = A_{\|}A_\perp A_z$.

Now consider the problem of coherent control. The goal is to excite a spatiotemporal waveform at the thick edge of the wedge in such a way that the propagating SPP rays converge at an arbitrary nanofocus at the sharp edge where an ultrashort pulse is formed. To solve this problem, we use the idea of back-propagation or time-reversal.[79–81] We generate rays at the nanofocus as an ultrashort pulse containing just several oscillations of the optical field. Propagating these rays, we find amplitudes and phases of the fields at the thick edge at each frequency as given by the eikonal $\Phi(\boldsymbol{\rho})$. Then we complex conjugate the amplitudes of frequency components, which corresponds to the time reversal. We also multiply these amplitudes by $\exp(2\mathrm{Im}\,\Phi)$ which pre-compensates for the losses. This provides the required phase and amplitude modulation at the thick edge of the wedge.

We show an example of such calculations in Color Plate 3. Panel (a) displays the trajectories of SPPs calculated according to Eq. (69). The trajectories for different frequencies are displayed by colors corresponding to their visual perception. There is a very significant spectral dispersion: trajectories with higher frequencies are much more curved. The spatial-frequency modulation that we have found succeeds in bringing all these rays (with different frequencies and emitted at different x points) to the same nanofocus at the sharp edge.

The required waveforms at different x points of the thick edge of the wedge are shown in Color Plates 3(b)-(d) where the corresponding longitudinal electric fields are shown. The waves emitted at large x, i.e., at points more distant from the nanofocus, should be emitted significantly earlier to pre-compensate for the longer propagation times. They should also have different amplitudes due to the differences in A. Finally, there is clearly a negative chirp (gradual decrease of frequency with time). This is due to the fact that the higher frequency components propagate more slowly and

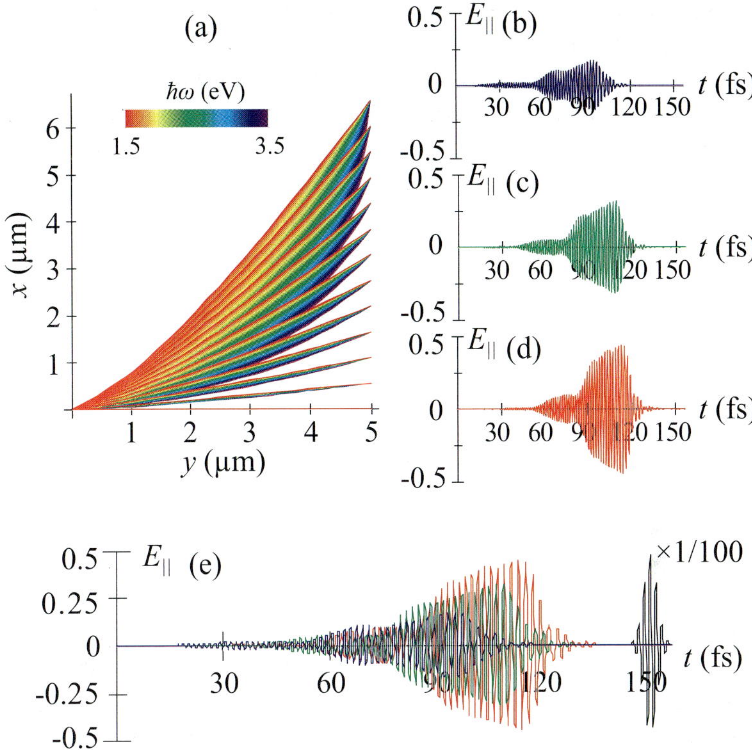

Color Plate 3. (a) Trajectories (rays) of SPP packets propagating from the thick edge to the nanofocus displayed in the xy plane of the wedge. The frequencies of the individual rays in a packet are indicated by color as coded by the bar at the top. (b)-(d) Spatiotemporal modulation of the excitation pulses at the thick edge of the wedge required for nanofocusing. The temporal dependencies (waveforms) of the electric field for the phase-modulated pulses for three points at the thick edge boundary: two extreme points and one at the center, as indicated, aligned with the corresponding x points at panel (a). (e) The three excitation pulses of panels (b)-(d) (as shown by their colors), superimposed to elucidate the phase shifts, delays, and shape changes between these pulses. The resulting ultrashort pulse at the nanofocus is shown by the black line. The scale of the electric fields is arbitrary but consistent throughout the figure.

therefore must be emitted earlier to form a coherent ultrashort pulse at the nanofocus.

In Color Plate 3(e) we display together all three of the representative waveforms at the thick edge to demonstrate their relative amplitudes and positions in time. The pulse at the extreme point in x (shown by blue) has the longest way to propagate and therefore is the most advanced in time. The pulse in the middle point (shown by green) is intermediate, and the pulse at the center ($x = 0$, shown by red) is last. One can notice also a counterintuitive feature: the waves propagating over longer trajectories are smaller in amplitude though one may expect the opposite to compensate for the larger losses. The explanation is that the losses are actually insignificant for the frequencies present in these waveforms, and the magnitudes are determined by adiabatic concentration factor A.

Color Plate 3(e) also shows the resulting ultrashort pulse in the nanofocus. This is a transform-limited, Gaussian pulse. The propagation along the rays completely compensates the initial phase and amplitude modulation, exactly as intended. As a result, the corresponding electric field of the waveform is increased by a factor of 100. Taking the other component of the electric field and the magnetic field into account, the corresponding increase of the energy density is by a factor $\sim 10^4$ with respect to that of the SPPs at the thick edge.

Consider the efficiency of the energy transfer to the nanoscale. This is primarily determined by the cross section σ_{SPP} for scattering of photons into SPPs. For instance, for a metal sphere of radius R at the surface of the wedge, one can obtain an estimate $\sigma_{SPP} \sim R^6/(d_m^3 \lambda)$, where λ is the reduced photon wavelength. Setting $R \sim d_m$, we estimate $\sigma_{SPP} \sim 3$ nm^2. Assuming optical focusing into a spot of ~ 300 nm radius, this yields the energy efficiency of conversion to the nanoscale of $\sim 10^{-3}$. Taking into account the adiabatic concentration of energy by a factor of 10^4, the optical field intensity at the nanofocus is enhanced by one order of magnitude with respect to that of the incoming optical wave.

The criterion of applicability of the WKB approximation is $\partial k^{-1}/\partial y \ll 1$. Substituting $k = k_0 n$ and Eq. (70), we obtain a condition $d_m/(n_a L) << 1$. This condition is satisfied everywhere including the nanofocus since $n_a \sim 1$ and $d_m \ll L$ for adiabatic grading. The minimum possible size of the wavepacket at the nanofocus in the direction of propagation, Δx, is limited by the local SPP wavelength: $\Delta x \sim 2\pi/k \approx 2\pi d_f/n_a$. The minimum transverse size a (waist) of the SPP beam at the nanofocus can be calculated as the radius of the first Fresnel zone: $a = \pi/k_x \geq \pi/(k_0 n_x)$. Because n_x is constant along a trajectory, one can substitute its value at the thick edge (the launch site), where from Eq. (70) we obtain $n_x \approx n = n_a/d_m$. This results in $a \approx \pi d_m/n_a$; thus a is on order of the maximum thickness of the wedge, which is assumed also to be on the nanoscale.

To briefly conclude, we have theoretically described an approach to full coherent control of spatiotemporal energy localization on the nanoscale. From the point of view of electromagnetics this concept is a nanoplasmonic counterpart of the active phased array radar (APAR) and can be called NAPAR. As we described, in NAPAR from the thick edge of a plasmonic metal nanowedge, SPPs are launched, whose phases and amplitudes are independently modulated for each constituent frequency of the spectrum and at each spatial point of the excitation. This pre-modulates the departing SPP wave packets in such a way that they reach the required point at the sharp edge of the nanowedge in phase, with equal amplitudes forming a nanofocus where an ultrashort pulse with required temporal shape is generated. Of course, a similar control can be achieved on a nanoplasmonic slab of a constant thickness, but then the benefit of the adiabatic nanofocusing would be lost. The system described (NAPAR) constitutes a "nanoplasmonic portal" connecting the incident light field, whose features are shaped on the microscale, with the required point or features at the nanoscale.

Acknowledgments

This work was supported by grants from the Chemical Sciences, Biosciences and Geosciences Division of the Office of Basic Energy Sciences, Office of Science, U.S. Department of Energy, a grant CHE-0507147 from NSF, and a grant from the US-Israel BSF. The author is grateful to K. Nelson, M. Durach, and A. Rusina for useful discussions and help.

References

1. L. Novotny and B. Hecht, *Principles of Nano-Optics.* (Cambridge University Press, Cambridge, New York, 2006).
2. Near-field optics and surface plasmon polaritons. In ed. S. Kawata, *Topics in Applied Physics*, vol. 81. Springer Verlag, Berlin, New York, (2001).
3. M. I. Stockman, S. V. Faleev, and D. J. Bergman, Localization versus delocalization of surface plasmons in nanosystems: Can one state have both characteristics?, *Phys. Rev. Lett.* **87**, 167401–1–4, (2001).
4. M. I. Stockman, M. F. Kling, U. Kleineberg, and F. Krausz, Attosecond nanoplasmonic field microscope, *Nature Photonics.* **1**, 539–544, (2007).
5. T. Klar, M. Perner, S. Grosse, G. von Plessen, W. Spirkl, and J. Feldman, Surface-plasmon resonances in single metallic nanoparticles, *Phys. Rev. Lett.* **80**, 4249–4252, (1998).
6. J. Lehmann, M. Merschdorf, W. Pfeiffer, A. Thon, S. Voll, and G. Gerber, Surface plasmon dynamics in silver nanoparticles studied by femtosecond time-resolved photoemission, *Phys. Rev. Lett.* **85**, 2921–2924, (2000).
7. J. Bosbach, C. Hendrich, F. Stietz, T. Vartanyan, and F. Trager, Ultrafast dephasing of surface plasmon excitation in silver nanoparticles: Influence of

particle size, shape, and chemical surrounding, *Phys. Rev. Lett.* **89**, 257404–1–4, (2002).
8. C. Hendrich, J. Bosbach, F. Stietz, F. Hubenthal, T. Vartanyan, and F. Trager, Chemical interface damping of surface plasmon excitation in metal nanoparticles: A study by persistent spectral hole burning, *Appl. Phys. B.* **76**, 869–875, (2003).
9. T. Zentgraf, A. Christ, J. Kuhl, and H. Giessen, Tailoring the ultrafast dephasing of quasiparticles in metallic photonic crystals, *Phys. Rev. Lett.* **93**, 243901–1–4, (2004).
10. M. I. Stockman, Nanofocusing of optical energy in tapered plasmonic waveguides, *Phys. Rev. Lett.* **93**, 137404–1–4, (2004).
11. A. Ono, J. Kato, and S. Kawata, Subwavelength optical imaging through a metallic nanorod array, *Phys. Rev. Lett.* **95**, 267407–1–4, (2005).
12. G. Shvets, S. Trendafilov, J. B. Pendry, and A. Sarychev, Guiding, focusing, and sensing on the subwavelength scale using metallic wire arrays, *Phys. Rev. Lett.* **99**, 053903–1–4, (2007).
13. J. B. Pendry, Perfect cylindrical lenses, *Opt. Expr.* **11**, 755–760, (2003).
14. Z. Liu, H. Lee, Y. Xiong, C. Sun, and X. Zhang, Far-field optical hyperlens magnifying sub-diffraction-limited objects, *Science.* **315**, 1686–1686, (2007).
15. M. I. Stockman, S. V. Faleev, and D. J. Bergman, Coherent control of femtosecond energy localization in nanosystems, *Phys. Rev. Lett.* **88**, 067402–1–4, (2002).
16. D. J. Tannor and S. A. Rice, Control of selectivity of chemical reaction via control of wave packet evolution, *J. Chem. Phys.* **83**, 5013–5018, (1985).
17. In eds. P. Brumer and M. Shapiro, *Principles of the Quantum Control of Molecular Processes.* Wiley, New York, (2003).
18. R. S. Judson and H. Rabitz, Teaching lasers to control molecules, *Phys. Rev. Lett.* **68**, 1500, (1992).
19. G. Kurizki, M. Shapiro, and P. Brumer, Phase-coherent control of photocurrent directionality in semiconductors, *Phys. Rev. B.* **39**, 3435–3437, (1989).
20. T. C. Weinacht, J. Ahn, and P. H. Bucksbaum, Controlling the shape of a quantum wavefunction, *Nature.* **397**, 233–235, (1999).
21. P. Brumer and M. Shapiro, Laser control of molecular processes, *Ann. Rev. Phys. Chem.* **43**, 257–282, (1992).
22. H. Rabitz, R. de Vivie-Riedle, M. Motzkus, and K. Kompa, Chemistry - whither the future of controlling quantum phenomena?, *Science.* **288**, 824–828, (2000).
23. J. M. Geremia and H. Rabitz, Optimal identification of hamiltonian information by closed-loop laser control of quantum systems, *Phys. Rev. Lett.* **89**, 263902–1–4, (2002).
24. N. A. Nguyen, B. K. Dey, M. Shapiro, and P. Brumer, Coherent control in nanolithography: Rydberg atoms, *J. Phys. Chem. A.* **108**, 7878–7888, (2004).
25. M. Shapiro and P. Brumer, Quantum control of bound and continuum state dynamics, *Physics Reports.* **425**, 195–264, (2006).
26. A. Assion, T. Baumert, M. Bergt, T. Brixner, B. Kiefer, V. Seyfried,

M. Strehle, and G. Gerber, Control of chemical reactions by feedback-optimized phase-shaped femtosecond laser pulses, *Science*. **282**, 919–922, (1998).
27. R. Bartels, S. Backus, E. Zeek, L. Misoguti, G. Vdovin, I. P. Christov, M. M. Murnane, and H. C. Kapteyn, Shaped-pulse optimization of coherent emission of high-harmonic soft x-rays, *Nature*. **406**, 164–166, (2000).
28. N. Dudovich, D. Oron, and Y. Silberberg, Single-pulse coherently controlled nonlinear raman spectroscopy and microscopy, *Nature*. **418**, 512–514, (2002).
29. T. Brixner, G. Krampert, T. Pfeifer, R. Selle, G. Gerber, M. Wollenhaupt, O. Graefe, C. Horn, D. Liese, and T. Baumert, Quantum control by ultrafast polarization shaping, *Phys. Rev. Lett.* **92**, 208301–1–4, (2004).
30. M. I. Stockman, D. J. Bergman, and T. Kobayashi, Coherent control of nanoscale localization of ultrafast optical excitation in nanosystems, *Phys. Rev. B*. **69**, 054202–1–10, (2004).
31. M. I. Stockman and P. Hewageegana, Nanolocalized nonlinear electron photoemission under coherent control, *Nano Lett.* **5**, 2325–2329, (2005).
32. M. Sukharev and T. Seideman, Phase and polarization control as a route to plasmonic nanodevices, *Nano Lett.* **6**, 715–719, (2006).
33. A. Kubo, K. Onda, H. Petek, Z. Sun, Y. S. Jung, and H. K. Kim, Femtosecond imaging of surface plasmon dynamics in a nanostructured silver film, *Nano Lett.* **5**, 1123–1127, (2005).
34. M. Aeschlimann, M. Bauer, D. Bayer, T. Brixner, F. J. G. d. Abajo, W. Pfeiffer, M. Rohmer, C. Spindler, and F. Steeb, Adaptive subwavelength control of nano-optical fields, *Nature*. **446**, 301–304, (2007).
35. M. Bauer, C. Wiemann, J. Lange, D. Bayer, M. Rohmer, and M. Aeschlimann, Phase propagation of localized surface plasmons probed by time-resolved photoemission electron microscopy, *Appl. Phys. A*. **88**, 473–480, (2007).
36. M. Sukharev and T. Seideman, Coherent control of light propagation via nanoparticle arrays, *J. Phys. B: At. Mol. Opt. Phys.* pp. S283–S298, (2007).
37. M. Durach, A. Rusina, K. Nelson, and M. I. Stockman, Toward full spatiotemporal control on the nanoscale, *Nano Lett.* **7**, 3145–3149, (2007).
38. A. A. Mikhailovsky, M. A. Petruska, K. Li, M. I. Stockman, and V. I. Klimov, Phase-sensitive spectroscopy of surface plasmons in individual metal nanostructures, *Phys. Rev. B*. **69**, 085401–1–6, (2004).
39. A. A. Mikhailovsky, M. A. Petruska, M. I. Stockman, and V. I. Klimov, Broadband near-field interference spectroscopy of metal nanoparticles using a femtosecond white-light continuum, *Optics Lett.* **28**, 1686–1688, (2003).
40. T. Kawazoe, T. Shimizu, and M. Ohtsu, Second-harmonic generation in a near-field optical-fiber probe, *Opt. Lett.* **26**, 1687–1689, (2001).
41. H. G. Frey, F. Keilmann, A. Kriele, and R. Guckenberger, Enhancing the resolution of scanning near-field optical microscopy by a metal tip grown on an aperture probe, *Appl. Phys. Lett.* **81**, 5030–5032, (2002).
42. F. Keilmann, Scattering-type near-field optical microscopy, *J. Electron. Microsc.* **53**, 187–192, (2004).

43. F. Keilmann and R. Hillenbrand, Near-field microscopy by elastic light scattering from a tip, *Philos. Trans. Roy. Soc. A.* **362**, 787–805, (2004).
44. M. I. Stockman. Delivering energy to nanoscale: Rapid adiabatic transformation, concentration, and stopping of radiation in nano-optics. In eds. N. J. Halas and T. R. Huser, *Plasmonics: Metallic Nanostructures and Their Optical Properties II*, vol. 5512, pp. 38–49. SPIE, Denver, Colorado, (2004).
45. D. K. Gramotnev, Adiabatic nanofocusing of plasmons by sharp metallic grooves: Geometrical optics approach, *J. Appl. Phys.* **98**, 104302–1–11, (2005).
46. S. I. Bozhevolnyi, V. S. Volkov, E. Devaux, J.-Y. Laluet, and T. W. Ebbesen, Channel plasmon subwavelength waveguide components including interferometers and ring resonators, *Nature.* **440**, 508–511, (2006).
47. L. D. Landau and E. M. Lifshitz, *Quantum Mechanics: Non-Relativistic Theory.* (Pergamon Press, Oxford and New York, 1965).
48. L. D. Landau and E. M. Lifshitz, *Electrodynamics of Continuous Media.* (Pergamon, Oxford and New York, 1984).
49. C. Cohen-Tannoudji, B. Diu, and F. Laloe, *Quantum Mechanics.* (Wiley, New York, 1977).
50. D. E. Aspnes, S. M. Kelso, R. A. Logan, and R. Bhat, Optical properties of $Al_xGa_{1-x}As$, *J. Appl. Phys.* **60**, 754–767, (1986).
51. P. B. Johnson and R. W. Christy, Optical constants of noble metals, *Phys. Rev. B.* **6**, 4370–4379, (1972).
52. E. N. Economou, Surface plasmons in thin films, *Phys. Rev.* **182**, 539–554, (1969).
53. R. Ruppin, Effect of non-locality on nanofocusing of surface plasmon field intensity in a conical tip, *Phys. Lett. A.* **340**, 299–302, (2005).
54. V. N. Pustovit and T. V. Shahbazyan, Quantum-size effects in sers from noble-metal nanoparticles, *Microelectronics J.* **36**, 559–563, (2005).
55. V. N. Pustovit and T. V. Shahbazyan, Finite-size effects in surface-enhanced raman scattering in noble-metal nanoparticles: A semiclassical approach, *J. Opt. Soc. Am. A.* **23**, 1369–1374, (2006).
56. V. N. Pustovit and T. V. Shahbazyan, Surface-enhanced Raman scattering on the nanoscale: A microscopic approach, *J. Optics A.* **8**, S208–S212, (2006).
57. V. N. Pustovit and T. V. Shahbazyan, SERS from molecules adsorbed on small ag nanoparticles: A microscopic model, *Chem. Phys. Lett.* **420**, 469–473, (2006).
58. V. N. Pustovit and T. V. Shahbazyan, Microscopic theory of surface-enhanced raman scattering in noble-metal nanoparticles, *Phys. Rev. B.* **73**, 085408–1–7, (2006).
59. E. Kretschmann and H. Raether, Radiative decay of nonradiative surface plasmons excited by light, *Z. Naturforsch.* **A 23**, 2135–2136, (1968).
60. H. Raether, *Surface Plasmons on Smooth and Rough Surfaces and on Gratings.* (Springer Verlag, Berlin New York, 1988).
61. J. A. Kong, *Electromagnetic Wave Theory.* (Wiley, New York, 1990).
62. F. Keilmann, Surface-polariton propagation for scanning near-field optical microscopy application, *J. Micros.* **194**, 567–570, (1999).

63. A. Bouhelier, J. Renger, M. R. Beversluis, and L. Novotny, Plasmon-coupled tip-enhanced near-field optical microscopy, *J. Microsc. - Oxford.* **210**, 220–224, (2003).
64. I. A. Larkin, M. I. Stockman, M. Achermann, and V. I. Klimov, Dipolar emitters at nanoscale proximity of metal surfaces: Giant enhancement of relaxation in microscopic theory, *Phys. Rev. B.* **69**, 121403(R)–1–4, (2004).
65. A. Liebsch, Screening properties of a metal surface at low frequencies and finite wave vectors, *Phys. Rev. Lett.* **54**, 67–70, (1985).
66. K. Li, M. I. Stockman, and D. J. Bergman, Self-similar chain of metal nanospheres as an efficient nanolens, *Phys. Rev. Lett.* **91**, 227402–1–4, (2003).
67. E. Verhagen, L. Kuipers, and A. Polman, Enhanced nonlinear optical effects with a tapered plasmonic waveguide, *Nano Lett.* **7**, 334–337, (2007).
68. E. Verhagen, A. Polman, and L. Kuipers, Nanofocusing in laterally tapered plasmonic waveguides, *Opt. Expr.* **16**, 45–57, (2008).
69. C. Ropers, C. C. Neacsu, T. Elsaesser, M. Albrecht, M. B. Raschke, and C. Lienau, Grating-coupling of surface plasmons onto metallic tips: A nano-confined light source, *Nano Lett.* **7**, 2784–2788, (2007).
70. C. Ropers, D. R. Solli, C. P. Schulz, C. Lienau, and T. Elsaesser, Localized multiphoton emission of femtosecond electron pulses from metal nanotips, *Phys. Rev. Lett.* **98**, 043907, (2007).
71. A. Apolonski, P. Dombi, G. G. Paulus, M. Kakehata, R. Holzwarth, T. Udem, C. Lemell, K. Torizuka, J. Burgdoerfer, T. W. Hansch, and F. Krausz, Observation of light-phase-sensitive photoemission from a metal, *Phys. Rev. Lett.* **92**, 073902–1–4, (2004).
72. D. Zeidler, A. Staudte, A. B. Bardon, D. M. Villeneuve, R. Dorner, and P. B. Corkum, Controlling attosecond double ionization dynamics via molecular alignment, *Phys. Rev. Lett.* **95**, 203003–1–4, (2005).
73. A. Kubo, N. Pontius, and H. Petek, Femtosecond microscopy of surface plasmon polariton wave packet evolution at the silver/vacuum interface, *Nano Lett.* **7**, 470–475, (2007).
74. M. M. Wefers and K. A. Nelson, Programmable phase and amplitude femtosecond pulse shaping, *Opt. Lett.* **18**, 2032–2034, (1993).
75. T. Feurer, J. C. Vaughan, and K. A. Nelson, Spatiotemporal coherent control of lattice vibrational waves, *Science.* **299**, 374–377, (2003).
76. W. Nomura, M. Ohtsu, and T. Yatsui, Nanodot coupler with a surface plasmon polariton condenser for optical far/near-field conversion, *Appl. Phys. Lett.* **86**, 181108–1–3, (2005).
77. L. L. Yin, V. K. Vlasko-Vlasov, J. Pearson, J. M. Hiller, J. Hua, U. Welp, D. E. Brown, and C. W. Kimball, Subwavelength focusing and guiding of surface plasmons, *Nano Lett.* **5**, 1399–1402, (2005).
78. R. J. Mailloux, *Phased Array Antenna Handbook.* (Artech House, Boston, 2005).
79. G. Lerosey, J. de Rosny, A. Tourin, A. Derode, G. Montaldo, and M. Fink, Time reversal of electromagnetic waves, *Phys. Rev. Lett.* **92**, 193904 –1–3, (2004).

80. G. Lerosey, J. de Rosny, A. Tourin, A. Derode, and M. Fink, Time reversal of wideband microwaves, *Appl. Phys. Lett.* **88**, 154101 –1–3, (2006).
81. G. Lerosey, J. de Rosny, A. Tourin, and M. Fink, Focusing beyond the diffraction limit with far-field time reversal, *Science.* **315**, 1120–1122, (2007).

Chapter 13

NANOPLASMONICS: COMPONENTS, DEVICES AND CIRCUITS

Mark L. Brongersma, Jon A. Schuller, Justin White, Young Chul Jun

Geballe Laboratory for Advanced Materials, Stanford University
476 Lomita Mall, Stanford, CA 94305, USA
E-mail: Brongersma@stanford.edu

Sergey I. Bozhevolnyi

Department of Sensors, Signals and Electrotechnics, University of Southern Denmark, Niels Bohrs Allé 1E, 5230 Odense M, Denmark
E-mail:seib@sense.sdu.dk

Thomas Søndergaard

Department of Physics and Nanotechnology, Aalborg University
Skjernvej 4A, DK-9220 Aalborg Øst, Denmark

Rashid Zia

Devision of Engineering, Brown University
Box D, 182 Hope St., Providence, RI 02912

This chapter reviews some of the key developments in the rapidly emerging field of nanoplasmonics. The use of metallic nanostructures has enabled the creation of nanoscale optical components, devices, and circuits that offer new types of functionalities that are either very hard or impossible to obtain with dielectrics or semiconductors. A tremendous synergy could be attained by integrating plasmonic, electronic, and conventional dielectric photonic devices on the same chip and taking advantage of the strengths of each technology. A number of challenges need to be addressed before this dream can be realized.

1. The value of plasmonics as a new device technology

The ever increasing demand for faster information transport and processing capabilities is undeniable. Our data hungry society has driven the enormous progress in the Si electronics industry and we have witnessed a continuous progression towards smaller, faster, and more efficient electronic devices over the last five decades. Device scaling has also brought about a myriad of challenges. Currently, two of the most daunting problems preventing significant increases in processor speed are thermal and RC delay time issues associated with electronic interconnection.[1,2,3] Optical interconnects, on the other hand, possess an almost unimaginably large data carrying capacity and may offer interesting new solutions for circumventing these problems.[4] Optical alternatives may be particularly attractive for future chips with more distributed architectures in which a multitude of fast electronic computing units (cores) need to be connected via high speed links. Unfortunately, their implementation is hampered by the large size mismatch between electronic and dielectric photonic components. Dielectric photonic devices are limited in size by the fundamental laws of diffraction to about half a wavelength of light and tend to be at least 1 or 2 orders of magnitude larger than their nanoscale electronic counterparts. This obvious size mismatch between electronic and photonic components presents a major challenge for interfacing these technologies. It thus appears that further progress will require the development of a radically new chip-scale device technology that can facilitate information transport between nanoscale devices at optical frequencies and bridge the gap between the world of nanoscale electronics and microscale photonics. In the following, we discuss a candidate technology that recently emerged[5,6] and was termed "plasmonics".[7] This device technology exploits the unique optical properties of nanoscale metallic structures to route and manipulate light at the nanoscale. An increased synergy could be attained by integrating plasmonic, electronic, and conventional photonic devices on the same chip and taking advantage of the strengths of each technology.[8] In the following, we present some recent studies on plasmonic structures and conclude by providing an assessment of the potential opportunities and limitations for chip-scale plasmonics.

Noble metal nanostructures may possess exactly the right combination of electronic and optical properties to tackle the interconnect crisis outlined above and realize the dream of significantly faster processing speeds. The metals commonly used in electrical interconnection such as copper (Cu) and aluminum (Al) allow for the excitation of surface plasmon-polaritons (SPPs). SPPs are electromagnetic waves that propagate along a metal-dielectric interface and are coupled to the free electrons in the metal, as discussed in detail in Chapter 1 and several papers and books.[9,10,11,12,13]

From an engineering standpoint a SPP can be viewed as a special type of light wave and the metallic interconnects that support such waves thus serve as tiny optical waveguides, termed plasmonic waveguides. The notion that the optical mode ("light beam") diameter normal to the metal interface can be significantly smaller than the wavelength of light,[14] has generated significant excitement and sparked the dream that one day we will be able to interface nanoscale electronics with similar sized optical (plasmonic) devices.

2. Diffraction-limited plasmonic waveguides

The dramatic advances in nanofabrication and full-field electromagnetic (EM) simulation techniques have enabled the realization of a wide variety of plasmonic waveguide structures. The transport properties of these structures strongly depend on the topology of the metallic surface and the optical properties of the metal and dielectric environment. Interestingly, the boundary conditions at metal\dielectric interfaces give rise to very different behavior than is commonly observed in dielectric optical components. This different behavior has been employed in many applications that are truly unique to metallic nanostructures and has led to the field of "Plasmonics". During the development of this field it has become eminently clear that one should not blindly apply dielectric waveguide concepts to metallic waveguide structures. In particular, the possibility to obtain non-diffraction limited transport in metal nanostructures has been of great interest and numerous groups are pursuing this exciting area. In the following, several of the most relevant waveguide structures for information transport are discussed and their relative strengths and weaknesses commented upon.

Plasmonic stripe waveguides provide a natural starting point for this discussion as such stripes very closely resemble conventional metal interconnects on computer chips. Colour plate 1 (a) shows an optical micrograph of a set of Au stripes generated by electron beam lithography on a SiO_2 glass cover slip. The stripes are 55 nm thick and their widths range from 5 μm to 50 nm. Au stripes are ideal for fundamental waveguide transport studies as they are easy to fabricate, do not oxidize, and exhibit a qualitatively similar plasmonic response as Cu and Al.[15] The stripes are attached to an extended Au pad from which SPPs can be launched into a stripe of choice. A scanning electron microscopy (SEM) image of a 250 nm wide stripe is shown as an inset. The red arrow shows pictorially how light is launched from a focused laser spot into a 1 μm wide stripe. A photon scanning tunneling microscope (PSTM)[16] has been the tool of choice to characterize SPP propagation along extended films as well as metal stripe waveguides.[17,18,19] In such microscopes, SPPs are typically excited using the well-known Kretschmann geometry that enables phase matching of a free space excitation beam and the SPP.[20] In the PSTM used for this study, an apertured cantilever-probe, as shown schematically in Colour plate 3(a), is used to locally tap into the guided SPP waves and scatter light towards a photodetector in the far-field. The detected signal provides a measure of the local field intensity right underneath the tip and the propagation of SPPs can be imaged by scanning the tip over the patterned metal structure.

Colour plates 1 (b), (c), and (d) show PSTM images of SPPs excited at a wavelength of 785 nm and propagating along 3.0 μm, 1.5 μm, and 0.5 μm wide Au stripes. The 3.0 μm wide stripe can be used to propagate signals over several tens of microns. Similar to previous far-field measurements along silver (Ag) stripes[21] it is clear that the propagation distance of SPPs decreases with decreasing stripe width. A better understanding of this behavior can be obtained from full-field simulations and a recently developed intuitive dielectric waveguide models for plasmonic waveguides.[22,23,24,25] A selected set of these simulation results is presented next, followed by a discussion of the potential uses for these relatively short propagation distance waveguides.

Recent numerical work demonstrated that the modal solutions of plasmonic stripe waveguides are hybrid transverse electric-transverse magnetic (TE-TM) modes, and therefore, proper analysis requires

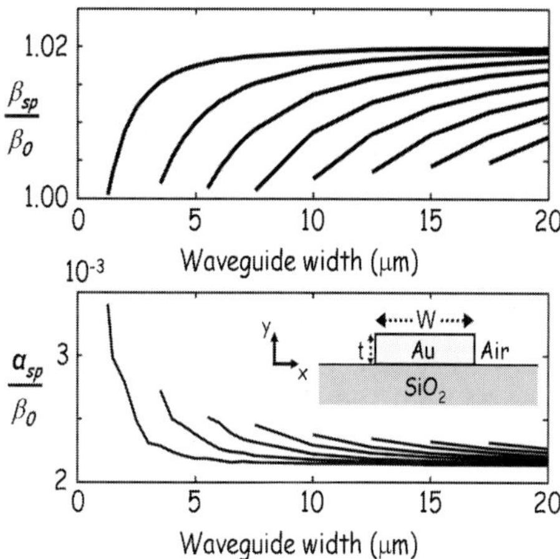

Fig. 1. The calculated complex propagation constants (β_{sp} + iα_{sp}) for the 8 lowest order leaky, quasi-transverse magnetic (TM) surface plasmon-polariton modes of varying width Au stripe waveguides. In these calculations the Au stripe thickness was taken to be t = 55 nm and the free space excitation wavelength was chosen as λ = 800nm. The magnitudes of β_{sp} and α_{sp} were normalized to the real part of the free space propagation constant, β_0. The inset shows the simulated geometry (Adapted from reference 8).

numerical solution of the full vectorial wave equation.[26,27,28] In the following, we describe EM simulations that solve Helmholtz equation using a full vectorial magnetic field-finite difference method.[29] This is a frequency domain method and has clear advantages over the very popular finite difference time domain methods. One particularly attractive feature is that realistic, frequency dependent dielectric constants for the metals and dielectrics can be used as inputs to the simulation. Colour plate 2 shows two modal solutions obtained with this method for a 55 nm thick and 3.5 mm wide Au stripe on glass at a wavelength of 800 nm. It is worth noticing that SPP modes are supported on both the top and bottom Au surface. These waves can simultaneously carry information without interacting. The mode propagating along the top metal/air interface is called a leaky mode and the mode at the bottom metal/glass interface is called a bound mode. The field associated with these modes have a large H_x component, which is reminiscent of the purely TM nature of the SPP modes on infinite metal films. For this reason they are often called quasi-TM modes.

In addition to calculating the field intensity distributions of SPP modes, the FVH-FDM also provides values for the real and imaginary parts of the propagation constants. Figure 1 shows the complex propagation constants (β_{sp} + iα_{sp}) determined for the lowest order leaky, quasi-TM modes supported by 55 nm thick Au stripes of various widths, W, at an excitation wavelength of 800 nm.[24] The inset in the bottom graph shows the geometry used in the simulations. For this simulation the dielectric properties of Au (ε_{Au} = -26.14 + 1.85 i) at the excitation wavelength of 800 nm were used.[30]

Several important trends can be discerned from these plots. Similar to dielectric waveguides, larger structures tend to support an increased number of modes. Higher order SPP modes exhibit a higher number of maxima in the transverse magnetic field, H_x, along the x-direction. As the stripe width is decreased to a couple times the excitation wavelength, the wave number, β_{sp}, for all modes starts to decrease from the value of an infinitely wide stripe (β_{sp} = 1.02). The decrease of the propagation constant with decreasing stripe width first results in a reduced confinement of the modes and finally cutoff occurs for a width of ~1.3μm. At this width the SPP propagation constant has become equal to the propagation constant in air, β_0. It is worth noting that the width of the stripes needs to be chosen quite a bit bigger than the surface plasmon wavelength to avoid cutoff. The diminished confinement for narrow stripes results in a concomitant increase in the radiation losses into the high index SiO_2 substrate. This explains the larger α_{sp} observed for small stripe width. The 0.5 μm wide waveguide in Colour plate 1(d) is below cutoff and does not support a quasi-TM mode. It appears, however, that there is a finite propagation length, which can be explained by taking into account the contribution to the measured field intensity from the radiation continuum (i.e. non-guided waves).[25] It is interesting to note that simulations also predict cutoff for the bound modes, albeit at a slightly narrower stripe width.[24]

Recently, a SPP analogue to Young's experiment was employed to reveal the strong analogy between SPP propagation and light propagation in conventional dielectric optics.[31] This analogy enabled the construction of a unified theory for SPP propagation, diffraction, and interference. Colour plate 3(a) shows a schematic of the experimental configuration used in this experiment. Using electron-beam lithography, 48 nm thick Au films were patterned on glass substrates such that two metal stripe waveguides protrude from an extended metal film region, termed the launchpad. The two stripes play a role that is equivalent to the role of the two slits in the original experiment. The stripes are 2 μm wide, separated by a distance of 2 μm, and terminated into another extended thin film region, named the termination pad. As before, a PSTM was used to visualize the SPP propagation. The near-field intensity recorded by the PSTM is presented in Colour plate 3 (b). The light intensity distribution closely resembles the well-known diffraction and interference pattern of

Young's original double-slit experiment. This experiment and the associated theory predict an effective diffraction limit for the lateral confinement on metal stripe waveguides and justify the use of well-developed dielectric photonics concepts in the design of new plasmonic functionalities.

The type of knowledge presented on the propagation behavior of SPPs along plasmonic interconnects (mode size, propagation length, operation bandwidth, and cutoff) is essential for chip-designers and process engineers. It is clear that the short propagation distances found for plasmonic stripe waveguides preclude direct competition with low-loss dielectric waveguide components in many situations. However, plasmonic structures can add new types of functionality to chips that cannot be obtained with dielectric photonics, as will be shown in the following sections.

Photonic band gap (PBG) materials provide another natural approach towards miniaturization of optical circuitry. These materials make use of the PBG effect that is essentially a manifestation of Bragg reflection of propagating waves (in any direction) because of periodic modulation of the refractive index.[32] Indeed, efficient waveguiding along straight and sharply bent line defects in two-dimensional (2D) PBG structures (i.e., planar waveguide structures with periodic arrays of holes to control the light propagation in the waveguide plane) has been demonstrated for light wavelengths inside the PBG.[33] It became clear that these PBG structures, when properly designed and realized, might be advantageous for miniature photonic circuits allowing for an unprecedented level of integration.[34] Furthermore, one can conjecture that other (quasi) 2D waves, e.g., SPPs, might be employed for the same purpose (see Fig. 2). In fact, the PBG effect for SPPs excited along a specific direction was first observed more than 30 years ago in the experiments on light diffraction by periodically corrugated metal surfaces.[35] The SPPBG effect for all directions in the surface plane of a silver film having a 2D periodic surface profile has also been reported.[36] It should be emphasized that the SPP interaction with a periodic surface corrugation, similarly to the interaction of a waveguide mode with a periodic array of holes, inevitably produces scattered waves propagating away from the surface (See Fig. 2(b)). This (unwanted) process results in an additional propagation loss and has to be taken into account when considering the SPPBG structures.

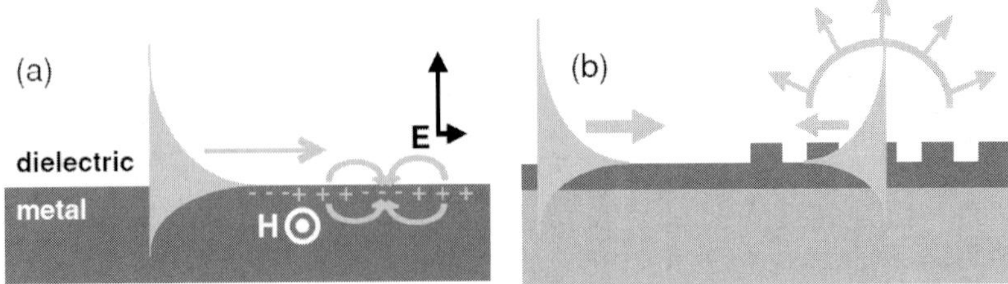

Fig. 2. (a) Schematic representation of an SPP propagating along a metal-dielectric interface and orientations of electric and magnetic field components. (b) Illustration of the SPP reflection by a periodic array of surface scatterers, showing also scattered field components propagating away from the metal surface.

SPP guiding along line defects in SPPBG structures was first demonstrated with ~ 45-nm-high and 200-nm-wide gold bumps arranged in a 410-nm-period triangular lattice on the surface of a 45-nm-thick gold film.[37] The efficient SPP reflection by such an area and SPP guiding along channels free from scatterers was observed, as well as significant deterioration of these effects at wavelengths outside the bandgap, indicating the occurrence of a SPPBG effect in these structures. The observations were made with a stand-alone scanning near-field optical microscope (SNOM) used to collect the radiation scattered by an uncoated sharp fiber tip into fiber modes and an arrangement for SPP excitation in the usual Kretschmann configuration (See Fig. 3(a)). Similar experiments have followed revealing that the SPP guiding is strongly damped in narrow channels[38] contrary to what was expected from similar studies on (conventional) photonic BG structures.[34] It has also been found that the SPP guiding around sharp corners exhibited a bend loss increasing quadratically with the bend angle.[39] In fact, the bend loss was found to be significant for bend angles exceeding 15° (Fig. 3), an observation that was again somewhat unexpected. Although further progress in this area can be expected, these structures are limited in their size like the stripe waveguides. Other waveguide structures have recently emerged that can guide EM energy well below the diffraction limit and may enable the ultimate scaling of photonic circuits. These structures are the topic of the next section.

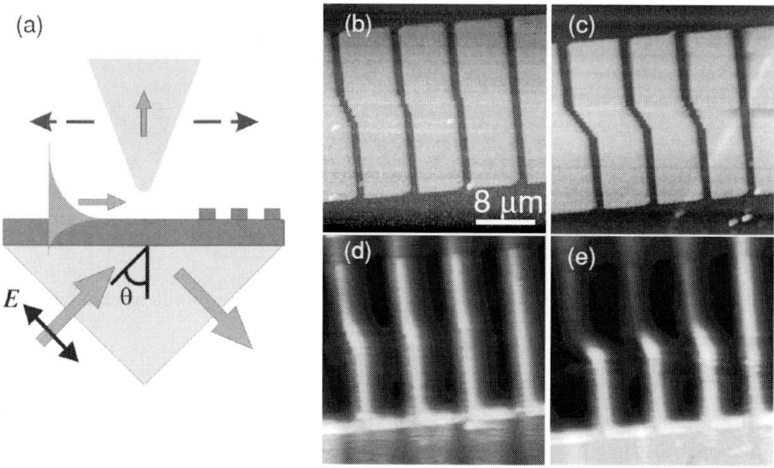

Fig. 3. (a) Schematic representation of the excitation of SPPs and their detection with a fiber tip of a collection SNOM. The polarization of the incident laser beam is also indicated. Gray-scale (b, c) topographical and (d, e) near-field optical images (32x32 μm^2) for SPPBG structures (made of gold) at a wavelength of 737 nm with (b, d) and (c, e) taken simultaneously. Periodic surface structures are of ΓM orientation with the following parameters: the period is 410 nm, the bump height and width are 45 nm and 200 nm, respectively. The structures contain ~ 2-μm-wide straight defects followed (from right to left) by double-bent defects with (b, d) 5°, 10°, and 15° and (c, e) 20°, 25°, and 30°. (Adapted from reference 39).

3. Plasmonic waveguides affording nanoscale mode confinement

Whereas weakly guided stripe waveguides and SPPBG structures cannot achieve deep sub-wavelength confinement, there exist alternative strongly guiding geometries which can provide markedly better confinement. For the discussion on novel interconnection schemes, this category of structures is of great interest. For example, Takahara's original paper on SPP modes of a metal cylinder shows that subwavelength mode diameters are possible and propagation over short distances can be realized.[14] The key to designing and realizing deeply sub-wavelength waveguides is to find geometries like this one wherein the SPP momentum increases with decreasing lateral dimensions.

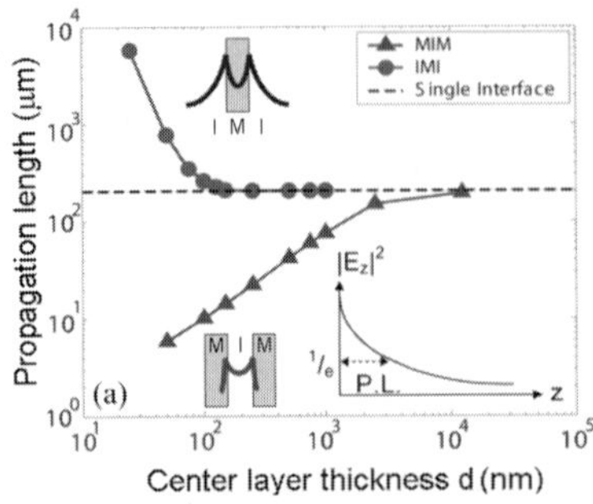

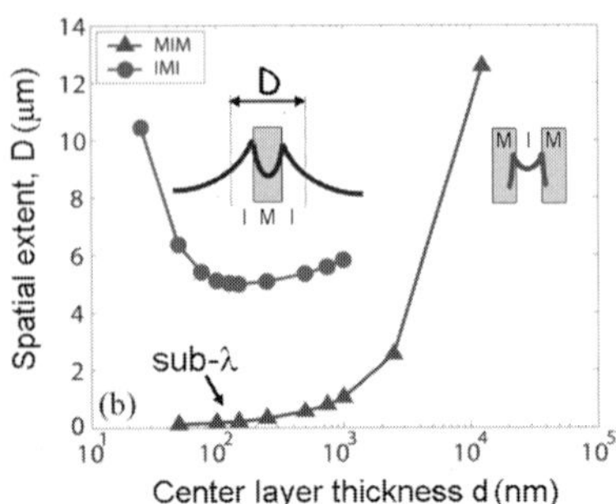

Fig. 4. (a) Plot of the SPP propagation length as a function of the center layer thickness for the case of a metal\insulator\metal (MIM) and a insulator\metal\insulator (IMI) plasmonic waveguide. The calculations were performed of the (a) propagation length and (b) spatial extent of the SPP modes as a function of decreasing center layer thickness. The insets graphically illustrate plotted terms. The reflection pole method was used at $\lambda = 1.55\mu m$ and Au was taken for the metal and air as the insulator (Adapted from reference 15).

Building upon this idea, a myriad of alternative guiding geometries were explored, including arrays of metal nanoparticles,[40,41] metal wedges,[42,43] plasmonic slot waveguides,[44,45,46,47] and channels or grooves in metal films.[48,49,50,51,52] These structures have been the topic of considerable theoretical and experimental investigation.

In the following, we will consider metallic waveguide structures consisting of two closely spaced metals. They provide a simple model system that combines propagation distances of a few to a few tens of micron with deep sub-wavelength confinement.[15] These structures provide a natural geometry to perform simultaneous electrical and optical functions (see e.g sections 5 and 6). For this reason, they are ideally suited for integration with nanoscale electronic circuitry.

Figure 4 (a) shows a comparison of the propagation lengths (point where the exponential decay in $|E_z|^2$ falls to the 1/e point) for planar metal\insulator\metal (MIM) waveguides and waveguides consisting of a metal film sandwiched between two insulators (IMI waveguides). These calculations where performed using the well-established reflection pole method[53] at the important telecom wavelength of 1.55 μm.[8] This technique, based on the transfer matrix formalism, monitors the phase of the reflection coefficient denominator, and can be used to solve for the complex propagation constants of both bound and leaky modes in lossy waveguides. We again used Au (ε_{Au} = -95.92 + i 10.97 at λ = 1.55 μm) as the metal and air as the insulator.[30] For a sufficiently large center layer thickness the propagation lengths for these two types of waveguides converge to the propagation length found for a single interface (dashed line in Fig. 4 (a)). This is reasonable since the SPP modes on the two metal surfaces decouple when the spacing between them becomes large. As the center layer thickness decreases and the SPPs at the two interfaces start to interact, the propagation length along MIM structures decreases while it increases along IMI structures. In fact, IMI waveguides can reach centimeter propagation distances for very thin metal films. For obvious reasons, these are termed long range SPP (LRSPP) modes.[54,55] These "large" propagation distances can be understood by realizing that the spatial extent of the modes at these extremely thin metallic film thicknesses becomes as large as 10 μm, as shown in Fig. 4(b). In that case the SPP waves are almost entirely propagating in the lossless air region with very little field intensity inside the lossy (resistive) metal.

The MIM modes exhibit a continuous decrease in the propagation length as the center insulating layer thickness is reduced. However, Fig. 4(b) shows that by "pushing" the metals closer together it is feasible to realize deep sub-wavelength mode diameters without running into problems with cutoff. For example, the spatial extent decreases to about 100nm ($< \lambda/15$) for metal to metal spacing of 50nm. It thus seems possible to transport information in a deep sub-wavelength mode over short (~ 1 μm) distances, which is impossible with conventional dielectric components. For CMOS compatible Cu and Al plasmonic waveguides similar numbers are found and were discussed in a recent publication.[15] It is not needed to have a planar structure and slot in metallic films of finite thickness also support MIM-like modes. For example, Colour plate 4 shows the field distribution for a 200 nm wide slot waveguide fabricated in a 400 nm thick film of Ag[56]. The surrounding dielectric was chosen to be silica glass and the free space operation wavelength is 1.5 μm. Similar to the planar MIM structures, micronscale propagation lengths can be achieved and the mode does not exhibit cutoff when the width of the slot is decreased. The propagation losses do increase substantially when the slot width is decreased due to an increased overlap of the mode with the metal. As with every plasmonic waveguide, there is thus a trade-off between mode confinement and propagation length. Interestingly, within the short propagation distances MIM structures do allow for routing of electromagnetic energy around sharp corners and split signals in T-junctions.[45] These unique features can be used to realize truly nanoscale photonic functionality and circuitry,[51,57] although the maximum size of such circuits will be limited in size by the SPP propagation length. It is important to recognize that for every type of waveguide, there is a clear, but different, trade-off between confinement and propagation distance (loss). The use of one type of waveguide over another will thus depend on application specific constraints.

Based on the data presented above, it seems that the propagation lengths for plasmonic waveguides are too short to propagate SPPs with high confinement over the length of an entire Si chip (~ 1cm). Although, the manufacturability of long-range SPP waveguides may well be straightforward within a CMOS foundry, in many applications such waveguides will have a hard time competing with well-established, low-loss and high-confinement dielectric waveguides made from Si or Si_3N_4.[58,59,60] However, it is possible to create new capabilities by capitalizing on an additional strongpoint of metallic nanostructures: unparalleled subwavelength light concentration. In the following section, we will show how short, wavelength-scale sections of IMI and MIM waveguide structures can serve as miniature optical antennas and resonators. These structures are short compared to the decay length of the SPPs and reasonable quality factors can be attained.

4. Plasmonic antenna and resonator structures

Metal nanostructures have the ability to concentrate light into nanoscale volumes. For this reason, many applications of metallic nanostructures have been found in the fields of optical characterization, sensing, and nanoscale structure manipulation.[61] Extreme light focusing has also been employed to enhance a diversity of non-linear optical phenomena, such as surface enhanced Raman scattering (SERS)[62,63,64,65] and two-photon-excited photoluminescence.[66] Probably the most intensively investigated pathway to produce large field enhancements is directed towards the exploitation of (electrostatic) resonances in electron oscillations in metal nanostructures that are small compared to the wavelength of light. Much research has been devoted to analyze the properties of particles with different shapes and configurations, ranging from individual pointed particles[67,68] to their pairs[69,70,] and ensembles.[71,72] the requirements on curvature radii and inter-particle gaps involved are rather stringent resulting in large variations in the resonant effects observed for nominally identical structures. Enhancement factors on the order of 100 have been predicted and observed for spherical particles. Even greater enhancements can be obtained near carefully engineered metal optical antenna structures that basically resemble scaled down versions of a car antenna.[73] Such antennas have recently even enabled single molecule studies by SERS and white-light supercontinuum generation.[74,75,76,77]

Recently, an alternative to electrostatic resonances was proposed to reach significant and robust field enhancements.[78] It aims at exploiting retardation-based resonances involving (slow) SPPs supported by wavelength-scale structures. For the long-wavelength part of the visible spectrum and the infrared, SPPs are quite close to the light line (i.e., to free propagating light waves) and unwieldy to be directly used for constructing strongly localized and enhanced fields. However, the placement of two (or more) metal-dielectric interfaces close to each other induces a coupling between the SPPs of individual interfaces, which in turn slows down the appropriate resulting SPP-mode. The substantial slowing of these waves allows for nm-sized field localization at any wavelength.[9] Unlike conventional SPPs on thick films, strongly confined slow SPPs efficiently reflect from the structure terminations. This provides an opportunity to realize very simple resonator designs, such as simple stripes, sets of stripes, stripes with nanoscale gaps, etc.

Recently, it has indeed demonstrated that the scattering resonances of metallic stripes are related to the excitation of standing wave resonances composed of forward and backward propagating slow SPPs on the strip.[79] In particular, it was shown that the resonance wavelengths, λ, of a strip with width, w, are related by the equation:

$$w\frac{2\pi}{\lambda}n_{SPP} = m\pi - \varphi \ , \tag{1}$$

where n_{SPP} is the real part of the slow SPP mode index, φ is the phase acquired upon reflection of a SPP at the strip terminations, and m is an integer. Colour plate 5(a) shows the cross-sectional field distributions for the lowest order resonance ($m = 1$) of a 96-nm-wide silver strip resonator in water. It is seen that the largest fields are found at the corners of the strip with the field maximum being limited by the 2-nm-radius of corner rounding. Equation 1 shows that an increase of the width of the resonator with an integer number of half wavelengths of the slow SPP will again result in a new resonator with the same resonance wavelength. For these resonators, the nm-sized mode confinement is achieved at the cost of a large increase in the propagation loss. A compromise is thus to be sought

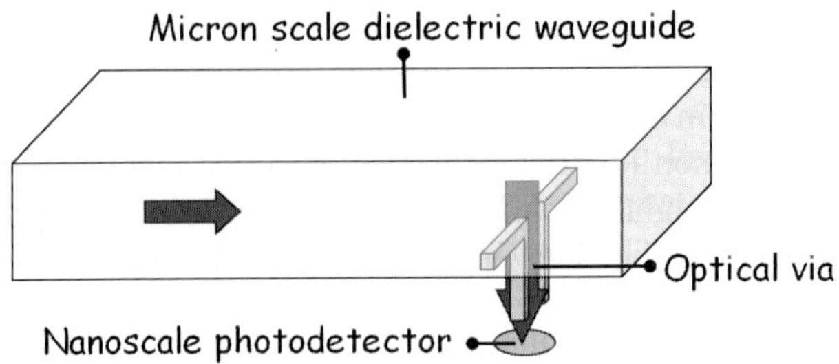

Fig. 5. Schematic illustration how a nanoscale antenna structure can serve as a bridge between microscale dielectric components and nanoscale electronic devices (Adapted from reference 8).

be designed for large scale manufacturing. The ability to concentrate light into a small (a few tens of nm) detector is critical to ensure a small capacitance and thus low-noise, low-power operation. For future CMOS circuitry photodetector capacitances should not exceed a few fF in order to ensure operation at a few fJ/bit (assuming 1V voltage swing for operation). The ability to concentrate light into deep subwavelength photodetectors also offers a substantial increase in speed. Carrier transit times are in semiconductor structures are typically limited to about 10 ps/micron. Conventional photodetectors with absorption depths of about 1 micron thus have operating speeds limited to at least a few times 10 ps.

Several types of plasmon-enhanced detectors have recently emerged for mid-infrared[83] detection and optical interconnection schemes.[84,85] Each of these devices employs a nanopatterned metallic structure capable of concentrating light into a subwavelength region. Colour plate 7 shows an example of a patterned metal film (also termed a plasmonic lens) with a grating and subwavelength slit (MIM waveguide) capable of concentrating light into a deep subwavelength Si detector, which sits at the bottom of the slit.[86] The grating has a set of 60 nm wide and 50 nm deep grooves, which are spaced by 570 nm. In the center of the grating is a 60 nm wide slit that goes down from the metal surface and terminates on the Si substrate. The grating serves to couple a normally incident free space beam to SPPs propagating along the metal. If light hits a grating with a grating constant, G, it can gain a momentum, $2\pi n/G$, in the direction of the periodicity, where n is an integer. This additional "momentum kick" enables the coupling of normally incident light (with no in-plane momentum) to SPPs and the condition $\beta_{sp} = 2\pi/G$ is satisfied.[11] It has also been shown by Leveque[87] and Ditlbacher[88] et al.

that an efficient coupling of about 20% can be achieved using a finite size grating through a careful choice of the grating dimensions. Light that is not directly incident on the detector can thus be "funneled" towards the slit with reasonable efficiencies, enabling collection from a larger area. The optical properties of the slit are similar to those of the MIM resonator shown Colour plate 6. For this reason, it will enable a resonant enhancement of the fields in the detector at the bottom of the slit. The full-field simulation in Colour plate 7 indeed shows the large field concentration at the bottom of the slit. When the slit and grating resonances are designed to occur at the same wavelength, an over 20-fold absorption enhancement can be attained in a 50 nm wide and 200 nm deep detector right behind the slit when an illumination wavelength of 632 nm is used. Even large absorption enhancements can be attained at longer wavelengths where the SPP propagation losses are lower and the natural absorption depth of light in Si is larger.

Developments in this area are very rapid and similar plasmon-enhanced photodetectors have already been used for high-speed optical clock distribution (10–20 GHz).[89] By using metallic nanostructures as a bridge between photonics and electronics, one plays to the strengths of the metallic nanostructures (concentrating fields and subwavelength guiding), the dielectric waveguides (low loss information transport) and the nanoscale electronic components (high speed information processing).[8] These unique features of metallic nanostructures play a critical role in the emerging new class of active plasmonic devices. Active plasmonic structures naturally allow for the realization of devices in which electrical signals can control optical ones and vice versa. In the following section we will investigate a couple of such devices in more detail.

6. Recent developments in active plasmonics

In the previous text, we have already discussed the possibility to introduce gain materials in metallic structures and to concentrate light into deep subwavelength volumes. In the near term, this may enable the realization of fast, nanoscale optical sources and detectors using plasmonics. For the realization of a complete chip-scale optical link, we also need high-speed and power-efficient optical modulators[90] and their development has recently experienced a number of important advances.[91,92,93] Unfortunately, many of the traditional CMOS modulator approaches run into problems as the weak non-linear optical effects in Si (e.g. free carrier dispersion) and Si compatible materials exhibit poor scaling characteristics. Pioneering work on electrically-pumped plasmonic modulators employed a linear electro-optic effect for fast modulation of SPP waveguide components.[94] The changes in the refractive index for this effect are limited to ~ 10^{-3}, implying that the interaction length (and thereby the appropriate SPP mode propagation length) needed for efficient modulation should be at least ~ 10^3 wavelengths of light. As discussed in the preceding chapters, such a requirement is extremely difficult to meet.

In this context we should mention that, for long-range SPP modes (LRSPPs) found in diffraction-limited stripe waveguides,[22,27,28] one can realize low-loss propagation and consequently efficient modulation, e.g., by using thermo-optical material properties. In fact, LRSPP-based thermo-optical modulators and switches were the first plasmonic components, in which the same metal circuitry was used for both guiding optical radiation and transmitting electrical signals controlling its guidance. Thin gold stripes embedded in polymer were heated by electrical currents causing changes in the effective index of LRSPP modes supported by these stripes. Using well-known modulation schemes, interferometric modulators (a device that modulates the transmission intensity by playing on interferences between different optical paths) and directional-coupler switches (a device in which mode coupling between parallel waveguides is controlled by the relative phase), operating at telecom wavelengths have been realized featuring low driving powers (~ 10 mW), high extinction ratios (> 30 dB), moderate response times (~ 1 ms) and quite low in- and out-coupling losses (~ 0.5 dB per coupling) when using single-mode polarization-maintained optical fibers.[98] It is worth noticing that the achieved driving powers in the devices are considerably lower than those of their conventional (dielectric waveguide based) counterparts because the heating electrodes are positioned exactly at the field maxima (in lateral cross sections) of the LRSPP, thereby maximizing induced changes in the effective index. Despite remarkable results obtained with LRSPP-based devices, their application potential is severely limited due to diffraction limited operation which precludes their usage in high density photonic circuits.

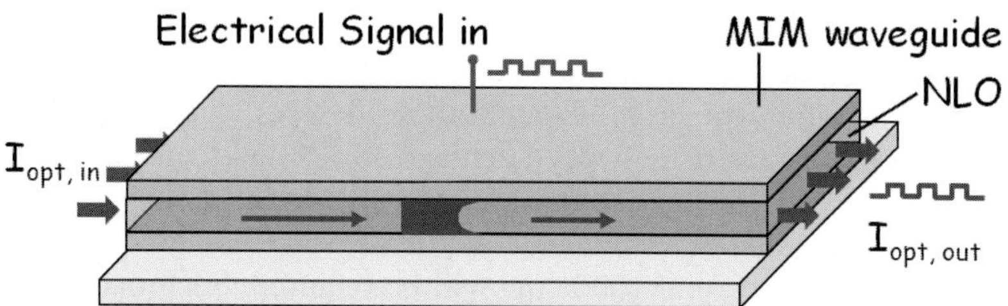

Fig. 6. Schematic of a plasmonic modulator capable of encoding electrical signals into an optical data stream. A continuous wave optical signal is injected into an MIM plasmonic waveguide from the left. The waveguide is filled with a non-linear optical medium which exhibits a large change in an optical property (typically an absorption change) in response to an electrically applied field applied across the two metal plates making up the waveguide. The excellent overlap of the MIM modes with the active medium allows for efficient, low-power switching.

There are several exciting opportunities for further performance enhancements and miniaturization in plasmonic modulators as they can nicely capitalize on two of the most useful properties of plasmonic nanostructures. First, the high electrical conductivity of metallic structures combined with their ability to support SPP waves enables them to perform simultaneous electrical and optical functions. More specifically, 1-dimensional and 2-dimensional metallic structures can be used as electrical contacts and act as plasmonic waveguides carrying an optical information stream. Second, metallic structures posses an unparalleled ability to concentrate and manipulate light at the nanoscale. The resulting field enhancements can be utilized to boost non-linear effects and enable an excellent mode overlap with active media. These concepts are illustrated in Fig. 6, which shows a schematic of a model nanoscale plasmonic modulator. This device is capable of encoding electrical signals into an optical data stream. A continuous wave optical signal is injected into an MIM plasmonic waveguide from the left. The waveguide is filled with a non-linear optical medium which exhibits a large change in an optical property (typically absorption) in response to an electric field applied across the two metal plates making up the waveguide. As MIM waveguides do not exhibit cut-off, the metal-to-metal spacing can be very small (a few tens of nm) and low voltage swings can cause large swings in the applied electric fields. Furthermore, the excellent overlap of the MIM modes with the active medium allows for efficient, low-power switching. Whereas the applicability of high confinement plasmonic waveguide structures is severely limited by the rapid attenuation of the SPP waves, there may well be a range of viable applications in active plasmonic devices, such as modulators. As long as the non-linearities are strong enough, these devices can be made small compared to characteristic decay lengths of SPPs and their performance parameters should not suffer too much from the unavoidable resistive losses in the metals. For modulators, optically induced absorption depth swings exceeding several 1000 cm^{-1} will be required. Such materials are currently tested in all optical switching configurations and electrically controlled devices may be soon to follow.[95,96,97,98]

7. Integration of plasmonic components

Many different types of SPP waveguides and various SPP-based components have been discussed above and in preceding chapters. However, we are still facing tremendous challenges in accomplishing true integration of these components into complete plasmonic circuits dedicated to particular purposes. In other words, SPP-based devices, in particular those utilizing SPP waveguides which allow deep sub-wavelength confinement, are yet to be developed even if in laboratory realization. It is very difficult to combine system configurations that support nanoscale SPP modes (e.g. the gap waveguides discussed in Chapters 4 and 5) with the terminations that can be readily coupled to (single-mode) optical fibers as discussed in Chapter 6. Another formidable challenge is connected with the typically large propagation losses that severely limit the available propagation lengths. The latter implies not only that the number of waveguide elements (bends, splitters etc.) that can be used for composing a device is limited. It is even more problematic that the waveguide length available for influencing the propagating SPP mode is often too small to actually observe this influence (and thus to realize a given functionality). This point was directly evident from the earlier discussion on plasmonic modulators.

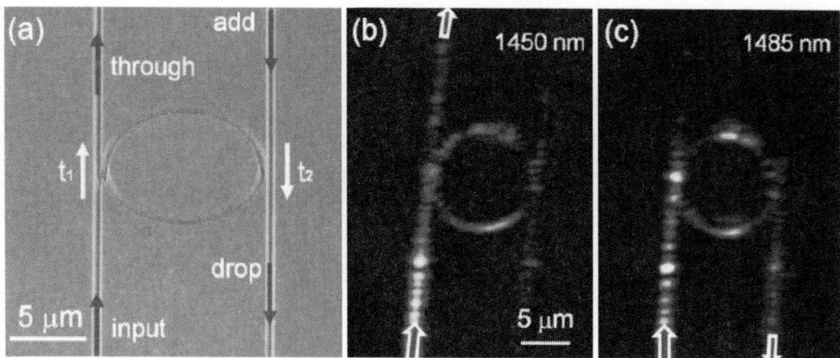

Fig. 7. SEM (a) and near-field optical images (b,c) of the CPP-based add-drop multiplexer. Note that the two different wavelengths, 1450 nm and 1485 nm, exit in different arms of the device so as to be, for example, correspondingly transmitted further in an optical communication network or dropped to a local receiver port (Adapted from reference 99).

Future integrated optical information and communication systems will heavily rely on wavelength division multiplexing (WDM). In WDM components light signals at specific wavelength can be selected and routed to different locations on a chip. For such systems SPP propagation loss is also detrimental to the overall performance and limits the achievable bandwidth. Even within this limitation, wavelength selection with a minimum of insertion losses is a very challenging problem, since any modification of the waveguide geometry (needed to introduce its sensitivity to the radiation wavelength) would inevitably lead to the occurrence of additional absorption and scattering of an SPP mode propagating through the structure. Considering development of various SPP-based chip-scale components, probably the most rapid and impressive progress (Reviewed in Chapter 11) has been with plasmonic waveguides formed by V-grooves cut into gold that support the propagation of channel plasmon-polaritons (CPPs). The CPP waveguides were shown to exhibit both strong confinement and acceptable propagation loss,[50] features that were advantageously exploited for realization of ultra-compact and low-loss bends, splitters, Mach-Zehnder interferometers and waveguide-ring resonators operating at telecom wavelengths.[51] The waveguide-ring resonator being sensitive in its operation to the light wavelength was later used to realize a simple add-drop multiplexer, which is one of the basic devices required in any telecommunication network.[99] The investigated add-drop multiplexer comprised two straight waveguides serving for input and output of radiation (in the form of CPP modes) exchanged the power via a 5-μm-radius ring waveguide inserted between and coupled to these waveguides (Fig. 7). Depending on the phase accumulated by the CPP travelling around the ring, which is determined by the CPP wavelength and ring circumference, the input radiation (end-fired with a tapered optical fiber) appeared at the through or drop port of the multiplexer. Note that the demonstrated suppression of the dropped wavelength in the through channel was quite large (on average, ~ 5) allowing for efficient wavelength filtering with a bandwidth of ~ 40 nm. In principle, the radiation at the dropped wavelength can also be added into the through port by end-firing the add port, allowing for example to exchange the information content at the dropped wavelength. Finally, it should be mentioned that the fabricated device operating at telecom wavelengths featured an insertion loss of < 3 dB and a footprint of only 100 μm^2.

It is important to realize that with the latest advances in CMOS compatible fabrication techniques a variety of the functional, nanoscale plasmonic structures described above can be fabricated in a Si foundry right now. Current silicon-based integrated circuit technology is capable of making nanoscale metallic structures such as the copper and aluminum interconnects that route electronic signals between the transistors on a chip. This mature processing technology can thus also be used to our advantage to integrate plasmonic devices with their electronic and dielectric photonic counterparts.

Concluding this section, we would like to emphasize that, even though a large number of elements necessary for a complete SPP miniature circuit have been realized, a lot of work is still needed in realizing complete integrated plasmonic circuitry. [8,100] Because of the complexity of the tasks, various parallel paths and options are being and will continue to be explored and compared. Out of this diversity, the most promising possibilities will naturally emerge for further development. In this context, it is impossible to overestimate the importance of the interplay between theoretical and experimental work that has already been very important in the progress made so far. In SPP circuits, all light-matter interactions are occurring at the nanoscale bringing many unexpected physical effects into play. Therefore elaborate characterization techniques, such as near-field imaging with subwavelength resolution and extensive computations using modern software tools are precious guides for designing devices. This point was already clearly demonstrated in the preceding chapters.

8. Plasmon-enhanced photon sources and quantum plasmonics

The operation of a large class of plasmonic devices can simply be predicted by well-developed classical electromagnetic simulation tools. In contrast, the design of future, plasmon-enhanced photon sources will rely on a deep understanding of the interactions between molecular- or nanoscale emitters and surrounding electromagnetic fields. This is an important area for future developments in plasmonics as solid state light emitters are rapidly gaining importance in a number of applications such as traffic lights, room lighting, displays, and the telecommunications industry.[101] The main reasons for their success are compactness, spectral and spatial control over the emission, high efficiency, and the availability of relatively inexpensive fabrication routes. From an electronics viewpoint it is desirable to inject current into such devices using low resistivity metal contacts. However, more resistive transparent conductors such as Indium Tin Oxide are often used for the electrode as they allow for more efficient light extraction than a reflective metal contact. Moreover, metal contacts have always reduced device efficiency by allowing emitters, such as excited molecules or semiconductor nanoparticles, to quickly decay into non-radiative SPP modes.[102,103] The non-radiative SPPs possess a high momentum (wave vector) component in the metal-insulator plane that prevents coupling to low momentum free space photons (See Chapter 1). The energy of such SPPs is typically dissipated in the metal, lowering device efficiency and making SPP excitation undesirable. Recently, several attempts have been made to effectively recover the energy of SPPs and to use nanostructured metals in light emitting devices to increase the out-coupling efficiency, to decrease the radiative lifetime of emitters, and to modify their angular distribution.[104,105,106] In contrast to sophisticated dielectric cavity structures which have been widely used for light emission enhancements, simple metal structures such as films,[102] nanowires,[107] or nanoparticles.[108,109] can easily provide large spontaneous emission enhancements due to the very large density of states near the surface plasmon resonance frequency. By incorporating antennas or suitable out-coupler structures, the surface plasmon waves can be out-coupled efficiently into the free-space or into dielectric waveguides.

Because of their small mode volume, MIM structures may provide another promising geometry for achieving large emission enhancement and efficient coupling into a well-defined SPP waveguide mode.[110] Colour plate 8 illustrates this point for a single dipolar emitter which is oriented perpendicular (⊥) to the Ag\SiO_2 interfaces of a planar MIM waveguide and radiating at 800 nm. The calculation methodology that was used to determine the spontaneous emission enhancement factor, F_P, for such an emitter is similar to the approach outlined in work by Chance and co-workers.[102] In this work, F_P is determined by considering the work done on the dipole by its own reflected field from the metal surfaces. The frequency-dependent dielectric constant of Ag was obtained from the literature[111] and we assume an internal quantum efficiency 1 for the emitter.

Colour plate 8(a) shows the calculated value of F_P as a function of the free-space wavelength for the ⊥ dipole emitter in the center of a MIM waveguide with a 30 nm metal-to-metal spacing (solid black line). In addition to a resonant enhancement peak around 400 nm (which is close to surface plasmon resonance wavelength of silver), there is a strong off-resonant enhancement which increases linearly with wavelength. For comparison, F_P, is also plotted for a ⊥ dipole emitter spaced by the same distance from a single metal surface (dashed red line), which clearly lacks such off-resonant enhancement. It is clear that the decay rate enhancements afforded by the MIM waveguide are substantially higher than those obtained for the single metal surface due to the strong enhancement of SPP excitations in small gap sizes. This strong enhancement is obtained off-resonance and over a broad wavelength regime.

Colour plate 8(b) shows the dependences of the total decay rates of such an emitter as a function of the metal-to-emitter spacing, when placed in the center of an Ag/SiO_2/Ag MIM waveguide or near a SiO_2/Ag interface. In both cases, the total decay rate increases rapidly when the emitter-to-metal spacing is reduced below 200 nm because of the enhanced surface plasmon excitation rates. In the single interface case, the rate increases roughly exponentially with decreasing spacing as the coupling to the (exponentially decaying) SPP mode is increased. In the case of an MIM waveguide, the decay rate increases faster than exponential with decreasing metal-to-emitter spacings as the mode volume, the phase velocity, and the group velocity all decrease and give rise to larger enhancements.[110] When the emitters are placed within a few tens of nanometers, undesired decay into large wavevector excitations, termed ‘lossy surface waves’, becomes important as well and even further increases the total decay rate.[102,110] Interestingly, we find that the strong coupling to gap SPPs in MIM slab structures can enable it to be the dominant decay channel for a wide range of metal-emitter distances at the considered off-resonant wavelength. Only when the metal-emitter distance becomes larger than ~ 250 nm, the fraction of gap SPP excitation decreases due to higher-order TM waveguide modes (which are similar to conventional waveguide modes). This behavior is in stark contrast to that observed for an emitter near a single metal surface, which exhibits relatively stronger coupling to LSW and free space modes, as shown in Colour plate 8(d).

The differences in the behavior of the MIM and single plasmonic waveguides are rooted in the more tightly confined SPP modes between the two metallic plates. Comparing the coupling of a dipole emitter to the main SPP mode of a metal nanowire[112,113] and MIM waveguide, it should be borne in mind that, with subwavelength MIM waveguides, one can reduce the number of emission channels so that only the emission into the gap SPP mode (being the only propagating mode survived for small gaps) would be allowed. Moreover, the polarization properties of the gap SPP mode ensure a better overlap with the dipole source field as compared to the fundamental SPP mode of a nanowire.[114]

Because of these beneficial properties MIM waveguide structures and resonators with embedded light emitters may find applications in creating electrically pumped, nanoscale local light sources or in generating guided single plasmons (or single photons) on an integrated optical circuit. Recently, spontaneous emission enhancements near nanometallic structures have already been utilized to realize single photon sources.[112] The ability of simple metallic structures to induce strong, coherent interactions between individual quantum emitters and photons may have important consequences for the fields of quantum computation and quantum information science.[113]

9. Conclusions

Plasmonics has the potential to play a unique and important role in enhancing the processing speed and capabilities of future integrated circuitry. The field of plasmonics has witnessed an explosive growth over the last few years and our knowledge base in this area is rapidly expanding. As a result, the role of plasmonic devices on a chip is also becoming more well-defined and can be captured in the graph shown in Fig. 8. This graph shows the operating speeds and critical dimensions of different chip-scale device technologies. In the past devices were

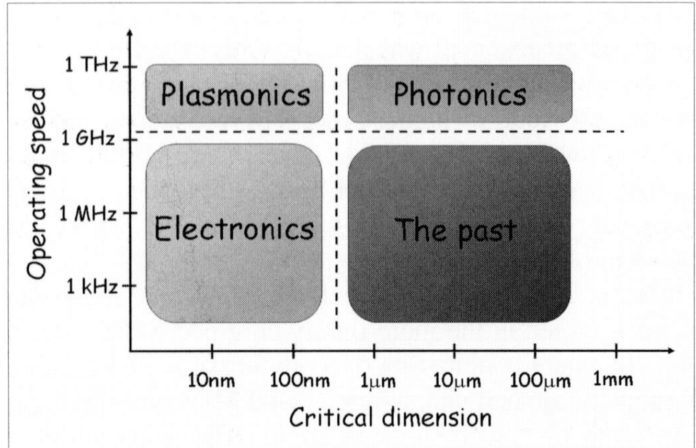

Fig. 8. Graph of the operating speeds and critical dimensions of different chip-scale device technologies. This graph can be used to highlight the strengths of different chip-scale device technologies (Adapted from reference 8).

relatively slow and bulky. The semiconductor industry has performed an incredible job at scaling electronic devices to nanoscale dimensions. Unfortunately, interconnect delay time issues may limit processing speeds of purely electronic integrated circuits. In stark contrast, photonic devices possess an enormous data carrying capacity (bandwidth). Unfortunately, dielectric photonic components are limited in their size by the fundamental laws of diffraction, preventing the same scaling that was possible for electronics. Finally, plasmonics offers precisely what electronics and photonics do not have: The size of electronics and the speed of photonics. Plasmonic devices might therefore naturally interface with similar speed photonic devices and with similar size electronic components. For these reasons, plasmonics may well serve as the missing link between two device technologies that have a difficult time communicating today. By increasing the synergy between these technologies, plasmonics may be able to unleash the full potential of nanoscale functionality and become the next wave of chip-scale technologies.

Acknowledgments

The authors would like to thank the students and postdoctoral scholars in the groups for valuable discussions on the topics of this chapter. This work at Stanford was supported by a DoD MURI sponsored by the AFOSR (F49550-04-1-0437), an NSF Career Award (ECS-0348800), the Center for Probing the Nanoscale, an NSF NSEC (PHY-0425897), and the Intel Corporation. The research in Denmark was carried out in the framework of the European Network of Excellence, PLASMO-NANO-DEVICES (FP6-2002-IST-1-507879), and supported by the Danish Research Agency through the NABIIT project (Contract No. 2106-05-033) and the FTP project (Contract No. 274-07-0258).

References

1. M.T. Bohr, *Tech. Dig. IEDM*, 241, (1995).
2. T.-Y. Chiang, B. Shieh, and K.C. Saraswat, *IEEE Symposium on VLSI Circuits, Digest of Technical Papers*, 38, (2002).
3. K. Banerjee, A.A., G. Dixit, and C. Hu, *Tech. Dig. IEDM*, 65 (1996).
4. D.A.B. Miller, *Proceedings of the IEEE*, **88**, 728 (2000).
5. W.L. Barnes, A. Dereux, and T.W. Ebbesen, *Nature*, **424**, 824 (2003).
6. E. Ozbay, *Science,* **311**, 189 (2006).
7. M.L. Brongersma, J.W. Hartman, and H.H. Atwater, Symposium, 29 Nov.-2 Dec. 1999, Boston, MA, USA. 1999: Warrendale, PA, USA : Mater. Res. Soc, 2001, This reference contains the first occurrence of the word "Plasmonics" in the title, subject, or abstract in the Inspec ® database.
8. R. Zia, J.A. Schuller and M.L. Brongersma, *Materials Today* **9**, 20-27 (2006).
9. E.N. Economou, *Phys. Rev.*, **182**, 539 (1969).
10. J.J. Burke, G.I. Stegeman, and T. Tamir, *Phys. Rev. B* **33**, 5186 (1986).
11. H. Raether, *Surface plasmons on smooth and rough surfaces and on gratings*. New York: Springer-Verlag 1988.
12. M.L Brongersma and P.G. Kik (Editors), *Surface Plasmon Nanophotonics,* Springer Verlag, New York, 2007.
13. S.A. Maier, *Plasmonics—Fundamentals and Applications,* Springer Verlag, New York, 2007.
14. Takahara, J., et al., Guiding of a one-dimensional optical beam with nanometer diameter. *Opt. Lett.*, 1997. 22(7): p. 475-7.
15. Rashid Zia, Mark D. Selker, Peter B. Catrysse, and Mark L. Brongersma, *J. Opt. Soc. of Am. A***21**, 2442 (2004).
16. R.C. Reddick, R.J. Warmack, and T.L. Ferrell, *Phys. Rev. B***39**, 767 (1989).
17. J.C. Weeber, et al., *Phys. Rev. B***64**, 045411 (2001).
18. J.C. Weeber, Y. Lacroute, and A. Dereux, *Phys. Rev. B***68**, 115401-1 (2003).
19. J.R. Krenn et al., *Europhys. Lett.* **60**, 663 (2002).
20. E. Kretschmann, *Zeitschrift fur Physik A***241**, 313 (1971).
21. B. Lamprecht, B., et al., *Appl. Phys. Lett.* **79**, 51 (2001).
22. A. Boltasseva, T. Nikolajsen, K. Leasson, K. Kjaer, M.S. Larsen, and S.I. Bozhevolnyi, *J. Lightwave Techn.* **23**, 413 (2005).
23. R. Zia, A. Chandran, and M.L. Brongersma, *Opt. Lett.* **30**, 1473 (2005).
24. R. Zia, M.D. Selker, and M.L. Brongersma, *Phys. Rev. B***71**, 165431-1 (2005).

25. R. Zia, J.A. Schuller, and M.L. Brongersma, *Phys. Rev. B*74, 165415 (2006).
26. S.J. Al-Bader, *IEEE Journal of Quantum Electronics* **40**, 325 (2004).
27. P. Berini, *Phys. Rev. B*61, 10484 (2000).
28. P. Berini, *Phys. Rev.B*63, 125417-15 (2001).
29. P. Lusse, et al., *J. of Lightw. Techn.* **12**, 487 (1994).
30. E.D. Palik, *Handbook of Optical Constants of Solids*. 1985: Orlando, FL, USA : Academic Press, 1985. xviii, 804p.
31. Rashid Zia and Mark L. Brongersma, *Nature Nanotechnology*, 2, 426 (2007).
32. J. D. Joannopoulos, R. D. Meade, and J. N. Winn, *Photonic Crystals* (Princeton University Press, Princeton, 1995).
33. T. F. Krauss and R. M. De La Rue, *Prog. Quant. Elect.* **23**, 51 (1999).
34. C. M. Soukoulis, ed., *Photonic Crystals and Light Localization in the 21st Century*, (Kluwer, Dordrecht, 2001).
35. R. H. Ritchie, E. T. Arakawa, J. J. Cowan, and R. N. Hamm, *Phys. Rev. Lett.* **21**, 1530 (1968).
36. S. C. Kitson, W. L. Barnes, and J. R. Sambles, *Phys. Rev. Lett.* **77**, 2670 (1996).
37. S.I. Bozhevolnyi, J. Erland, K. Leosson, P.M.W. Skovgaard, and J.M. Hvam, *Phys. Rev. Lett.* **86**, 3008 (2001).
38. S. I. Bozhevolnyi, V.S. Volkov, K. Leosson, and J. Erland, *Opt. Lett.* **26**, 734 (2001).
39. S.I. Bozhevolnyi, V.S. Volkov, K. Leosson, and A. Boltasseva, *Appl. Phys. Lett.* **79**, 1076 (2001).
40. M. Quinten, A. Leitner, J.R. Krenn, F.R. Aussenegg, *Opt. Lett.* **23**, 1331(1998).
41. M.L. Brongersma, J.W. Hartman, H.A. Atwater, *Phys. Rev. B* **62**, R16356-16359 (2000).
42. D.F.P. Pile, et al., *Appl. Phys. Lett.* **87**, 1-3 (2005).
43. D.F.P. Pile, D.K. Gramotnev, M. Haraguchi, T. Okamoto, and M. Fukui, *J. of Appl. Phys.* **100**, 013101 (2006).
44. G. Veronis, G. and S.H. Fan Opt. Lett. 30, 3359-3361 (2005).
45. G. Veronis, and S.H. Fan, *Appl. Phys. Lett.* **87,** 131102-1 (2005).
46. J.A.Dionne, L.A. Sweatlock, H.A. Atwater, and A. Polman, *Phys. Rev. B*73, 035407 (2006).
47. L. Chen, J. Shakya, and M. Lipson, *Opt. Lett.* **31**, 2133-2135 (2006).
48. D.F.P. Pile, D.K. Gramotnev, *Opt. Lett.***29**, 1069 (2004).
49. I.V. Novikov, A.A. Maradudin, *Phys. Rev. B* **66**, 035403 (2002).
50. S.I. Bozhevolnyi, V.S. Volkov, E. Devaux, T.W. Ebbesen, *Phys. Rev. Lett.* **95**, 046802 (2005).
51. S.I. Bozhevolnyi, V.S. Volkov, E. Devaux, J.Y. Laluet, and T.W. Ebbesen, Nature **440**, 508 (2006).
52. E. Moreno, F.J. Garcia-Vidal, S.G. Rodrigo, L. Martin-Moreno, and S.I. Bozhevolnyi, *Opt. Lett.* **31**, 3447 (2006).
53. E. Anemogiannis, E.N. Glytsis, and T.K. Gaylord, *J. of Lightw. Techn.***17**, 929 (1999).
54. D. Sarid, *Phys. Rev. Lett.***47**, 1927 (1981).
55. F.Z. Yang, G.W. Bradberry, and J.R. Sambles, *Phys. Rev. Lett.***66**, 2030 (1991).
56. Ning-Ning Feng, Mark L. Brongersma, and Luca Dal Negro, *IEEE Journ. Of Quant. Electron.* **43**, 479- 485 (2007).

57. K. Tanaka, and M. Tanaka, *Appl. Phys. Lett.***82**, 1158 (2003).
58. K.K. Lee, et al., *Opt. Lett.***26**, 1888 (2001).
59. N. Daldosso, et al., *J. of Lightw. Techn.***22**, 1734 (2004).
60. D.K. Sparacin, S.J. Spector, and L.C. Kimerling, *J. of Lightw. Techn.***23**, 2455 (2005).
61. V.M. Markel and T.F. George (Eds), "Optics of Nanostructured Materials," (Wiley, New York, 2001).
62. N. Felidj, N., *Appl. Phys. Lett.***82**, 3095 (2003).
63. K.E. Shafer-Peltier, *J. Am. Chem. Soc.***125**, 588 (2003).
64. C.L. Haynes and R.P. Van Duyne, *J. Phys. Chem.B***107**, 7426 (2003).
65. V.P. Drachev, et al., *J. Phys. Chem. B***108**, 18046 (2004).
66. P. J. Schuck, D.P. Fromm, A. Sundaramurthy, G.S. Kino, and W.E. Moerner, *Phys. Rev.Lett.* **94**, 017402 (2005).
67. G.T. Boyd, Th. Rasing, J.R.R. Leite, and Y.R. Shen, *Phys. Rev. B***30**, 519 (1984), and references therein.
68. J. Sánchez, L. Novotny, and X. S. Xie, *Phys. Rev. Lett.* **82**, 4014(1999).
69. Kneipp, H. Kneipp, I. Itzkan, R.R. Dasari, and M.S. Feld, *J. Phys. Condens. Matter* **14**, R597 (2002).
70. P. J. Schuck, D.P. Fromm, A. Sundaramurthy, G.S. Kino, and W.E. Moerner, Phys. Rev.Lett. 94, 017402 (2005).
71. A.K. Sarychev and V.M. Shalaev, Phys. Rep. 335, 275 (2000).
72. M.I. Stockman, in *Optics of Nanostructured Materials*, Ref. 1, p. 313, and references therein.
73. K.B. Crozier, et al., *J. Appl. Phys.***94**, 4632 (2003).
74. S.M. Nie, and S.R. Emery, *Science* **275**,1102 (1997).
75. K. Kneipp, K., et al., *Phys. Rev. Lett.***78**, 1667 (1997).
76. D.P. Fromm, et al., *J. Chem. Phys.***124**, 6 (2006).
77. P. Mühlschlegel, H.-J. Eisler, O.J.F. Martin, B. Hecht, D.W. Pohl, Science 308, 1607 (2005).
78. T. Søndergaard and S. I. Bozhevolnyi, *Phys. Rev. B* **75**, 073402 (2007).
79. T. Søndergaard and S. I. Bozhevolnyi, *Opt. Express* **15**, 4198 (2007).
80. T. Søndergaard and S. I. Bozhevolnyi, *Phys. stat. sol. (b)***245**, 9 (2008). T. Søndergaard and S. I. Bozhevolnyi, *Phys. stat. sol. (b)***245**, 9 (2008).
81. M. A. Noginov, G. Zhu, M. Bahoura, J. Adegoke, C. E. Small, B. A. Ritzo, V. P. Drachev, and V. M. Shalaev, *Opt. Lett.* **31**, 3022 (2006).
82. S. I. Bozhevolnyi and T. Søndergaard, *Opt. Express* **15**, 10869 (2007).
83. Z. Yu, G. Veronis, S. Fan, and M.L. Brongersma, *Opt. Lett.***89**, 151116-1-3 (2006).
84. T. Ishi, J. Fujikata, K. Makita, T. Baba, and K. Ohashi, *Jpn. J. Appl. Phys.***44**, L364 (2005).
85. L. Tang, D.A.B. Miller, A.K. Okyay, J.A. Matteo, Y. Yuen, K.C. Saraswat, L. Hesselink, *Opt. Lett.* **31**, 1519 (2006).
86. J. White, E. Barnard, M.L. Brongersma, Unplublished.
87. G. Leveque and O.J.F. Martin, J. Appl. Phys. **100**, 124301-1-6 (2006).
88. H. Ditlbacher, J. R. Krenn, A. Hohenau, A. Leitner, and F. R. Aussenegg, *Appl. Phys. Lett.* **83**, 3665 (2003).
89. J. Fujikata, K. Nishi, A. Gomyo, J. Ushida, T. Ishi, H. Yukawa, D. Okamoto, M. Nakada, T. Shimizu, M. Kinoshita, K, Nose, M. Mizuno, T. Tsuchizawa, T.

Watanabe, K. Yamada, S. Itabashi, K. Keishi, *IEICE Trans. Electron.* **E91–C**, (2008).
90. L. Pavesi "*A review of the various efforts to a silicon laser,*" *in Photonics Packaging and Integration III*" edited by R. A. Heyler, D. J. Robbins and G. E. Jabbour, *Proceedings of SPIE* 4997, 206-220, (2003).
91. M. Lipson, *Nanotechnology*, 15, S622-627 (2004).
92. A. Liu, R. Jones, L. Liao, D. Samara-Rubio, D. Rubin, O. Cohen, R. Nicolaescu and M. Paniccia, *Nature* **427**, 615-618 (2004).
93. Rohan D. Kekatpure and Mark L. Brongersma, *Opt. Lett.* **30**, 2149-2151 (2005).
94. T. Nikolajsen, K. Leosson, S. I. Bozhevolnyi, Appl. Phys. Lett. 85, 5833 (2004).
95. A.V. Krasavin, N.I Zheludev, *Appl. Phys. Lett.* **84**, 1416-18 (2004).
96. Dintinger, J. Robel, I. Kamat, P. V. Genet, C. Ebbesen, T. W., *Advanced Materials*, **18**, 1645 (2006).
97. D. Pacifici, H.J. Lezec, and H.A. Atwater, *Nature Photonics*, **1** (2007).
98. R.A. Pala, K.T. Shimizu, N.A. Melosh, M.L. Brongersma, *Nano Lett.***8**, 1506-1510 (2008).
99. V. S. Volkov, S. I. Bozhevolnyi, E. Devaux, J.-Y. Laluet, and T. W. Ebbesen, *Nano Lett.* **7**, 880 (2007).
100. T.W. Ebbesen, C. Genet, and S.I. Bozhevolnyi, Physics Today, p.44, May 2008.
101. E.F. Schubert, J.K. Kim, *Science* **308**, 1274 (2005).
102. R.R. Chance, A. Prock, R. Silbey, *Adv. Chem. Phys.* **37**, 1 (1978).
103. W.L. Barnes, *J. Modern Optics* **45**, 661 (1998).
104. W.L. Barnes, *J. Lightwave Technol.* **17**, 2170 (1999).
105. K. Okamoto, I. Niki, A. Shvartser, Y. Narukawa, T. Mukai, A. Scherer, *Nature* Materials **3**, 601 (2004).
106. John S.Q. Liu and Mark L. Brongersma, *Appl. Phys. Lett.* **90**, 091116-1-3 (2007).
107. D.E. Chang et al., *Phys. Rev. Lett.* **97**, 053002 (2006).
108. H. Mertens, A.F. Koenderink, and A. Polman, *Phys. Rev. B* **76**, 115123 (2007).
109. J.A. Dionne et al., *Phys. Rev. B* **73**, 035407 (2006).
110. Y.C. Jun, R.D. Kekapture, J.S. White, and M.L. Brongersma, *Submitted* 2008.
111. P.B. Johnson and R.W. Christy, *Phys. Rev. B* **6**, 4370 (1972).
112. A.V. Akimov et al., *Nature* **450**, 402 (2007).
113. D.E. Chang et al., *Nature Physics 3*, 807 (2007).
114. S.I. Bozhevolnyi, unpublished.

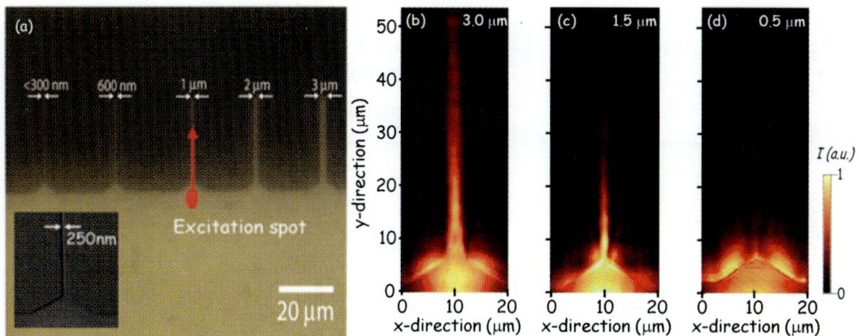

Colour Plate 1. (a) Optical microscopy image of a SiO_2 substrate with an array of Au stripes attached to a large launchpad generated by electron beam lithography. The red arrow illustrates the launching of a surface plasmon-polariton (SPP) into a 1 μm wide stripe. (b), (c) , and (d) show near-field optical microscopy images of SPPs excited at λ = 785 nm and propagating along 3.0 μm, 1.5μm, and 0.5 μm wide Au stripes, respectively (Adapted from reference 8).

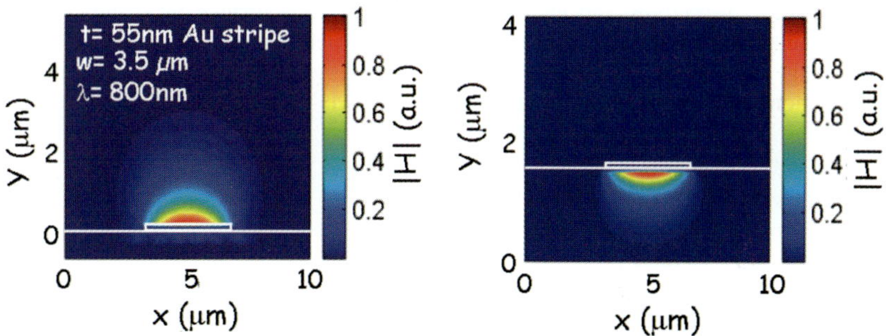

Colour Plate 2. Simulated SPP mode profiles appropriate to a 55 nm thick and 3.5 μm wide Au stripe on a SiO_2 glass substrate. It shows both the leaky (left) and bound (right) SPP modes propagating at the top air/metal and bottom glass/ metal interfaces. Both modes can be employed simultaneously for information transport (Adapted from reference 25).

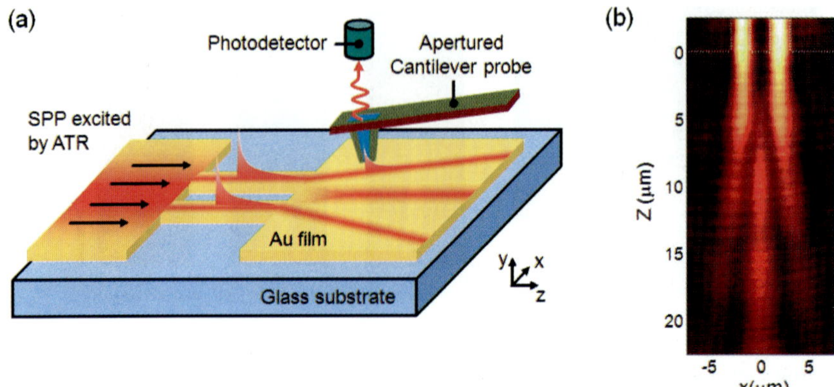

Colour Plate 3. (a) Schematic of an experimental configuration in which photon scanning tunneling microscopy (PSTM) is used to visualize Young's double-slit experiment for SPPs. SPPs are first launched onto the surface of a patterned Au film (on the left) by an attenuated total reflection (ATR) technique. These SPPs then excite the guided polariton modes supported by two adjacent metal stripe waveguides. These stripes perform an analogous role to the slits in Young's traditional double-slit experiment. An apertured cantilever-probe is used to locally tap into the guided SPP waves and scatter light towards a photodetector in the far-field. The detected signal provides a measure of the local field intensity right underneath the tip and the propagation, diffraction, and interference of SPPs can be imaged by scanning the tip over the patterned metal. (b) Experimental near-field image taken with a photon scanning tunneling microscope demonstrating guided polariton propagation, diffraction, and interference for the structure shown in (a) (Adapted from reference 31).

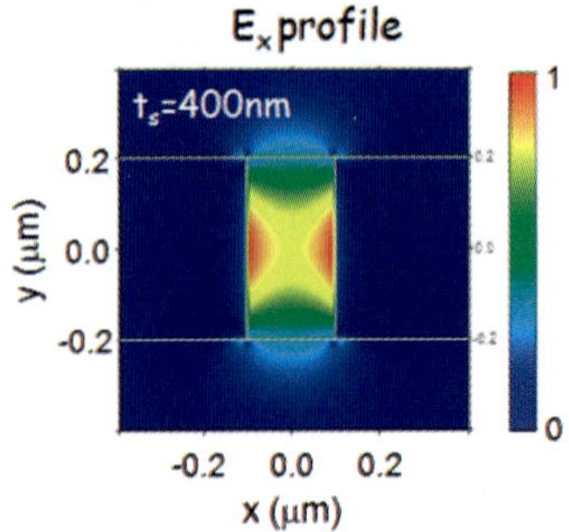

Colour Pate 4. The E_x field distribution for a 200 nm wide slot waveguide fabricated in a 400 nm thick film of Ag. The surrounding dielectric was chosen to be silica glass and the free space operation wavelength is 1.5 μm (Adapted from reference 56).

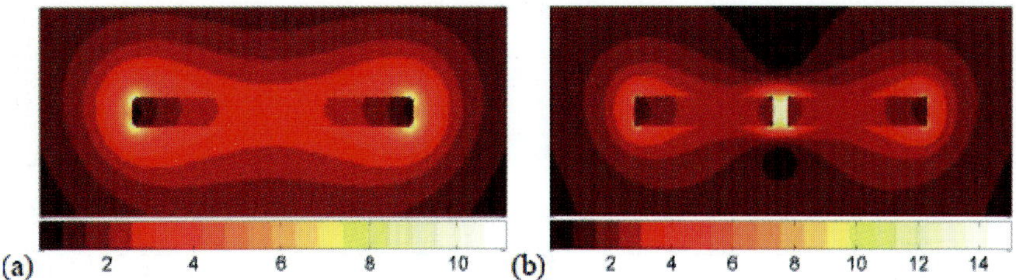

Colour Plate 5. Electric field magnitude distributions for 10-nm-thin silver strips embedded in water and bottom-illuminated. (a) Single strip with w=96nm, λ=672nm and (b) two strips with w=48nm separated by a 5-nm-wide gap, λ=580nm (Adapted from reference 79).

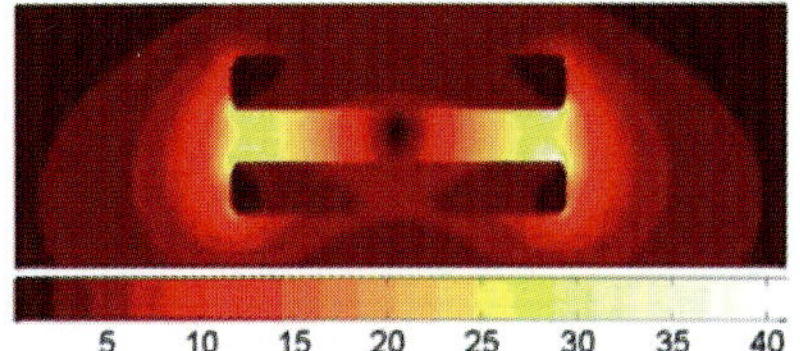

Colour Plate 6. Electric field magnitude at resonance (wavelength ~700 nm) for two silver strips of thickness 20 nm and width 130 nm separated by 20 nm. The strips are bottom-illuminated with a plane wave (Adapted from reference 80).

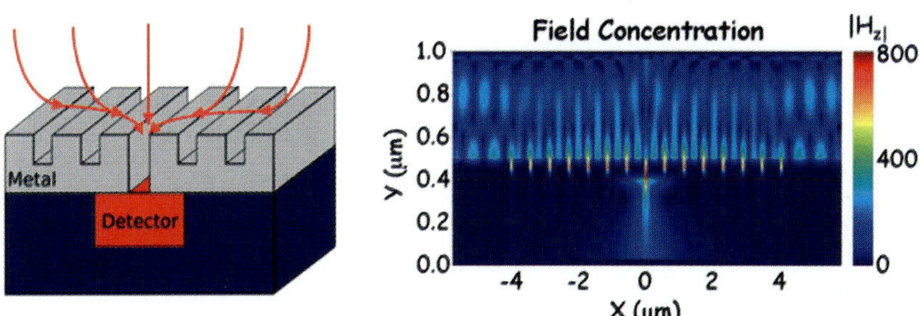

Colour Plate 7. Schematic of a plasmon-enhanced detector employing grating and slit (MIM waveguide) resonances to concentrate light from a large illumination area into a deep subwavelength photodetector (left). Full field electromagnetic simulation of a "plasmon lens" capable of concentrating light into a 60 nm wide Si photodiode. The plasmon lens consists of a 60 nm wide central slit surrounded by a larger grating structure. The grating has a set of 60 nm wide and 50 nm deep grooves, which are spaced by 570nm. The illumination wavelength was chosen to be 632 nm.

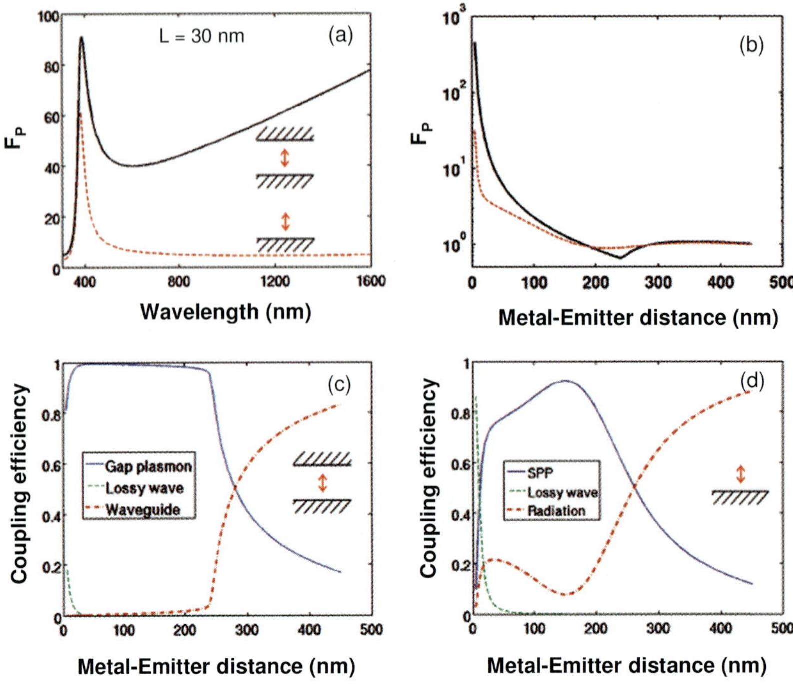

Colour Plate 8. The spontaneous emission behavior of an excited dipolar emitter inside an Ag/SiO_2/Ag MIM slab waveguide or near a Ag/SiO_2 interface. The dipole is oriented perpendicular to the metal\dielectric interfaces and its quantum efficiency in free space was assumed to be 1. (a) Spontaneous emission enhancement factor, F_P, for a dipolar emitter as a function of free-space wavelength for the MIM slab waveguide (solid black line) vs. the single metal-dielectric interface case (dashed red line). In the MIM case, the emitter was in the center of the gap. (b) The dependences of the total decay rate of a single emitter placed in the center of an MIM waveguide (solid black line) and near a metal surface (dashed red line) as a function of the metal-to-emitter spacing. The emission wavelength was chosen to be $\lambda = 800$ nm. (c) Fraction of dissipated energy to the different decay pathways described in the text as a function of metal-emitter distance. Again, the emitter was in the middle of the gap and the emission wavelength was $\lambda = 800$ nm. (d) Same as (c) but for the single metal/dielectric interface case (Adapted from reference 110).

INDEX